Environmental Contaminants and Medicinal Plants Action on Female Reproduction

Environmental Contaminants and Medicinal Plants Action on Female Reproduction

Alexander V. Sirotkin
Professor,
Constantine the Philosopher University in Nitra, Slovakia

Adriana Kolesarova
Professor,
Slovak University of Agriculture in Nitra, Slovakia

Academic Press is an imprint of Elsevier
125 London Wall, London EC2Y 5AS, United Kingdom
525 B Street, Suite 1650, San Diego, CA 92101, United States
50 Hampshire Street, 5th Floor, Cambridge, MA 02139, United States
The Boulevard, Langford Lane, Kidlington, Oxford OX5 1GB, United Kingdom

Copyright © 2022 Elsevier Inc. All rights reserved.

No part of this publication may be reproduced or transmitted in any form or by any means, electronic or mechanical, including photocopying, recording, or any information storage and retrieval system, without permission in writing from the publisher. Details on how to seek permission, further information about the Publisher's permissions policies and our arrangements with organizations such as the Copyright Clearance Center and the Copyright Licensing Agency, can be found at our website: www.elsevier.com/permissions.

This book and the individual contributions contained in it are protected under copyright by the Publisher (other than as may be noted herein).

Notices

Knowledge and best practice in this field are constantly changing. As new research and experience broaden our understanding, changes in research methods, professional practices, or medical treatment may become necessary.

Practitioners and researchers must always rely on their own experience and knowledge in evaluating and using any information, methods, compounds, or experiments described herein. In using such information or methods they should be mindful of their own safety and the safety of others, including parties for whom they have a professional responsibility.

To the fullest extent of the law, neither the Publisher nor the authors, contributors, or editors, assume any liability for any injury and/or damage to persons or property as a matter of products liability, negligence or otherwise, or from any use or operation of any methods, products, instructions, or ideas contained in the material herein.

Library of Congress Cataloging-in-Publication Data
A catalog record for this book is available from the Library of Congress

British Library Cataloguing-in-Publication Data
A catalogue record for this book is available from the British Library

ISBN: 978-0-12-824292-6

For information on all Academic Press publications visit our website at
https://www.elsevier.com/books-and-journals

Publisher: Andre G. Wolff
Acquisitions Editor: Kattie Washington
Editorial Project Manager: Sam W. Young
Production Project Manager: Kiruthika Govindaraju
Cover Designer: Christian Bilbow

Typeset by TNQ Technologies

Contents

Preface	xv

1. Female reproductive system and its regulation

1.1	Introduction	1
1.2	The ovary	1
	1.2.1 Folliculogenesis	2
	1.2.2 Preovulatory changes and ovulation	4
	1.2.3 Luteogenesis and luteolysis	5
	1.2.4 Oocytes	6
1.3	Fallopian tubes	6
1.4	Uterus	7
1.5	Extra- and intracellular regulators of female reproductive processes	7
1.6	Some female reproductive disorders	14
1.7	Conclusions and possible direction of future studies	15
	References	15

2. Environmental contaminants and their influence on health and female reproduction

2.1	Heavy metals	21
	2.1.1 Introduction	21
	2.1.2 Provenance and properties	22
	2.1.3 Physiological actions	23
	2.1.4 Mechanisms of action	26
	2.1.5 Effects on female reproductive processes	27
	2.1.5.1 Effect on ovaries	27
	2.1.5.2 Effect on uterus and pregnancy	29
	2.1.6 Mechanism of action on female reproductive processes	30
	2.1.7 Application in reproductive biology and medicine	31
	2.1.8 Conclusions and possible direction of future studies	32
	References	32
2.2	Mycotoxins	38
	2.2.1 Introduction	38
	2.2.2 Provenance and properties	38

v

vi Contents

2.2.3	Physiological actions	43
2.2.4	Mechanism of action	43
2.2.5	Effects on female reproductive processes	44
	2.2.5.1 Effect on ovarian cell functions	44
	2.2.5.2 Effects on oocytes and embryos	45
2.2.6	Mechanisms of action on female reproductive processes	46
2.2.7	Application in reproductive biology and medicine	48
2.2.8	Conclusion	49
References		49

2.3. The oil-related environmental contaminants ... 54

2.3.1	Introduction	54
2.3.2	Provenance and properties	54
2.3.3	Physiological actions	55
2.3.4	Mechanisms of action	56
2.3.5	Effects on female reproductive processes	57
	2.3.5.1 Effects on the CNS	57
	2.3.5.2 Effects on the ovarian and reproductive state	57
	2.3.5.3 Effect on ovarian cell functions	58
	2.3.5.4 Effects on oocytes and embryos	59
	2.3.5.5 Effects on oviducts	60
	2.3.5.6 Effects on cytogenetics of somatic and generative cells	60
	2.3.5.7 Effects on reproductive hormones	60
2.3.6	Mechanisms of action on female reproductive processes	61
2.3.7	Application in reproductive biology and medicine	62
2.3.8	Conclusions and possible direction of future studies	64
References		66

2.4. Tobacco smoking ... 70

2.4.1	Introduction	70
2.4.2	Provenance and properties	70
2.4.3	Physiological actions	71
2.4.4	Mechanism of action	73
2.4.5	Effects on female reproductive processes	73
	2.4.5.1 The effect on ovarian and reproductive state	73
	2.4.5.2 Effect on reproductive hormones	74
	2.4.5.3 Effect on embryos	74
2.4.6	Mechanisms of action on female reproductive processes	75
2.4.7	Application in reproductive biology and medicine	75
2.4.8	Conclusions and possible direction of future studies	76
References		77

3. Food/medicinal herbs and their influence on health and female reproduction

3.1	Apricot (*Prunus armeniaca* L.)	81
3.1.1	Introduction	81
3.1.2	Provenance and properties	82
3.1.3	Physiological and therapeutic actions	83

Contents **vii**

	3.1.4 Mechanisms of action	86
	3.1.5 Effects on female reproductive processes	87
	3.1.5.1 Effect on reproductive hormones	87
	3.1.5.2 Effect on ovarian follicular cell functions	88
	3.1.6 Mechanisms of action on female reproductive processes	88
	3.1.7 Application in reproductive biology and medicine	89
	3.1.8 Conclusions and possible direction of future studies	90
	References	90
3.2	**Black elder (*Sambucus nigra* L.)**	96
	3.2.1 Introduction	96
	3.2.2 Provenance and properties	96
	3.2.3 Physiological and therapeutic actions	98
	3.2.4 Mechanisms of action	99
	3.2.5 Effects on female reproductive processes	100
	3.2.5.1 Effect on ovarian cell functions	100
	3.2.5.2 Effect on embryo	101
	3.2.6 Mechanisms of action on female reproductive processes	101
	3.2.7 Application in reproductive biology and medicine	102
	3.2.8 Conclusions and possible direction of future studies	102
	References	103
3.3	**Buckthorn (*Hippophae rhamnoides* L.)**	108
	3.3.1 Introduction	108
	3.3.2 Provenance and properties	109
	3.3.3 Physiological and therapeutic actions	110
	3.3.4 Mechanisms of action	111
	3.3.5 Effects on female reproductive processes	113
	3.3.5.1 Effect on ovarian follicular cell functions	113
	3.3.5.2 Effect on vagina and uterus	114
	3.3.6 Mechanisms of action on female reproductive processes	114
	3.3.7 Potential for application in reproductive biology and medicine	115
	3.3.8 Conclusions and possible direction of future studies	116
	References	116
3.4	**Buckwheat (*Fagopyrum tataricum*, L., *Fagopyrum esculentum* Moench)**	121
	3.4.1 Introduction	121
	3.4.2 Provenance and properties	121
	3.4.3 Physiological and therapeutical actions	122
	3.4.4 Mechanisms of action	123
	3.4.5 Effects on female reproductive processes	125
	3.4.5.1 Effect of on ovarian and reproductive state	125
	3.4.5.2 Effect on ovarian cell functions	125
	3.4.5.3 Effect on reproductive hormones	125
	3.4.5.4 Effect on ovarian cell response to environmental contaminants	126

viii Contents

	3.4.6	Mechanisms of action on female reproductive processes	126
	3.4.7	Application in reproductive biology and medicine	127
	3.4.8	Conclusions and possible direction of future studies	127
	References		128
3.5	**Curcuma/turmeric (*Curcuma longa* L., *Curcuma zedoaria* (Christm.) Roscoe)**		**130**
	3.5.1	Introduction	130
	3.5.2	Provenance and properties	130
	3.5.3	Physiological and therapeutic actions	131
	3.5.4	Mechanisms of action	132
	3.5.5	Effect on female reproductive processes	132
		3.5.5.1 Effect on ovarian and reproductive state	132
		3.5.5.2 Effect on ovarian cell functions	133
		3.5.5.3 Effect on reproductive hormones	133
		3.5.5.4 Mechanisms of action on female reproductive processes	134
	3.5.6	Application in reproductive biology and medicine	135
	3.5.7	Conclusion and possible directions of further studies	136
	References		137
3.6	**Flaxseed (*Linum usitatissimum* L.)**		**141**
	3.6.1	Introduction	141
	3.6.2	Provenance and properties	141
	3.6.3	Physiological action	142
	3.6.4	Mechanisms of action	143
		3.6.4.1 Flaxseed constituents responsible for its physiological effects	143
		3.6.4.2 Mediators of flaxseed effects	144
	3.6.5	Effects on female reproductive processes	145
		3.6.5.1 Effect on ovarian and reproductive state	145
		3.6.5.2 Effect on oocytes and embryos	146
		3.6.5.3 Effect on reproductive hormones	147
	3.6.6	Mechanisms of action on female reproductive processes	148
		3.6.6.1 Flaxseed constituents responsible for its effects on female reproductive processes	148
		3.6.6.2 Mediators of flaxseed effects on female reproductive processes	149
	3.6.7	Application in reproductive biology and medicine	150
	3.6.8	Conclusions and possible direction of future studies	151
	References		152
3.7	**Ginkgo (*Ginkgo biloba*, L.)**		**156**
	3.7.1	Introduction	156
	3.7.2	Provenance and properties	156
	3.7.3	Physiological and therapeutic actions	157
	3.7.4	Mechanisms of action	159
		3.7.4.1 Ginkgo constituents responsible for particular effects	159

Contents **ix**

	3.7.4.2	Mediators of ginkgo and its constituents' effects	160
3.7.5		Effects on female reproductive processes	162
	3.7.5.1	Effect on ovarian and reproductive state	162
	3.7.5.2	Effect on ovarian cell functions	163
	3.7.5.3	Effect on oocytes and embryos	163
	3.7.5.4	Effect on reproductive hormones	163
3.7.6		Mechanisms of action on female reproductive processes	164
3.7.7		Application in reproductive biology and medicine	165
3.7.8		Conclusions and possible direction of future studies	165
References			166
3.8		**Grape (*Vitis vinifera* L.)**	**170**
3.8.1		Introduction	170
3.8.2		Provenance and properties	171
3.8.3		Physiological and therapeutic actions	171
3.8.4		Mechanisms of action	172
3.8.5		Effects on female reproductive processes	172
	3.8.5.1	Effect on ovaries	173
	3.8.5.2	Effect on uterus	174
3.8.6		Mechanisms of action on female reproductive processes	174
3.8.7		Application in reproductive biology and medicine	177
3.8.8		Conclusions and possible direction of future studies	178
References			178
3.9		**Pomegranate (*Punica granatum* L.)**	**183**
3.9.1		Introduction	183
3.9.2		Provenance and properties	183
	3.9.2.1	Bioavailability, metabolism and pharmacokinetics	184
3.9.3		Physiological and therapeutic actions	185
3.9.4		Mechanisms of action	186
3.9.5		Effects on female reproductive processes	189
3.9.6		Mechanisms of action on female reproductive processes	191
3.9.7		Application in reproductive biology and medicine	191
3.9.8		Conclusions and possible direction of future studies	192
References			192
3.10		**Puncture vine (*Tribulus terrestris* L.)**	**199**
3.10.1		Introduction	199
3.10.2		Provenance and properties	199
3.10.3		Physiological and therapeutical actions	200
3.10.4		Mechanisms of action	201
3.10.5		Effect on male reproductive processes	203
3.10.6		Mechanisms of effects on male reproductive processes	204
3.10.7		Effects on female reproductive processes	205

x Contents

	3.10.8	Mechanisms of effects on female reproductive processes	207
	3.10.9	Application in reproductive biology and medicine	208
	3.10.10	Conclusions and possible direction of future studies	208
	References		209
3.11	Rooibos (*Aspalathus linearis* (Burm.f.) R.Dahlgren)		213
	3.11.1	Introduction	213
	3.11.2	Provenance and properties	213
	3.11.3	Physiological actions	214
	3.11.4	Mechanisms of action	215
	3.11.5	Effects on female reproductive processes	216
	3.11.6	Mechanisms of action on female reproductive processes	217
	3.11.7	Application in reproductive biology and medicine	217
	3.11.8	Conclusions and possible direction of future studies	218
	References		218
3.12	Tea (*Camelia sinensis* L.)		220
	3.12.1	Introduction	220
	3.12.2	Provenance and properties	220
	3.12.3	Physiological actions	221
	3.12.4	Mechanisms of action	222
	3.12.5	Effects on female reproductive processes	223
		3.12.5.1 Effect on ovarian and reproductive state	223
		3.12.5.2 Effect on ovarian cell functions	224
		3.12.5.3 Effect on oocytes and embryos	224
		3.12.5.4 Effect on reproductive hormones	225
	3.12.6	Mechanisms of action on female reproductive processes	226
	3.12.7	Application in reproductive biology and medicine	227
	3.12.8	Conclusions and possible direction of future studies	228
	References		229
3.13	Vitex (*Vitex agnus-castus* L.)		232
	3.13.1	Introduction	232
	3.13.2	Provenance and properties	232
	3.13.3	Physiological actions	233
	3.13.4	Mechanisms of action	234
	3.13.5	Effects on female reproductive processes	236
		3.13.5.1 Effect on ovarian and reproductive state	236
		3.13.5.2 Effect on ovarian cell functions	236
		3.13.5.3 Effect on reproductive hormones	237
	3.13.6	Mechanisms of action on female reproductive processes	237
	3.13.7	Application in reproductive biology and medicine	238
	3.13.8	Conclusions and possible direction of future studies	240
	References		240

Contents **xi**

4. Plant molecules and their influence on health and female reproduction

4.1	Amygdalin	245
	4.1.1 Introduction	245
	4.1.2 Provenance and properties	246
	4.1.3 Physiological and therapeutic actions	248
	4.1.4 Mechanisms of action	249
	4.1.5 Effects on female reproductive processes	251
	4.1.5.1 Effect on ovarian follicular cell functions	251
	4.1.5.2 Effect on reproductive hormones	251
	4.1.5.3 Effect on reproductive state	252
	4.1.6 Mechanisms of action on female reproductive processes	252
	4.1.7 Application in reproductive biology and medicine	253
	4.1.8 Conclusions and possible direction of future studies	253
	References	254
4.2	Apigenin	259
	4.2.1 Introduction	259
	4.2.2 Provenance and properties	259
	4.2.3 Physiological and therapeutical actions	260
	4.2.4 Mechanisms of action	262
	4.2.5 Effects on female reproductive processes	263
	4.2.5.1 Effect on ovarian and reproductive state	263
	4.2.6 Effect on ovarian and uterine cell functions	264
	4.2.7 Effect on oocytes and embryos	265
	4.2.8 Effect on reproductive hormones	265
	4.2.9 Effect on response to adverse external factors	266
	4.2.10 Mechanisms of action on female reproductive processes	267
	4.2.11 Application in reproductive biology and medicine	270
	4.2.12 Conclusions and possible direction of future studies	271
	References	272
4.3	Berberine	276
	4.3.1 Introduction	276
	4.3.2 Provenance and properties	276
	4.3.3 Physiological and therapeutic actions	277
	4.3.4 Mechanisms of action	278
	4.3.5 Effects on female reproductive processes	280
	4.3.5.1 Effect on ovaries	280
	4.3.5.2 Effect of vagina and uterus	281
	4.3.6 Mechanisms of action on female reproductive processes	281
	4.3.6.1 Proliferation and apoptosis	281
	4.3.6.2 Oxidative stress	282
	4.3.6.3 Prostaglandins	282
	4.3.6.4 Hormones and growth factors	282
	4.3.7 Application in reproductive biology and medicine	283
	4.3.8 Conclusions and possible direction of future studies	283
	References	284

xii Contents

4.4 Capsaicin 287
  4.4.1 Introduction 287
  4.4.2 Provenance and properties 287
  4.4.3 Physiological and therapeutic actions 288
  4.4.4 Mechanisms of action 290
  4.4.5 Effects on female reproductive processes 293
        4.4.5.1 Effect on ovarian and reproductive state 293
        4.4.5.2 Effect on ovarian cell functions 294
        4.4.5.3 Effect on reproductive hormones 294
  4.4.6 Mechanisms of action on female reproductive processes 294
  4.4.7 Application in reproductive biology and medicine 295
  4.4.8 Conclusions and possible direction of future studies 296
  References 296
4.5 Daidzein 303
  4.5.1 Introduction 303
  4.5.2 Provenance and properties 303
  4.5.3 Physiological and therapeutical actions 305
  4.5.4 Mechanisms of action 306
  4.5.5 Effect on female reproductive processes 307
        4.5.5.1 Effect on ovarian and reproductive state 307
        4.5.5.2 Effect on ovarian cell functions 308
        4.5.5.3 Effect on reproductive hormones 308
        4.5.5.4 Mechanisms of action on female reproductive processes 309
        4.5.5.5 Application in reproductive biology and medicine 311
  4.5.6 Conclusion and possible direction of further studies 312
  References 313
4.6 Diosgenin 317
  4.6.1 Introduction 317
  4.6.2 Provenance and properties 317
  4.6.3 Physiological action 318
  4.6.4 Mechanisms of action 320
  4.6.5 Effects on female reproductive processes 320
        4.6.5.1 Effect on ovarian and reproductive state 320
  4.6.6 Effect on ovarian cell functions 321
  4.6.7 Effect on oocytes and embryos 321
  4.6.8 Effect on reproductive hormones 321
  4.6.9 Mechanisms of action on female reproductive processes 322
  4.6.10 Application in reproductive biology and medicine 323
  4.6.11 Conclusions and possible direction of future studies 324
  References 325
4.7 Isoquercitrin 328
  4.7.1 Introduction 328
  4.7.2 Provenance and properties 328

Contents **xiii**

4.7.3	Physiological and therapeutic actions	330
4.7.4	Mechanisms of action	331
4.7.5	Effects on female reproductive processes	332
4.7.6	Mechanisms of action on female reproductive processes	333
4.7.7	Application in reproductive biology and medicine	334
4.7.8	Conclusions and possible direction of future studies	334
References		334
4.8	**Punicalagin**	338
4.8.1	Introduction	338
4.8.2	Provenance and properties	338
4.8.3	Physiological and therapeutic actions	340
4.8.4	Mechanisms of action	340
4.8.5	Effects on female reproductive processes	342
	4.8.5.1 Effect on reproductive health and fertility	342
	4.8.5.2 Effect on ovarian follicular cell functions	343
	4.8.5.3 Effect on uterus and pregnancy	343
4.8.6	Mechanisms of action on female reproductive processes	343
4.8.7	Application in reproductive biology and medicine	344
4.8.8	Conclusions and possible direction of future studies	345
References		345
4.9	**Quercetin**	349
4.9.1	Introduction	349
4.9.2	Provenance and properties	349
4.9.3	Physiological actions	351
4.9.4	Mechanisms of action	352
4.9.5	Effects on female reproductive processes	355
	4.9.5.1 Effect on ovarian and reproductive state	355
	4.9.5.2 Effect on ovarian cell functions	355
	4.9.5.3 Effect on oocytes and embryos	356
	4.9.5.4 Effect on reproductive hormones	357
	4.9.5.5 Effect on ovarian disfunctions	358
	4.9.5.6 Effect on response to hazardous factors	359
4.9.6	Mechanisms of action on female reproductive processes	360
4.9.7	Application in reproductive biology and medicine	362
4.9.8	Conclusions and possible direction of future studies	363
References		364
4.10	**Resveratrol**	371
4.10.1	Introduction	371
4.10.2	Provenance and properties	371
4.10.3	Physiological and therapeutic actions	372
4.10.4	Mechanisms of action	374
4.10.5	Effects on female reproductive processes	374
	4.10.5.1 Effect on ovarian and reproductive state and on ovarian cell functions	374

xiv Contents

		4.10.5.2 Effect on reproductive hormones	375
		4.10.5.3 Possible causes of variability in resveratrol effects	376
	4.10.6	Mechanisms of action on female reproductive processes	377
	4.10.7	Application in reproductive biology and medicine	378
	4.10.8	Conclusion and possible directions of the further studies	380
	References		382
4.11	**Rutin**		387
	4.11.1	Introduction	387
	4.11.2	Provenance and properties	387
	4.11.3	Physiological actions	389
	4.11.4	Mechanisms of action	390
	4.11.5	Effects on female reproductive processes	391
		4.11.5.1 Effect on ovarian and reproductive state	391
		4.11.5.2 Effect on ovarian cell functions	392
		4.11.5.3 Effect on oocytes and embryos	392
	4.11.6	Effect on reproductive hormones	392
	4.11.7	Mechanisms of action on female reproductive processes	393
	4.11.8	Application in reproductive biology and medicine	394
	4.11.9	Conclusions and possible direction of future studies	395
	References		396

Conclusion	401
Index	403

Preface

From the biological point of view, the sense of life is to ensure the survival of particular genes and their transfer to the next generation, i.e., reproduction. Sexual reproduction provides a good opportunity for generation and selection of particular gene combinations. From the viewpoint of structure and regulation, female reproductive processes are more complicated, than the male ones. Female reproductive system should ensure not only generation of gametes, but also their fertilization, resulted embryogenesis and offspring care. Understanding the internal and external regulators of female reproduction should help to characterize, to predict and to control they, as well as to prevent and to treat reproductive disorders. Currently, the impact of environmental contaminants and food/medicinal herbs on female reproductive health and disease risk is poorly understood. It is clear, however, that nutrition spanning the entire developmental lifespan plays an integral role. Improved insight into the plant molecules and their mechanism of action is likely to have significant implications for our current understanding of female reproductive disorders, and therefore for the health and reproductive potential of future generations.

The occurrence of such disorders is currently growing in all mammalian species (Canipari et al., 2020). Worldwide, the inability to have children affects 10%−20% of all couples. Subfertility affects much more people (Roudebush et al., 2008; Wojsiat et al., 2017; Smith et al., 2018). Females are more susceptible to these disorders because the number of germ cells is limited, and they are not renewable (Canipari et al., 2020). The most common female reproductive disorders are ovarian cancer, polycystic ovarian syndrome, endometriosis and premature ovarian failure. All these disorders jeopardize not only women's fertility, but its health and life.

Pollution of environment is considered as the cause of the current reproductive problems. Pollutants can induce and dysfunctions of endocrine and intracellular regulation of reproductive processes, ovarian state and fertility in humans (Canipari et al., 2020) and domestic animals (Cortinovis et al., 2013). These dysfunctions can be due to ability of pollutant to induce oxidative stress—accumulation of reactive oxygen species, which are destructive for DNA and proteins, oogenesis and embryogenesis (Wojsiat et al., 2017). Oxidative stress and resulted decline in function and quality of ovarian follicles, oocytes and embryos could be prevented by antioxidants (Budani et al., 2020; Timóteo-Ferreira et al., 2021). The most natural, accessible and inexpensive source of

xvi Preface

antioxidants, which can prevent numerous reproductive and nonreproductive disorders are plant extracts and plant constituents—flavonoids (quercetin, daidzein, isoquercitrin, and rutin), ellagitannins (punicalagin, ellagic acid, etc.), glycosides (amygdalin, etc.), sapogenins (diosgenin), stilbenoids (resveratrol) with potentially high antioxidant and curative activity and the ability to eliminate reprotoxicity induced by environmental stress. On the other hand, these natural substances are able to involve to regulation of processes of follicullogenesis, luteogenesis, oogenesis, embryogenesis, and significantly regulate female reproductive functions. Medicinal and functional food plants and their constituents could represent a simple, cheap and efficient tool to control female reproductive processes and to prevent and to treat reproductive disorders induced by environmental contaminants and oxidative stress (Akkol et al., 2020; Mihanfar et al., 2021; Sirotkin et al., 2021).

Therefore, knowledge concerning character, properties, effects and mechanisms of action of environmental contaminants, plants, and plant molecules on the cells of female reproductive system could be useful for management of human and animal reproduction and prevention and treatment of its disorders. Nevertheless, to our knowledge, this knowledge was not summarized in one book yet.

The present book tries to address the following queries:

- How some common environmental contaminants affect female reproductive processes, and how could be mechanisms of their effects?
- How some food or medical plants affect female reproductive processes, and how could be mechanisms of their effects, and what for their molecules are responsible for these plant effects?
- Whether some food or medical plants and plant molecules could be useful as a regulators of female reproduction, drugs for treatment of female reproductive disorders and natural protector against contaminant effects?

This book is an attempting to summarize the available knowledge concerning effect of most common environmental contaminants (heavy metals, mycotoxins, oil-related contaminants, tobacco smoking), food/medicinal herbs (apricot, seeds, black elder, buckthorn, buckwheat, curcuma/turmeric, flaxseed, ginkgo, grape, pomegranate, puncture vine, rooibos, tea, vitex) and plant constituents (amygdalin, apigenin, barberine, capsaicin, daidzein, diosgenin, isoquercitrin, punicalagin, quercetin, resveratrol, and rutin). Their provenance and properties, nonreproductive physiological and therapeutic actions, and their mechanisms are shortly outlined. More detailed is described and analyzed their influence on various targets and processes in female reproductive systems, possible mechanisms of their effects, possible areas of application in reproductive biology and medicine, as well as the limitations of the existing knowledge and possible areas of their further studies. For better understanding the targets and mechanisms of action of contaminants and plant preparations, the book is started with a short description of basic principles and regulators of female reproductive processes. Each chapter is supported with the key literary sources.

Search for literature was performed in agreement with the PRISMA (= preferred reporting items for systematic review) criteria (Moher et al., 2009). To search the related articles, MedLine/Pubmed, Web of Science, SCOPUS databases between the year 1995 and 2021 were used. In cases of repeated or conflicting information or references, more recent and full sources have been preferred.

The authors hope, that this book could be interested for scientists working in areas of biology, physiology, cellular and molecular biology, ecology, for specialists in pharma industry, human and veterinary medicine, agriculture/animal production, food production, toxicologists, pharmacologists, pharmaceutical scientists, pharmacists, nutritionists, medicinal and natural product chemists, specialists in phytotherapy and naturopathy, healthcare professionals—pharmacists, doctors, and nurses—with an interest in herbal therapeutics and functional food, as well as the common public.

Finally, the authors would like to take this opportunity to thank MSc. Simona Baldovska, PhD. and MSc. Michal Mihal for their assistance and collection of data, as well as PaedDr. Ingrid Sulanova, MSc and Dr. Timea Sommerova for text language editing. The results of the research, which was financially supported by the Ministry of Education, Science, Research and Sport of the Slovak Republic (projects APVV-18-0312, APVV-15-0296, DS-FR-19-0049, VEGA 1/0266/20, VEGA 1/0327/16, VEGA 1/0392/17, KEGA 033SPU-4/2021), and the Excellent scientific team "Center of Animal Reproduction (CeRA)," constituted the significant basis for writing this book.

References

Akkol EK, Dereli FTG, Sobarzo-Sánchez E, Khan H. Roles of medicinal plants and constituents in gynecological cancer therapy: current literature and future directions. Curr Top Med Chem 2020;20(20):1772—90. https://doi.org/10.2174/1568026620666200416084440.

Budani MC, Tiboni GM. Effects of supplementation with natural antioxidants on oocytes and preimplantation embryos. Antioxidants 2020;9(7):612. https://doi.org/10.3390/antiox9070612.

Canipari R, De Santis L, Cecconi S. Female fertility and environmental pollution. Int J Environ Res Public Health 2020;17(23):8802. https://doi.org/10.3390/ijerph17238802.

Cortinovis C, Pizzo F, Spicer LJ, Caloni F. Fusarium mycotoxins: effects on reproductive function in domestic animals-a review. Theriogenology 2013;80(6):557—64. https://doi.org/10.1016/j.theriogenology.2013.06.018.

Mihanfar A, Nouri M, Roshangar L, Khadem-Ansari MH. Polyphenols: natural compounds with promising potential in treating polycystic ovary syndrome. Reprod Biol 2021;21(2):100500. https://doi.org/10.1016/j.repbio.2021.100500.

Moher D, Liberati A, Tetzlaff J, Altman DG. Preferred reporting items for systematic reviews and meta-analyses: the PRISMA statement. PLoS Med 2009;6(6):e1000097.

xviii Preface

Roudebush WE, Kivens WJ, Mattke JM. Biomarkers of ovarian reserve. Biomark Insights 2008;3:259−68. https://doi.org/10.4137/bmi.s537.

Sirotkin AV, Alwasel SH, Harrath AH. The influence of plant isoflavones daidzein and equol on female reproductive processes. Pharmaceuticals 2021;14(4):373. https://doi.org/10.3390/ph14040373.

Smits RM, Mackenzie-Proctor R, Fleischer K, Showell MG. Antioxidants in fertility: impact on male and female reproductive outcomes. Fertil Steril 2018;110(4):578−80. https://doi.org/10.1016/j.fertnstert.2018.05.028.

Timóteo-Ferreira F, Abreu D, Mendes S, Matos L, Rodrigues AR, Almeida H, Silva E. Redox imbalance in age-related ovarian dysfunction and perspectives for its prevention. Ageing Res Rev 2021;68:101345. https://doi.org/10.1016/j.arr.2021.101345.

Wojsiat J, Korczyński J, Borowiecka M, Żbikowska HM. The role of oxidative stress in female infertility and in vitro fertilization. Postepy Hig Med Dosw 2017;71(0):359−66. https://doi.org/10.5604/01.3001.0010.3820.

Chapter 1

Female reproductive system and its regulation

1.1 Introduction

The most important system for regulating female reproductive endocrine function is the hypothalamus-pituitary-ovary axis. Its dysfunction leads to the development of endocrine-related disease, such as congenital hypogonadotropic hypogonadism, polycystic ovarian syndrome (PCOS), and premature ovarian failure, anovulation, and others that are global causes of infertility. Oxidative stress and related inflammatory processes are causal factors and key promoters of all kinds of reproductive disorders including cancer that acts by dysregulating the expression of related genes. Novel extra- and intracellular regulators of reproductive functions due to their effects on secretory activity, cell proliferation, differentiation, and apoptosis in a variety of cells and tissues in the body, have gained attention in the field of reproductive sciences including female sexual maturation. Thus, understanding the signaling mechanisms that underlie the control of ovarian development is important for control of reproduction and the diagnosis of and intervention for female reproductive disorders.

1.2 The ovary

The ovary is a central organ of the female reproductive system. It has two main functions: the production of gametes (oocytes) and the secretion of signaling and regulatory substances. The secretions of the ovary influence other parts of the reproductive system, including those which control female maturation, cyclicity, gamete production, and behavior (brain, pituitary gland, and the ovary itself), and those which support embryo development, gestation, and lactation (reproductive tract, uterus, mammary gland). The ovary has two structural regions: the outer cortex contains follicles at various stages of development together with structures derived from follicles, while the inner medullar consists mainly of stromal tissues and vascular elements. Ovarian structures and their cyclic changes are described in special reviews and studies (Gougeon, 2004; Oktem and Oktay, 2008; Peter et al., 2009; Aerts and Bols,

Environmental Contaminants and Medicinal Plants Action on Female Reproduction
https://doi.org/10.1016/B978-0-12-824292-6.00024-6
Copyright © 2022 Elsevier Inc. All rights reserved.

1

2010a,b; Kolesarova et al., 2015). Genotypic females (XX) develop two ovaries that sit adjacent to the uterine horns. Each ovary is anchored at the medial pole by the utero-ovarian ligament to the uterus. The lateral ovarian pole is anchored to the pelvic sidewall by the infundibulopelvic ligament (i.e., suspensory ligament of the ovary), which carries the ovarian artery and vein. Each ovary contains one to two million primordial follicles that each contain primary oocytes (i.e., eggs) that can supply that female with enough follicles until she reaches her fourth or fifth decades of life. These primordial follicles are arrested in Prophase I of meiosis until the onset of puberty. At the onset of pubescence, the gonadotropic hormones began to induce the maturation of the primordial follicle allowing for completion of Meiosis I forming a secondary follicle. The secondary follicle begins Meiosis II, but this phase will not be completed unless that follicle is fertilized. With each ovulatory cycle, the number of follicles decreases eventually leading to the onset of menopause or the cessation of ovulatory function. Per each ovulation cycle, the average ovary loses 1000 follicles to the process of selecting a dominant follicle that will be released. This process accelerates in an age-dependent manner as well. It is also a common thought that the right and left ovaries alternate follicular releases each month (Holesh et al., 2021).

1.2.1 Folliculogenesis

Each follicle represents the basic functional unit of the ovary. From its very initial stage of development, the follicle consists of an oocyte surrounded by somatic cells. The oocyte grows and matures to become fertilizable, and the somatic cells proliferate and differentiate into the major suppliers of steroid sex hormones as well as generators of other local regulators. The process by which a follicle forms, proceeds through several growing stages, develops to eventually release the mature oocyte, and turns into a *corpus luteum* is known as "folliculogenesis" (Gershon and Dekel, 2020). Follicle growth and oocyte maturation are associated with dynamic transcriptional regulation in both oocyte and granulosa cells compartments of the follicle (Sánchez and Smitz, 2012). The formation of the primordial follicles is the first stage of folliculogenesis, a complex but well-organized physiological process. Activated primordial follicles, having a single layer of flattened granulosa cells surrounding the primordial oocytes, sequentially differentiate into primary, secondary, and eventually antral (Graafian) follicles (McGee and Hsueh, 2000; Atwood, 2016; Shah et al., 2018).

Much of the structure of the female reproductive tract develops prenatally (Cunha et al., 2018); however, the early ovarian developmental process occurs during the embryonic and postnatal period and is regulated through a series of molecular signaling events (Wear et al., 2017). The dynamic transcriptional regulation and interactions of human germlines and surrounding somatic cells during folliculogenesis remain unknown (Zhang et al., 2018). In women, the

Female reproductive system and its regulation **Chapter | 1** **3**

nongrowing population of follicles that comprise the ovarian reserve is determined at birth and serves as the reservoir for future fertility. This reserve of dormant, primordial follicles, and the mechanisms controlling their selective activation which constitute the committing step into folliculogenesis are essential for determining fertility outcomes in women (Ford et al., 2020).

Primordial germ cells (precursors of oocytes) appear within the embryonic ovary and multiply by mitosis during prenatal ontogenesis. During postnatal ontogenesis, several thousands of germ cells cease their mitosis and begin meiosis, becoming oogonia and then oocytes. Neighboring ovarian cells differentiate into follicular somatic cells, under the influence of oogonia/oocyte-derived factors (Oktem and Oktay, 2008; Tingen et al., 2009).

Each oocyte becomes enclosed by one layer of flat follicular cells to form a primordial follicle. Subsequent proliferation of follicular cells leads to the formation of a **primary follicle** surrounded by several layers of cuboidal, epithelial cells. Under the influence of the oocyte, which is now in a state of meiotic arrest, the cells differentiate and produce a membrane made of protein and glycoprotein called the *zona pellucida*; they also supply the oocyte with nutrients and secrete signaling substances, including hormones such as estrogen, which regulate oogenesis and other components of the reproductive system. On the other hand, oocyte through oocyte-specific secretes, promotes development of surrounding follicle (Fortune, 2003; Craig at al., 2007; Hutt and Albertini, 2007; McLaughlin and McIver, 2009). The single layer of cells immediately adjacent to the oocyte, the *corona radiata*, has long processes which penetrate the zona pellucida and connect with the oocyte by gap junctions. The layers of more elongated cells surrounding these are called granulosa cells (and form the *zona granulosa* or *stratum granulosum*) and they initially fill the space between the corona cells and the outer basement membrane (*basal lamina* or *membrana basalis*) of the follicle. The outermost (mural) layer of granulosa cells is columnar rather than cuboidal and have intimate contact with the extracellular matrix components which make up the membrane (including collagen Type IV, laminin, and heparan sulfate proteoglycan). The basement membrane is synthesized by the granulosa cells, working in cooperation with cells on its outer side, which makes up the theca interna. The theca interna is comprised of several different types of cells including fibroblasts, capillary endothelial cells, and endocrine cells. The latter secrete a range of steroid and protein hormones, some of which may be transported across the basement membrane to regulate the activity of the granulosa cells and the oocyte and to be converted into other hormones. For example, in many species thecal androgen is converted by granulosa cells into estrogens which is then carried into the general circulation. A further external layer of tissue, the theca externa, consists of structural cells, connective tissue and capillaries. The latter supply the whole follicle with nutrients, gases and extraovarian signaling molecules. They reach into the theca interna but do not penetrate the follicular basement membrane. This means that the internal parts

4 Contaminants and Plants Action on Female Reproduction

of the follicle (granulosa, corona, oocyte) are avascular and must receive blood-born nutrients, gases, and other substances by transfer across the follicle wall (Gougeon, 2004).

During early follicular growth, granulosa cells divide and fill the expanding follicle. At a certain stage of growth, the volume of the follicle increases more rapidly than the total volume of the proliferating granulosa cells. Some inner layers of cells may also die or fail to replicate, possibly because of limited nutrient availability. At this stage, the differential volume of the expanding follicle fills with a fluid (follicular fluid or *liquor folliculi*) and a chamber or *antrum* is formed, which largely separates the outer granulosa cell layers from the oocyte and corona. Some granulosa cells may remain contiguous with the corona cells, forming a short stalk called the *cumulus oophorus*. Follicles reaching this stage of development are called **secondary or antral follicles** (Gougeon, 2004; Oktem and Oktay, 2008; Aerts and Bols, 2010a).

At each stage of folliculogenesis, from the primordial and primary stages and through each level of antral growth, there is an intensive selection and loss of follicles. As a result, less than 1% of follicles ovulate and more than 99% of the primordial follicles originally present in the ovary are subjected to degeneration (known as *atresia*) due to programmed cell death (*apoptosis*). Final destiny of follicle depends on balance between apoptosis and proliferation rate in follicular cells (Craig et al., 2007; Webb and Campbell, 2007; Adams et al., 2008; Zhou et al., 2019). Autophagy and apoptosis are involved in the regulation of both primordial follicular development as well as atresia (Zhou et al., 2019). In growing follicles, only a subset of oocytes is capable of maturation, fertilization, and early embryo development. This proportion of competent oocytes is increasing along with size of oocyte and follicular and decreasing along with occurrence of follicular atresia/apoptosis (Feng et al., 2007; Mermillod et al., 2008). In one ovarian cycle one distinguishes from one to four (mainly two or three) so-called follicular waves, during whose particular cohort of follicles are entrained to recruitment, rapid growth, and atresia (Adams et al., 2008; Mihm and Evans, 2008; Peter et al., 2009). Approximately two-thirds of women develop two follicle waves throughout an interovulatory interval and the remainder exhibit three waves of follicle development. Major and minor waves of follicle development have been observed. Major waves are those in which a dominant follicle develops; dominant follicles either regress or ovulate. In minor waves, physiologic selection of a dominant follicle is not manifest (Baerwald and Pierson, 2020).

1.2.2 Preovulatory changes and ovulation

Several of the follicles undergo atretic degeneration while a few of them, under the stimulation of cyclic gonadotropins during puberty, reach the preovulatory stage. In females who have reached the reproductive age, the preovulatory/Graafian follicles are the main source of ovarian estrogens

(Zeleznik, 2001; Hsueh et al., 2015; Atwood, 2016). Ovulation is the appropriately timed release of a mature, developmentally competent oocyte from the ovary into the oviduct, where fertilization occurs. Importantly, ovulation is tightly linked with oocyte maturation, demonstrating the interdependency of these two parallel processes, both essential for female fertility. Initiated by pituitary gonadotropins, the ovulatory process is mediated by intrafollicular paracrine factors from the theca, mural, and cumulus granulosa cells, as well as the oocyte itself. The result is the induction of cumulus expansion, proteolysis, angiogenesis, inflammation, and smooth muscle contraction, which are each required for follicular rupture (Robker et al., 2018). Under the influence of the preovulatory gonadotropin surge during each reproductive cycle, the dominant Graafian follicle releases the mature oocyte into the fimbriae of the fallopian tube while the remaining theca and granulosa cells differentiate into the progesterone secreting *corpus luteum*. Cyclic changes induced in the endometrium lead to menstruation in the female organism. The total number of ovarian follicles is determined early in life, and depletion of this pool culminates into reproductive senescence. The fate of each follicle is controlled by diverse autocrine and paracrine factors (Hsueh et al., 2015; Atwood, 2016; Shah et al., 2018), including the activins and inhibins (Adu-Gyamfi et al., 2020).

1.2.3 Luteogenesis and luteolysis

After ovulation, the follicle wall in the majority of mammalian species develops into a new tissue called the *corpus luteum*. Its structure, life span and regulators are described in several reviews (Niswender et al., 2000, 2002; Meidan et al., 2005; Skarzynski et al., 2008). The *corpus luteum* formed from the luteinized somatic follicular cells following ovulation, vasculature cells, and immune cells is critical for progesterone production and maintenance of pregnancy. Follicular theca cells differentiate into small luteal cells that produce progesterone in response to luteinizing hormone (LH), and granulosa cells luteinize to become large luteal cells that have a high rate of basal production of progesterone. The formation and function of the *corpus luteum* rely on the appropriate proliferation and differentiation of both granulosa and theca cells. If any aspect of granulosa or theca cell luteinization is perturbed, then the resulting luteal cell populations (small luteal cells, large luteal cells, vascular, and immune cells) may be reduced and compromise progesterone production. Thus, many factors that affect the differentiation/lineage of the somatic cells and their gene expression profiles can alter the ability of a *corpus luteum* to produce the progesterone critical for pregnancy (Abedel-Majed et al., 2019).

The *corpus luteum* of its luteal phase produces progesterone, which inhibits endometrial proliferation and determines endometrial receptivity (Mesen and Young, 2015). As well as conditioning the uterus to support implantation and pregnancy, progestagens and other hormones from the *corpus luteum* limit the development of other follicles, thereby preventing maturation and ovulation until the *corpus luteum* regresses (luteolysis) (Meidan et al., 2005; Skarzynski et al., 2008). Luteolysis consists of two phases, functional luteolysis, and structural luteolysis. A rapid functional regression of *corpus luteum* is characterized by a decrease of progesterone production, which induce structural regression (Teeli et al., 2019). During structural luteolysis, luteal cells undergo apoptosis and autophagy. Histologically, the first indication of this is the apoptotic shrinkage of the large luteal cells. By contrast the small cells appear selectively hyperstimulated during early luteolysis. At the later stages of luteolysis both cell types become apoptotic and are eventually destroyed, leading to the morphological regression of the *corpus luteum*. All that remains is a nodule of dense, functionless connective tissue called the *corpus albicans* (Niswender et al., 2000; Teeli et al., 2019).

1.2.4 Oocytes

Oocytes production is the most important generative function of the ovary. The oocyte is located within ovarian follicles surrounded by a particular group of granulosa cells—cumulus oophorus. Oocyte is maturing together with its ovarian follicle. Oocyte in the growing follicles is on the diplotene-dictionhene phase of meiosis, which is initiated before ovulation and accomplished after penetration of oocyte by spermatozoa. To be able to be fertilized and to generate embryo, oocyte should accomplish meiosis (nuclear maturation), which is finished by generation of haploid oocyte. Fusing of haploid spermatozoa with haploid oocyte generates diploid zygote, which, after transfer through fallopian tubes, is implanting into uterus where prenatal ontogenesis occurs.

To enable fertilization and the first stages of embryogenesis, oocyte should accomplish also so-called cytoplasmic maturation—accumulation and redistribution of structures and molecules necessary for completion of meiosis, first divisions of zygote, and embryo development before start of embryonal own biosynthetic processes. Both nuclear and cytoplasmic maturation of oocytes are under control of hormones and other factors of pituitary and ovarian origin. On the other hand, growth factors produced by oocyte promote ovarian follicle development (Sirotkin, 2014; Lonergan and Fair, 2016).

1.3 Fallopian tubes

The anatomy of the fallopian tube is complex starting from its embryological development and continuing with its vascular supply and ciliated microstructure, that is the key of the process of egg transport to the site of fertilization (Briceag et al., 2015). The Fallopian tubes are muscular conduits connecting the ovaries with the uterus and are divided into the following

Female reproductive system and its regulation Chapter | 1   7

regions: fimbriated infundibulum, ampulla and isthmus. At puberty, the extrauterine portion of the tube measures approximately 11 cm and the intramural portion is 1.5–2 cm long, these dimensions are concealed by the convolutions of the duct in situ within its supporting mesentery, the mesosalpinx (Moore and Dalley, 2006).

There are many strongly documented causes of tubal infertility: infections (Chlamydia Trachomatis, Gonorrhea, and genital tuberculosis), intrauterine contraceptive devices, endometriosis, and complications after abdominal surgery. Around 30% of the infertile women worldwide have associated fallopian tubes pathology. Unfortunately, for a long time, this aspect of infertility has been neglected due to the possibility of bypassing this deadlock through *in vitro* fertilization (Briceag et al., 2015).

1.4 Uterus

The anatomy of the uterus is defined with the angles of the vagina, cervix, and uterine corpus. Hereunder there are angles of version and flexion (Fidan et al., 2017). Each month the endometrium becomes inflamed, and the luminal portion is shed during menstruation. The subsequent repair is remarkable, allowing implantation to occur if fertilization takes place. Aberrations in menstrual physiology can lead to common gynecological conditions, such as heavy or prolonged bleeding. Increased knowledge of the processes involved in menstrual physiology may also have translational benefits at other tissue sites. Menstruation occurs naturally in very few species. Human menstruation is thought to occur as a consequence of preimplantation decidualization, conferring embryo selectivity and the ability to adapt to optimize function. The study of menstruation, in both normal and abnormal scenarios, is essential for the production of novel, acceptable medical treatments for common gynecological complaints (Maybin and Critchley, 2015).

Endometriosis is a chronic inflammatory condition defined by the presence of glands and stroma outside the uterine cavity. It occurs in 7%–10% of all women of reproductive age and may present as pain or infertility. The pelvic pain may be in the form of dysmenorrhea, dyspareunia, or pelvic pain itself (Brown et al., 2012). Furthermore, endometriosis is considered as a cause of ovarian cancer (Shih et al., 2021; see also below).

1.5 Extra- and intracellular regulators of female reproductive processes

The hypothalamic-pituitary-ovarian axis is a tightly regulated system controlling female reproduction.

Gonadotropin-releasing hormone (GnRH) is a tropic peptide hormone made and secreted by the hypothalamus. It is a releasing hormone that stimulates the release of follicle-stimulating hormone (FSH) and LH from the anterior pituitary gland through variations in GnRH pulse frequency. Low-frequency GnRH pulses are responsible for FSH secretion whereas high-frequency pulses are responsible for LH secretion. Furthermore, GnRH is able to affect the number of LH/hCG (human chorionic gonadotropin) receptors and the response of granulosa cells and their response to gonadotropins. Furthermore, GnRH can directly regulate the release of hormones by ovarian cells and promote oocyte maturation (Sirotkin, 2014).

The release of GnRH and its pulses are controlled by the feedback mechanisms including ovarian steroid hormones—estrogens (see later in the chapter). During the follicular phase of the ovarian cycle, granulosa cells autonomously increase they own production of estrogen contributing to elevation in estrogen serum levels. This elevation is communicated to the hypothalamus and contributes to the increase in GnRH pulse frequency eventually stimulating the LH surge that eventually induces the follicular rupture and release from the *corpus luteum* and luteinization of the granulosa cells enabling the synthesis of progesterone in place of estrogen. Finally, the low levels of LH following the surge restarts the FSH production by the slow-pulsation frequency of GnRH release (Sirotkin, 2014; Holesh et al., 2021).

Kisspeptin is considered as a physiological promoter of GnRH production, as well as promoter or various reproductive processes. The hypothalamic kisspeptin neurons have receptors for adipokine leptin and estrogen receptor α (ERα) which suggest that kisspeptin is a hormone-mediating effect of metabolism and estrogens on GnRH secretion. In addition, kisspeptin is a direct autocrine and paracrine up-regulator of follicle development, oocyte maturation, embryo blastocysts and trophoblast development, and implantation and placentation. Its potential benefits for promotion of puberty and ovarian folliculogenesis and cyclicity, hypothalamo-pituitary-ovarian axis, embryo survival, development, trophoblast attachment, implantation and pregnancy can be proposed (D'Occhio et al., 2020).

Follicle-Stimulating Hormone. Gonadotropins FSH and LH are heterodimeric glycoproteins with alpha/beta subunits. The hormone-specific FSHβ-subunit is noncovalently associated with the common α-subunit that is also present in the LH, another gonadotrophic hormone secreted by gonadotrophs and thyroid-stimulating hormone (TSH) secreted by thyrotrophs (Das and Kumar, 2018). FSH is produced by pituitary gonadotrophs, which release it in response to slow-frequency pulsatile GnRH (Holesh et al., 2021).

FSH receptors are G-protein coupled receptors are found in the granulosa cells in developing ovarian follicles. FSH stimulates ovarian follicular growth, as well as growth and maturation of immature oocytes in mature (Graafian) secondary follicles before ovulation. Furthermore, it promotes generation of the LH receptors in this follicle that, because of its FSH-dependent maturation, is capable of ovulation and forming a corpus luteum in response to the mid-cycle surge of LH (Sirotkin, 2014).

Luteinizing Hormone is a gonadotropin synthesized and secreted by the anterior pituitary gland in response to high-frequency GnRH release. LH is responsible for inducing ovulation, preparation for fertilized oocyte uterine implantation, and the ovarian production of progesterone through stimulation of theca cells and luteinized granulosa cells. Prior to the LH surge, LH interacts with theca cells that are adjacent to granulosa cells in the ovary. These cells produce androgens that diffuse into the granulosa cells and convert to estrogen for follicular development. The LH surge creates the environment for follicular eruption by increasing the activity of the proteolytic enzymes that weaken the ovarian wall allowing for passage of the oocyte. After the oocyte is released, the follicular theca and luteinized granulosa cells are transformed to luteal cells. Their function is now to produce mainly progesterone which is the hormone responsible for maintaining the uterine environment that can accept a fertilized embryo (Holesh et al., 2021).

The relationship between FSH and LH hormones is responsible for the process that induces follicular development, rupture, release, and endometrial reception or shedding. Disruption in the hormonal communication between the gonadotropin-releasing hormones, gonadotropic hormones, and their receptors can lead to anovulation or amenorrhea leading to various pathologic sequelae as a consequence (Holesh et al., 2021).

Prolactin. This peptide hormone is produced in a number of tissues including ovarian granulose cell, but the most known source of prolactin is anterior pituitary. Prolactin was reported to affect release of ovarian steroid and peptide hormones, as well as the number of macrophages, regulated state, and promoting lysis of *corpus luteum*. The involvement of prolactin in control of fecundity is not demonstrated yet, although it can be involved in so-called lactational infertility during suckling, which is associated with increased prolactin release (Sirotkin, 2014). Despite the temporal association of high prolaction release with ovarian inactivity and the influence of prolactin on ovarian functions listed above, no association between polymorphysm in prolactin genes and pig fertility rate was found (Kątska-Książkiewicz et al., 2006).

Steroid hormones progestins, androgens, and estrogens are synthesized from cholesterol by the ovary in a sequential manner, with each serving as substrate for the subsequent steroid in the pathway. The two-cell, two gonadotrophin model describes the role of theca and granulosa cells in the production of steroids, highlighting the cooperation between the two-cell types, which is necessary for estrogen production. According to this model, in the theca, under the influence of LH, cholesterol is converted to pregnenolone and metabolized through a series of substrates ending in androgen production. Androgens produced by the theca cells transported to the

10 Contaminants and Plants Action on Female Reproduction

granulosa cells where they are aromatized to estrogens under the influence of FSH (Sirotkin, 2014).

Progesterone is a steroid produced primarily by the *corpus luteum* and the placenta (Henderson, 2018) and an essential hormone in the process of reproduction (Di Renzo et al., 2016). Progesterone can inhibit mitosis of granulosa cells, transition of primordial follicles to primary follicle, and suppress apoptosis in granulosa cells. Progesterone is necessary for induction of ovulation of Graafian follicle by two ways: induction of preovulatory LH surge and promotion of the production of proteolytic enzymes important for the rupture of follicles at ovulation, either directly, or by enhancing endometrial relaxin production, which is thought to stimulate the release of proteases by granulosa cells (Sirotkin, 2014). Progesterone is involved in the control of menstrual cycle: it is responsible for preparing the endometrium for the uterine implantation of the fertilized egg and maintenance of pregnancy (Di Renzo et al., 2016; Holesh et al., 2021). If a fertilized egg implants, the corpus luteum secretes progesterone in early pregnancy until the placenta develops and takes over progesterone production for the remainder of the pregnancy (Holesh et al., 2021). Therefore, it is used in gynecology as a drug for the maintenance of pregnancy (Di Renzo et al., 2016).

Androgens and estrogens are known to be critical regulators of mammalian physiology and development. While these two classes of steroids share similar structures (in general, estrogens are derived from androgens via the enzyme aromatase), they subserve markedly different functions via their specific receptors. In the past, estrogens such as estradiol were thought to be most important in the regulation of female biology, while androgens such as testosterone and dihydrotestosterone were believed to primarily modulate development and physiology in males. However, the emergence of patients with deficiencies in androgen or estrogen hormone synthesis or actions, as well as the development of animal models that specifically target androgen- or estrogen-mediated signaling pathways, have revealed that estrogens and androgens regulate critical biological and pathological processes in both males and females. While estrogens are considered to be the dominant sex steroid in women, in fact, serum androgen levels in women are higher than estrogen levels most of the time. The exception is during the preovulatory and midluteal phases of the menstrual cycle, when androgen and estrogen levels are similar. Therefore, it is reason-able to consider that androgens might have important physiologic effects in women. With regard to steroid metabolism, most androgen actions are likely mediated by intracellular conversion to dihydrotestosterone; thus, it is unclear whether serum testosterone levels truly reflect active androgen levels. In addition, testosterone is readily converted to 17-β-estradiol (E2) by aromatase in most tissues; therefore, observations associated with high testosterone levels might really reflect estrogen actions (Hammes and Levin, 2019).

Androgens are promoting proliferation of follicular cells, recruitment and development of ovarian follicles up to preovulatory stage, either stimulate or suppress development of Graafian follicles and their ovulation, increase apoptosis and follicular atresia, at different stages of folliculogenesis and promotes oocyte nuclear maturation. Testosterone treatments altered release of progesterone, estradiol and other hormones by ovarian cells. Furthermore, they increase the number of FSH and the insulin-like growth factor I (IGF-I) receptors in primary and periovulatory follicles and the response of follicles to these hormones. In contrast to FSH receptors, androgens are reported to inhibit FSH-stimulated LH receptor expression by granulosa cells, which could explain the inhibitory action of androgens on late, LH-dependent, stages of folliculogenesis (Sirotkin, 2014).

Estrogens are steroid hormones which are responsible for the growth and regulation of the female reproductive system and secondary sex characteristics. Estrogens are produced mainly by the granulosa cells of the developing follicle and exerts negative feedback on LH production in the early part of the menstrual cycle. However, once estrogen levels reach a critical level as oocytes mature within the ovary in preparation for ovulation, estrogen begins to exert positive feedback on LH production, leading to the LH surge through its effects on GnRH pulse frequency. The granulosa cells —derived estrogen needed to maturate the developing dominant follicle. Estrogen also has many other effects that are important for bone health, cardiovascular health, and other nonreproductive processes in premenopausal patients. The preovulatory FSH and LH surge causes ovulation of ovarian follicle, luteinization of the granulosa cells and generation LH receptors. This transition enables granulosa cells to respond to LH and produce progesterone (Holesh et al., 2021).

Thyroid hormones are vital for the proper functioning of the female reproductive system, since they modulate the metabolism and development of ovarian, uterine, and placental tissues. Therefore, hypo- and hyperthyroidism may result in subfertility or infertility in both women and animals. Other well-documented sequelae of maternal thyroid dysfunctions include menstrual/estral irregularity, anovulation, abortion, preterm delivery, preeclampsia, intrauterine growth restriction, postpartum thyroiditis, and mental retardation in children (Silva et al., 2018).

The **renin-angiotensin system** is widely expressed in the ovary, uterus, vagina, and placenta. Angiotensin II (Ang II), angiotensin-converting enzyme 2 (ACE2) and Ang-(1—7) regulate follicle development and ovulation, modulate luteal angiogenesis and degeneration, and also influence the regular changes in endometrial tissue and embryo development (Jing et al., 2020). Novel coronavirus (2019-nCoV) virus invades the target cell by binding to ACE2 and modulates the expression of ACE2 in host cells. ACE2, a pivotal component of the renin-angiotensin system, exerts its physiological functions by modulating the levels of Ang II and Ang-(1—7). ACE2 is widely expressed in the ovary, uterus, vagina and placenta. Ang II, ACE2, and Ang-(1—7) regulate follicle development and ovulation, modulate luteal angiogenesis and degeneration, and also influence the regular changes in endometrial tissue and embryo development. Taking these functions into account, 2019-nCoV may disturb the female reproductive functions through regulating ACE2 (Jing et al., 2020).

Insulin signaling acts cooperatively with gonadotropins in mammals and lower vertebrates to mediate various aspects of ovarian development, mainly owing to evolution of the endocrine system in vertebrates (Das and Arur, 2017). In addition, insulin shares its receptors and, therefore, interacts with insulin-like growth factors, the important regulators of ovarian cell proliferation, apoptosis and fecundity (Sirotkin, 2014).

The insulin-like growth factors (IGFs) (IGF-I and IGF-II) are polypeptides with high sequence similarity to insulin. IGF-I has probably no effects on primordial follicle development, but both IGF-I and IGF-II stimulate growth of secondary follicles. In antral follicles, these IGFs stimulate granulosa cell proliferation and steroidogenesis and inhibit apoptosis. Furthermore, IGFs augment expression of FSH and LH receptors and response of granulosa and theca cells and oocytes to gonadotropins. Moreover, IGFs are considered as main local mediators of gonadotropin action in the ovary. IGF-I and −II are potent promoters of ovarian secretory activity: they stimulate release of progesterone, testosterone, estradiol and other hormones by ovarian cells. Both IGF-I and IGF-II promote maturation of oocytes and embryogenesis (Sirotkin, 2014).

Vascular endothelial growth factor (VEGF) is a potent angiogenic factor produced by various cells including ovarian cells. It is mitogen and survival factor for the vascular endothelium, which induces formation of new capillaries in target organs. Regulation of follicular angiogenesis has been shown to be important for the development of ovulatory follicles and *corpus luteum.* Administration of VEGF has been shown to stimulate preantral follicular growth, increase the number of preovulatory follicles and supports maintenance of *corpus luteum.* Manipulation of the angiogenic process may affect ovarian folliculogenesis and embryogenesis. Furthermore, VEGF can be involved in control of some reproductive disorders. At least, women with polycystic ovarian syndrome, ovarian hyperstimulating syndrome, and ovarian tumors has increased VEGF expression in ovaries. On the other hand, VEGF blockers can inhibit ovarian cancerogenesis (Sirotkin et al., 2014).

Prostaglandins (PGs) F and E groups are produced by various tissues, but mainly by myometrium and endometrium of the uterus. LH and ovarian steroids are involved in regulating endometrial PG production in many species (Piotrowska-Tomala et al., 2020).

PGs are involved in control of reproduction at the level of the pituitary, ovary, uterus, placenta. They play an important role in maternal recognition of pregnancy, implantation, maintenance of gestation, microbial-induced abortion, parturition, postpartum uterine and ovarian infections, and resumption of postpartum ovarian cyclicity. In the ovary, the most known functions of PGs are control of ovulation, luteogenesis, luteolysis, and stimulation of oocyte maturation and cumulus oophorus expansion. During *Corpus luteum* development, prostaglandin F2α can increase number of both small and large luteal cells, but in the second half of luteal cycle, prostaglandin F2α is potent luteolytic agens. Prostaglandin E₂ is considered as a luteotropic. Moreover, prostaglandin E₂ can protect *corpus luteum* from the luteolytic effect of prostaglandin F2α. Prostaglandins are involved in the control of release of some ovarian hormones (Sirotkin, 2014).

Cytokines are originally referred as immunomodulating signaling substances that are secreted by certain cells of the immune system. Some cytokines are however secreted by ovarian follicular cells and oocytes. They are involved in control of follicular development, ovulation, uterine receptivity, decidualization and placentation (Sirotkin, 2014). Inflammation promotes the production of inflammatory cytokines, which orchestrates the antiinfectious innate immune response. However, an overzealous production, leading up to a cytokine storm, can be deleterious and contributes to suppression of reproductive processes and even mortality consecutive to sepsis or toxic shock syndrome (Cavaillon, 2018).

Mammalian target of rapamycin (mTOR) regulators can affect ovarian functions. The mTOR system regulates healthy human ovarian cell proliferation and apoptosis and indicates that the action of mTOR regulators on ovarian cell apoptosis can be mediated by the transcription factor p53 (Sirotkin et al., 2019).

Oxidative stress and related inflammatory processes are factor and key promoter of all kinds of reproductive disorders related to granulosa cell apoptosis that acts by dysregulating the expression of related genes (Zhang et al., 2019; Marí-Alexandre et al., 2019). The occurrence of oxidative stress is due to the excessive production of reactive oxygen species (ROS). ROS are a double-edged sword; they do not only play an important role as secondary messengers in many intracellular signaling cascades, but they also exert indispensable effects on pathological processes involving the female genital tract. ROS and antioxidants join in the regulation of reproductive processes in both animals and humans. Imbalances between prooxidants and antioxidants could lead to a number of female reproductive diseases including polycystic ovarian syndrome, endometriosis, cancer, preeclampsia, and so on (Lu et al., 2018).

Some other hormones including **leptin, ghrelin, oxytocin, arginine-vasotocin, endothelin** (ET-1), **adrenocorticotropic hormone** (ACTH), **growth factors such as insulin-like growth factor I** (IGF-I), **epidermal growth factor** (EGF), **nuclear peroxisome proliferator-activated receptor gamma** (PPARγ), pharmacological regulators of some protein kinases such as **protein kinase A** (PKA), **mitogen-activated protein** (MAP) kinase, **cell division cycle protein two homolog** (CDC2 kinase, CDK), **tyrosine kinases**, several **transcription factors** and **plant molecules** (resveratrol, rapamycin) and **small RNAs** (miRNAs, siRNAs, and cycloRNAs) on the functions of ovarian cells (proliferation, apoptosis, secretory activity, expression of some protein kinases, inflammation, malignant transformation, and other dysfunctions) and reproductive end points (blood level of reproductive hormones, ovarian morphology, number of ovulations, embryo yield and quality, number and viability of offspring), and their possible interrelationships and practical application in control of reproductive processes was found (Sirotkin, 2014; Sirotkin et al., 2014). Sexual maturation in females is influenced by puberty-related changes in some ovarian signaling substances: apoptosis regulator Bcl-2, apoptotic protein Bax, protein kinase A, cAMP responsive element binding protein 1 (CREB-1), T tyrosine kinase, mitogen-activated protein kinases (MAPK), cyclin B1, and protein kinase G (Kolesarova et al., 2015).

1.6 Some female reproductive disorders

The disfunction of this axis can lead to female reproductive disorders. The World Health Organization (WHO) classified they into three categories. Group I ovulation disorders involve hypothalamic failure characterized as hypogonadotropic hypogonadism. Group II disorders display an eugonadal state commonly associated with a wide range of endocrinopathies. Finally, group III constitutes hypergonadotropic hypogonadism secondary to depleted ovarian function (Mikhael et al., 2019). It is known that inadequate regulation or co-ordination of recruitment and development of primordial follicles can induce pathological states. For example, exhaustion of follicles in ovary induces **premature ovarian failure** (POF) and **infertility** (Skinner, 2005; Broekmans et al., 2006; Kawamura et al., 2016), which affects up to 20% of reproductive aged couples (Roudebush et al., 2008). But some of ovaries still contain residual dormant follicles. A new infertility treatment and named it as *in vitro* activation was developed, which enables POF patients to conceive using their own eggs by activation of residual dormant follicles (Kawamura et al., 2016).

On the contrary, intensive formation of small preantral follicles and early growing follicles, not reaching later stages of their development and ovulation are characteristics of **polycystic ovarian syndrome** (PCOS) (Franks et al., 2008). PCOS is defined by a combination of signs of endocrine and ovarian dysfunction - reduced release of FSH and estrogens and increased output of LH and androgens, suppression of ovarian follicular growth and ovulation and resulted infertility. PCOS is frequently associated with abdominal adiposity, insulin resistance, obesity, metabolic disorders, cardiovascular risk factors and inflammatory processes (Abraham Gnanadass et al., 2021). Mounting evidence suggests that PCOS might be a complex multigenic disorder with strong epigenetic and environmental influences, including diet and lifestyle factors (Escobar-Morreale, 2018).

Endometriosis is an often painful disorder in which tissue similar to the tissue that normally lines the inside of the uterus—the endometrium—grows outside your uterus. Endometriosis most commonly involves the ovaries, fallopian tubes and the tissue lining pelvis or spread beyond pelvic organs. It seems a multigenic disorder with several forms and associated with several pathogenic changes (Rolla, 2019).

Ovarian cancer and endometriosis belong to the most common reproductive disorders, which have common biological characteristics: increased cell proliferation, inhibition of apoptosis, promotion of angiogenesis, inflammation, oxidative stress, and invasion of surrounding tissue, which are controlled by common genes and miRNAs. Furthermore, their functional interrelationships have been suggested. Unlike other human cancers, in which all primary tumors arise de novo, ovarian epithelial cancers are primarily imported from either endometrial or fallopian tube epithelium endometriotic lesions and bleeding during irregular menstruation induce oxidative stress, malignant transformation of fallopian tube and endometrium, migration of cancer cells and resulted development of ovarian cancer (Marí-Alexandre et al., 2019; Shih et al., 2021).

1.7 Conclusions and possible direction of future studies

The evidence accumulated by contemporarily reproductive biology demonstrate that female reproduction is performed by functional interrelationships between various reproductive organs and their regulators. Some (but not all) extracellular regulators (hormones hypothalamo-pituitary-ovarian origin, as well as hormones, cytokines and growth factors produced by other cells) have been listed here. Furthermore, intracellular regulators and mediators of extracellular factors on target cells include receptors, postreceptor protein kinase, transcription factors, small RNAs, regulators of cell cycle and apoptosis. Irregularity in production, release, metabolism, reception and mediators of effects of these extra- and intracellular regulators and in their interrelationships induces some animal and human reproductive disorders.

Studies of the functional interrelationships between reproductive organs, processes and their regulators, as well as of character and mechanisms of intervention of environmental contaminants and plant molecules could enable better understanding mechanisms of reproductive processes, diagnostics and prediction of reproductive state and fertility and their disfunctions, as well as search for the new (or old, but forgotten) approaches to protect, to prevent and to treat reproductive disorders. The next chapters will describe sources, general physiological and specific reproductive effects of the most common and known environmental contaminants, medicinal and functional food plants and their molecules, mechanisms of their action and finally their application for control of reproductive processes and prevention and treatment of their disorders.

References

Abedel-Majed MA, Romereim SM, Davis JS, Cupp AS. Perturbations in lineage specification of granulosa and theca cells may alter corpus luteum formation and function. Front Endocrinol 2019;10:832. https://doi.org/10.3389/fendo.2019.00832.

Abraham Gnanadass S, Divakar Prabhu Y, Valsala Gopalakrishnan A. Association of metabolic and inflammatory markers with polycystic ovarian syndrome (PCOS): an update. Arch Gynecol Obstet 2021;303(3):631−43. https://doi.org/10.1007/s00404-020-05951-2.

16 Contaminants and Plants Action on Female Reproduction

Adams GP, Jaiswal R, Singh J, Malhi P. Progress in understanding ovarian follicular dynamics in cattle. Theriogenology 2008;69(1):72−80. https://doi.org/10.1016/j.theriogenology.2007.09.026.

Appiah Adu-Gyamfi E, Tanam Djankpa F, Nelson W, Czika A, Kumar Sah S, Lamptey J, Ding YB, Wang YX. Activin and inhibin signaling: from regulation of physiology to involvement in the pathology of the female reproductive system. Cytokine 2020;133:155105. https://doi.org/10.1016/j.cyto.2020.155105.

Aerts JM, Bols PE. Ovarian follicular dynamics: a review with emphasis on the bovine species. Part I: folliculogenesis and pre-antral follicle development. Reprod Domest Anim 2010a;45(1):171−9. https://doi.org/10.1111/j.1439-0531.2008.01302.x.

Aerts JM, Bols PE. Ovarian follicular dynamics. A review with emphasis on the bovine species. Part II: antral development, exogenous influence and future prospects. Reprod Domest Anim 2010b;45(1):180−7. https://doi.org/10.1111/j.1439-0531.2008.01298.x.

Atwood CS, Meethal SV. The spatiotemporal hormonal orchestration of human folliculogenesis, early embryogenesis and blastocyst implantation. Mol Cell Endocrinol 2016;430:33−48. https://doi.org/10.1016/j.mce.2016.03.039.

Baerwald A, Roger Pierson R. Ovarian follicular waves during the menstrual cycle: physiologic insights into novel approaches for ovarian stimulation. Fertil Steril 2020;114(3):443−57. https://doi.org/10.1016/j.fertnstert.2020.07.008.

Briceag I, Costache A, Purcarea VL, Cergan R, Dumitru M, Briceag I, Sajin M, Ispas AT. Fallopian tubes–literature review of anatomy and etiology in female infertility. J Med Life 2015;8(2):129−31.

Broekmans FJ, Kwee J, Hendriks DJ, Mol BW, Lambalk CB. A systematic review of tests predicting ovarian reserve and IVF outcome. Hum Reprod Update 2006;12:685−718. https://doi.org/10.1093/humupd/dml034.

Brown J, Kives S, Akhtar M. Progestagens and anti-progestagens for pain associated with endometriosis. Cochrane Database Syst Rev 2012;2012(3):CD002122. https://doi.org/10.1002/14651858.CD002122.pub2.

Cavaillon JM. Exotoxins and endotoxins: inducers of inflammatory cytokines. Toxicon 2018;149:45−53. https://doi.org/10.1016/j.toxicon.2017.10.016.

Craig J, Resaca M, Wang H, Resaca S, Thompson W, Zhu C, Kostunica F, Tsang BK. Gonadotropin and intra-ovarian signals regulating follicle development and atresia: the delicate balance between life and death. Front Biosci 2007;12:3628−39.

Cunha GR, Robboy SJ, Kurita T, Isaacson D, Shen J, Cao M, Baskin LS. Development of the human female reproductive tract. Differentiation 2018;103:46−65. https://doi.org/10.1016/j.diff.2018.09.001.

D'Occhio MJ, Campanile G, Baruselli PS. Peripheral action of kisspeptin at reproductive tissues-role in ovarian function and embryo implantation and relevance to assisted reproductive technology in livestock: a review. Biol Reprod 2020;103(6):1157−70. https://doi.org/10.1093/biolre/ioaa135.

Das D, Arur S. Conserved insulin signaling in the regulation of oocyte growth, development, and maturation. Mol Reprod Dev 2017;84(6):444−59. https://doi.org/10.1002/mrd.22806.

Das N, Kumar TR. Molecular regulation of follicle-stimulating hormone synthesis, secretion and action. J Mol Endocrinol 2018;60(3):R131−55. https://doi.org/10.1530/JME-17-0308.

Di Renzo GC, Giardina I, Clerici G, Brillo E, Gerli S. Progesterone in normal and pathological pregnancy. Horm Mol Biol Clin Invest 2016;27(1):35−48. https://doi.org/10.1515/hmbci-2016-0038.

Gershon E, Nava D. Newly identified regulators of ovarian folliculogenesis and ovulation. Molecular Cell Int J Mol Sci 2020;21(12):4565. https://doi.org/10.3390/ijms21124565.

Escobar-Morreale HF. Polycystic ovary syndrome: definition, aetiology, diagnosis and treatment. Nat Rev Endocrinol 2018;14(5):270−84. https://doi.org/10.1038/nrendo.2018.24.

Feng WG, Sui HS, Han ZB, Chang ZL, Zhou P, Liu DJ, Bao S, Tan JH. Effects of follicular atresia and size on the developmental competence of bovine oocytes: a study using the well-in-drop culture system. Theriogenology 2007;67:1339−50. https://doi.org/10.1016/j.theriogenology.2007.01.017.

Fidan U, Keskin U, Ulubay M, Öztürk M, Bodur S. Value of vaginal cervical position in estimating uterine anatomy. Clin Anat 2017;30(3):404−8. https://doi.org/10.1002/ca.22854.

Ford EA, Beckett EL, Roman SD, McLaughlin EA, Sutherland JM. Advances in human primordial follicle activation and premature ovarian insufficiency. Reproduction 2020;159(1):R15−29. https://doi.org/10.1530/REP-19-0201.

Fortune JE. The early stages of follicular development: activation of primordial follicles and growth of preantral follicles. Anim Reprod Sci 2003;78:135 63. https://doi.org/10.1016/s0378-4320(03)00088-5.

Franks S, Stark J, Hardy K. Follicle dynamics and anovulation in polycystic ovary syndrome. Hum Reprod Update 2008;14(4):367−78. https://doi.org/10.1093/humupd/dmn015.

Gougeon A. Dynamics of human follicular growth: morphologic, dynamic, and functional aspects. In: Leung PCK, Adash EY, editors. The ovary. 2nd ed. Amsterdam: Elsevier-Academic Press; 2004. p. 25−43.

Hammes SR, Levin ER. Impact of estrogens in males and androgens in females. J Clin Invest 2019;129(5):1818−26. https://doi.org/10.1172/JCI125755.

Henderson VW. Progesterone and human cognition. Climacteric 2018;21(4):333−40. https://doi.org/10.1080/13697137.2018.1476484.

Holesh JE, Bass AN, Lord M. Physiology, ovulation. 2020. In: StatPearls [internet]. Treasure Island (FL: StatPearls Publishing; 2021.

Hsueh AJW, Kawamura K, Cheng Y, Fauser BCJM. Intraovarian control of early folliculogenesis. Endocr Rev 2015;36(1):1−24. https://doi.org/10.1210/er.2014-1020.

Hutt KJ, Albertini DF. An oocentric view of folliculogenesis and embryogenesis. Reprod Biomed Online 2007;14:758−64. https://doi.org/10.1016/s1472-6483(10)60679-7.

Jing Y, Run-Qian L, Hao-Ran W, Hao-Ran C, Ya-Bin L, Yang G, Fei C. Potential influence of COVID-19/ACE2 on the female reproductive system. Mol Hum Reprod 2020;26(6):367−73. https://doi.org/10.1093/molehr/gaaa030.

Katska-Ksiazkiewicz L, Lechniak-Cieślak D, Korwin-Kossakowska A, Alm H, Ryńska B, Warzych E, Sosnowski J, Sender G. Genetical and biotechnological methods of utilization of female reproductive potential in mammals. Reprod Biol 2006;6(Suppl 1):21−36.

Kawamura K, Kawamura N, Hsueh AJ. Activation of dormant follicles: a new treatment for premature ovarian failure? Curr Opin Obstet Gynecol 2016;28(3):217−22. https://doi.org/10.1097/GCO.0000000000000268.

Kolesarova A, Sirotkin AV, Mellen M, Roychoudhury S. Possible intracellular regulators of female sexual maturation. Physiol Res 2015;64:379−86. https://doi.org/10.33549/physiolres.932838.

Lonergan P, Fair T. Maturation of oocytes in vitro. Ann Rev Anim Biosci 2016;4:255−68. https://doi.org/10.1146/annurev-animal-022114-110822.

Lu J, Wang Z, Cao J, Chen Y, Dong Y. A novel and compact review on the role of oxidative stress in female reproduction. Reprod Biol Endocrinol 2018;16(1):80. https://doi.org/10.1186/s12958-018-0391-5.

18 Contaminants and Plants Action on Female Reproduction

Marí-Alexandre J, Carcelén AP, Agababyan C, Moreno-Manuel A, García-Oms J, Calabuig-Fariñas S, Gilabert-Estellés J. Interplay between MicroRNAs and oxidative stress in ovarian conditions with a focus on ovarian cancer and endometriosis. Int J Mol Sci 2019;20(21):5322. https://doi.org/10.3390/ijms20215322.

Maybin JA, Critchley HO. Menstrual physiology: implications for endometrial pathology and beyond. Hum Reprod Update 2015;21(6):748−61. https://doi.org/10.1093/humupd/dmv038.

McGee EA, Hsueh AJW. Initial and cyclic recruitment of ovarian follicles. Endocr Rev 2000;21:200−14. https://doi.org/10.1210/edrv.21.2.0394.

McLaughlin EA, McIver SC. Awakening the oocyte: controlling primordial follicle development. Reproduction 2009;137(1):1−11. https://doi.org/10.1530/REP-08-0118.

Meidan R, Levy N, Kisliouk T, Podlovny L, Rusiansky M, Klipper E. The yin and yang of corpus luteum-derived endothelial cells: balancing life and death. Domest Anim Endocrinol 2005;29(2):318−28. https://doi.org/10.1016/j.domaniend.2005.04.003.

Mesen TB, Young SL. Progesterone and the luteal phase: a requisite to reproduction. Obstet Gynecol Clin N Am 2015;42(1):135−51. https://doi.org/10.1016/j.ogc.2014.10.003.

Mermillod P, Dalbiès-Tran R, Uzbekova S, Thélie A, Traverso JM, Perreau C, Papillier P, Monget P. Factors affecting oocyte quality: who is driving the follicle? Reprod Domest Anim 2008;43(2):393−400. https://doi.org/10.1111/j.1439-0531.2008.01190.x.

Mihm M, Evans AC. Mechanisms for dominant follicle selection in monovulatory species: a comparison of morphological, endocrine and intraovarian events in cows, mares and women. Reprod Domest Anim 2008;43(2):48−56. https://doi.org/10.1111/j.1439-0531.2008.01142.x.

Mikhael S, Punjala-Patel A, Gavrilova-Jordan L. Hypothalamic-pituitary-ovarian Axis disorders impacting female fertility. Biomedicines 2019;7(1):5. https://doi.org/10.3390/biomedicines7010005.

Moore KL, Dalley AF. Clinically oriented anatomy. 5th ed. Philadelphia: LWW; 2006. p. 1−300.

Niswender GD, Juengel JL, Silva PJ, Rollyson MK, McIntush EW. Mechanisms controlling the function and life span of the corpus luteum. Physiol Rev 2000;80(1):1−29. https://doi.org/10.1152/physrev.2000.80.1.1.

Niswender GD. Molecular control of luteal secretion of progesterone. Reproduction 2002;123(3):333−9. https://doi.org/10.1530/rep.0.1230333.

Oktem O, Oktay K. The ovary: anatomy and function throughout human life. Ann N Y Acad Sci 2008;1127:1−9. https://doi.org/10.1196/annals.1434.009.

Peter AT, Levine H, Drost M, Bergfelt DR. Compilation of classical and contemporary terminology used to describe morphological aspects of ovarian dynamics in cattle. Theriogenology 2009;71:1343−57.

Piotrowska-Tomala KK, Jonczyk AW, Skarzynski DJ, Szóstek-Mioduchowska AZ. Luteinizing hormone and ovarian steroids affect in vitro prostaglandin production in the equine myometrium and endometrium. Theriogenology 2020;153:1−8. https://doi.org/10.1016/j.theriogenology.2020.04.039.

Rebecca L, Robker R, Hennebold JD, Russell DL. Coordination of ovulation and oocyte maturation: a good egg at the right time. Endocrinology 2018;159(9):3209−18. https://doi.org/10.1210/en.2018-00485.

Rolla E. Endometriosis: advances and controversies in classification, pathogenesis, diagnosis, and treatment. F1000Research 2019;8:F1000. https://doi.org/10.12688/f1000research.14817.1. Faculty Rev-529.

Roudebush WE, Kivens WJ, Mattke JM. Biomarkers of ovarian reserve. Biomark Insights 2008;3:259−68. https://doi.org/10.4137/bmi.s537.

Sánchez F, Smitz J. Molecular control of oogenesis. Biochim Biophys Acta 2012;1822:1896—912. https://doi.org/10.1016/j.bbadis.2012.05.013.

Shah JS, Sabouni R, Vaught KCC, Owen CM, Albertini DF, Segars JH. Biomechanics and mechanical signaling in the ovary: a systematic review. J Assist Reprod Genet 2018;35:1135—48. https://doi.org/10.1007/s10815-018-1180-y.

Shih IM, Wang Y, Wang TL. The origin of ovarian cancer species and precancerous landscape. Am J Pathol. 2021;191(1):26—39. https://doi.org/10.1016/j.ajpath.2020.09.006.

Silva JF, Ocarino NM, Serakides R. Thyroid hormones and female reproduction. Biol Reprod 2018;99(5):907—21. https://doi.org/10.1093/biolre/ioy115.

Sirotkin AV. Regulators of ovarian functions. 2nd ed., 194. New York: Nova Science Publishers, Inc.; 2014, ISBN 978-1-62948-574-4.

Sirotkin AV, Adamcova E, Rotili D, Mai A, Mlyncek M, Mansour L, Alwasel S, Harrath AH. Comparison of the effects of synthetic and plant-derived mTOR regulators on healthy human ovarian cells. Eur J Pharmacol 2019;854:70—8. https://doi.org/10.1016/j.ejphar.2019.03.048.

Sirotkin AV, Chrenek P, Kolesarová A, Parillo F, Zerani M, Boiti C. Novel regulators of rabbit reproductive functions. Anim Reprod Sci 2014;148(3—4):188—96. https://doi.org/10.1016/j.anireprosci.2014.06.001.

Skarzynski DJ, Ferreira-Dias G, Okuda K. Regulation of luteal function and corpus luteum regression in cows: hormonal control, immune mechanisms and intercellular communication. Reprod Domest Anim 2008;43(2):57—65. https://doi.org/10.1111/j.1439-0531.2008.01143.x.

Skinner MK. Regulation of primordial follicle assembly and development. Hum Reprod Update 2005;11:461—71. https://doi.org/10.1093/humupd/dmi020.

Teeli AS, Leszczyński P, Krishnaswamy N, Ogawa H, Tsuchiya M, Śmiech M, Skarzynski D, Taniguchi H. Possible mechanisms for maintenance and regression of corpus luteum through the ubiquitin-proteasome and autophagy system regulated by transcriptional factors. Front Endocrinol (Lausanne). 2019;10:748. https://doi.org/10.3389/fendo.2019.00748.

Tingen C, Kim A, Woodruff TK. The primordial pool of follicles and nest breakdown in mammalian ovaries. Mol Hum Reprod 2009;15(12):795—803. https://doi.org/10.1093/molehr/gap073.

Wear HM, Eriksson A, Yao HH, Watanabe KH. Cell-based computational model of early ovarian development in mice. Biol Reprod 2017;97(3):365—77. https://doi.org/10.1093/biolre/iox089.

Webb R, Campbell BK. Development of the dominant follicle: mechanisms of selection and maintenance of oocyte quality. Soc Reprod Fertil Suppl 2007;64:141—63. https://doi.org/10.5661/rdr-vi-141.

Zeleznik AJ. Follicle selection in primates: "many are called but few are chosen". Biol Reprod 2001;65:655—9. https://doi.org/10.1095/biolreprod65.3.655.

Zhang Y, Yan Z, Qin Q, Nisenblat V, Chang HM, Yu Y, et al. Transcriptome landscape of human folliculogenesis reveals oocyte and granulosa cell interactions. Mol Cell 2018;72(6):1021—1034.e4. https://doi.org/10.1016/j.molcel.2018.10.029.

Zhang JQ, Wang XW, Chen JF, Ren QL, Wang J, Gao BW, Shi ZH, Zhang ZJ, Bai XX, Xing BS. Grape seed procyanidin B2 protects porcine ovarian granulosa cells against oxidative stress-induced apoptosis by upregulating let-7a expression. Oxid Med Cell Longev 2019:1076512. https://doi.org/10.1155/2019/1076512.

Zhou J, Peng X, Mei S. Autophagy in ovarian follicular development and atresia. Int J Biol Sci 2019;15(4):726—37. https://doi.org/10.7150/ijbs.30369.

Chapter 2

Environmental contaminants and their influence on health and female reproduction

Chapter 2.1

Heavy metals

2.1.1 Introduction

Heavy metal pollution is one of the major environmental problems faced today (Rahman and Singh 2019; Fu and Xi, 2020; Ajiboye et al., 2021). In recent years, high concentrations of heavy metals in different natural systems including atmosphere, pedosphere, hydrosphere, and biosphere have become a global issue (Rahman and Singh 2019). The coexistence of heavy metals and organics in industrial effluents is a prevalent problem (Ajiboye et al., 2021). Several industrial activities and some natural processes are responsible for their high contamination in the environment with the effects on various microorganisms, plants, and animals (Rahman and Singh 2019). Humans can be exposed to heavy metals through multiple routes. Health concerns relating to the exposure to organic contaminants have been raised that can be detected at low concentrations in drinking water have been raised, especially in regard to reproduction (Martínez-Sales et al., 2015). Drinking water contaminated with heavy metals namely: arsenic, cadmium, nickel, mercury, chromium, zinc, and lead is becoming a major health concern (Rehman et al., 2018). Certain heavy metals, such as arsenic, cadmium, chromium, mercury, and lead, are nonthreshold toxins and can exert toxic effects at very low concentrations (Rahman and Singh 2019). Heavy metals may react with chlorine and oxygen in the human body, and exert their toxic effects (Rusyniak et al., 2010). The general mechanism of heavy metal toxicity is through the production of reactive oxygen species (ROS), the appearance of oxidative damage, and

Environmental Contaminants and Medicinal Plants Action on Female Reproduction
https://doi.org/10.1016/B978-0-12-824292-6.00031-3
Copyright © 2022 Elsevier Inc. All rights reserved.

22 Contaminants and Plants Action on Female Reproduction

subsequent adverse effects on health. Therefore, water contaminated with heavy metals causes high morbidity and mortality worldwide (Fu and Xi, 2020).

2.1.2 Provenance and properties

Heavy metals can be found naturally in Earth's crust. The main problem is that heavy metals can accumulate in different parts of plants thus causing various harmful effects on human health (Rehman et al., 2018). However, it is a fact that it is impossible to have an environment totally free of heavy metals. Entry of these heavy metals into the human body could occur in various ways such as through consumption of contaminated food, drinking water, or air (Rehman et al., 2018).

A major source of heavy metal pollution is the improper dumping and discarding of industrialized waste products through which they are disposed directly into water and land areas (Dixit et al., 2015). The main source of human exposure to heavy metals is from contaminated drinking water. As technology continues to advance, heavy metals in drinking water have exceeded recommended limits from regulators around the world (Fu and Xi, 2020). This poses harmful health risks to humans, aquatic lives and the entire ecosystem, because majority of these mixed pollutants amass in water in concentrations, which are more than the permissible discharge limits in the environment (Ajiboye et al., 2021).

Heavy metals can disturb the body's metabolic functions in various ways. Moreover, they may accumulate in vital body organs such as the liver, heart, kidney, and brain disturbing normal biological functioning (Singh, 2007). Accumulation of lead, cadmium, mercury, nickel, and zinc in muscle, liver, and kidney (Kolesarova et al., 2008; Kalafova et al., 2012) can cause several health issues (Chandra et al., 2011). The worst consequences are observed for females since the number of germ cells present in the ovary is fixed during fetal life, and the cells are not renewable. This means that any pollutant affecting hormonal homeostasis and/or the reproductive apparatus inevitably harms reproductive performance (Canipari et al., 2020). The heavy metals such pose potential risks to sustainability of environment and thus to our future generations (Kolesarova et al., 2010). On other hand some of them are essential trace elements and play an important role in cell functions (Kolesarova et al., 2011b).

The impact of several heavy metals is the focus of study by many researchers. For example, mercury is a naturally occurring chemical element found on the Earth's crust, existing in three different chemical forms: elemental, inorganic, and organic (EPA, 2021). In humans, mercury exposure occurs predominantly through the consumption of seafood or raw fish, but also dental amalgams, button cell batteries, broken thermometers, and compact fluorescent light bulbs, and skin-lightning creams. Mercury exposure also depends on the geographical location/region, education level, and type of job (Al-Saleh et al., 2011). With increasing incidences of nickel contamination over recent years, the toxicity of nickel has also attracted more attention (Zhang et al., 2019). Nickel has extensive applications in various modern industrial fields, such as refining, electroplating, welding, electroforming, and the production of nickel-cadmium batteries (Binet et al., 2018). Rizvi et al. (2020) has suggested that nickel exposure can result in various adverse health effects including developmental toxicity, genotoxicity, immunotoxicity, hematotoxicity, reprotoxicity, and neurotoxicity.

2.1.3 Physiological actions

Heavy metals can express toxicity (Martin et al., 2009; Rusyniak et al., 2010; Chandra et al., 2011; Bernhoft, 2012; O'Neal and Zheng, 2015; Rzymski et al., 2015; Rahman and Singh 2019; Fu and Xi, 2020; Rizvi et al., 2020), genotoxicity, immunotoxicity, hematotoxicity (Rizvi et al., 2020), carcinogenic (Bhasin et al., 2002; Genestra, 2007; García-Esquinas et al., 2013; Huy et al., 2014; Bjørklund et al., 2018; Rehman et al., 2018; Engwa et al., 2019), mutagenic (Genestra, 2007), and neuromodulatory effects (Burton and Guilarte, 2009; Bernhoft, 2012; Brown et al., 2012; O'Neal and Zheng, 2015; Milatovic et al., 2017; Engwa et al., 2019; Debnath et al., 2019; Yang et al., 2020; Rizvi et al., 2020), and can induce cardiovascular disorders (Klos et al., 2006; Martin et al., 2009; Huy et al., 2014; Rehman et al., 2018; Debnath et al., 2019), neuronal damage (Martin et al., 2009; Huy et al., 2014; Rehman et al., 2018), renal injuries (Martin et al., 2009; Rehman et al., 2018), diabetes (Huy et al., 2014; Rehman et al., 2018), osteoporosis (Nishijo et al., 2017), mental retardation, dyslexia, autism, psychosis, allergies (Martin et al., 2009), and reproductive toxicity (Borja-Aburto et al., 1999; Debnath et al., 2019; Rizvi et al., 2020). On the other hand some metals in low doses are essential for the human body such as iron (Pantopoulos, 2012; Anderson and Frazer, 2017; Yiannikourides and Latunde-Dada, 2019; Mintz et al., 2020), zinc (Steinbrenner and Klotz, 2020), and others.

Mercury and its compounds affect the nervous system. Increased exposure to mercury can alter brain functions and lead to tremors, shyness, irritability, memory problems, and changes in hearing or vision. Short-term exposure to metallic mercury vapors at higher levels can lead to vomiting, nausea, skin rashes, diarrhea, lung damage, and high blood pressure, while short-term exposure to organic mercury poisoning can lead to depression, tremors, headache, fatigue, memory problems, and hair loss (Engwa et al., 2019). Exposure to chronic levels of mercury can lead to tremor of the hands, memory loss, insomnia, erethism and timidity. Researchers also noted that occupational exposure to mercury can be associated with measurable declines in the performance of visual scanning, neurobehavioral tests of motor speed, visuomotor coordination, verbal, and visual memory. Dimethylmercury is a very toxic compound. It can penetrate skin even through latex gloves and even low dosage exposure can cause the degeneration of the central nervous system and eventually death (Bernhoft, 2012). Induction of apoptosis in cerebellar and differentiated human neurons are observed at low organic mercury concentrations, while higher concentrations of organic mercury result in necrosis (Lohren et al., 2015).

Cadmium and its compounds have several health effects on humans. The main problem is the inability of human body to excrete cadmium. What is worse is that cadmium is even reabsorbed by the kidney thereby limiting its excretion. Long-term exposure to cadmium leads to its deposition in bones and lungs (Bernard, 2008). Cadmium can cause bone mineralization. This can lead to osteoporosis, as many studies on humans or animals have revealed. The most known disease caused by cadmium is called "Itai-itai" disease. This disease caused an epidemic of bone fractures in Japan due to cadmium contamination (Nishijo et al., 2017). Increased risk of bone fractures in women as well as decreased bone density and height loss in both men and women, have been linked to increased cadmium toxicity in the population. Cadmium is especially toxic to the kidneys. It is accumulated in the proximal tubular cells in higher concentration. According to this fact, we can assume that cadmium exposure can cause renal dysfunction and kidney disease. It has been found that it can also cause the formation of renal stones and hypercalciuria and disturbances in calcium metabolism (Mudgal et al., 2010a,b).

Lead poisoning can be divided into acute and chronic. Acute exposure to lead can lead to headaches, loss of appetite, fatigue, hallucinations, abdominal pain, vertigo, and arthritis. Chronic exposure can cause mental retardation, dyslexia, autism, psychosis, allergies, paralysis, hyperactivity, kidney damage, brain damage, and coma. Severe poisoning can cause even death (Martin et al., 2009).

Arsenic exposure can lead to acute or chronic toxicity. Acute arsenic poisoning can lead to the destruction of blood vessels, gastrointestinal tissue and can affect the heart and the brain (Martin et al., 2009). Long-term exposure can even lead to pulmonary disease, peripheral vascular disease, diabetes mellitus, hypertension, neurological problems, and cardiovascular disease. Chronic arsenic exposure can result in irreversible changes in vital organs which can lead to death. Also, it can promote the development of a number of cancers, for example, cancers of the bladder, liver, lung, skin, and possibly kidney and colon cancers (Huy et al., 2014).

Iron is essential for human health. Iron plays a crucial role in oxygen transport, cellular proliferation, oxidative metabolism, and many catalytic reactions. Same as other essential heavy metals, to be beneficial, the amount of iron in the human body needs to be maintained in an ideal range. The main consumer of iron in body is bone marrow whereas liver is the key component of iron homeostasis. It is responsible for synthetic, storing, and regulatory functions in homeostasis (Yiannikourides and Latunde-Dada, 2019). In the human body, iron is required as a cofactor for hemoproteins and nonhem iron-containing proteins. The most important hemoproteins include hemoglobin and myoglobin, which are part of oxygen binding and transport (Pantopoulos, 2012). However, some forms of iron can cause serious health problems. Iron toxicity can lead to clinical symptoms and is characterized by hypotension, shocks, lethargy, hepatic necrosis, tachycardia, metabolic acidosis, and may sometimes lead to death (Hillman, 2001). Iron is also known to generate free radicals which are suggested to be responsible for asbestos related cancer (Bhasin et al., 2002).

Zinc is essential trace element, and an inadequate dietary intake has been implicated in the decline of immune and cognitive functions in aged persons and in the pathogenesis of age-related disorders. The micronutrient is often marketed as "antioxidants" in mineral supplements; however, zinc is not antioxidants per se but it may exert beneficial effects as component of enzymes and other proteins that catalyze redox reactions and/or are involved in the maintenance of redox homeostasis. It should also be noted that the observed deficiencies in micronutrients may not necessarily be attributable to inadequate dietary intake as the absorption and distribution within the body might also be influenced by factors such as medications or interaction with other food ingredients (Steinbrenner and Klotz, 2020).

Toxic effect of **copper** nanoparticles has been evident more in male mice than in females. Adverse effect of copper on male reproductive functions has been indicated by the decrease in spermatozoa parameters such as concentration, viability and motility. Copper nanoparticles are capable of generating oxidative stress *in vitro* thereby leading to reproductive toxicity (Roychoudhury et al., 2016).

Manganese has begun to be a metal of global concern when methylcyclopentadienyl manganese tricarbonyl (MMT), which was known to be toxic, was introduced as a gasoline additive. It has been claimed that MMT is linked with the development of Parkinson's disease-like syndrome of tremor, gait disorder, postural instability, and cognitive disorder (O'Neal and Zheng, 2015). Exposure to high levels of manganese can result in neurotoxicity. It can cause a neurological disease called manganism. It is characterized by rigidity, action tremor, gait disturbances, bradykinesia, memory and cognitive dysfunction, and also mood disorder (Klos et al., 2006). Often, manganism is mistaken for Parkinson's disease because the symptoms are very similar. The main difference between manganism and Parkinson's disease is the insensitivity of manganism to levodopa administration and also the progression of the disease is different (Guilarte, 2010).

2.1.4 Mechanisms of action

Several principal extra- and intracellular mechanisms of toxic effect of heavy metals on the cells have been proposed, although the evidence for each of this mechanism is limited, and the interrelationships between these mechanisms are poorly investigated yet. The following mechanisms of heavy metals action are suggested:

(1) production of free radicals leads to oxidative damage of proteins and DNA (Bhasin et al., 2002; Genestra, 2007; Fu and Xi, 2020), oxidative stress and consequential destruction of healthy cells (Flora et al., 2012),

(2) generation of hydroxyl radical OH• by the Fenton reaction (Engwa et al., 2019) and its reaction with biological molecules, for example, lipids, proteins, and DNA (Valko et al., 2004),

(3) alterations of histones and DNA methylation (Bjørklund et al., 2018),

(4) binding to glutathione's sulfhydryl group, inactivates glutathione, and increases oxidative stress. It blocks the activity of enzymes (δ-amino-levulinic acid dehydratase, glutathione reductase, glutathione peroxidase, and glutathione-S-transferase) and further reduces the glutathione levels (Kim et al., 2015),

(5) destabilization of the cellular membrane through lipid peroxidation, which can cause hemolytic anemia (Kim et al., 2015),

(6) the activation of transcription factors that are redox-sensitive whereas another function involves its role as a mitogenic signal (Genestra, 2007),

(7) the activation of apoptosis through caspase as well as ultrastructural changes in the hepatocytes (Renu et al., 2021), myogenic cells (Usuki et al., 2008), and murine T and B lymphoma cells (Kim and Sharma, 2003),

(8) alteration of the expression of p53 protein and also decreased expression of p21, binding DNA-binding proteins and at the same time disrupt DNA repair processes thereby increasing the risk of carcinogenesis (García-Esquinas et al., 2013),

(9) inflammation involving tumour necrosis factor-α (TNF-α), proin-flammatory cytokines, a mitogen-activated protein kinase (MAPK), the extracellular signal-regulated kinases (ERK) pathways in the event of heavy metal hepatotoxicity (Renu et al., 2021),

(10) disrupting the intracellular second messenger systems and alter the functioning of the central nervous system (Brown et al., 2012),

(11) disruption of ATP synthesis by inhibiting the F1/F0 ATP synthase, in-hibition ATP synthesis at two sites in the brain of mitochondria which are either the glutamate/aspartate exchanger or the complex II (succinate dehydrogenase) depending on the mitochondrial energy source (Gunter et al., 2010), and

Environmental contaminants and their influence Chapter | 2 **27**

(12) dopamine reactive species taken up the dopamine transporter which can cause dopaminergic neurotoxicity (Benedetto et al., 2010).

2.1.5 Effects on female reproductive processes

The female reproductive system is sensitive to heavy metal exposure (Kolesarova et al., 2010; Rzymski et al., 2014, 2015; Roychoudhury et al., 2016) but some potential effects of heavy metals on the reproductive system is still not examined fully. Heavy metals can be toxic even at low doses but there is no clear answer as to which range of doses can be considered dangerous and may affect female reproductive system. On the other hand, some metals are essential for human health (Yiannikourides and Latunde-Dada, 2019). Heavy metals cause the alteration in ovarian follicular cell functions (Motta et al., 2021), uterus (Höfer et al., 2009; Mintz et al., 2020), and induce infertility (Rzimsky et al., 2015).

2.1.5.1 Effect on ovaries

Mercury affects fertility, disrupts estrous cycle and, impairs ovarian follicular development (Schuurs et al., 1999). Mercury mostly causes menstrual cycle abnormalities such as changes in bleeding patterns and cycle length (Davis et al., 2001). Mercury chloride treatment of hamsters disrupts their estrus cycles, reduces plasma and luteal progesterone levels, suppresses follicular maturation, and disrupts hypothalamus-pituitary gonadotropin secretion. It was observed that in rats, mercury lengthens the estrous cycle and induces morphological changes in the *corpus luteum* (Davis et al., 2001). However, existing data indicate that mercury may cause infertility. Although, the mechanisms behind mercury-induced infertility needs to be understood (Rzimsky et al., 2015).

Cadmium can inhibit the process of steroidogenesis (Kawai et al., 2002). *In vivo* cadmium treatment on rat ovaries exhibits suppression of progesterone, estradiol and testosterone production (Piasek and Laskey, 1994). Cadmium induces oocyte degeneration, no matter the stage of development. In apparently healthy oocytes, changes in cytoplasm, cortical alveoli, and/or chorion carbohydrates composition are observed. Cadmium also induces significant changes in the localization of progesterone and estrogen receptors β (ERβ), a result that well correlates with the observed increase in ovarian metals concentrations. The acute modifications detected are suggestive of a significantly impaired fecundity and of a marked endocrine disrupting effects of cadmium in this teleost species (Motta et al., 2021).

Lead exposure significantly reduced the maturation and fertilization of oocytes *in vivo*, leading to a decrease in the fertility (Jiang et al., 2021). Moreover, lead exposure caused histopathological and ultrastructural changes in oocytes and ovaries (Jiang et al., 2021). The dietary lead exposure significantly enhanced the follicular atresia rate (Ma et al., 2020). A circulation of luteinizing (LH) and follicle-stimulating hormone (FSH) was suppressed by lead exposure in monkeys. Estradiol level was also reduced without affecting progesterone. This caused overt signs of menstrual irregularity (Foster, 1992). Lead-induced the inhibition of insulin-like growth factor I (IGF-I) release by granulosa cells isolated from ovaries of nonpregnant gilts, but progesterone release was not changed (Kolesarova et al., 2010). In case of pregnant gilts, lead administration promoted progesterone release by ovarian granulosa cells (Kolesarova et al., 2010). Both prenatal and neonatal lead exposure in mice resulted in the suppression of ovarian homogenate-4 androgen production (Taupeau et al., 2001).

Cobalt addition stimulated IGF-I release by rat ovarian fragments. Progesterone release by rat ovarian fragments was inhibited after cobalt sulfate addition. However, sulfate addition did not cause any change in the release of 17β-estradiol by rat ovarian fragments *in vitro* (Roychoudhury et al., 2014b,c).

Short-term **copper** administration has been found to exert deleterious effect on intracellular organelles of rat ovarian cells *in vivo*. *In vitro* administration in porcine ovarian granulosa cells releases IGF-I, steroid hormone progesterone, and induces expression of peptides related to proliferation and apoptosis (Roychoudhury et al., 2014a, 2016). Similarly, as cadmium, copper induces oocyte degeneration in about one third of the previtellogenic oocytes, no matter the stage of development. In apparently healthy oocytes, changes in cytoplasm, cortical alveoli and/or chorion carbohydrates composition are observed. On the other hand, copper did not induce changes in the localization of progesterone and beta-estrogen receptors. The acute modifications detected are suggestive of a significantly impaired fecundity and of a marked endocrine disrupting effects of copper in this teleost species (Motta et al., 2021).

Molybdenum addition decreased IGF-I release by porcine ovarian granulosa cells (Kolesarova et al., 2011b). Progesterone release by rat ovarian fragments was not affected by molybdenum. However, addition of ammonium molybdate was found to cause dose-dependent decrease in 17β-estradiol release by ovarian fragments. Also, addition of ammonium molybdate inhibited IGF-I release. Results suggest ammonium molybdate induced inhibition in the release of growth factor IGF-I and its dose-dependent effect on secretion of steroid hormone 17β-estradiol but not progesterone (Roychoudhury et al., 2014b).

The data listed here demonstrated the ability of many heavy metals to affect and to disrupt ovarian cycle, release of ovarian hormones and growth factors, ovarian folliculogenesis, oogenesis, ovulation and fecundity.

2.1.5.2 Effect on uterus and pregnancy

All rapidly dividing cells require **iron,** and the cells of endometrium are no exception (Anderson and Frazer, 2017). Endometrial tissue is known to express divalent metal transporter 1. This is used for transporting and storing iron as ferritin (Yanatori and Kishi, 2019). The main problem for women health is that iron is lost through each menstrual cycle because of excretion of endometrial cells, and iron-rich hemoglobin. This can cause anemia in women with heavy menstrual bleeding (Mintz et al., 2020).

Mercury compounds were found in maternal blood and infant hair which is suggestive of its effect on pregnancy outcome (Lee et al., 2010). Mercury can cause fetotoxicity because of its ability to cross the placenta and blood-brain barrier thereby causing neuronal dysfunction of the fetus (Verma et al., 2018). Mercury affects impairs embryo implantation. All forms of mercury-induced reproductive disturbances, such as spontaneous abortions, congenital malformations, and stillbirth are reported (Schuurs et al., 1999). Still, there is limited evidence of the direct impact of mercury on the human reproductive system (Rzimsky et al., 2015).

Cadmium accumulation in endometrial tissue is higher in female smokers. Levels of lead were higher depending on age, which can be related to the release of **lead** that has accumulated in bones during life (Rzymski et al., 2014). Low levels of cadmium in blood correlated with the metal content in uterus which significantly decreased estradiol concentration (Nasiadek et al., 2011). The effect of cadmium is associated with hypertrophy, hyperplasia of the endometrium (Höfer et al., 2009), and an increase in uterine weight and thickness of the epithelium (Rzymski et al., 2014), endometriosis, spontaneous abortions, and endometrial cancers (Borja-Aburto et al., 1999). When pregnant women are exposed to high cadmium levels, there is an increase in the possibility of premature delivery or even miscarriage probably because of cadmium-induced compromise of placental function. Reduction in progesterone production in human trophoblast cells, mediated by cadmium, indicated that cell death is not because of apoptosis. It was found out that there is a specific block of P450 side-chain cleavage expression and activity (Kawai et al., 2002). There are some pieces of evidence of pregnancy-related metabolic changes that can increase lead mobilization from bones and can expose the fetus to endogenous metal content (Rzymski et al., 2015).

Sodium **arsenite** can induce disruption in embryo implantation, disrupt synthesis of female sex hormones and induce female infertility (Mehta and Hundal, 2016).

These observations suggest the ability of some heavy metals to affect functions of uterus, embryo implantation and development, pregnancy and labor.

2.1.6 Mechanism of action on female reproductive processes

Heavy metals influence reproduction at various regulatory levels via extra- and intracellular signaling pathways by induction of oxidative stress, modulation of signaling pathway of steroidogenesis, proliferation, and apoptosis:

(1) Dietary lead exposure could induce oxidative stress by impairing the **nuclear factor erythroid 2-related factor 2 (Nrf2)/Kelch-like ECH-associated protein 1 (Keap1) signaling pathway** in the ovaries. The oxidative stress could be due to reduction in antioxidant enzyme activities. At least, after dietary lead exposure, **superoxide dismutase (SOD) and glutathione peroxidase** (GSH-Px) activities and **glutathione** contents were decreased quadratically, and there were decreases in the activities of **catalase** (CAT) and **glutathione reductase** (GR), whereas **malondialdehyde** content was linearly increased. In addition, except for manganese superoxide dismutase, the gene expressions of copper-zinc superoxide dismutase, CAT, and GR were significant decreased. In addition, there were quadratic decreases in the mRNA expressions of GSH-Px and Nrf2. By way of contrast, Keap1 gene expression was increased (Ma et al., 2020),

(2) lead can affect the pathway of proliferation and apoptosis of porcine ovarian granulosa cells through intracellular substances such as **cyclin B1** and **caspase-3** (Kolesarova et al., 2010),

(3) lead activates Nrf2/Keap1 pathway and impairs oocyte maturation and fertilization by inducing oxidative stress, lead exposure triggered oxidative stress with a decreased increased amount of reactive oxygen species, and abnormal mitochondrial distribution. This action can be due to lead-induced decrease in the activities of **catalase, glutathione peroxidase, total superoxide dismutase**, and **glutathione-S transferase**, and increases in the levels of **malonaldehyde** in mouse ovaries (Jiang et al., 2021),

(4) lead exposure activated the Nrf2 signaling pathway to protect oocytes against oxidative stress (Jiang et al., 2021),

(5) lead reduces **mRNA** and protein levels of **estrogen receptors** and **P450 aromatase** in cultured human ovarian granulosa cells *in vitro* (Taupeau et al., 2001),

(6) arsenic produces oxidative stress in reproductive organs, this generation of oxidative stress in arsenic toxicity may be due to excess of free radical formation, autooxidation of accumulated δ-aminolevulinic acid, inhibition of antioxidant enzymes and enhanced inflammatory response (Flora and Agrawal, 2017),

(7) in the ovary, zinc oxide (ZnO) nanoparticles (NPs) induce cell apoptosis in **Sonic hedgehog (Shh) pathway** activated ovary cells, and affect the synthesis of steroid hormones (Kuang et al., 2021),

(8) ZnO NPs activate **mitochondrial-mediated signaling pathway** and induce **caspase** depend damage that ultimately injured the uterus (Kuang et al., 2021),

(9) cadmium stimulates **estrogen receptors α and β** and also upregulates **progesterone receptors** (Borja-Aburto et al., 1999; Rzymski et al., 2015). Therefore, cadmium might be considered as endocrine disrupter,

(10) similarly, interference of copper in the pathway apoptosis and proliferation of porcine ovarian granulosa cells through hormonal and intracellular peptide **cyclin B1** was suggested,

(11) copper effect on ovarian cell proliferation could be mediated by **IGF-I** and **cyclin B1** (Roychoudhury et al., 2014a),

(12) cobalt sulfate decreased the expression of both the of markers of apoptosis **Bax** and **caspase-3** in rat ovarian fragments (Roychoudhury et al., 2014c),

(13) the mechanism of action of molybdenum (Kolesarova et al., 2011) and silver (Kolesarova et al., 2011) on porcine ovarian functions such as proliferation and apoptosis of granulosa cells is through hormonal and intracellular substances such as **cyclin B1 and caspase-3** (Kolesarova et al., 2011),

(14) cancer cells exhibit an iron-seeking phenotype achieved through dysregulation of iron metabolic proteins. These changes are mediated, at least in part, by oncogenes and tumor suppressors. The dependence of cancer cells on iron has implications in a number of cell death pathways, including **ferroptosis**, an **iron-dependent form of cell death** (Torti et al., 2018).

These observations suggest multiple mechanisms of heavy metal actions on female reproductive systems including changes in extra- and intracellular regulators of oxidative processes, proliferation, apoptosis, secretion and reception of steroid hormones and growth factors.

2.1.7 Application in reproductive biology and medicine

Reactive oxygen species (ROS) are a double-edged sword—they serve as key signal molecules in physiological processes but also induce pathological processes from oocyte maturation to fertilization, embryo development and pregnancy—age-dependent reproductive insufficiency, complications in pregnancy and labor, gynecological cancers. These disorders can be prevented by antioxidants (Agarwal et al., 2005). Heavy metals can be both oxidants and antioxidants. For example, both iron excess and iron depletion can be utilized in anticancer therapies (Torti et al., 2018). Therefore, in some conditions presence of optimal amount of heavy metals in environment could be essential for organism. For example, iron, zinc, selenium and copper that play a crucial role in oxygen transport, cellular proliferation, oxidative metabolism, and many catalytic reactions including in female reproductive system (see above). Therefore, the optimal supply of organism with these metals could be beneficial for health and reproduction.

On the other hand, the excess heavy metals could induce numerous reproductive disorders listed above via promotion of oxidative stress, which can affect hormones, regulators of cell proliferation, apoptosis, metabolism and other important cellular processes. The adverse influence of heavy metals could be prevented by monitoring the environment, including food and water. The heavy metal-induced oxidative stress could be prevented by plant molecules with antioxidative properties. Some of such plants and their molecules are listed in this book.

2.1.8 Conclusions and possible direction of future studies

The currently obtained data listed here showed that many heavy metals (mercury, lead, arsenic, cadmium, and others) are able affect and to disrupt the main female reproductive events—ovarian cycle, release of ovarian hormones and growth factors, ovarian cell proliferation, apoptosis, ovarian folliculogenesis, oogenesis, ovulation, fecundity, functions of uterus, embryo implantation, and development, pregnancy, and labor. These heavy metal actions on female reproductive systems can be mediated by changes in extra- and intracellular regulators of oxidative processes, proliferation, apoptosis, secretion, and reception of steroid hormones and growth factors.

Although understanding the effects of heavy metals on reproduction is very important from both theoretical and practical viewpoints, the current available knowledge is relatively poor and superficial. The mechanisms of heavy metals are usually stated on the basis of signaling molecules affected by these metals, but their mediatory role and hierarchic interrelationships with other signaling molecules are remain usually not fully determined. These mechanisms are usually studied on *in vitro* models, which should not fully reflect the behavior of heavy metals in the whole organism. The direct strong clinical evidence concerning character, mechanisms and consequences of heavy metals on human reproductive processes are usually missing. Nevertheless, the importance of heavy metals for maintenance of health and function of female reproductive system could be a good motivation for further studies in this area.

References

Agarwal A, Gupta S, Sharma RK. Role of oxidative stress in female reproduction. Reprod Biol Endocrinol 2005;3:28. https://doi.org/10.1186/1477-7827-3-28.

Ajiboye TO, Oyewo OA, Onwudiwe DC. Simultaneous removal of organics and heavy metals from industrial wastewater: a review. Chemosphere 2021;262:128379. https://doi.org/10.1016/j.chemosphere.2020.128379.

Al-Saleh I, Shinwari N, Mashhour A, Mohamed GED, Rabah A. Heavy metals (lead, cadmium and mercury) in maternal, cord blood and placenta of healthy women. Int J Hyg Environ Health 2011;214(2):79−101. https://doi.org/10.1016/j.ijheh.2010.10.001.

Environmental contaminants and their influence Chapter | 2 **33**

Anderson GJ, Frazer DM. Current understanding of iron homeostasis. Am J Clin Nutr 2017;106(6):1559S−66S. https://doi.org/10.3945/ajcn.117.155804.

Benedetto A, Au C, Avila DS, Milatovic D, Aschner M. Extracellular dopamine potentiates mn-induced oxidative stress, lifespan reduction, and dopaminergic neurodegeneration in a BLI-3−dependent manner in *Caenorhabditis elegans*. PLoS Genet 2010;6(8):e1001084. https://doi.org/10.1371/journal.pgen.1001084.

Bernard A. Cadmium & its adverse effects on human health. Indian J Med Res 2008;128(4):557. http://hdl.handle.net/2078.1/35901.

Bernhoft RA. Mercury toxicity and treatment: a review of the literature. J Environ Public Health 2012. https://doi.org/10.1155/2012/460508.

Bhasin G, Kauser H, Athar M. Iron augments stage-I and stage-II tumor promotion in murine skin. Canc Lett 2002;183(2):113−22. https://doi.org/10.1016/S0304-3835(02)00116-7.

Bjørklund G, Aaseth J, Chirumbolo S, Urbina MA, Uddin R. Effects of arsenic toxicity beyond epigenetic modifications. Environ Geochem Health 2018;40(3):955−65. https://doi.org/10.1007/s10653-017-9967-9.

Binet MT, Adams MS, Gissi F, Golding LA, Schlekat CE, Garman ER, Stauber JL. Toxicity of nickel to tropical freshwater and sediment biota: a critical literature review and gap analysis. Environ Toxicol Chem 2018;37(2):293−317. https://doi.org/10.1002/etc.3988.

Borja-Aburto VH, Hertz-Picciotto I, Lopez MR, Farias P, Rios C, Blanco J. Blood lead levels measured prospectively and risk of spontaneous abortion. Am J Epidemiol 1999;150(6):590−7. https://doi.org/10.1093/oxfordjournals.aje.a010057.

Brown MJ, Margolis S. Lead in drinking water and human blood lead levels in the United States. The morbidity and mortality weekly report (MMWR). Washington, DC: Center for Disease Control and Prevention (CDC); 2012.

Burton NC, Guilarte TR. Manganese neurotoxicity: lessons learned from longitudinal studies in nonhuman primates. Environ Health Perspect 2009;117(3):325−32. https://doi.org/10.1289/ehp.0800035.

Canipari R, De Santis L, Cecconi S. Female fertility and environmental pollution. Int J Environ Res Publ Health 2020;17(23):8802. https://doi.org/10.3390/ijerph17238802.

Chandra R, Bharagava RN, Kapley A, Purohit HJ. Bacterial diversity, organic pollutants and their metabolites in two aeration lagoons of common effluent treatment plant (CETP) during the degradation and detoxification of tannery wastewater. Bioresour Technol 2011;102(3):2333−41. https://doi.org/10.1016/j.biortech.2010.10.087.

Davis BJ, Price HC, O'Connor RW, Fernando RRAS, Rowland AS, Morgan DL. Mercury vapor and female reproductive toxicity. Toxicol Sci 2001;59(2):291−6. https://doi.org/10.1093/toxsci/59.2.291.

Debnath B, Singh WS, Manna K. Sources and toxicological effects of lead on human health. Indian J Med Specialities 2019;10(2):66. https://doi.org/10.4103/INJMS.INJMS_30_18.

Dixit R, Malaviya D, Pandiyan K, Singh UB, Sahu A, Shukla R, Paul D. Bioremediation of heavy metals from soil and aquatic environment: an overview of principles and criteria of fundamental processes. Sustainability 2015;7(2):2189−212. https://doi.org/10.3390/su7022189.

Engwa GA, Ferdinand PU, Nwalo FN, Unachukwu MN. Mechanism and health effects of heavy metal toxicity in humans. Poison Modern World N Tricks Old Dog 2019;10.

EPA. Basic information about mercury. 2021. https://www.epa.gov/mercury/basic-information-about-mercury.

Flora G, Gupta D, Tiwari A. Toxicity of lead: a review with recent updates. Interdiscipl Toxicol 2012;5(2):47−58. https://doi.org/10.2478/v10102-012-0009-2.

34 Contaminants and Plants Action on Female Reproduction

Flora SJ, Agrawal S. Arsenic, cadmium, and lead. Reprod Develop Toxicol 2017:537−66. https://doi.org/10.1016/B978-0-12-804239-7.00031-7.

Foster WG. Reproductive toxicity of chronic lead exposure in the female cynomolgus monkey. Reprod Toxicol 1992;6(2):123−31. https://doi.org/10.1016/0890-6238(92)90113-8.

Fu Z, Xi S. The effects of heavy metals on human metabolism. Toxicol Mech Methods 2020;30(3):167−76. https://doi.org/10.1080/15376516.2019.1701594.

García-Esquinas E, Pollán M, Umans JG, Francesconi KA, Goessler W, Guallar E, Navas−Acien A. Arsenic exposure and cancer mortality in a US-based prospective cohort: the strong heart study. Cancer Epidemiol Prevent Biomark 2013;22(11):1944−53. https://doi.org/10.1158/1055-9965.EPI-13-0234-T.

Genestra M. Oxyl radicals, redox-sensitive signalling cascades and antioxidants. Cell Signal 2007;19(9):1807−19. https://doi.org/10.1016/j.cellsig.2007.04.009.

Guilarte TR. Manganese and Parkinson's disease: a critical review and new findings. Environ Health Perspect 2010;118(8):1071−80. https://doi.org/10.1289/ehp.0901748.

Gunter TE, Gerstner B, Lester T, Wojtovich AP, Malecki J, Swarts SG, Gunter KK. An analysis of the effects of Mn2+ on oxidative phosphorylation in liver, brain, and heart mitochondria using state 3 oxidation rate assays. Toxicol Appl Pharmacol 2010;249(1):65−75. https://doi.org/10.1016/j.taap.2010.08.018.

Hillman RS. Hematopoietic agents: growth factors, minerals, and vitamins. In: Hardman JG, Limbird LE, Gilman AG, editors. Goodman & Gilman's the pharmacological basis of therapeutics. 10th ed. New York: McGraw-Hill; 2001. p. 1487−518.

Höfer N, Diel P, Wittsiepe J, Wilhelm M, Degen GH. Dose-and route-dependent hormonal activity of the metalloestrogen cadmium in the rat uterus. Toxicol Lett 2009;191(2−3):123−31. https://doi.org/10.1016/j.toxlet.2009.08.014.

Huy TB, Tuyet-Hanh TT, Johnston R, Nguyen-Viet H. Assessing health risk due to exposure to arsenic in drinking water in Hanam Province, Vietnam. Int J Environ Res Publ Health 2014;11(8):7575−91. https://doi.org/10.3390/ijerph110807575.

Jiang X, Xing X, Zhang Y, Zhang C, Wu Y, Chen Y, Meng R, Jia H, Cheng Y, Zhang Y, Su J. Lead exposure activates the Nrf2/Keap1 pathway, aggravates oxidative stress, and induces reproductive damage in female mice. Ecotoxicol Environ Saf 2021;207:111231. https://doi.org/10.1016/j.ecoenv.2020.111231. Epub 2020 Sep 8. PMID: 32916527.

Kalafova A, Kovacik J, Capcarova M, Kolesarova A, Lukac N, Stawarz R, Formicki G, Laciak T. Accumulation of zinc, nickel, lead and cadmium in some organs of rabbits after dietary nickel and zinc inclusion. J Environ Sci Health A Tox Hazard Subst Environ Eng 2012;47(9):1234−8. https://doi.org/10.1080/10934529.2012.672073.

Kawai M, Swan KF, Green AE, Edwards DE, Anderson MB, Henson MC. Placental endocrine disruption induced by cadmium: effects on P450 cholesterol side-chain cleavage and 3β-hydroxysteroid dehydrogenase enzymes in cultured human trophoblasts. Biol Reprod 2002;67(1):178−83. https://doi.org/10.1095/biolreprod67.1.178.

Kim HC, Jang TW, Chae HJ, Choi WJ, Ha MN, Ye BJ, Hong YS. Evaluation and management of lead exposure. Ann Occup Environ Med 2015;27(1):1−9. https://doi.org/10.1186/s40557-015-0085-9.

Kim SH, Sharma RP. Cytotoxicity of inorganic mercury in murine T and B lymphoma cell lines: involvement of reactive oxygen species, Ca2+ homeostasis, and cytokine gene expression. Toxicol Vitro 2003;17(4):385−95. https://doi.org/10.1016/S0887-2333(03)00040-7.

Klos KJ, Chandler M, Kumar N, Ahlskog JE, Josephs KA. Neuropsychological profiles of manganese neurotoxicity. Eur J Neurol 2006;13(10):1139−41. https://doi.org/10.1111/j.1468-1331.2006.01407.x.

Kolesarova A, Capcarova M, Sirotkin AV, Medvedova M, Kalafova A, Filipejova T, Kovacik J. In vitro assessment of molybdenum-induced secretory activity, proliferation and apoptosis of porcine ovarian granulosa cells. J Environ Sci Health A Tox Hazard Subst Environ Eng 2011;46(2):170−5. https://doi.org/10.1080/10934529.2011.532430.

Kolesarova A, Roychoudhury S, Slivkova J, Sirotkin A, Capcarova M, Massanyi P. In vitro study on the effects of lead and mercury on porcine ovarian granulosa cells. J Environ Sci Health A Tox Hazard Subst Environ Eng 2010;45(3):320−31. https://doi.org/10.1080/10934520903467907.

Kolesarova A, Slamecka J, Jurcik R, Tataruch F, Lukac N, Kovacik J, Capcarova M, Valent M, Massanyi P. Environmental levels of cadmium, lead and mercury in brown hares and their relation to blood metabolic parameters. J Environ Sci Health A Tox Hazard Subst Environ Eng 2008;43(6):646−50. https://doi.org/10.1080/10934520801893741.

Kuang H, Zhang W, Yang L, Aguilar ZP, Xu H. Reproductive organ dysfunction and gene expression after orally administration of ZnO nanoparticles in murine. Environ Toxicol 2021;36(4):550−61. https://doi.org/10.1002/tox.23060.

Lee BE, Hong YC, Park H, Ha M, Koo BS, Chang N, Roh YM, Kim BN, Kim YJ, Kim BM, Jo SJ, Ha EH. Interaction between GSTM1/GSTT1 polymorphism and blood mercury on birth weight. Environ Health Perspect 2010;118(3):437−43. https://doi.org/10.1289/ehp.0900731.

Lohren H, Blagojevic L, Fitkau R, Ebert F, Schildknecht S, Leist M, Schwerdtle T. Toxicity of organic and inorganic mercury species in differentiated human neurons and human astrocytes. J Trace Elem Med Biol 2015;32:200−8. https://doi.org/10.1016/j.jtemb.2015.06.008.

Ma Y, Shi YZ, Wu QJ, Wang YQ, Wang JP, Liu ZH. Effects of varying dietary intoxication with lead on the performance and ovaries of laying hens. Poultry Sci 2020;99(9):4505−13. https://doi.org/10.1016/j.psj.2020.06.015. Epub 2020 Jul 2.

Martin S, Griswold W. Human health effects of heavy metals. Environ Sci Technol Brief Citizens 2009;15:1−6.

Martínez-Sales M, García-Ximénez F, Espinós FJ. Zebrafish as a possible bioindicator of organic pollutants with effects on reproduction in drinking waters. J Environ Sci 2015;33:254−60. https://doi.org/10.1016/j.jes.2014.11.012.

Mehta M, Hundal SS. Effect of sodium arsenite on reproductive organs of female Wistar rats. Arch Environ Occup Health 2016;71(1):16−25. https://doi.org/10.1080/19338244.2014.927346.

Mintz J, Mirza J, Young E, Bauckman K. Iron therapeutics in women's health: past, present, and future. Pharmaceuticals 2020;13(12):449. https://doi.org/10.3390/ph13120449.

Milatovic D, Gupta RC, Yin Z, Zaja-Milatovic S, Aschner M. Manganese in reproductive and developmental toxicology. 2017. p. 567−81. https://doi.org/10.1016/B978-0-12-804239-7.00032-9.

Motta CM, Simoniello P, Di Lorenzo M, Migliaccio V, Panzuto R, Califano E, Santovito G. Endocrine disrupting effects of copper and cadmium in the oocytes of the Antarctic Emerald rockcod *Trematomus bernacchii*. Chemosphere 2021;268:129282. https://doi.org/10.1016/j.chemosphere.2020.129282. Epub 2020 Dec 18.PMID: 33360142.

Mudgal V, Madaan N, Mudgal A, Singh RB, Mishra S. Effect of toxic metals on human health. Open Nutraceuticals J 2010;3:94−9. https://doi.org/10.2174/18763960010030100188.

Nasiadek M, Swiatkowska E, Nowinska A, Krawczyk T, Wilczynski J, Sapota A. The effect of cadmium on steroid hormones and their receptors in women with uterine myomas. Arch Environ Contamin Toxicol 2011;60(4):734−41. https://doi.org/10.1007/s00244-010-9580-8.

Nishijo M, Nakagawa H, Suwazono Y, Nogawa K, Kido T. Causes of death in patients with Itai-itai disease suffering from severe chronic cadmium poisoning: a nested case−control analysis of a

36 Contaminants and Plants Action on Female Reproduction

follow-up study in Japan. BMJ Open 2017;7(7):e015694. https://doi.org/10.1136/bmjopen-2016-015694.

O'Neal SL, Zheng W. Manganese toxicity upon overexposure: a decade in review. Curr Environ Health Rep 2015;2(3):315−28. https://doi.org/10.1007/s40572-015-0056-x.

Pantopoulos K, Porwal SK, Tartakoff A, Devireddy L. Mechanisms of mammalian iron homeostasis. Biochemistry 2012;51(29):5705−24. https://doi.org/10.1021/bi300752r.

Piasek M, Laskey JW. Acute cadmium exposure and ovarian steroidogenesis in cycling and pregnant rats. Reprod Toxicol 1994;8(6):495−507. https://doi.org/10.1016/0890-6238(94)90032-9.

Rahman Z, Singh VP. The relative impact of toxic heavy metals (THMs) (arsenic (As), cadmium (Cd), chromium (Cr)(VI), mercury (Hg), and lead (Pb)) on the total environment: an overview. Environ Monit Assess 2019;191(7):419. https://doi.org/10.1007/s10661-019-7528-7.

Rehman K, Fatima F, Waheed I, Akash MSH. Prevalence of exposure of heavy metals and their impact on health consequences. J Cell Biochem 2018;119(1):157−84. https://doi.org/10.1002/jcb.26234.

Renu K, Chakraborty R, Myakala H, Koti R, Famurewa AC, Madhyastha H, Vellingiri B, George A, Valsala Gopalakrishnan A. Molecular mechanism of heavy metals (lead, chromium, arsenic, mercury, nickel and cadmium) - induced hepatotoxicity - a review. Chemosphere 2021;271:129735. https://doi.org/10.1016/j.chemosphere.2021.129735.

Rizvi A, Parveen S, Khan S, Naseem I. Nickel toxicology with reference to male molecular reproductive physiology. Reprod Biol 2020;20(1):3−8. https://doi.org/10.1016/j.repbio.2019.11.005.

Roychoudhury S, Bulla J, Sirotkin AV, Kolesarova A. In vitro changes in porcine ovarian granulosa cells induced by copper. J Environ Sci Health A Tox Hazard Subst Environ Eng 2014a;49(6):625−33. https://doi.org/10.1080/10934529.2014.865404.

Roychoudhury S, Detvanova L, Sirotkin AV, Toman R, Kolesarova A. In vitro changes in secretion activity of rat ovarian fragments induced by molybdenum. Physiol Res 2014b;63(6):807−9. https://doi.org/10.33549/physiolres.932836.

Roychoudhury S, Sirotkin AV, Toman R, Kolesarova A. Cobalt-induced hormonal and intracellular alterations in rat ovarian fragments in vitro. J Environ Sci Health B 2014c;49(12):971−7. https://doi.org/10.1080/03601234.2014.951586.

Roychoudhury S, Nath S, Massanyi P, Stawarz R, Kacaniova M, Kolesarova A. Copper-induced changes in reproductive functions: in vivo and in vitro effects. Physiol Res 2016;65(1):11−22. https://doi.org/10.33549/physiolres.933063.

Rusyniak DE, Arroyo A, Acciani J, Froberg B, Kao L, Furbee B. Heavy metal poisoning: management of intoxication and antidotes. Mol Clin Environ Toxicol 2010:365−96. https://doi.org/10.1007/978-3-7643-8338-1_11.

Rzymski P, Rzymski P, Tomczyk K, Niedzielski P, Jakubowski K, Poniedziałek B, Opala T. Metal status in human endometrium: relation to cigarette smoking and histological lesions. Environ Res 2014;132:328−33. https://doi.org/10.1016/j.envres.2014.04.025.

Rzymski P, Tomczyk K, Poniedzialek B, Opala T, Wilczak M. Impact of heavy metals on the female reproductive system. Ann Agric Environ Med 2015;22(2). https://doi.org/10.5604/12321966.1152077.

Schuurs AHB. Reproductive toxicity of occupational mercury. A review of the literature. J Dent 1999;27(4):249−56. https://doi.org/10.1016/S0300-5712(97)00039-0.

Singh MR. Impurities-heavy metals: IR perspective. Int J Phys Sci 2007;5(4):1045−58.

Steinbrenner H, Klotz LO. Selenium and zinc: "antioxidants" for healthy aging? Z Gerontol Geriatr 2020;53(4):295−302. https://doi.org/10.1007/s00391-020-01735-0.

Taupeau C, Poupon J, Nomé F, Lefèvre B. Lead accumulation in the mouse ovary after treatment-induced follicular atresia. Reprod Toxicol 2001;15(4):385−91. https://doi.org/10.1016/S0890-6238(01)00139-3.

Torti SV, Manz DH, Paul BT, Blanchette-Farra N, Torti FM. Iron and cancer. Annu Rev Nutr 2018;38:97−125. https://doi.org/10.1146/annurev-nutr-082117-051732.

Usuki F, Fujita E, Sasagawa N. Methylmercury activates ASK1/JNK signaling pathways, leading to apoptosis due to both mitochondria-and endoplasmic reticulum (ER)-generated processes in myogenic cell lines. Neurotoxicology 2008;29(1):22−30. https://doi.org/10.1016/j.neuro.2007.08.011.

Valko M, Izakovic M, Mazur M, Rhodes CJ, Telser J. Role of oxygen radicals in DNA damage and cancer incidence. Mol Cell Biochem 2004;266(1):37−56. https://doi.org/10.1023/B:MCBI.0000049134.69131.89.

Verma R, Vijayalakshmy K, Chaudhiry V. Detrimental impacts of heavy metals on animal reproduction: a review. J Entomol Zoo Stud 2018;6:27−30. ISSN: 2320-7078.

Yanatori I, Kishi F. DMT1 and iron transport. Free Radic Biol Med 2019;133:55−63. https://doi.org/10.1016/j.freeradbiomed.2018.07.020.

Yang L, Zhang Y, Wang F, Luo Z, Guo S, Strähle U. Toxicity of mercury: molecular evidence. Chemosphere 2020;245:125586. https://doi.org/10.1016/j.chemosphere.2019.125586.

Yiannikourides A, Latunde-Dada GO. A short review of iron metabolism and pathophysiology of iron disorders. Medicines 2019;6(3):85. https://doi.org/10.3390/medicines6030085.

Zhang X, Gan X, Zhang EQ, Ye Y, Cai Y, Han A, Tian M, Wang Y, Wang C, Su L, Liang C. Ameliorative effects of nano-selenium against NiSO4-induced apoptosis in rat testes. Toxicol Mech Methods 2019;29(7):467−77. https://doi.org/10.1080/15376516.2019.1611979.

38 Contaminants and Plants Action on Female Reproduction

Chapter 2.2

Mycotoxins

2.2.1 Introduction

Mycotoxins are a relatively large, diverse group of naturally occurring, and fungal toxins, many of which have been strongly implicated as chemical agents of toxic disease in humans and animals (Barac, 2019). Main problem with mycotoxins is that contamination generally affects cereals/cereal products (Bryła et al., 2018), and a large number of raw materials and finished feed intended for animal production. Mycotoxins can impact human health via animal products such as eggs, milk, and meat when consumed from intoxicated animals (Binder et al., 2007). According to Food and Drug Organization there are more than 25% of food production in the world that is to some degree contaminated with mycotoxins. On a global level, 30%−100% of feed and food samples are contaminated with at least one mycotoxin (Rodrigues and Naehrer, 2012). Growth, distribution, and mycotoxin production in fungi can be significantly affected by climate. Increasing change of climate has the potential to also increase the risk that mycotoxigenic fungi pose to feed and food safety (Pinotti et al., 2016). Mycotoxins can have a disruptive or even fatal impact on both human and animal health. Among many others, mycotoxins can cause infertility and reproductive alterations. In this chapter we are reviewing knowledge about how mycotoxins can affect human health and the female reproductive system.

This chapter could be important because mycotoxins represent serious reproductive health problem now. The previous related reviews are relatively old, and they can be and may be marked out of date. There are not many reports about the direct effect of mycotoxins on human ovarian cells but a lot of questions about mycotoxins mechanism of action that are yet to be answered now. With this review we are trying to point out the need for further research.

2.2.2 Provenance and properties

Mycotoxins are produced by fungi in grains if they parasitize the host plants during growth. Fungi can also produce mycotoxins if they grow as saprophytes on grains during storage and harvest (Fletcher and Blaney, 2016). Mycotoxins are natural secondary metabolites of fungi for example *Aspergillus, Fusarium,*

Alternaria, or *Penicillium* (Ukwuru et al., 2017; Haque et al., 2020). They are structurally diverse group of toxic and low molecular weight compound (Ukwuru et al., 2017; Haque et al., 2020). Mycotoxins are grouped according to their toxicity (mutagenic, teratogenic, or carcinogenic) (Ostry et al., 2017). Mycotoxins are considered most toxigenic to public health, agriculture and animal husbandry (Marin et al., 2013).

Deoxynivalenol (DON, $C_{15}H_{20}O_6$, Fig. 2.2.1) is a type B trichothecene, produced by the *Fusarium* species. Exposure to DON might cause disruptive effects such as reduced weight gain, neuroendocrine changes and immune modulation in animals and humans (Urbanek et al., 2018). **Zearalenone** (ZEA, $C_{18}H_{22}O_5$, Fig. 2.2.2) is classified as a xenoestrogen, an exogenous compound which resembles the structure of naturally occurring estrogens with its chemical structure. This property of ZEA determines its ability to bind to estrogen receptors of cell and its bioaccumulation. This leads to disorders of the hormonal balance of the body, which in consequence may lead to numerous diseases of reproductive system such as prostate, ovarian, cervical or

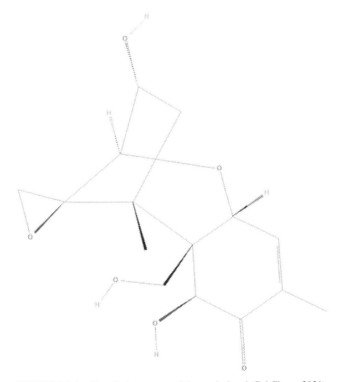

FIGURE 2.2.1 Chemical structure of deoxynivalenol (PubChem, 2021).

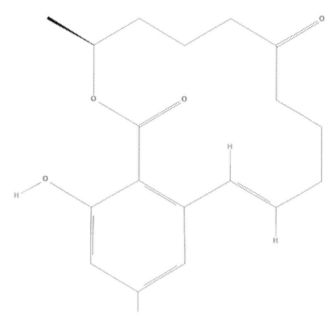

FIGURE 2.2.2 Chemical structure of zearalenone (PubChem, 2021).

breast cancers (Rogowska et al., 2019). **T-2 toxin** (T-2, $C_{24}H_{34}O_9$, Fig. 2.2.3) is a type A trichothecene produced by *Fusarium* species, and the most cytotoxic mycotoxin of the group (Taroncher et al., 2020). For the last decade, it has garnered considerable attention due to its potent neurotoxicity. Worryingly, T-2 toxin can cross the blood-brain barrier and accumulate in the central nervous system (CNS) to cause neurotoxicity (Dai et al., 2019). **HT2 toxin** (HT2, $C_{22}H_{32}O_8$, Fig. 2.2.4) is one of the type A trichothecene mycotoxins has exerted various toxic effects on human and livestock, as it induces lesions in multiple tissues including reproductive system (Zhang et al., 2019).

The examples of the harmful mycotoxins are ochratoxins, aflatoxins, trichothecenes, zearalenone, fumonisins, citrinin, and also patulin. Human and animal intoxication by mycotoxins are possible via food and feed (Sun et al., 2020). Main problem is, when animals are fed with contaminated feed, mycotoxins can accumulate in their tissues, which also affects eggs, meat, and milk. These products are then also mycotoxin-contaminated and can be a potential hazard for human health (Hou et al., 2020). Mycotoxins can be dangerous both to humans and animals upon inhalation, ingestion, or skin contact. Disease caused by mycotoxins is called mycotoxicosis (Marin et al., 2013). In organism, mycotoxin is absorbed in the small intestine and transferred to the bloodstream, and then is transported by plasma proteins and red blood cells directly to the liver. There is the toxin metabolized by microsomal-mixed function oxidase enzymes. Then toxin is converted to more toxic and

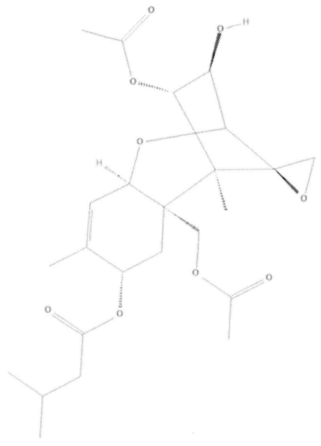

FIGURE 2.2.3 Chemical structure of T-2 toxin (PubChem, 2021).

highly reactive metabolites that have a high affinity to bind to the DNA and then lead to DNA mutations. The metabolites also bind to other macromolecules like RNA and proteins, which leads to inhibition of RNA, DNA, protein synthesis, and cellular function disorders. In addition, in the endoplasmic reticulum of liver cells, toxin is also hydroxylated to toxic metabolites (Rotimi et al., 2019).

Fungi can produce toxins during cultivation, harvesting or, even when we are storing food, which leads to contamination. Production of mycotoxins is a problem in many countries around the world where production technologies and storage conditions are inefficient, which can lead to economic losses. Production can be affected by temperature, water activity, oxygen, pH and substrate composition (Thanushree et al., 2019). Growth, distribution, and mycotoxin production in fungi can be significantly

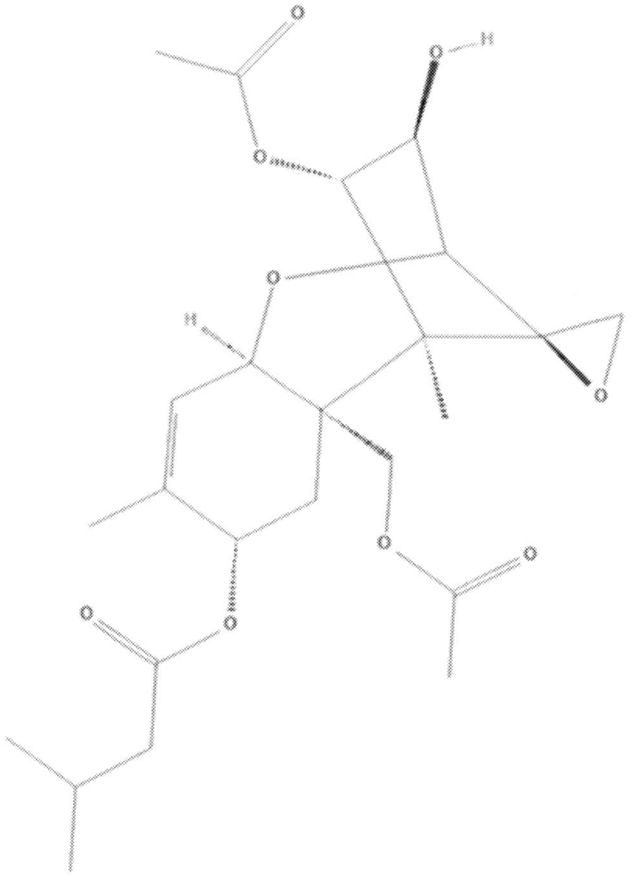

42 Contaminants and Plants Action on Female Reproduction

FIGURE 2.2.4 Chemical structure of HT2 toxin (PubChem, 2021).

affected by climate. Increasing change of climate has the potential to also increase the risk that mycotoxigenic fungi pose to feed and food safety (Pinotti et al., 2016). Mycotoxin contamination is considered to be an unpredictable problem. Main problem is the resistance of mycotoxins to high temperatures and chemical or physical treatments. According to this, standard cooking is not a suitable option for mycotoxin elimination (Alshannaq and Yu, 2017). Since the initial discovery of mycotoxins, new methods have been found to detect mycotoxins in food and feed. Main problems are difficulties in detecting low-level mycotoxin contamination, the great diversity of mycotoxin chemical structures, complex food matrices in which the mycotoxin contamination occurs necessitating complicated extraction processes and the cooccurrence of mycotoxins (Stroka and Maragos, 2016).

2.2.3 Physiological actions

Mycotoxins can express toxic, immunotoxic (Bondy and Pestka, 2000; Voss et al., 2006; Solcan et al., 2015; Urbanek et al., 2018), mutagenic (Voss et al., 2006; Ostry et al., 2017; Rotimi et al., 2019; Do et al., 2020), carcinogenic (Ostry et al., 2017; Do et al., 2020) including induction of liver cancer (Matsuda et al., 2011), teratogenic (Ostry et al., 2017), growth and development retarding (Voss et al., 2006), and neuromodulatory (Matsuda et al., 2011; Urbanek et al., 2018; Dai et al., 2019) effects, can directly induce inflammation (Maresca and Fantini, 2010), and lead to enterocyte apoptosis (Solcan et al., 2015), and immune-related changes such as thymic aplasia, delayed cutaneous hypersensitivity, inhibition of phagocytosis by macrophages, leukocyte migration and also lymphocyte proliferation (Bondy and Pestka, 2000).

2.2.4 Mechanism of action

Several principal extra- and intracellular mechanisms of toxic effect of mycotoxins on the cells have been proposed, although the evidence of each of this mechanism is limited, and the interrelationships between these mechanisms are poorly investigated yet. The following mechanisms of mycotoxins action are suggested:

(1) generation of reactive oxygen species (ROS), decreasing the function of the antioxidant system (Huang et al., 2019), the dysfunction of mitochondria and release free radicals which include ROS, induction of lipid peroxidation leading to changes in membrane integrity, cellular redox signaling, and in the antioxidant status of the cells (Wu et al., 2014),

(2) inhibition of the key signaling transduction pathways, including a mitogen-activated protein kinase (MAPK) and Janus kinase (JAK)/signal transducers and activators of transcription (STAT), by triggering the ribosomal stress response, thereby regulating signaling molecules involved in apoptosis, such as interleukins IL-6, IL-1β, and tumor necrosis factor-α (TNF-α; Wu et al., 2017),

(3) suppression of nucleic acid synthesis (Rotimi et al., 2019), induction of DNA damage and mutagenesis (Voss et al., 2006; Do et al., 2020),

(4) zinc deficiency, which then causes growth faltering and immune deficiency (Turner et al., 2007),

(5) inhibition of RNA, DNA (Rotimi et al., 2019), protein synthesis, which indicates that ribosomes may be an important target for mycotoxins (Wu et al., 2014), and cellular function disorders (Wu et al., 2014; Rotimi et al., 2019), and

(6) reduction in the intestinal TCR1, TCR2, CD4+, CD8+ lymphocyte population which led to enterocyte apoptosis (Solcan et al., 2015).

2.2.5 Effects on female reproductive processes

The most dangerous effects of mycotoxins could be their influence on the most important biological process—reproduction, especially female reproduction and its regulators including steroidogenesis (Medvedova et al., 2011; Kolesarova et al., 2012; Santos et al., 2013; Maruniakova et al., 2014, 2015; Halenar et al., 2015; Capcarova et al., 2015; Kolesarova et al., 2017a,b; Urbanek et al., 2018; Rogowska et al., 2019).

2.2.5.1 Effect on ovarian cell functions

Mycotoxins among others can be teratogenic, and they also can cause problems in female reproduction. *Fusarium* mycotoxins, such as DON, ZEA, and fumonisin B1, which are mainly produce by molds *Fusarium* genus, can cause serious reproductive and health problems to both livestock and humans (Milićević et al., 2010). Although recent studies about *Fusarium* mycotoxins have shown that it can be potential regulator of intracellular steroidogenesis, there are many studies on reproductive toxicity that was caused by DON and fumonisin B1 that was based on *in vitro* models (Santos et al., 2013; Urbanek et al., 2018). The *in vitro* experiments revealed the direct action of mycotoxins on basic ovarian follicular cell functions—viability, proliferation, apoptosis, secretory activity, and the response to the endocrine regulators.

DON and ZEA, *Fusarium* mycotoxins is frequently occurring in feed of pigs together (Kolesarova et al., 2017a,b). DON is a potential regulator of intracellular steroidogenesis (Urbanek et al., 2018). The stimulatory effect of DON on the progesterone (Medvedova et al., 2011; Halenar et al., 2015; Kolesarova et al., 2017a,b) and 17β-estradiol (Kolesarova et al., 2017b) secretion by porcine ovarian granulosa cells was found. These findings suggest a possible involvement of DON into the processes of steroidogenesis and appear to be endocrine modulator of porcine ovaries (Halenar et al., 2015). On the other hand, the effects of ZEA on progesterone and estradiol secretion by the cells were not confirmed. DON in combination with the other fusariotoxin ZEA may impair steroidogenesis. The results also demonstrate different toxicological effects of fusariotoxins on follicle-stimulating hormone-induced secretion of progesterone and estradiol. All these results taken together suggest that fusariotoxin and their interactions can impact ovarian steroidogenesis, thereby demonstrating their potential reproductive effects on pigs (Kolesarova et al., 2017b).

Next, *in vitro* results suggest that reproductive toxicity of animals induced by a mycotoxin - DON can be inhibited by a protective natural substance—resveratrol (Kolesarova et al., 2012). Resveratrol in combination with DON stimulated the progesterone release by porcine ovarian granulosa cells. On the other hand, the stimulatory effect of resveratrol in combination with DON was lower in comparison with alone DON effect. The results indicate (1) the dose-depended stimulatory effects of resveratrol, DON and combination of resveratrol with DON on release of steroid hormone progesterone and (2) reduction of the stimulatory effect of DON by resveratrol (Kolesarova et al., 2012). Frizzell et al. (2011) demonstrated effects of ZEA on the endocrine system. Zearalenone and its metabolites can induce progesterone, cortisol, testosterone and estradiol in the human adrenal gland-derived cells. We can assume that this is indirectly caused by an inhibition of apoptosis and also an increased cell proliferation (Khosrokhavar et al., 2009).

T-2 and HT2 toxin are one of the most toxic mycotoxins of type A-trichothecenes, which are produced mainly by *Fusarium* species (Kolesarova et al., 2017a). Feeding of mice with mycotoxin-contaminated diet containing DON, ZEA, aflatoxin reduced ovarian size, i.e., suppressed ovarian growth and folliculogenesis (Hou et al., 2014). HT2 toxin decreased 17β-estradiol secretion by ovarian fragments. On the other hand, 17β-estradiol secretion was not affected by HT2 toxin exposure combined with insulin-like growth factor I (IGF-I), metabolic hormones leptin and ghrelin. HT2 toxin has potent direct dose-dependent effects on ovarian steroidogenesis in rabbits. These direct effects of HT2 mycotoxin on ovarian steroidogenesis could impact negatively on the reproductive performance of rabbits (Kolesarova et al., 2017a). Alone IGF-I addition stimulates progesterone release by cultured porcine ovarian granulosa cells, while either T-2 or HT2 toxins could prevent the IGF-I-induced progesterone secretion (Maruniakova et al., 2014). In cultured rabbit ovarian fragments, trichothecene as T-2 toxin, but not HT2 toxin prevented the stimulatory action of IGF-I on progesterone secretion, On the other hand, T-2 and HT2 toxins did not modify the action of leptin or ghrelin on progesterone secretion by cultured rabbit ovarian fragments (Maruniakova et al., 2015).

2.2.5.2 Effects on oocytes and embryos

Hou et al. (2014) submitted a study about oocyte quality in mice after it was affected by mycotoxin-contaminated diet by DON, ZEA, aflatoxin. Their results showed that diet contaminated with naturally occurring mycotoxins has resulted in decrease in size of oocytes and appearance of unclear oocyte cytoplasm. Oocytes from these mice exhibited low developmental competence with reduced germinal vesicle breakdown and polar body extrusion rates. Embryo developmental competence also showed a similar pattern, and the majority of embryos could not develop to the morula stage. Moreover, a large percentage of oocytes derived from mice that were fed a mycotoxin-contaminated diet exhibited aberrant spindle morphology, a loss of the cortical granule-free domain, and abnormal mitochondrial distributions, which further supported the decreased oocyte quality. Thus, the results demonstrate that mycotoxins are toxic to the mouse reproductive system by affecting oocyte quality (Hou et al., 2014). Oocyte activity might be decreased after the mycotoxin exposure (Hou et al., 2014). DON exposure reduced porcine

46 Contaminants and Plants Action on Female Reproduction

oocytes maturation capability through affecting cytoskeletal dynamics, cell cycle, autophagy/apoptosis and epigenetic modifications (Han et al., 2016). DON is the most frequent contaminant of grains. It has been found out that DON can reduce reproductive performance in mammals by impairing oocyte maturation and embryo development (Lu et al., 2018). *In vitro* studies on DON have shown that this mycotoxin can alter fetus growth and cause bone malformation. Although the effects of DON are known, the potential effects of DON on placental function and embryogenesis are to this day not yet adequately evaluated (Yu et al., 2017). Schoevers et al. (2012) submitted a study about ZEA effect on pigs. ZEA reduced the number of follicles with normal morphology during fetal and early postnatal period. Prepubertal gilts, which were exposed to ZEA, did not have their development potential of oocytes affected. The reduction of normal follicles during the fetal and early postnatal period can lead to premature exhaustion of the follicle pool.

Taken together, the results of the preformed studies demonstrate direct influence of mycotoxins on basic ovarian cell functions including hormones release and response to hormonal stimulators, quality of oocytes and development of embryos. The results demonstrate that mycotoxins are toxic to the female reproductive system in various species.

2.2.6 Mechanisms of action on female reproductive processes

The most evident mechanisms of adverse effects of mycotoxins on female reproduction and fecundity can be ability of mycotoxins to act on ovarian cell steroidogenesis, proliferation and apoptosis through intracellular regulators (Khosrokhavar et al., 2009; Medvedova et al., 2011; Halenar et al., 2015; Kolesarova et al., 2017a,b; Gerez et al., 2021). Such effects were demonstrated for DON, ZEA, T2 and HT2 toxin and others in previous studies (Medvedova et al., 2011; Halenar et al., 2015; Kolesarova et al., 2017a,b). At present, there is limited information about the influence of mycotoxins exposure on mammalian reproduction.

ZEA determines its ability to bind to **estrogen receptors** of cell and its bioaccumulation (Rogowska et al., 2019).

IGF-I secretion by porcine ovarian granulosa cells was inhibited by DON, while the expression of a regulatory protein involved in mitosis **cyclin B1** and **proliferating cell nuclear antigen (PCNA)** expression was stimulated by DON. On the other hand, marker of apoptosis caspase-3 expression was not influenced by DON treatment. This and previous results (Medvedova et al., 2011; Halenar et al., 2015; Kolesarova et al., 2017b) results indicate (1) a direct effect of DON on secretion of growth factor IGF-I and steroids hormones, (2) expression of markers of proliferation (cyclin B1 and PCNA) but not on the (3) expression of marker of apoptosis (caspase-3) in porcine ovarian granulosa cells. This *in vitro* study suggests the dose-dependent association of DON on porcine ovarian functions (Medvedova et al., 2011).

Similar to DON, fumonisin B1 is a common contaminant of maize that often occurs with other *Fusarium* toxins (Eskola et al., 2019). Cortinovis et al. (2014) stated that fumonisin B1 induces inhibitory effects on porcine granulosa **cell proliferation** and also delays attainment of **sexual maturity**. Even with this knowledge, its mechanism of action on ovarian tissue is unknown to this day (Gerez et al., 2021).

DON inhibits porcine **oocyte maturation** and disrupted meiotic spindle by reducing phosphorylated **phospho-mitogen-activated protein kinase (p-MAPK)** protein level, which caused retardation of cell cycle progression. In addition, up-regulated **LC3 protein expression** and aberrant **Lamp2, LC3 and mTOR mRNA** levels were observed with DON exposure, these results indicated that DON treatment induces **autophagy/apoptosis** in porcine oocytes. DON exposure increases **DNA methylation** level in porcine oocytes through altering DNA methyltransferase 3A (DNMT3A) mRNA levels. Histone methylation levels were also changed showing with increased H3K27me3 and H3K4me2 protein levels, and mRNA levels of their relative methyltransferase genes, indicating that epigenetic modifications were affected. Taken together, the results suggested that DON exposure reduced porcine oocytes maturation capability through affecting cytoskeletal dynamics, cell cycle, autophagy/apoptosis and epigenetic modifications (Han, 2016). Similarly, as ZEA, DON was found to have a potent effect on *in vitro* porcine oocyte maturation, significantly decreasing the proportion of oocytes reaching metaphase II. Deoxynivalenol at high concentration can inhibit cumulus expansion and induce cumulus **cell death** (Schoevers et al., 2010). Without cumulus cell expansion and induction of cumulus cell death, the oocyte maturation can be negatively influenced by altering the glutathione levels in the oocyte (Qian et al., 2003). It needs to be emphasized that in many experimental approaches high, nonrealistic concentrations of mycotoxins have been used. This can be helpful to identify mechanisms and vulnerable phases in oocyte maturation (Santos et al., 2013). Many researchers stated that oxidative stress is considered to be an important mechanism of toxicity of mycotoxins. T-2 toxin developed stress reaction in porcine ovarian granulosa cells and increased **generation of ROS** (Capcarova et al., 2015). Although the mechanism of fumonisin B_1 is not yet fully understood, Gbore et al. (2012) studied the effect of fumonisin B_1 on rats' ovaries. They noted an increase in primordial and growing follicles with type II degeneration. When we compare results of Gbore et al. (2012) and results of Gerez et al. (2021) we can suggest that fumonisin B_1 has higher toxicity then DON. Rats fed by fumonisin B_1 contaminated diet for 14 days have shown no histological abnormalities in their ovaries. Although no histological changes were observed, there was a significant reduction on the serum follicle-stimulating hormone (FSH) and luteinizing hormone (LH) levels (Gbore et al., 2012). HT2 exposure impaired mouse embryo development by

48 Contaminants and Plants Action on Female Reproduction

inducing **oxidative stress, mitochondria dysfunction,** and **DNA damage** (Zhang et al., 2019).

These observations indicate that mycotoxins can affect reproductive processes including steroidogenesis, proliferation, apoptosis via changes in generation of ROS, the release of growth factors and ovarian hormones and in response of ovarian cells to these hormones (probably due to mycotoxin-induced changes in hormone receptors). Mycotoxins such as aflatoxin, zearalenone, deoxynivalenol, ochratoxin, fumonisin, and trichothecenes affect oocyte quality by disruption of intracellular signaling pathway. On the other hand, there is still no sufficient direct evidence of mediatory role of hormones in mycotoxin action on reproduction.

2.2.7 Application in reproductive biology and medicine

The general and reproductive toxicity of mycotoxins demonstrate that they can disrupt female reproductive processes at different levels. Studies on animal models have shown that mycotoxins especially fumonisin B1, DON and ZEA, T-2 and HT2 toxin can affect ovarian **folliculogenesis**, release of hormones, quality, and maturation of oocyte and also embryo development, as well as the response of reproductive system to upstream hormonal regulators.

Character and intensity of mycotoxins influence on animal and human health depend on numerous interrelated factors including:

(1) mycotoxin production, storage, and consumption;
(2) mycotoxin monitoring in external and internal environment;
(3) limitation of mycotoxins release, distribution, and contents in the agricultural products; and
(4) characterization, prediction, mitigation, and prevention of adverse mycotoxin effects on organisms.

It means that prevention of harmful mycotoxins action on female reproductive processes could be achieved by control and reduction in mycotoxin generation and accumulation in environment and food. To this day, new improvements in the analytical methodology in a variety of food matrix are needed to prevent further mycotoxin intoxication. It is necessary to support the enforcement of mycotoxin regulations, to protect consumer's health, and also support the agriculture industry, and facilitate international food trade (Shephard, 2016).

The second approach to reduce the negative impact of mycotoxins on reproduction could be search for drugs, which could mitigate or prevent their effects. The testing of some hormonal preparations did not show the ability of IGF-I, leptin and ghrelin to prevent mycotoxin action on ovarian cells (Maruniakova et al., 2014, 2015). One of the mechanisms of mycotoxin toxic effect is generation of ROS (see earlier in the chapter). Therefore, it is not to

Environmental contaminants and their influence Chapter | 2 **49**

be excluded, that some plant preparation or plant molecules with antioxidant activity could be promising tool to neutralize adverse influence of mycotoxins on reproduction and health.

2.2.8 Conclusion

In this chapter we focused mainly on the female reproductive system. The few performed studies demonstrated the toxic influence of various mycotoxins on general health and specifically to female reproductive processes (release of hormones, ovarian cell proliferation, apoptosis, oogenesis and embryogenesis).

The main problem is that only limited number of reports concerning mycotoxin influence on female reproduction has been published yet. In addition, all the reported studies were done on laboratory animals or cell culture, while in these experiments the nonrealistic too high dosages of mycotoxins, which the organisms cannot be exposed in real life, were tested. We can only assume that based on the animal and *in vitro* models, the human female reproductive system can have a similar reaction to mycotoxins intoxication. Confirmation is needed by *in vivo* studies or epidemiological surveys relating mycotoxin exposure to adverse effects on humans and animals (Santos et al., 2013). With the help of this review, we hope to encourage new research focusing on this topic.

References

Alshannaq A, Yu JH. Occurrence, toxicity, and analysis of major mycotoxins in food. Int J Environ Res Publ Health 2017;14(6):632. https://doi.org/10.3390/ijerph14060632.

Barac A. Mycotoxins and human disease. Clin Relevant Mycos 2019:213—25. https://doi.org/10.1007/978-3-319-92300-0_14.

Binder EM, Tan LM, Chin LJ, Handl J, Richard J. Worldwide occurrence of mycotoxins in commodities, feeds and feed ingredients. Anim Feed Sci Technol 2007;137(3—4):265—82. https://doi.org/10.1016/j.anifeedsci.2007.06.005.

Bondy GS, Pestka JJ. Immunomodulation by fungal toxins. J Toxicol Environ Health B Crit Rev 2000;3(2):109—43. https://doi.org/10.1080/109374000281113.

Bryła M, Waśkiewicz A, Ksieniewicz-Woźniak E, Szymczyk K, Jedrzejczak R. Modified Fusarium mycotoxins in cereals and their products-metabolism, occurrence, and toxicity: an updated review. Molecules 2018;23(4):963. https://doi.org/10.3390/molecules23040963.

Capcarova M, Petruska P, Zbynovska K, Kolesarova A, Sirotkin AV. Changes in antioxidant status of porcine ovarian granulosa cells after quercetin and T-2 toxin treatment. J Environ Sci Health B 2015;50(3):201—6. https://doi.org/10.1080/03601234.2015.982425.

Cortinovis C, Caloni F, Schreiber NB, Spicer LJ. Effects of fumonisin B1 alone and combined with deoxynivalenol or zearalenone on porcine granulosa cell proliferation and steroid production. Theriogenology 2014;81(8):1042—9. https://doi.org/10.1016/j.theriogenology.2014.01.027.

Dai C, Xiao X, Sun F, Zhang Y, Hoyer D, Shen J, Tang S, Velkov T. T-2 toxin neurotoxicity: role of oxidative stress and mitochondrial dysfunction. Arch Toxicol 2019;93(11):3041—56. https://doi.org/10.1007/s00204-019-02577-5.

50 Contaminants and Plants Action on Female Reproduction

Do TH, Tran SC, Le CD, Nguyen HBT, Le PTT, Le HHT, Thai-Nguyen HT. Dietary exposure and health risk characterization of aflatoxin B1, ochratoxin A, fumonisin B1, and zearalenone in food from different provinces in Northern Vietnam. Food Control 2020;112:107108. https://doi.org/10.1016/j.foodcont.2020.107108.

Eskola M, Kos G, Elliott CT, Hajšlová J, Mayar S, Krska R. Worldwide contamination of food-crops with mycotoxins: validity of the widely cited 'FAO estimate' of 25%. Crit Rev Food Sci Nutr 2019;60(16):2773−89. https://doi.org/10.1080/10408398.2019.1658570.

Fletcher MT, Blaney BJ. Mycotoxins. Encyclop Food Grains 2016;2:290−6. https://doi.org/10.1016/b978-0-12-394437-5.00112-1.

Frizzell C, Ndossi D, Verhaegen S, Dahl E, Eriksen G, Sørlie M, Ropstadb E, Mullerf M, Elliott CT, Connolly L. Endocrine disrupting effects of zearalenone, alpha-and beta-zearalenol at the level of nuclear receptor binding and steroidogenesis. Toxicol Lett 2011;206(2):210−7. https://doi.org/10.1016/j.toxlet.2011.07.015.

Gbore FA, Owolawi TJ, Erhunwunsee M, Akele O, Gabriel-Ajobiewe RA. Evaluation of the reproductive toxicity of dietary fumonisin B 1 in rats. Jordan J Biol Sci 2012;5(3):183−90. ISSN: 1995-6673.

Gerez JR, Camacho T, Marutani VHB, de Matos RLN, Hohmann MS, Júnior WAV, Bracarense APF. Ovarian toxicity by fusariotoxins in pigs: does it imply in oxidative stress? Theriogenology 2021;165:84−91. https://doi.org/10.1016/j.theriogenology.2021.02.003.

Halenar M, Medvedova M, Maruniakova N, Kolesarova A. Assessment of a potential preventive ability of amygdalin in mycotoxin-induced ovarian toxicity. J Environ Sci Health B 2015;50(6):411−6. https://doi.org/10.1080/03601234.2015.1011956.

Han J, Wang QC, Zhu CC, Liu J, Zhang Y, Cui XS, Sun SC. Deoxynivalenol exposure induces autophagy/apoptosis and epigenetic modification changes during porcine oocyte maturation. Toxicol Appl Pharmacol 2016;300:70−6. https://doi.org/10.1016/j.taap.2016.03.006.

Haque MA, Wang Y, Shen Z, Li X, Saleemi MK, He C. Mycotoxin contamination and control strategy in human, domestic animal and poultry: a review. Microb Pathog 2020;142:104095. https://doi.org/10.1016/j.micpath.2020.104095.

Hou S, Ma J, Cheng Y, Wang H, Sun J, Yan Y. One-step rapid detection of fumonisin B1, dexyonivalenol and zearalenone in grains. Food Control 2020;117:107107. https://doi.org/10.1016/j.foodcont.2020.107107.

Hou YJ, Xiong B, Zheng WJ, Duan X, Cui XS, Kim NH, Sun SC. Oocyte quality in mice is affected by a mycotoxin-contaminated diet. Environ Mol Mutagen 2014;55(4):354−62. https://doi.org/10.1002/em.21833.

Huang D, Cui L, Dai M, Wang X, Wu Q, Hussain HI, Yuan Z. Mitochondrion: a new molecular target and potential treatment strategies against trichothecenes. Trends Food Sci Technol 2019;88:33−45. https://doi.org/10.1016/j.tifs.2019.03.004.

Khosrokhavar R, Rahimifard N, Shoeibi S, Hamedani MP, Hosseini MJ. Effects of zearalenone and α-Zearalenol in comparison with Raloxifene on T47D cells. Toxicol Mech Methods 2009;19(3):246−50. https://doi.org/10.1080/15376510802455347.

Kolesarova A, Capcarova M, Maruniakova N, Lukac N, Ciereszko RE, Sirotkin AV. Resveratrol inhibits reproductive toxicity induced by deoxynivalenol. J Environ Sci Health A Tox Hazard Subst Environ Eng 2012;47(9):1329−34. https://doi.org/10.1080/10934529.2012.672144.

Kolesarova A, Maruniakova N, Kadasi A, Halenar M, Marak M, Sirotkin AV. The effect of HT-2 toxin on ovarian steroidogenesis and its response to IGF-I, leptin and ghrelin in rabbits. Physiol Res 2017a;66(4):705−8. https://doi.org/10.33549/physiolres.933610.

Kolesarova A, Medvedova M, Halenar M, Sirotkin AV, Bulla J. The influence of deoxynivalenol and zearalenone on steroid hormone production by porcine ovarian granulosa cells in vitro. J Environ Sci Health B 2017b;52(11):823−32. https://doi.org/10.1080/03601234.2017.1356175.

Lu Y, Zhang Y, Liu JQ, Zou P, Jia L, Su YT, Sun SC. Comparison of the toxic effects of different mycotoxins on porcine and mouse oocyte meiosis. PeerJ 2018;6:e5111. https://doi.org/10.7717/peerj.5111.

Maresca M, Fantini J. Some food-associated mycotoxins as potential risk factors in humans predisposed to chronic intestinal inflammatory diseases. Toxicon 2010;56(3):282−94. https://doi.org/10.1016/j.toxicon.2010.04.016.

Marin S, Ramos AJ, Cano-Sancho G, Sanchis V. Mycotoxins: occurrence, toxicology, and exposure assessment. Food Chem Toxicol 2013;60:218−37. https://doi.org/10.1016/j.fct.2013.07.047.

Maruniakova N, Kadasi A, Sirotkin AV, Bulla J, Kolesarova A. T-2 toxin and its metabolite HT-2 toxin combined with insulin-like growth factor-I modify progesterone secretion by porcine ovarian granulosa cells. J Environ Sci Health A Tox Hazard Subst Environ Eng 2014;49(4):404−9. https://doi.org/10.1080/10934529.2014.854650.

Maruniakova N, Kadasi A, Sirotkin AV, Leśniak A, Ferreira AM, Bulla J, Kolesarova A. Assessment of T-2 toxin effect and its metabolite HT-2 toxin combined with insulin-like growth factor I, leptin and ghrelin on progesterone secretion by rabbit ovarian fragments. J Environ Sci Health B 2015;50(2):128−34. https://doi.org/10.1080/03601234.2015.975622.

Matsuda Y, Ichida T, Fukumoto M. Hepatocellular carcinoma and liver transplantation: clinical perspective on molecular targeted strategies. Med Mol Morphol 2011;44(3):117. https://doi.org/10.1007/s00795-011-0547-2.

Medvedova M, Kolesarova A, Capcarova M, Labuda R, Sirotkin AV, Kovacik J, Bulla J. The effect of deoxynivalenol on the secretion activity, proliferation and apoptosis of porcine ovarian granulosa cells in vitro. J Environ Sci Health B 2011;46(3):213−9. https://doi.org/10.1080/03601234.2011.540205.

Milićević DR, Škrinjar M, Baltić T. Real and perceived risks for mycotoxin contamination in foods and feeds: challenges for food safety control. Toxins 2010;2(4):572−92. https://doi.org/10.3390/toxins2040572.

National Center for Biotechnology Information. PubChem Compound Summary for CID 40024, Deoxynivalenol. PubChem, https://pubchem.ncbi.nlm.nih.gov/compound/Deoxynivalenol. Accessed June 1, 2021.

National Center for Biotechnology Information. PubChem Compound Summary for CID 5281576, Zearalenone. PubChem, https://pubchem.ncbi.nlm.nih.gov/compound/Zearalenone. Accessed June 1, 2021.

National Center for Biotechnology Information. PubChem Compound Summary for CID 6711181, T-2 Mycotoxin. PubChem, https://pubchem.ncbi.nlm.nih.gov/compound/T-2-Mycotoxin. Accessed June 1, 2021.

National Center for Biotechnology Information. PubChem Compound Summary for CID 322238, Mycotoxin HT 2. PubChem, https://pubchem.ncbi.nlm.nih.gov/compound/Mycotoxin-HT-2. Accessed June 1, 2021.

Ostry V, Malir F, Toman J, Grosse Y. Mycotoxins as human carcinogens—the IARC Monographs classification. Mycotoxin Res 2017;33(1):65−73. https://doi.org/10.1007/s12550-016-0265-7.

Pinotti L, Ottoboni M, Giromini C, Dell'Orto V, Cheli F. Mycotoxin contamination in the EU feed supply chain: a focus on cereal byproducts. Toxins 2016;8(2):45. https://doi.org/10.3390/toxins8020045.

52 Contaminants and Plants Action on Female Reproduction

Qian Y, Shi WQ, Ding JT, Sha JH, Fan BQ. Predictive value of the area of expanded cumulus mass on development of porcine oocytes matured and fertilized in vitro. J Reprod Dev 2003;49(2):167−74. https://doi.org/10.1262/jrd.49.167.

Rodrigues I, Naehrer K. A three-year survey on the worldwide occurrence of mycotoxins in feedstuffs and feed. Toxins 2012;4(9):663−75. https://doi.org/10.3390/toxins4090663.

Rogowska A, Pomastowski P, Sagandykova G, Buszewski B. Zearalenone and its metabolites: effect on human health, metabolism and neutralisation methods. Toxicon 2019;162:46−56. https://doi.org/10.1016/j.toxicon.2019.03.004.

Rotimi OA, Rotimi SO, Goodrich JM, Adelani IB, Agbonihale E, Talabi G. Time-course effects of acute aflatoxin B1 exposure on hepatic mitochondrial lipids and oxidative stress in rats. Front Pharmacol 2019;10:467. https://doi.org/10.3389/fphar.2019.00467.

Santos RR, Schoevers EJ, Roelen BAJ, Fink-Gremmels J. Mycotoxins and female reproduction: in vitro approaches. World Mycotoxin J 2013;6(3):245−53. https://doi.org/10.3920/WMJ2013.1596.

Schoevers EJ, Fink-Gremmels J, Colenbrander B, Roelen BAJ. Porcine oocytes are most vulnerable to the mycotoxin deoxynivalenol during formation of the meiotic spindle. Theriogenology 2010;74(6):968−78. https://doi.org/10.1016/j.theriogenology.2010.04.026.

Schoevers EJ, Santos RR, Colenbrander B, Fink-Gremmels J, Roelen BA. Transgenerational toxicity of Zearalenone in pigs. Reprod Toxicol 2012;34(1):110−9. https://doi.org/10.1016/j.reprotox.2012.03.004.

Shephard GS. Current status of mycotoxin analysis: a critical review. J AOAC Int 2016;99(4):842−8. https://doi.org/10.5740/jaoacint.16-0111.

Solcan C, Pavel G, Floristean V, Chiriac I, Slencu B, Solcan G. Effect of ochratoxin A on the intestinal mucosa and mucosa-associated lymphoid tissues in broiler chickens. Acta Vet Hung 2015;63(1):30−48. https://doi.org/10.1556/avet.2015.004.

Stroka J, Maragos CM. Challenges in the analysis of multiple mycotoxins. World Mycotoxin J 2016;9(5):847−61. https://doi.org/10.3920/WMJ2016.2038.

Sun Z, Xu J, Wang G, Song A, Li C, Zheng S. Hydrothermal fabrication of rectorite based biocomposite modified by chitosan derived carbon nanoparticles as efficient mycotoxins adsorbents. Appl Clay Sci 2020;184:105373. https://doi.org/10.1016/j.clay.2019.105373.

Taroncher M, Rodríguez-Carrasco Y, Ruiz MJ. T-2 toxin and its metabolites: characterization, cytotoxic mechanisms and adaptive cellular response in human hepatocarcinoma (HepG2) cells. Food Chem Toxicol 2020;145:111654. https://doi.org/10.1016/j.fct.2020.111654.

Thanushree MP, Sailendri D, Yoha KS, Moses JA, Anandharamakrishnan C. Mycotoxin contamination in food: an exposition on spices. Trends Food Sci Technol 2019;93:69−80. https://doi.org/10.1016/j.tifs.2019.08.010.

Ukwuru MU, Ohaegbu CG, Muritala A. An overview of mycotoxin contamination of foods and feeds. J Biochem Microbial 2017;1:101.

Urbanek KA, Habrowska-Górczyńska DE, Kowalska K, Stańczyk A, Domińska K, Piastowska-Ciesielska AW. Deoxynivalenol as potential modulator of human steroidogenesis. J Appl Toxicol 2018;38(12):1450−9. https://doi.org/10.1002/jat.3623.

Voss KA, Gelineau-vanWaes JB, Riley RT. Fumonisins: current research trends in developmental toxicology. Mycotoxin Res 2006;22:61−9.

Wu QH, Wang X, Yang W, Nüssler AK, Xiong LY, Kuča K, Yuan ZH. Oxidative stress-mediated cytotoxicity and metabolism of T-2 toxin and deoxynivalenol in animals and humans: an update. Arch Toxicol 2014;88(7):1309−26. https://doi.org/10.1007/s00204-014-1280-0.

Wu Q, Wang X, Nepovimova E, Miron A, Liu Q, Wang Y, Kuca K. Trichothecenes: immuno-modulatory effects, mechanisms, and anti-cancer potential. Arch Toxicol 2017;91 (12):3737–85. https://doi.org/10.1007/s00204-017-2118-3.

Yu M, Chen L, Peng Z, Nüssler AK, Wu Q, Liu L, Yang W. Mechanism of deoxynivalenol effects on the reproductive system and fetus malformation: current status and future challenges. Toxicol Vitro 2017;41:150–8. https://doi.org/10.1016/j.tiv.2017.02.011.

Zhang L, Li L, Xu J, Pan MH, Sun SC. HT-2 toxin exposure induces mitochondria dysfunction and DNA damage during mouse early embryo development. Reprod Toxicol 2019;85:104–9. https://doi.org/10.1016/j.reprotox.2019.02.011.

Chapter 2.3

The oil-related environmental contaminants

2.3.1 Introduction

The low-weight aromatic hydrocarbons benzene, toluene, ethylbenzene, m/p-xylene, and o-xylene (BTEX) are among the most common and hazardous sources of environmental contamination.

In comparison with other toxic environmental contaminants, BTEX are especially dangerous because of their (1) multiple sources of contamination in the environment (e.g., oil production, oil refrigeration, oil transportation, the production of petroleum, solvents, coal-derived products, traffic, tobacco smoking, voluntary inhalation as drugs of abuse, etc.); (2) relatively high solubility in water and air and therefore easy migration and distribution in the environment and easy transport into the cells; (3) high solubility in lipids and thus ready absorption from environment in gastrointestinal tract and accumulation in the adipose tissue; (4) ability to be trapped for long periods of time in the soil minerals of contaminated sites, and subsequently bioaccumulated, due to their hydrophobic and stable chemical properties; (5) low physico-chemical and biological degradation in ecosystems; and (6) multiple toxic influences on living organisms (WHO, 2005; Varjani et al., 2017).

This chapter outlines provenance, properties, general physiological effects, and mechanisms of action of BTEX. It discusses in details BTEX action on female reproductive cycle and fecundity, state of the ovary, oviduct and uterus, basic ovarian cell functions, oogenesis and embryogenesis, endocrine regulators of reproduction, and intracellular mechanisms of action. Finally, the possible approaches to mitigate and eliminate adverse effects of BTEX, as well as the limitations of the current knowledge concerning BTEX action on female reproduction have been outlined.

2.3.2 Provenance and properties

Aromatic hydrocarbons including benzene, toluene, ethyl benzene, xylene (BTEX, Fig. 2.3.1) occur naturally in crude oil and can be found in sea water in the vicinity of natural gas and petroleum deposits. Other natural sources of BTEX compounds include gas emissions from volcanoes and forest fires. The primary human-caused releases of BTEX compounds are through emissions from motor vehicles and aircrafts, and cigarette smoke. Furthermore, BTEX compounds are created and used during the processing of petroleum products

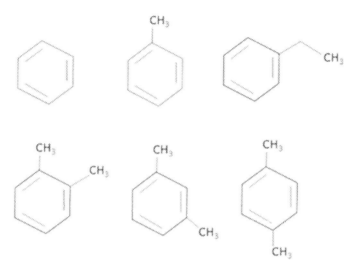

FIGURE 2.3.1 Chemical structure of BTEX. Upper line: benzene, toluene and ethylbenzene. Lower line: orto-, meta-, and paraxylene. *From Montero-Montoya et al. (2018).*

and during the production of consumer goods such as paints and lacquers, thinners, rubber products, adhesives, inks, cosmetics, and pharmaceutical products. BTEX compounds are among the most abundantly produced chemicals in the world. Oil production and processing and the subsequent production and consumption of petroleum products represent the basis of a modern economy. Unfortunately, leaks and spills of oil products (e.g., crude oil, petroleum, petrochemical products, aromatic hydrocarbons, etc.) are among the most common hazardous sources of environmental contamination. The most common sources of exposure to BTEX compounds are from breathing contaminated air, particularly in areas of heavy motor vehicle traffic and petrol stations, and through cigarette smoke. Exposure to BTEX from water contributes only a small percentage of the total daily intake, compared with inhaled air and dietary sources. Emissions from the combustion of gasoline and diesel fuels are the largest contributors to atmospheric BTEX concentrations. However, levels indoors (where people spend greater than 83% of their time) can be many times greater than outdoors. The exposition of people to BTEX in the air is several times higher in the urban and industrial areas as in the countryside, but newly polluted areas are gaining importance in countries where accelerated industrialization is taking place in suburban or rural settings (https://environment.des.qld.gov.au/management/activities/non-mining/fraccing/btex-chemicals; Bolden et al., 2015; Montero-Montoya et al., 2018).

2.3.3 Physiological actions

BTEX can express toxic, immunotoxic (Maronpot, 1987; WHO, 2005; Faber et al., 2006; Roberts et al., 2007; Hannigan and Bowen, 2010; Lin et al., 2011),

56 Contaminants and Plants Action on Female Reproduction

mutagenic (Apie t al., 1995; Kennel et al., 2000; Rana and Verma, 2005), carcinogenic (Maronpot, 1987; Rana and Verma, 2005; Fernández-Navarro et al., 2017), embryotoxic and teratogenic (Hood and Ottley, 1985; Waldner, 2008; Bioven et al., 2009; Aguilera et al., 2010; Hannigan et al., 2010; Webb et al., 2014; Bolden et al., 2015), growth retarding (Ungváry et al., 1981; Hood and Ottley, 1985; Faber et al., 2006; Roberts et al., 2007; Jarosz et al., 2008; Aguilera et al., 2010), metabolic (Faber et al., 2006; Roberts et al., 2007; Jarosz et al., 2008), and neuromodulatory (Krishek et al., 1994; Cruz et al., 2000; Burmistrov et al., 2001; Bale et al., 2002; Rana and Verma, 2005; Cruz, 2011) effects. Health effects of BTEX are significantly associated with ambient level exposure included sperm abnormalities, cardiovascular disease, respiratory dysfunction, asthma, sensitization to common antigens, and more (Bolden et al., 2015; Varjani et al., 2017).

2.3.4 Mechanisms of action

Several principal extra- and intracellular mechanisms of toxic effect of BTEX on the cells have been proposed, although the evidence for each of this mechanism is limited, and the interrelationships between these mechanisms are poorly investigated yet. The following mechanisms of BTEX action are suggested:

(1) Inhibition of cell growth, cell viability and cell protein content (Shen, 1998).
(2) Influence level of polyamines spermidine, spermine, and putrescine, which are known to be important in cell proliferation (Shen, 1998).
(3) Suppresses the cell cycle by p53-mediated overexpression of p21, an inhibitor of cyclin-dependent kinase, and a promoter of mitosis (Yoon et al., 2001).
(4) Genomic and epigenomic effect on chromatin structure and promotion the DNA damage (Morgan et al., 2005).
(5) Promotion of generation of reactive oxygen species, which in turn stimulate apoptosis (Ávila et al., 2016).
(6) BTEX are endocrine disrupters—phenoles (see Fig. 2.3.1), which possess the similar chemical and biological properties as steroid hormones. BTEX can bind to steroid hormones (estrogen and androgen) receptors and either up- or down regulate they and the steroid hormones-dependent processes (Bolden et al., 2015, 2018).
(7) BTEX can affect production of various hormones including estrogens, androgens, glucocorticoids, insulin, and serotonin, which are involved in control of various physiological processes and heath (Bolden et al., 2018).

It is noteworthy, that the toxic effects of BTEX can be due to not (or not only) BTEX theirselves, but by their metabolites 3-methylcatechol, 4-methylcatechol, 4-hydroxybenzoic acid, and 4-hydroxy-3-methoxybenzoic acid (Shen, 1998).

2.3.5 Effects on female reproductive processes

The most dangerous effects of BTEX could be their influence on the most important biological process—reproduction, especially female reproduction, and its regulators (Sharma et al., 2016). Female organisms accumulate 3.7—6.8 times more xylene than males, and ovaries are the primary accumulation site of these hydrocarbons (Suter-Eichenbeger et al., 1998). The developing fetus is particularly sensitive to BTEX action (Webb et al., 2014).

2.3.5.1 Effects on the CNS

The inhibitory effects of **benzene** (Rana and Verma, 2005), **toluene** (Krishek et al., 1994; Burmistrov et al., 2001; Bale et al., 2002; Cruz, 2011; Furlong et al., 2016), and **ethylbenzene** (Cruz et al., 2000; Cruz, 2011) on neuromediator receptors in the CNS are well known. In addition, toluene was able to exert its neurotoxic effect via an increase in brain free radicals (Burmistrov et al., 2001). Neuromediators play a key role in the control of sexual behavior and of neuromediator-dependent neurohormones, including gonadotropin-releasing hormones, oxytocin, etc., which in turn are the main regulators of the downstream pituitary-ovarian axis (Sirotkin, 2014). Therefore, the influence of BTEX on reproduction via CNS is possible, although there is still no direct and strong evidence for it.

2.3.5.2 Effects on the ovarian and reproductive state

Female workers' contact with **benzene** is associated with ovarian hypo- and hyperplasia, ovarian and uterine retardation (Maronpot, 1987), reduction in the duration of the luteal phase of the menstrual cycle (Chen et al., 2000, 2001), and the retardation of fetal growth during pregnancy (Aguilera et al., 2010; Webb et al., 2014) and increased incidence of ovarian cancer (Fernández-Navarro et al., 2017). Increase benzene concentration in human ovarian follicular fluid was associated with reduced oocyte and embryo production, although pregnancy rate was not linked with ovarian benzene level (Alviggi et al., 2014). Exposure of mice to benzene increased the incidence of ovarian granulosa cell tumors and ovarian benign mixed tumors (National Toxycology Program, 1986). Exposure of cows to benzene increases the incidence of odd calves (Waldner, 2008). **Toluene** exposure in rats suppresses the development of growing but not primordial ovarian follicles (Tap et al., 1996). The inhalation of toluene increases the incidence of maternal and fetal morbidity and embryonic malformations in women (Kuczkowski, 2007; Hannigan and Bowen, 2010), cows (Waldner, 2008), and rats (Bowen et al., 2006; Jarosz et al., 2008), although other studies have not detected any adverse effects on implantation or the number and viability of rat fetuses. Inhalation of **ethylbenzene** does not affect rat ovarian structure, function, and fertility (Faber et al., 2006). The accessible databases do not contain publications concerning

58 Contaminants and Plants Action on Female Reproduction

the reproductive effects of **xylenes,** although decreased embryonic growth (Ungváry et al., 1981) and increased prenatal mortality (Hood and Ottley, 1985) in rats exposed to xylene have been reported.

The available information demonstrated the adverse effect of all BTEX on animal and human female reproduction and fecundity.

2.3.5.3 Effect on ovarian cell functions

The *in vitro* experiments revealed the direct action of BTEX on basic ovarian follicular cell functions—viability, proliferation, apoptosis, secretory activity, and the response to the upstream regulators.

Benzene directly promoted both proliferation and apoptosis and reduced viability of cultured bovine (Sirotkin et al., 2020e) and porcine ovarian granulosa cells (Sirotkin et al., 2020a,c; Tarko et al., 2020). In other experiments, benzene promoted proliferation but not apoptosis and reduced viability of porcine granulosa cells (Sirotkin et al., 2020h). Furthermore, in porcine cells benzene addition stimulated oxytocin release and inhibited progesterone and prostaglandin F release (Tarko et al., 2020). In other experiments, benzene either promoted the release of progesterone, oxytocin and prostaglandin F (Sirotkin et al., 2020a,b), increased progesterone and estradiol and increased testosterone output (Sirotkin et al., 2020h), or reduced progesterone and estradiol release by these cells (Sirotkin et al., 2020c). In addition, benzene prevented the stimulatory effect of FSH on porcine granulosa cell proliferation (Sirotkin et al., 2020a). In cultured mice ovaries, benzene suppressed progesterone, and IGF-I release (Sirotkin et al., 2017). It did not influence production of progesterone, IGF-I, oxytocin, and prostaglandin F by cultured rabbit ovarian fragments (Földešiová et al., 2017).

Xylene, like benzene, was able to increase both proliferation and apoptosis and to reduce viability of cultured bovine (Sirotkin et al., 2020e) and porcine (Tarko et al., 2018; Sirotkin et al., 2020a) ovarian granulosa cells. Other experiments however reveal the suppressive action of xylene on not only on viability, but also on proliferation, apoptosis, progesterone and estradiol release by porcine cells (Sirotkin et al., 2020d). In bovine granulosa cells, xylene was able to suppress progesterone release (Sirotkin et al., 2020e), but in porcine granulosa cells it either inhibited (Tarko et al., 2018) or promoted progesterone secretion, and stimulated IGF-I and oxytocin release (Tarko et al., 2018; Sirotkin et al., 2020a). In cultured feline ovarian fragments, xylene inhibited oxytocin and stimulated prostaglandin F release (Sirotkin et al., 2020g). Furthermore, xylene prevented the stimulatory action of FSH on prostaglandin F output by cultured porcine granulosa cells (Sirotkin et al., 2020a).

Xylene promoted progesterone and testosterone P4 and T (but not IGF-I) in cultured mice ovaries (Sirotkin et al., 2017).

Toluene, promoted bovine (Sirotkin et al., 2020a) and porcine (Sirotkin et al., 2020b) ovarian granulosa cell apoptosis, but it did not influence their proliferation and viablity. Other studies (Sirotkin et al., 2020f) demonstrated the inhibitory influence of toluene on cell viability and proliferation, but no influence on apoptosis in porcine granulosa cells. The inhibitory action of toluene on viability, proliferation and apoptosis was observed in cultured mare granulosa cells (Tarko et al., 2020). Some studies demonstrated also the stimulatory influence of toluene on progesterone and inhibitory effect of toluene on estradiol, but not to testosterone secretion by porcine granulosa cells (Sirotkin et al., 2020f) or the suppressive action of toluene on progesterone and oxytocin and its stimulatory action on prostaglandin F release by these cells. In cultured feline ovarian fragments, toluene promoted both oxytocin and prostaglandin F output (Sirotkin et al., 2020g). Finally, toluene prevented the stimulatory action of FSH on progesterone and prostaglandin F output (Sirotkin et al., 2020b).

Taken together, the results of the preformed studies demonstrate direct influence of BTEX on basic ovarian cell functions. Despite some contradictions probably due to differences between model species and initial state of ovarian cells, they show, that BTEX can reduce cell viability of ovarian cells in various species probably via increase in their apoptosis and decrease in proliferation/apoptosis rate. Furthermore, they can affect a wide array of ovarian hormones—progestagen, androgen, estrogen, oxytocin, IGF-I, and prostaglandin F. Finally, they can prevent the response of ovarian cells to physiological stimulator—gonadotropin FSH.

2.3.5.4 Effects on oocytes and embryos

Xenopus oocytes are used as a common model for studying the effects of BTEX on membrane receptors. These oocytes demonstrate inhibitory effects of **benzene, m-xylene,** and **ethylbenzene** on membrane N-methyl-D-aspartate receptors (Cruz et al., 2000); of **benzene** on gamma-aminobutyric acid receptors (Krishek et al., 1994); of **toluene** on nicotinic acetylcholine receptors (Bale et al., 2002; Furlong et al., 2016) and ethanol-sensitive potassium channels (Del Re et al., 2006); and of **xylenes** on nicotinic acetylcholine receptors (van Kleef et al., 2008). Therefore, BTEX might also suppress reproduction via the inhibition of oocyte signaling systems. Nevertheless, there is practically no evidence concerning direct action of BTEX on mammalian oocytes.

60 Contaminants and Plants Action on Female Reproduction

BTEX can disrupt reproduction due to their embryotoxic and teratogenic actions. The suppressive effects of **benzene, m-xylene, ethylbenzene, and xylenes** on bovine (Waldner, 2008) and human (Hood and Ottley, 1985; Aguilera et al., 2010; Hannigan et al., 2010; Slama et al., 2009) embryogenesis and of **toluene** (Bowen et al., 2009) and **xylene** (Hood and Ottley, 1985) on rat embryonic growth have been reported. In addition, toluene is able to inhibit steroidogenic enzymes and testosterone synthesis in fetal rats (Tsukahara et al., 2009), suggesting the potential inhibitory influence of this aromatic hydrocarbon on gonadal steroidogenesis and related sexual maturation. There is some evidence for critical windows of vulnerability during prenatal and early postnatal development, during which BTEX exposures can cause potentially permanent damage to the growing embryo and fetus (Webb et al., 2014). Therefore, the adverse effect of BTEX on embryogenesis and related fecundity are evident.

2.3.5.5 Effects on oviducts

Fertility and embryo viability could be influenced by BTEX via an influence of oviduct and uterine functions. Reveles et al. (2005) reported that **benzene** exposure inhibits ciliary beat frequency, oocyte pickup rates, and infundibular smooth muscle contraction rates. The inhalation of para-**xylene** does not influence rat uterine and ovarian venous outflow (Ungváry et al., 1981). Therefore, BTEX influence on oviduct and uterus is possible, but this aspect of BTEX action on female reproduction remains poorly investigated yet.

2.3.5.6 Effects on cytogenetics of somatic and generative cells

Cytogenetic analysis demonstrated the mutagenic effects of **toluene** and its analogues on hamster ovarian cells (Apie t al., 1995; Kennel et al., 2000). **Benzene** increased the incidence of aneuploidy in mature mouse oocytes (Zeng et al., 2001). The destructive effects of BTEX on chromosomes in somatic ovarian, brain, and other cells and in oocytes could be due to the ability of BTEX to increase the level of mutagenic and proapoptotic free radicals in these cells. The inhalation of toluene increased the activities of glutathione peroxidase and catalase and the intensity of lipid peroxidation in rat ovaries and brain cortices (Burmistrov et al., 2001). Therefore, the adverse action of BTEX on chromosomes is possible, but it remains poorly documented yet.

2.3.5.7 Effects on reproductive hormones

Women with occupational exposures to various BTEX (**benzene, ethylbenzene, toluene, and xylenes**) have reduced preovulatory blood gonadotropin (follicle-stimulating hormone, FSH, luteinizing hormone, LH) and prostaglandin level (Chen et al., 2001; Reutman et al., 2002). High **benzene** level in woman ovarian follicular fluid was associated with increased FSH and decreased estradiol level in blood plasma (Alviggi et al., 2014). The inhalation of **toluene** by rats reduces the hypothalamic level of GnRH and plasma levels of gonadotropins, changes the FSH:LH rate, and increases the plasma level of antigonadotropin prolactin. These changes are associated with reduced blood progesterone and estradiol levels but not with their response to gonadotropin administration (Stepanov et al., 1990). Inhalation of **paraxylene** decreases plasma progesterone and estradiol levels but not the release of these hormones by ovaries in rats (Ungváry et al., 1981). These observations suggest that BTEX can reduce the output of ovarian steroid hormones via the inhibition of upstream hypothalamic gonadotropin-releasing hormone (GnRH)/pituitary gonadotropin production or the response of the ovary to gonadotropin. Taken together, the available literary data suggest the influence of BTEX on hormones-products of hypothalamo-hypophysial-ovarian axis.

2.3.6 Mechanisms of action on female reproductive processes

The most evident mechanisms of adverse effects of BTEX on female reproduction and fecundity can be ability of BTEX to promote ovarian cell **apoptosis**. Such effect was demonstrated for benzene (Sirotkin et al., 2020a,c,e; Tarko et al., 2020), xylene (Sirotkin et al., 2020a,e; Tarko et al., 2018) and toluene (Sirotkin et al., 2020a,b). Some studies however show lack (toluene: Sirotkin et al., 2020f) or even inhibitory action (xylene: Sirotkin et al., 2020d) of BTEX on ovarian cell apoptosis.

The ability of BTEX to induce cell apoptosis can explain their adverse influence on **viability** of ovarian cells (benzene: Sirotkin et al., 2020a,c,e; Tarko et al., 2020; xylene: Sirotkin et al., 2020e; Tarko et al., 2018; Sirotkin et al., 2020a; toluene: Sirotkin et al., 2020f) and embryos (benzene, m-xylene, ethylbenzene, and xylenes: Waldner, 2008; Hood and Ottley, 1985; Aguilera et al., 2010; Hannigan et al., 2010; Slama et al., 2009; toluene: Bowen et al., 2009; xylene: Hood and Ottley, 1985).

There are reports concerning stimulatory action of BTEX on ovarian cell **proliferation** (benzene: Sirotkin et al., 2020a,c,e; Tarko et al., 2020, xylene: Sirotkin et al., 2020e; Tarko et al., 2018; Sirotkin et al., 2020a). Other studies revealed no (toluene: Sirotkin et al., 2020a,b) and even inhibitory (xylene: Lin et al., 2011; Sirotkin et al., 2020d; toluene: Sirotkin et al., 2020f) action of BTEX on ovarian cell proliferation. The reduction in cell proliferation can explain ability of BTEX to down-regulate cell viability.

The influence of BTEX on reproductive processes can be mediated by **protein kinases**. For example, the possible mechanisms of inhibitory action of BTEX on ovarian cells are shown by Lin et al. (2011). In their experiments, the xylene analogue ($L^1 = \alpha,\alpha'$-diamino-p-xylene, $L^2 = 4,4'$-methylenedianiline) arrested the cycle of cultured human cancer cells at G2 or M phase, most likely via influences on the accumulation and phosphorylation of proliferation-related protein kinases Checkpoint kinase 1/2 (CHK1/2), extracellular signal-regulated kinases 1/2 (ERK1/2), and p38 subunit of a mitogen-activated protein kinase (p38 MAPK). It was suggested that CHK1/2, ERK1/2, and p38 MAPK can be intracellular mediators of some BTEX effects on ovarian cell proliferation. Furthermore, because these MAP kinases can be important regulators of oocyte maturation and ovarian hormone release (Sirotkin, 2014), these kinases might be mediators of BTEX in various ovarian processes.

The apoptosis: proliferation rate can be regulated by **transcription factor p53**. In nonovarian cells, the transcription factor p53 promotes p21, which via inhibition of PCNA and cyclin-dependent protein kinase blocks cell cycle and induces DNA repair (Karamian et al., 2016). Furthermore, p53 is a known promoter of cell apoptosis and senescence (Chen, 2016). In bovine ovarian cells benzene, xylene and toluene affected p53 accumulation, which was associated with the corresponding changes in accumulation of cell proliferation (PCNA) and apoptosis (bax) marker (Sirotkin et al., 2020e). These observations suggest, that transcription factor p53 can be mediator of BTEX action on both proliferation and apoptosis of ovarian cells.

Another putative mechanism of BTEX action on reproduction can be **oxidative stress**. At least, toluene was able to increase accumulation of reactive oxygen species and oxidative enzymes in rat brain and ovaries (Burmistrov et al., 2001). It is not to be excluded, that adverse influence of BTEX on DNA and chromosome structures mentioned above could be due to reactive oxygen species with mutagenic properties.

Changes in **hormones** release and action is the most probably extracellular mechanism of BTEX action on reproductive processes. The ability of BTEX to affect release of pituitary and ovarian peptide, steroid hormones and prostaglandins, the key regulators of reproductive processes (Sirotkin, 2014), as well as to mitigate the response of ovarian cells to gonadotropin action in various species has been described above. These observations indicate, that BTEX can affect reproductive processes via changes in both release of pituitary and ovarian hormones and in response of ovarian cells to these hormones (probably due to BTEX-induced changes in hormone receptors). On the other hand, there is still no sufficient direct evidence for mediatory role of hormones in BTEX action on reproduction.

2.3.7 Application in reproductive biology and medicine

Charakter and intensity of BTEX influence on animal and human health depend on numerous interrelated factors including

1. BTEX production, storage, and consumption;
2. BTEX monitoring in external and internal environment;
3. Limitation of BTEX release, distribution, and contents in the environment;
4. Prevention of contacts of living organisms with BTEX; and
5. Characterization, prediction, mitigation, and prevention of adverse BTEX effects on organisms.

Description and discussion of the points one through four are worth of several special books; therefore, they are out of scope of the present review. Moreover, reduction in BTEX content in the environment below the level considered as safe for health is sometimes not sufficient to make they really safe: these endocrine disrupters can affect reproductive processes at the concentrations far below this level (Bolden et al., 2015, 2018), and their adverse influence on puberty, genome and could be manifested only in future, even in the next generations.

The results of the occupational studies described in subchapter 2.3.5.2. suggest, that BTEX level in external and internal environment can be an index for detection and prediction of development of reproductive and developmental disorders. Furthermore, the ability of some BTEX to promote ovarian cell proliferation and to reduce their apoptosis can indicate their potential ability to induce ovarian cancer, which is characterized just by such changes in cell behavior. The cancerogenic action of BTEX has been previously described for nonreproductive systems (Maronpot, 1987; Rana and Verma, 2005). The increased incidence of ovarian cancer in regions contaminated by benzene (Fernández-Navarro et al., 2017) suggests, that BTEX can be cancerogenic also for the ovary too.

More important from the practical viewpoint is a search for approached which can mitigate and prevent adverse effects of BTEX. There are indications that BTEX action on ovarian cells can depend on metabolic state. At least in mices obesity was associated with increased responsibility of their ovaries to benzene and decreased responsibility to xylene (Sirotkin et al., 2017). Ovarian granulosa cells isolated from cows with tendency to emaciation have also increased responsibility to benzene, xylene, and toluene (Sirotkin et al., 2020e). Therefore, obesity could increase the adverse influence of BTEX to ovarian cells. The influence of metabolic state on reproduction is mediated by metabolic hormones including leptin and ghrelin. These hormones were able to mimic and promote xylene and toluene action on release of hormones by feline ovarian fragments (Sirotkin et al., 2020g). An other metabolic factor, gravidity, did not affect the response of feline ovarian fragments to xylene (Sirotkin et al., 2020g).

Search for drugs which can mitigate and prevent adverse effects of BTEX on reproduction is important from both theoretical and practical viewpoints. The information concerning mechanisms of BTEX action on the ovary reviewed in subchapter 2.3.6. could ne helpful for such search. If BTEX effects are mediated by their influence on regulators of apoptosis, proliferation, viability, protein kinases, transcription factor, oxidative processes, and hormones, BTEX action could be modified by targeting these mediators. It is known that some effects of stress on ovarian function could be prevented or neutralized with stem cell therapy (Bhartiya et al., 2016), plants containing antioxidants and other adaptogenes (Huang and Chen, 2008; Liang and Yin, 2010), hormones, growth factors, pharmacological and genomic regulators of

64 Contaminants and Plants Action on Female Reproduction

intracellular signaling molecules (Sirotkin, 2012, 2014). Modification some of these targets can really modify and even completely prevent BTEX action on ovarian cell functions. For example, addition of extract of fennel (*Foeniculum vulgare* Mill.) prevented the toluene effect on viability, proliferation, apoptosis and hormone release by cultured mare ovarian granulosa cells (Tarko et al., 2020). Addition of extracts of buckwheat (*Fagopyrum esculentum*), rooibos (*Aspalathus linearis*), but not of vitex (*Vitex Agnus-Castus*) prevented the adverse effect of benzene on viability of porcine ovarian cells, but all these extracts promoted benzene effects on proliferation, apoptosis and steroidogenesis in these cells (Sirotkin et al., 2020c). These extracts however were able to prevent all the effects of xylene on these cells (Sirotkin et al., 2020d).

The plant constituent resveratrol however did not prevent, but rather promoted influence of benzene (Tarko et al., 2019; Sirotkin et al., 2020b) and xylene (Tarko et al., 2018) on cultured porcine ovarian cells. Other plant molecules, apigenin, daidzein, or rutin were able to prevent some and to promote other changes in these cells induced by benzene (Sirotkin et al., 2020h) and toluene (Sirotkin et al., 2020f). Feeding of rabbit does with *Yucca schidigera* did not affect or even induced the influence of benzene on their ovarian cells (Földešiová et al., 2017). Taken together, these studies demonstrated, that some food and medical plants or their constituents can modify ovarian response to BTEX, and some of these plants be natural protectors against adverse effects of BTEX on ovarian cells.

Other mediators of BTEX action can be potential targets for protection of reproductive processes from negative action of these hydrocarbons, but the drugs with such activity are still waiting for their discovery.

2.3.8 Conclusions and possible direction of future studies

The analysis of the available literature demonstrates that BTEX can exert negative effects on various female reproductive sites, including the CNS-pituitary-ovarian axis, their signaling molecules and receptors, ovarian follicles and somatic follicular cells, oocytes, embryos, oviducts, ovarian cycle, fertility, and the viability of offspring. These effects could be due to the ability of BTEX to destroy chromosomes, to affect cell metabolism, including the accumulation of free radicals, and to affect release of hormonal regulators of reproductive processes and intracellular protein kinases, transcription factors and other regulators/markers of proliferation and apoptosis.

Although understanding the effects of BTEX on reproduction is very important from both theoretical and practical (e.g., medical, ecological, etc.) viewpoints, the current available knowledge is poorly understood and superficial. Not all BTEX hydrocarbons are studied in relation to the main reproductive processes. The contamination of nature with BTEX is growing, but there is no any information concerning BTEX action on reproduction in wild animals.

Unfortunately, due to ethical issues, there is no direct and reliable experimental approach to evaluate BTEX action on human reproduction. The available data concerning association between occupational exposure of humans to BTEX with reproductive problems are insufficiently and inconclusive due to lack of data obtained in controlled experiments with exclusion other factors which can affect reproductive healt (financial, social, health status, nutrition, smoking, etc.) and low number of the studied probands. All the available data concerning *in vivo* effects of BTEX on reproduction were obtained only on rats, BTEX action on ovarian follicular cells were obtained on a number of animal species (mice, cow, pig, mare, and cat) but not in humans, while BTEX action on oocytes was demonstrated only on frog oocytes. These results cannot however be automatically extrapolated to human health. This warning is based on differences between results concerning character of BTEX effects *in vivo* and *in vitro*, as well as on different results obtained during *in vitro* experiments on ovarian cells of different species. For example, BTEX suppressed the secretory activity of hypothhalamo-hypophysial-ovarian axis *in vivo*, but often stimulated it *in vitro* (see subchapters 2.3.5.1 and 2.3.5.7). Therefore, character, rate and possible mechanisms of BTEX action can be detected and predicted on the basis of *in vivo* and *in vitro* experiments, but the obtained presented data should be interpreted very carefully. Moreover, they require validation by the adequate human (at least *in vitro*) experiments, which looks have not been performed yet.

The available data suggest that biological action of BTEX, its mechanisms and possible consequences could be evaluated and predicted by using *in vivo* and *in vitro* studies. On the other hand, it remains to be established, what for parameter would be adequate for assessment the BTEX action. For example, no association between toluene action on ovarian cell proliferation, apoptosis (defining ovarian follicular growth, selection, and atresia; Sirotkin, 2014), release of ovarian hormones and viability was found (Sirotkin et al., 2020a,b,f; Tarko et al., 2020). This indicates, the independent action of this BTEX as nonspecific toxin, regulator of endocrine system and regulator of ovarian follicular growth and atresia, and that toluene could principally be toxic, but not affect reproduction or vice versa. The validation of this hypothesis, as well as the selection of the adequate marker of BTEX action require further studies.

The current efforts to protect animals and humans from adverse action of BTEX are currently limited mainly by limitation of presence of BTEX in environment and contacts of living organisms with them. It is less known how to protect reproduction if such contact occurs. The search for physiological factor, which could mitigate and prevent adverse effects of BTEX on female reproduction are limited with a few studies concerning association of BTEX effects with animal metabolic state and a series of studies aimed to find plant or plant molecules, which could be natural protectors against BTEX action.

66 Contaminants and Plants Action on Female Reproduction

Although some promising plant protectors have been identified, their action should be verified by *in vivo* studies on animals and humans. In addition, the plant molecules responsible for these plant effects are not found yet.

A better understanding of the physiological, genetic, endocrine, and intracellular mechanisms of BTEX action could advance the characterization and prediction of the biological effects of BTEX. Furthermore, understanding mediators of BTEX action could help to find the adequate, efficient and cheap physiological protectors against adverse influence of BTEX on living organisms including female reproduction.

References

Aguilera I, Garcia-Esteban R, Iñiguez C, Nieuwenhuijsen MJ, Rodríguez A, Paez M, Ballester F, Sunyer J. Prenatal exposure to traffic-related air pollution and ultrasound measures of fetal growth in the INMA Sabadell cohort. Environ Health Perspect 2010;118:705—11.

Alviggi C, Guadagni R, Conforti A, Coppola G, Picarelli S, De Rosa P, Vallone R, Strina I, Pagano T, Mollo A, Acampora A, De Placido G. Association between intrafollicular concentration of benzene and outcome of controlled ovaria stimulation in IVF/ICSI cycles: a pilot study. J Ovarian Res 2014;7:67.

Api AM, Ford RA, San RH. An evaluation of musk xylene in a battery of genotoxicity tests. Food Chem Toxicol 1995;33:1039—45.

Ávila J, González-Fernández R, Rotoli D, Hernández J, Palumbo A. Oxidative stress in granulosa-lutein cells from in vitro fertilization patients. Reprod Sci 2016;23:1656—61.

Bale AS, Smothers CT, Woodward JJ. Inhibition of neuronal nicotinic acetylcholine receptors by the abused solvent, toluene. Br J Pharmacol 2002;137:375—83.

Bhartiya D, Shaikh A, Anand S, Patel H, Kapoor S, Sriraman K, Parte S, Unni S. Endogenous, very small embryonic-like stem cells: critical review, therapeutic potential and a look ahead. Hum Reprod Update 2016;23:41—76.

Bolden AL, Kwiatkowski CF, Colborn T. Correction to new look at BTEX: are ambient levels a problem? Environ Sci Technol 2015;49(19):11984—9. https://doi.org/10.1021/acs.est.5b03462.

Bolden AL, Schultz K, Pelch KE, Kwiatkowski CF. Exploring the endocrine activity of air pollutants associated with unconventional oil and gas extraction. Environ Health 2018;17(1):26. https://doi.org/10.1186/s12940-018-0368-z.

Bowen SE, Irtenkauf S, Hannigan JH, Stefanski AL. Alterations in rat fetal morphology following abuse patterns of toluene exposure. Reprod Toxicol 2009;27:161—9.

Burmistrov SO, Arutyunyan AV, Stepanov MG, Oparina TI, Prokopenko VM. Effect of chronic inhalation of toluene and dioxane on activity of free radical processes in rat ovaries and brain. Bull Exp Biol Med 2001;132:832—6.

Chen H, Song L, Wang X, Wang S. Effect of exposure to low concentration of benzene and its analogues on luteal function of female workers. Wei Sheng Yan Jiu 2000;29:351—3 [in Chinese].

Chen H, Wang X, Xu L. Effects of exposure to low-level benzene and its analogues on reproductive hormone secretion in female workers. Zhonghua Yu Fang Yi Xue Za Zhi 2001;35:83—6 [in Chinese].

Chen J. The cell-cycle arrest and apoptotic functions of p53 in tumor initiation and progression. Cold Spring Harb Perspect Med 2016;6(3):a026104. https://doi.org/10.1101/cshperspect. a026104.

Cruz SL, Balster RL, Woodward JJ. Effects of volatile solvents on recombinant N-methyl-D-aspartate receptors expressed in Xenopus oocytes. Br J Pharmacol 2000;131:1303—8.

Cruz SL. The latest evidence in the neuroscience of solvent misuse: an article written for service providers. Subst Use Misuse 2011;46(Suppl 1):62—7.

Del Re AM, Dopico AM, Woodward JJ. Effects of the abused inhalant toluene on ethanol-sensitive potassium channels expressed in oocytes. Brain Res 2006;1087:75—82.

Faber WD, Roberts LS, Stump DG, Tardif R, Krishnan K, Tort M, Dimond S, Dutton D, Moran E, Lawrence W. Two generation reproduction study of ethylbenzene by inhalation in Crl-CD rats. Birth Defects Res B Dev Reprod Toxicol 2006;77:10—21.

Fernández-Navarro P, García-Pérez J, Ramis R, Boldo E, López-Abente G. Industrial pollution and cancer in Spain: an important public health issue. Environ Res 2017;159:555—63. https://doi.org/10.1016/j.envres.2017.08.049.

Földešiová M, Baláži A, Chrastinová Ľ, Pivko J, Kotwica J, Harrath AH, Chrenek P, Sirotkin AV. Yucca schidigera can promote rabbit growth, fecundity, affect the release of hormones in vivo and in vitro, induce pathological changes in liver, and reduce ovarian resistance to benzene. Anim Reprod Sci 2017;83.66- 76. https://doi.org/10.1016/j.anireprosci.2017.06.001.

Furlong TM, Duncan JR, Corbit LH, Rae CD, Rowlands BD, Maher AD, Nasrallah FA, Milligan CJ, Petrou S, Lawrence AJ, Balleine BW. Toluene inhalation in adolescent rats reduces flexible behaviour in adulthood and alters glutamatergic and GABAergic signalling. J Neurochem 2016;139:806—22.

Hannigan JH, Bowen SE. Reproductive toxicology and teratology of abused toluene. Syst Biol Reprod Med 2010;56:184—200.

Hood RD, Ottley MS. Developmental effects associated with exposure to xylene: a review. Drug Chem Toxicol 1985;8:281—97.

Huang ST, Chen AP. Traditional Chinese medicine and infertility. Curr Opin Obstet Gynecol 2008;20:211—5.

Jarosz PA, Fata E, Bowen SE, Jen KL, Coscina DV. Effects of abuse pattern of gestational toluene exposure on metabolism, feeding and body composition. Physiol Behav 2008;93:984—93.

Karimian A, Ahmadi Y, Yousefi B. Multiple functions of p21 in cell cycle, apoptosis and transcriptional regulation after DNA damage. DNA Repair 2016;42:63—71. https://doi.org/10.1016/j.dnarep.2016.04.008.

Kennel SJ, Foote LJ, Morris M, Vass AA, Griest WH. Mutation analyses of a series of TNT-related compounds using the CHO-hprt assay. J Appl Toxicol 2000;20(6):441—8. https://doi.org/10.1002/1099-1263(200011/12)20:6<441:aid-jat711>3.0.co;2-w.

Krishek BJ, Xie X, Bouchet MJ, Smart TG. m-sulphonate benzene diazonium chloride: a novel GABAA receptor antagonist. Neuropharmacology 1994;33:1125—30.

Kuczkowski KM. The effects of drug abuse on pregnancy. Curr Opin Obstet Gynecol 2007;19:578—85.

Liang ZH, Yin DZ. Preventive treatment of traditional Chinese medicine as antistress and anti-aging strategy. Rejuvenation Res 2010;13:248—52.

Lin M, Wang X, Zhu J, Fan D, Zhang Y, Zhang J, Guo Z. Cellular and biomolecular responses of human ovarian cancer cells to cytostatic dinuclear platinum(II) complexes. Apoptosis 2011;16:288—300.

Maronpot RR. Ovarian toxicity and carcinogenicity in eight recent National Toxicology Program studies. Environ Health Perspect 1987;73:125—30.

68 Contaminants and Plants Action on Female Reproduction

Montero-Montoya R, López-Vargas R, Arellano-Aguilar O. Volatile organic compounds in air: sources, distribution, exposure and associated illnesses in children. Ann Glob Health 2018;84(2):225−38. https://doi.org/10.29024/aogh.910.

Morgan GJ, Alvares CL. Benzene and the hemopoietic stem cell. Chem Biol Interact 2005;153-154:217−22. https://doi.org/10.1016/j.cbi.2005.03.025.

National Toxicology Program. NTP toxicology and carcinogenesis studies of benzene (CAS No. 71-43-2) in F344/N rats and B6C3F1 mice (Gavage studies). Natl Toxicol Progr Tech Rep 1986;289:1−277.

Rana SV, Verma Y. Biochemical toxicity of benzene. J Environ Biol 2005;26:157−68.

Reutman SR, LeMasters GK, Knecht EA, Shukla R, Lockey JE, Burroughs GE, Kesner JS. Evidence of reproductive endocrine effects in women with occupational fuel and solvent exposures. Environ Health Perspect 2002;110:805−11.

Riveles K, Roza R, Talbot P. Phenols, quinolines, indoles, benzene, and 2-cyclopenten-1-ones are oviductal toxicants in cigarette smoke. Toxicol Sci 2005;86:141−51.

Roberts LG, Nicolich MJ, Schreiner CA. Developmental and reproductive toxicity evaluation of toluene vapor in the rat II. Developmental toxicity. Reprod Toxicol 2007;23:521−31.

Sharma RP, Schuhmacher M, Kumar V. Review on crosstalk and common mechanisms of endocrine disruptors: scaffolding to improve PBPK/PD model of EDC mixture. Environ Int 2016;(16):30443. https://doi.org/10.1016/j.envint.2016.09.016. S0160-4120.

Shen Y. In vitro cytotoxicity of BTEX metabolites in HeLa cells. Arch Environ Contam Toxicol 1998;34(3):229−34. https://doi.org/10.1007/s002449900310.

Sirotkin AV. Application of RNA interference for the control of female reproductive functions. Curr Pharmaceut Des 2012;18:325−36.

Sirotkin AV. Regulators of ovarian functions. New York: Nova Science Publishers, Inc.; 2014. p. 194.

Sirotkin AV, Fabian D, Babel'ová Kubandová J, Vlčková R, Alwasel S, Harrath AH. Metabolic state can define the ovarian response to environmental contaminants and medicinal plants. Appl Physiol Nutr Metabol 2017;42(12):1264−9. https://doi.org/10.1139/apnm-2017-0262.

Sirotkin AV, Kadasi A, Baláži A, Kotwica J, Alrezaki A, Harrath AH. Mechanisms of the direct effects of oil-related contaminants on ovarian cells. Environ Sci Pollut Res Int 2020a;27(5):5314−22. https://doi.org/10.1007/s11356-019-07295-0.

Sirotkin A, Kádasi A, Balaží A, Kotwica J, Alwasel S, Harrath AH. The action of benzene, resveratrol and their combination on ovarian cell hormone release. Folia Biol 2020b;66(2):67−71.

Sirotkin AV, Macejková M, Tarko A, Fabova Z, Alrezaki A, Alwasel S, Harrath AH. Effects of benzene on gilts ovarian cell functions alone and in combination with buckwheat, rooibos, and vitex. Environ Sci Pollut Res Int 2020. https://doi.org/10.1007/s11356-020-10739-7.

Sirotkin AV, Macejková M, Tarko A, Fabova Z, Alwasel S, Harrath AH. Buckwheat, rooibos, and vitex extracts can mitigate adverse effects of xylene on ovarian cells in vitro. Environ Sci Pollut Res Int 2020. https://doi.org/10.1007/s11356-020-11082-7.

Sirotkin AV, Makarevich AV, Kubovicova E, Medvedova M, Kolesarova A, Harrath AH. Relationship between body conditions and environmental contaminants in bovine ovarian cells. Theriogenology 2020e;147:77−84. https://doi.org/10.1016/j.theriogenology.2020.02.022.

Sirotkin A, Záhoranska Z, Tarko A, Fabova Z, Alwasel S, Halim Harrath A. Plant polyphenols can directly affect ovarian cell functions and modify toluene effects. J Anim Physiol Anim Nutr 2020. https://doi.org/10.1111/jpn.13461. Epub ahead of print. PMID: 33058312.

Sirotkin AV, Tarko A, Kotwica J, Alrezaki A, Harrath AH. Interrelationships between metabolic hormones, leptin and ghrelin, and oil-related contaminants in control of oxytocin and prostaglandin F release by feline ovaries. Reprod Biol 2020g;20:254−8. https://doi.org/10.1016/j.repbio.2020.02.003.

Sirotkin A, Záhoranska Z, Tarko A, Popovska-Percinc, Alwasel S, Harrath AH. Plant isoflavones can prevent adverse effects of benzene on porcine ovarian activity: an in vitro study. Environ Sci Pollut Control Ser 2020h;27(23):29589—98. https://doi.org/10.1007/s11356-020-09260-8.

Slama R, Thiebaugeorges O, Goua V, Aussel L, Sacco P, Bohet A, Forhan A, Ducot B, Annesi-Maesano I, Heinrich J, Magnin G, Schweitzer M, Kaminski M, Charles MA, EDEN Mother-Child Cohort Study Group. Maternal personal exposure to airborne benzene and intrauterine growth. Environ Health Perspect 2009;117:1313—21.

Stepanov MG, Altukhov VV, Proĭmina FI, Savchenko ON, Danilova OA. Physiologic mechanisms of the reaction of the reproductive system in female rats to chronic exposure to low doses of toluene. Fiziol Zh SSSR Im I M Sechenova 1990;76:1096—102 [in Russian].

Suter-Eichenberger R, Altorfer H, Lichtensteiger W, Schlumpf M. Bioaccumulation of musk xylene (MX) in developing and adult rats of both sexes. Chemosphere 1998;56(13):2747—62.

Tap O, Solmaz S, Polat S, Mete UO, Ozbilgïn MK, Kaya M. The effect of toluene on the rat ovary: an ultrastructural study. J Submicr Cytol Pathol 1996;28:553—8.

Tarko A, Štochmalova A, Hrabovszka S, Vachanova A, Harrath AH, Alwasel S, Grossman R, Sirotkin AV. Can xylene and quercetin directly affect basic ovarian cell functions? Res Vet Sci 2018;119:308—12. https://doi.org/10.1016/j.rvsc.2018.07.010.

Tarko A, Štochmal'ová A, Jedličková K, Hrabovszká S, Vachanová A, Harrath AH, Alwasel S, Alrezaki A, Kotwica J, Baláži A, Sirotkin AV. Effects of benzene, quercetin, and their combination on porcine ovarian cell proliferation, apoptosis, and hormone release. Arch Anim Breed 2019;62(1):345—51. https://doi.org/10.5194/aab-62-345-2019.

Tarko A, Fabová Z, Kotwica J, Valocký I, Alrezaki A, Alwasel S, Harrath AH, Sirotkin AV. The inhibitory influence of toluene on mare ovarian granulosa cells can be prevented by fennel. Gen Comp Endocrinol 2020;295:113491. https://doi.org/10.1016/j.ygcen.2020.113491.

Tsukahara S, Nakajima D, Kuroda Y, Hojo R, Kageyama S, Fujimaki H. Effects of maternal toluene exposure on testosterone levels in fetal rats. Toxicol Lett 2009;185:79—84.

Ungváry G, Varga B, Horváth E, Tátrai E, Folly G. Study on the role of maternal sex steroid production and metabolism in the embryotoxicity of para-xylene. Toxicology 1981;19:263—8.

van Kleef RG, Vijverberg HP, Westerink RH. Selective inhibition of human heteromeric alpha9alpha10 nicotinic acetylcholine receptors at a low agonist concentration by low concentrations of ototoxic organic solvents. Toxicol Vitro 2008;22:1568—72.

Varjani SJ, Gnansounou E, Pandey A. Comprehensive review on toxicity of persistent organic pollutants from petroleum refinery waste and their degradation by microorganisms. Chemosphere 2017;188:280—91.

Waldner CL. The association between exposure to the oil and gas industry and beef calf mortality in Western Canada. Arch Environ Occup Health 2008;63:220—40.

Webb E, Bushkin-Bedient S, Cheng A, Kassotis CD, Balise V, Nagel SC. Developmental and reproductive effects of chemicals associated with unconventional oil and natural gas operations. Rev Environ Health 2014;29:307—18.

WHO. Petroleum products in drinking water. WHO guideliness for drinking water quality. Geneva, Switzerland: WHO; 2005.

Yoon BI, Hirabayashi Y, Kawasaki Y, Kodama Y, Kaneko T, Kim DY, Inoue T. Mechanism of action of benzene toxicity: cell cycle suppression in hemopoietic progenitor cells (CFU-GM). Exp Hematol 2001;29(3):278—85. https://doi.org/10.1016/s0301-472x(00)00671-8.

Zeng Q, Zheng L, Deng L. Study on frequencies of aneuploidy in mouse oocyte and female pronucleus of one cell zygote induced by benzene. Zhonghua Yu Fang Yi Xue Za Zhi 2001;35:87—9 [in Chinese].

Chapter 2.4

Tobacco smoking

2.4.1 Introduction

Tobacco smoking is one of the largest social and health problems. There is no rational explanation of the continued popularity of tobacco smoking. Smokers often acknowledge the harm and even report not to enjoy it, yet they keep smoking (Fidler and West, 2011). Strong urges to smoke are generated by nicotine from cigarettes which undermine and overwhelm concerns about the negative consequences of smoking, and the resolve not to smoke in those trying to stop (West and Shiffman, 2016). Although global tobacco use has fallen over the past two decades, the progress is still off track for achieving the World Health Organization's (WHO's) target of cutting tobacco use by 30% between 2010 and 2025 as part of the global efforts to reduce mortality from the four main noncommunicable diseases (cardiovascular diseases, cancer, chronic lung diseases and diabetes; WHO, 2020). The effects of smoking, alcohol consumption and drug addiction on female fertility have been heterogeneously investigated. The negative effects of smoking on female fertility have raised much interest in recent years, although most of the evidence is gathered from animal or human retrospective studies (de Angelis et al., 2020). Tobacco smoking is a major risk factor for six of the eight leading causes of premature mortality (ischemic heart disease, cerebrovascular disease, lower respiratory infections, chronic obstructive pulmonary disease, tuberculosis and cancer of the trachea, bronchus and lung). Moreover, smoking in pregnancy can lead to low birthweight and illness among infants (WHO, 2020). In addition, tobacco smoke is a reproductive hazard associated with premature reproductive senescence and reduced clinical pregnancy rates in female smokers. Despite an increased awareness of the adverse effects of cigarette smoke exposure on systemic health, many women remain unaware of the adverse effects of cigarette smoke on female fertility. This issue is compounded by our limited understanding of the molecular mechanisms behind cigarette smoke induced infertility (Sobinoff et al., 2013).

2.4.2 Provenance and properties

Nicotiana tabacum, L. common name tobacco is annual herb, shrub or small tree that is 0.9–1.50 m tall. Leaves of tobacco are elliptic and the flowers are clustered at end of the branches. Flowers of tobacco have a cylindrical calyx and are greenish or reddish in the upper part (El Zawi and Elyasseri, 2018).

Tobacco smoke contains over 7000 chemical compounds that are harmful for the smoker and nonsmoker alike (Matt et al., 2011). The main tobacco-specific nitrosamines, which are present in tobacco are nitrosoanabasine (NAB), nitrosoanatabine (NAT), 4-(methylnitrosamino)-1-(3-pyridyl)-1-butanone (NNK), and nitrosonornicotine (NNN). Other nitrosamines include 4-(methylnitrosamino)-1-(3-pyridyl)-1-butanol (NNAL), 4-(methylnitrosamino)-4-(3-pyridyl)-1-butanol (iso-NNAL), and 4-(methylnitrosamino)-4-3-pyridyl) butyric acid (iso-NNAC) (Moldoveanu et al., 2017). Tobacco kills more than eight million people each year. More than 7 million of those deaths are the result of direct tobacco use while around 1.2 million are the result of nonsmokers being exposed to second-hand smoke (WHO, 2020). Tobacco smoke is complex mixture of numerous mutagenic and carcinogenic substances. Vapor contains more than 500 compounds, for example polycyclic hydrocarbons, tobacco-specific N-nitrosamines, aromatic amines, ethylene oxide, and 1,3-butadiene (Feng et al., 2007). Main hazardous substance and lethal toxin in tobacco is nicotine. Nicotine is well known to be highly addictive. The other use of nicotine has been an insecticide (Mishra et al., 2015). It has been found out that main source of heavy metals intoxication, especially cadmium, is caused by tobacco smoking. Primary source of cadmium is soil enriched in tobacco and smoking (Lugon-Moulin et al., 2006). Cadmium is an accumulative toxin. It exhibits an extremely long biological half-life in the human body (20–40 years) (Klaassen et al., 2009). Benzo(a)pyrene is mutagen and carcinogen mainly found in smoke resulting from tobacco combustion. It is mainly present in tobacco tar (Luijten et al., 2014).

2.4.3 Physiological actions

Tobacco smoking and its chemical compounds can express toxic (Nishijo., 2017), mutagenic (Verma et al., 2012), carcinogen effects (Verma et al., 2012; Benowitz and Brunetta, 2016; Polosa et al., 2017) including lung (Verma et al., 2012; Oh and Kacker, 2014; Onor et al., 2017), pancreatic and esophageal cancer (Polosa et al., 2017), chronic pulmonary diseases (Romagna et al., 2013; Onor et al., 2017), neurotoxicity, hepatotoxicity, immunotoxicity, an placental toxicity (Verma et al., 2012), embryotoxic (Liu et al., 2007), gastrointestinal effects (Smith et al., 1992), tuberculosis (Oh and Kacker, 2014; Onor et al., 2017), cardiovascular (Oh and Kacker, 2014; Onor et al., 2017), including endothelial dysfunction, prothrombotic effects, inflammation, altered lipid metabolism, increased demand for myocardial oxygen and blood, and decreased supply of myocardial blood and oxygen, and insulin resistance (U.S. Department of Health and Human Services, 2014).

Tobacco smoking is currently the largest preventable cause of cancer-related deaths. Approximately 30% of **cancer** related to deaths are caused by tobacco smoking (Benowitz and Brunetta, 2016). Tobacco smoking together with exposure to second-hand smoking are major causes of coronary

72 Contaminants and Plants Action on Female Reproduction

heart disease, stroke, aortic aneurysm, and peripheral arterial disease (U.S. Department of Health and Human Services, 2014). Passive tobacco smokers have increased risk of developing serious diseases such as tuberculosis, cardiovascular disorders, lung cancer, emphysema (Oh and Kacker, 2014; Onor et al., 2017). In addition, it has been found out that passive smoking has also been linked with negative health consequences such as low-birth rate in offspring of mothers exposed to second-hand smoke, sudden infant death syndrome, and type 2 diabetes mellitus (Harris et al., 2016). The influence of selected tobacco constituents on health are listed below.

As we mentioned earlier, **nicotine** is one of the most toxic of all poisons and has a rapid onset of action. On direct application in humans, nicotine can cause irritation and burning sensation in the mouth and also throat. It also increases salivation, abdominal pain, nausea, vomiting and diarrhea (Smith et al., 1992). As can be seen in animal and human studies, predominant immediate effects consist of increase in blood pressure and pulse rate. Nicotine also increases hyperglycemia, plasma free fatty acids and also level of catecholamines in the blood. It has been observed reduced coronary blood flow, but on the other hand an increase in skeletal muscle blood flow (Dani and Heinemann, 1996). The increased rate of respiration causes a hypercoagulable state, hypothermia, increases the blood viscosity, and decreases skin temperature (Mishra et al., 2015).

Cadmium has several health effects in humans. The main problem is the inability of human body to excrete cadmium. What is worse, is that cadmium is even reabsorbed by the kidney thereby limiting its excretion. Long-term exposure to cadmium leads to its deposition in bones and lungs (Bernard, 2008). Cadmium can cause bone mineralization. This can lead to osteoporosis, as many researches on humans or animals have proven. The most known disease caused by cadmium is called "Itai-itai" disease. This disease causes an epidemic of bone fractures in Japan due to cadmium contamination (Nishijo et al., 2017). Increased risk of bone fractures in woman as well as decreased bone density and height loss in males and females has been linked to increased cadmium toxicity in population. Cadmium is especially toxic to kidneys. It is accumulated in the proximal tubular cells in higher concentration. According to this fact, we can assume that cadmium exposure can cause renal dysfunction and also kidney disease. It has been found out that it can also cause formation of renal stones and hypercalciuria and disturbances in calcium metabolism (Mudgal et al., 2010).

Benzo(a)pyrene (BaP) is environmental contaminant that can be absorbed by inhalation, oral and dermal routes (Walle et al., 2006). Respiratory epithelial cells are among the first to come in contact with BaP. Over the years it has been established that 60% of lung cancer were due to the mutations caused by BaP (Hecht, 1999). Among many other, BaP was found out to cause not only lung cancer, but it can contribute to many cancer-related diseases and also can cause neurotoxicity, hepatotoxicity, immunotoxicity, and placental toxicity. This is mainly because of BaP mutagen and carcinogen activity (Verma et al., 2012).

2.4.4 Mechanism of action

Several principal extra- and intracellular mechanisms of toxic effect of tobacco smoking on the cells have been proposed, although the evidence of each of this mechanism is limited, and the interrelationships between these mechanisms are poorly investigated yet. The following mechanisms of action of cigarette smoke and its chemical compounds are suggested:

(1) exhibition of cytotoxicity, oxidative stress, and alteration of the balance between cell proliferation and apoptosis, the reduction of antioxidants by ingredients, such as acrolein and aldehydes, undermines the defensive mechanisms of cells, and promotes cell damage (Romagna et al., 2013),

(2) a loss of cilia in the lungs, mucus gland hyperplasia, and overall inflammation resulting in the abnormal functioning of the lungs as well as injury (Onor et al., 2017),

(3) ganglionic transmission, nicotinic acetylcholine receptors on chromaffin cells via catecholamines and central nervous system stimulation of nicotinic acetylcholine receptors,

(4) an increase in oxidative stress and neuronal apoptosis, reactive oxygen species and lipid peroxide increase and also DNA damage by nicotine, an increase of activity in prefrontal cortex and visual system (Mishra et al., 2015), and

(5) a bind of cadmium in tobacco smoke on human DNA, resulting in DNA damage and gene mutations, genetic changes lead to uncontrolled cell growth and inhibit normal mechanisms that restrain cell growth and spread, resulting in cancer (Onor et al., 2017).

2.4.5 Effects on female reproductive processes

Tobacco smoking causes several reproductive alterations (Jennings et al., 2011; de Angelis et al., 2020; Zhan and Huang, 2021), changes in ovary (Jennings et al., 2011; de Angelis et al., 2020; Zhan and Huang, 2021) and in secretion of reproductive hormones (Barbieri et al., 2005; de Angelis et al., 2020). Moreover, it induces sudden infant death syndrome, premature births, and decreased fertility in women up to full infertility (Zhan and Huang, 2021).

2.4.5.1 The effect on ovarian and reproductive state

Tobacco smoking affects nearly all domains of the female reproductive function. Smoking is associated with early menopause and reduced levels of ovarian reserve markers, mediated by an impairment of antral follicle development, and growth resulting in cytotoxicity and production of poor quality oocytes (de Angelis et al., 2020). Direct or indirect tobacco smoking can affect the ovarian functions in many ways, for example impaired steroidogenesis and follicular development and follicles loss, an increase in zona pellucida thickness, a shorter pole-to-pole spindle length, but wider spindle equators, and manifested errors in chromosome alignment (Jennings et al., 2011). Also, tobacco smoking is associated with aneuploidy, because exposure to the harmful ingredients in smoking can induce meiotic spindle disturbance and chromosomal misalignment, leading to production of oocytes with abnormal chromosome numbers (Zhan and Huang, 2021). A smaller oocyte diameter in incipient antral follicles under smoke exposure and an increase of oocyte size after smoking cessation was noted. Granulosa cells count decreased after, but not during the smoke exposure period. According to this study, we may suggest that cigarette smoking elicits negative effects, which last after exposure withdrawal (Paixão et al., 2012). Various reproductive diseases can result from smoking-induced dysfunction at any corresponding stage of these processes, rendering to decreased female fecundity and even infertility (Zhan and Huang, 2021).

2.4.5.2 Effect on reproductive hormones

Tobacco smoking is associated with lower estrogens and progesterone and higher androgens blood levels. These effects are due to several ovarian and extraovarian factors including: increased levels of sex hormone binding globulin (SHBG); increased hepatic production of estrogens metabolites with minimal estrogenic activity by pushing the estrogens 2-hydrohylation pathway; inhibition of aromatase enzyme; increased expression of ovarian CYP1B1 enzyme; increased levels of adrenocorticotropic hormone (ACTH); inhibition of adrenal 21-hydrohylase enzyme (de Angelis et al., 2020). Higher level of testosterone and follicle-stimulating hormone (FSH) was found in female smokers (Barbieri et al., 2005).

2.4.5.3 Effect on embryos

Maternal cigarette (tobacco) smoking causes several reproductive abnormalities. Other toxins in tobacco smoke including nicotine, cadmium, lead, mercury, and polycyclic aromatic hydrocarbons, have been found to cause sudden infant death syndrome, premature births, and decreased fertility in women. More recent evidence indicates a causal relationship between maternal cigarette smoking and orofacial clefts and ectopic pregnancies (Benowitz and Brunetta, 2016). What is worse, declined quality of these oocytes compromised subsequent parthenogenetic development, with only about 5% of the resultant embryos developing to blastocysts that were almost aneuploid and hypogenetic with obviously decreased cell number (Liu et al., 2007).

Environmental contaminants and their influence Chapter | 2 **75**

2.4.6 Mechanisms of action on female reproductive processes

The most evident mechanisms of adverse effects of tobacco smoking on female reproduction and fecundity can be ability of tobacco smoking:

(1) increased reactive oxygen species (ROS) formation in smoke-exposed granulosa cells and in oocyte membranes, but also in follicles cultured *in vitro*, and in mice ovaries exposed to cigarette smoke components (Sobinoff et al., 2013),

(2) increased level of ROS induces granulosa cells-directed **oxidative stress** and DNA damage, resulting in cytotoxicity and production of poor quality oocytes (de Angelis et al., 2020), and initiation of lipid peroxidation (Paszkowski et al., 2002),

(3) influence on **antioxidative enzymes**: smoking affects mRNA levels of antioxidant enzymes in mural granulosa cells of smoker women undergoing an *In vitro* fertilization, and induce an overexpression of superoxide dismutase 2 (SOD2) and catalase in smoker women in comparison to nonsmokers (Budani et al., 2017),

(4) an induction of meiotic spindle disturbance and **chromosomal misalignment**, leading to aneuploidy, production of oocytes with abnormal chromosome numbers (Zhan and Huang, 2021),

(5) an effect on ovarian **steroidogenesis** and **follicle growth** (Budani and Tiboni, 2017), affect **folliculogenesis** and **oogenesis**, by means of direct ovotoxicity and central actions on the hypothalamus-pituitary-ovary axis (de Angelis et al., 2020),

(6) cigarette smoke alterations are interfering with several crucial reproductive events, including granulosa cells maturation, binding of gonadotropins to their receptors, and egg fertilizing capacity (Sobinoff et al., 2013),

(7) smoking role in **ovarian aging**, association between smoking and elevated levels of follicle-stimulating hormone and changes in antimüllerian hormone levels, a marker of ovarian reserve, or antral follicle count (de Angelis et al., 2020),

(8) smoking suppress **ovarian folliculogenesis**: it induces loss of large primary and secondary follicles, large follicles may be more sensitive to than primordial and small primary follicles, and increased primordial follicle recruitment incited by the loss of larger follicles (Madden et al., 2014).

2.4.7 Application in reproductive biology and medicine

Tobacco smoking causes several reproductive alterations of female reproductive system and changes in secretion of reproductive hormones regulated by the hypothalamus-pituitary-ovary axis, decreases fertility in women and induces infertility. For this reason and due to the occurrence of reproductive diseases and negative effects of tobacco smoking on human reproductive health is appropriate smoking cessation. Smoking cessation can be hard, especially because of nicotine addictiveness. One of the possible options for smoking cessation is nicotine replacement therapy. It is a temporarily replacement of the nicotine from the cigarettes. It can reduce motivation to smoke and also reduce nicotine withdrawal symptoms. Commercially available forms of nicotine replacement therapy are gums, nasal spray, transdermal patch, inhaler and sublingual tablets. These can increase rate of quitting by 50%–70%. The effectiveness of these therapy appears to be largely independent of the intensity of additional support (Stead et al., 2012). On the other hand, it is possible to look for alternative ways of replacing nicotine, possibly also in plants and their molecules. Moreover, it is not to be excluded, that some plant or plant molecules with antioxidant properties would be able to mitigate the oxidative stress induced by smoking and, therefore, its adverse consequence for reproduction and health. This could be a challenge for further research in this area.

2.4.8 Conclusions and possible direction of future studies

The analysis of the available literature demonstrates that tobacco smoking causes several reproductive abnormalities and can exert negative effects on various female reproductive functions, including the hypothalamus-pituitary-ovary axis, folliculogenesis, embryogenesis and their signaling molecules and receptors that lead to a decrease growth of ovarian follicles, ovarian aging, decrease of quality of oocytes, development of embryos, fertility, and induces infertility. These effects could be due to the ability of tobacco smoking to destroy chromosomes, to affect cell metabolism, including the accumulation of free radicals, DNA damage, resulting in cytotoxicity and production of poor quality oocytes and to affect release of hormonal regulators of reproductive processes and intracellular protein kinases, transcription factors, and other regulators/markers of proliferation and apoptosis.

Nevertheless, addressing some queries could help understanding character and mechanisms tobacco smoking action and to prevent its negative influence on reproduction and health. The action, mechanisms of action and interrelationships between numerous tobacco constituents require further clarification. The main studies stated the influence of smoking on some signaling molecules, but the hierarchical interrelationships between mediators of tobacco action are studied insufficiently. The search for the new approaches to reduce the attractivity of smoking for people and to replace or to neutralize the harmful effect of tobacco molecules mentioned above should be promising for reduction of smoking and its negative consequence for reproduction and health.

References

Barbieri RL, Sluss PM, Powers RD, McShane PM, Vitonis A, Ginsburg E, Cramer DC. Association of body mass index, age, and cigarette smoking with serum testosterone levels in cycling women undergoing in vitro fertilization. Fertil Steril 2005;83(2):302−8. https://doi.org/10.1016/j.fertnstert.2004.07.956.

Benowitz NL, Brunetta PG. Smoking hazards and cessation. In: Murray and Nadel's textbook of respiratory medicine. WB Saunders; 2016. p. 807−21.

Bernard A. Cadmium & its adverse effects on human health. Indian J Med Res 2008;128(4):557. http://hdl.handle.net/2078.1/35901.

Budani MC, Tiboni GM. Ovotoxicity of cigarette smoke: a systematic review of the literature. Reprod Toxicol 2017;72:164−81. https://doi.org/10.1016/j.reprotox.2017.06.184.

Budani MC, Carletti E, Tiboni GM. Cigarette smoke is associated with altered expression of antioxidant enzymes in granulosa cells from women undergoing in vitro fertilization. Zygote 2017;25(3):296−303. https://doi.org/10.1017/S0967199417000132.

Dani JA, Heinemann S. Molecular and cellular aspects of nicotine abuse. Neuron 1996;16(5):905−8.

de Angelis C, Nardone A, Garifalos F, Pivonello C, Sansone A, Conforti A, Pivonello R. Smoke, alcohol and drug addiction and female fertility. Reprod Biol Endocrinol 2020;18(1):1−26. https://doi.org/10.1186/s12958-020-0567-7.

El Zawi NR, Elyasseri AMA. Toxicity, teratogenicity and genotoxicity of nicotian tabacum L. and Nicotiana gluaca G: a review. Int J Photochem 2018;4:1−12.

Feng S, Plunkett SE, Lam K, Kapur S, Muhammad R, Jin Y, Roethig HJ. A new method for estimating the retention of selected smoke constituents in the respiratory tract of smokers during cigarette smoking. Inhal Toxicol 2007;19(2):169−79. https://doi.org/10.1080/08958370601052022.

Fidler JA, West R. Enjoyment of smoking and urges to smoke as predictors of attempts and success of attempts to stop smoking: a longitudinal study. Drug Alcohol Depend 2011;115(1−2):30−4. https://doi.org/10.1016/j.drugalcdep.2010.10.009.

Harris KK, Zopey M, Friedman TC. Metabolic effects of smoking cessation. Nat Rev Endocrinol 2016;12(5):299−308. https://doi.org/10.1038/nrendo.2016.32.

Hecht SS. Tobacco smoke carcinogens and lung cancer. J Natl Cancer Inst 1999;91(14):1194−210. https://doi.org/10.1093/jnci/91.14.1194.

Klaassen CD, Liu J, Diwan BA. Metallothionein protection of cadmium toxicity. Toxicol Appl Pharmacol 2009;238(3):215−20. https://doi.org/10.1016/j.taap.2009.03.026.

Liu Y, Li GP, White KL, Rickords LF, Sessions BR, Aston KI, Bunch TD. Nicotine alters bovine oocyte meiosis and affects subsequent embryonic development. Mol Reprod Dev 2007;74(11):1473−82.

Luijten M, Hernandez LG, Zwart EP, Bos PMJ, Van Steeg H, Van Benthem J. The sensitivity of young animals to benzo [a] pyrene-induced genotoxic stress. Environ Toxicol Chem 2014;20:102−6. http://hdl.handle.net/10029/311510.

Lugon-Moulin N, Martin F, Krauss MR, Ramey PB, Rossi L. Cadmium concentration in tobacco (Nicotiana tabacum L.) from different countries and its relationship with other elements. Chemosphere 2006;63(7):1074−86. https://doi.org/10.1016/j.chemosphere.2005.09.005.

Jennings PC, Merriman JA, Beckett EL, Hansbro PM, Jones KT. Increased zona pellucida thickness and meiotic spindle disruption in oocytes from cigarette smoking mice. Human Reprod 2011;26(4):878−84. https://doi.org/10.1093/humrep/deq393.

78 Contaminants and Plants Action on Female Reproduction

Madden JA, Hoyer PB, Devine PJ, Keating AF. Acute 7, 12-dimethylbenz [a] anthracene exposure causes differential concentration-dependent follicle depletion and gene expression in neonatal rat ovaries. Toxicol Appl Pharmacol 2014;276(3):179–87. https://doi.org/10.1016/j.taap.2014.02.011.

Matt GE, Quintana PJ, Destaillats H, Gundel LA, Sleiman M, Singer BC, Hovell MF. Thirdhand tobacco smoke: emerging evidence and arguments for a multidisciplinary research agenda. Environ Health Perspect 2011;119(9):1218–26. https://doi.org/10.1289/ehp.1103500.

Mishra A, Chaturvedi P, Datta S, Sinukumar S, Joshi P, Garg A. Harmful effects of nicotine. Indian J Med Paediat Oncol 2015;36(1):24. https://doi.org/10.4103/0971-5851.151771.

Moldoveanu SC, Zhu J, Qian N. Analysis of traces of tobacco-specific nitrosamines (TSNAs) in USP grade nicotine, e-liquids, and particulate phase generated by the electronic smoking devices. Beiträge zur Tabakforschung Int Contribut Tobacco Res 2017;27(6):86–96. https://doi.org/10.1515/cttr-2017-0009.

Mudgal V, Madaan N, Mudgal A, Singh RB, Mishra S. Effect of toxic metals on human health. Open Nutraceut J 2010;3:94–9. https://doi.org/10.2174/1876396001003010088.

Nishijo M, Nakagawa H, Suwazono Y, Nogawa K, Kido T. Causes of death in patients with Itai-itai disease suffering from severe chronic cadmium poisoning: a nested case–control analysis of a follow-up study in Japan. BMJ Open 2017;7(7):e015694. https://doi.org/10.1136/bmjopen-2016-015694.

Oh AY, Kacker A. Do electronic cigarettes impart a lower potential disease burden than conventional tobacco cigarettes?: review on e-cigarette vapor versus tobacco smoke. Laryngoscope 2014;124(12):2702–6. https://doi.org/10.1002/lary.24750.

Onor IO, Stirling DL, Williams SR, Bediako D, Borghol A, Harris MB, Sarpong DF. Clinical effects of cigarette smoking: epidemiologic impact and review of pharmacotherapy options. Int J Environ Res Publ Health 2017;14(10):1147. https://doi.org/10.3390/ijerph14101147.

Paixão LL, Gaspar-Reis RP, Gonzalez GP, Santos AS, Santana AC, Santos RM, Nascimento-Saba CCA. Cigarette smoke impairs granulosa cell proliferation and oocyte growth after exposure cessation in young Swiss mice: an experimental study. J Ovarian Res 2012;5(1):1–6. https://doi.org/10.1186/1757-2215-5-25.

Paszkowski T, Clarke RN, Hornstein MD. Smoking induces oxidative stress inside the Graafian follicle. Human Reprod 2002;17(4):921–5. https://doi.org/10.1093/humrep/17.4.921.

Polosa R, Cibella F, Caponnetto P, Maglia M, Prosperini U, Russo C, Tashkin D. Health impact of E-cigarettes: a prospective 3.5-year study of regular daily users who have never smoked. Sci Rep 2017;7(1):1–9. https://doi.org/10.1038/s41598-017-14043-2.

Romagna G, Allifranchini E, Bocchietto E, Todeschi S, Esposito M, Farsalinos KE. Cytotoxicity evaluation of electronic cigarette vapor extract on cultured mammalian fibroblasts (ClearStream-LIFE): comparison with tobacco cigarette smoke extract. Inhal Toxicol 2013;25(6):354–61. https://doi.org/10.3109/08958378.2013.793439.

Smith EW, Smith KA, Maïbach HI, Andersson PO, Cleary G, Wilson D. The local side effects of transdermally absorbed nicotine. Skin Pharmacol Physiol 1992;5(2):69–76. https://doi.org/10.1159/000211021.

Sobinoff AP, Beckett EL, Jarnicki AG, Sutherland JM, McCluskey A, Hansbro PM, McLaughlin EA. Scrambled and fried: cigarette smoke exposure causes antral follicle destruction and oocyte dysfunction through oxidative stress. Toxicol Appl Pharmacol 2013;271(2):156–67. https://doi.org/10.1016/j.taap.2013.05.009.

Stead LF, Perera R, Bullen C, Mant D, Hartmann-Boyce J, Cahill K, Lancaster T. Nicotine replacement therapy for smoking cessation. Cochrane Datab Syst Rev 2012;(11). https://doi.org/10.1002/14651858.CD000146.pub4.

U.S. Department of Health and Human Services. The health consequences of smoking—50 Years of progress: a report of the surgeon general. Atlanta, GA, USA: Department of Health and Human Services, Centers for Disease Control and Prevention, National Center for Chronic Disease Prevention and Health Promotion, Office on Smoking and Health; 2014. p. 1–36.

Verma N, Pink M, Rettenmeier AW, Schmitz-Spanke S. Review on proteomic analyses of benzo [a] pyrene toxicity. Proteomics 2012;12(11):1731–55. https://doi.org/10.1002/pmic.201100466.

Walle T, Walle UK, Sedmera D, Klausner M. Benzo [A] pyrene-induced oral carcinogenesis and chemoprevention: studies in bioengineered human tissue. Drug Metab Dispos 2006;34(3):346–50. https://doi.org/10.1124/dmd.105.007948.

West R, Shiffman S. Fast facts: smoking cessation. Karger Medical and Scientific Publishers; 2016, ISBN 978-1-903734-98-8.

WHO. Tobacco. In: Health at a glance: Asia/Pacific 2020: measuring progress towards universal health coverage. Paris: OECD Publishing; 2020. https://doi.org/10.1787/2c5c9396-en.

Zhan S, Huang J. Effects of cigarette smoking on preimplantation embryo development. Environment and Female Reproductive Health; 2021. p. 137–50. https://doi.org/10.1007/978-981-33-4187-6.

Chapter 3

Food/medicinal herbs and their influence on health and female reproduction

Chapter 3.1

Apricot (*Prunus armeniaca* L.)

3.1.1 Introduction

The apricot tree (*Prunus armeniaca* L.) is a member of the *Prunus* genus of the Rosaceae family (Bolarinwa et al., 2014; Lv et al., 2017; Saleem et al., 2018; Thodberg et al., 2018; Tanwar et al., 2018; Ayaz et al., 2020). It is one of the major worldwide preferable commercial crops in production and consumption (Ruiz et al., 2005; Ozturk et al., 2009; Lim, 2012; Kopcekova et al., 2018). Also, their seeds make an important contribution to the diet in many countries (Gezer et al., 2011). The fruit seeds of apricot trees are classified according to their taste into sweet apricot, semibitter apricot, and bitter apricot (Lee et al., 2013, 2014). Apricot seeds contain a wide variety of bioactive compounds, and that consumption of apricot seeds has been associated with a reduced risk of chronic diseases (Zhang et al., 2011; Kopcekova et al., 2018). As apricot seed contains a wide variety of bioactive components, it can be potentially useful in human nutrition (Kopcekova et al., 2018; Kolesárová et al., 2017). The use of apricot seeds for human nutrition is limited because of their content of the toxic, cyanogenic glycoside amygdalin, accompanied by minor amounts of prunasin (Kopcekova et al., 2018; Kolesárová et al., 2020). Due to the significant amounts of nutrients such as oils, proteins, amino acids, carbohydrates, soluble sugars, minerals, vitamins, and fibers in the apricot (see later), apricot seed exhibits a wide range of biological activity and is typically used in a variety of foods (Lee, 2014; Rai et al., 2016). There is a rising demand for the development of new foods with dietary properties. Apricot kernel flour could

Environmental Contaminants and Medicinal Plants Action on Female Reproduction
https://doi.org/10.1016/B978-0-12-824292-6.00003-9
Copyright © 2022 Elsevier Inc. All rights reserved.

82 Contaminants and Plants Action on Female Reproduction

offer an exciting potential as protein source in cereal products (Dhen et al., 2018). Despite rich source of protein and oil, wild (bitter) apricot (*Prunus armeniaca* L.) kernel is rarely exploited by food industries due to high amount of antinutrients and potentially toxic amygdalin (Tanwar et al., 2019).

In the present review we summarize the current (published after year 2000) knowledge concerning properties, physiological, dose-dependent effects, and benefit of apricot seeds on general physiological processes and health and on female reproductive processes at various regulatory levels.

3.1.2 Provenance and properties

Apricot seeds contain abundant compounds like carbohydrates, proteins, polyphenols, flavonoids, fiber, oil, mineral content and cyanogenic glycoside amygdalin (Alpaslan and Hayta, 2006; Zhou et al., 2016; Halenar et al., 2017; Tanwar et al., 2019; Kopcekova et al., 2018, 2021). Carbohydrate content of apricot kernel was reported variously 17.3%−27.9% (Alpaslan and Hayta, 2006). The reported protein content of apricot kernel ranged from 14.1% to 45.3% (Alpaslan and Hayta, 2006; Halenar et al., 2017; Kopcekova et al., 2018, 2021). Apricot kernel proteins contain 84.7% albumin, 7.65% globulin, 1.17% prolamin, and 3.54% glutelin. Nonprotein nitrogen is 1.17%, and other proteins are 1.85%. Essential amino acids in apricot kernel constituted 32% −34% (Alpaslan and Hayta, 2006) of the total amino acids. The composition and sequence of amino acids have a prominent influence on the antioxidant activities of peptides (Zhang et al., 2018). The major essential amino acids (mmol/100 g meal) were arginine (21.7−30.5) and leucine (16.2−21.6), and the predominant nonessential amino acid was glutamic acid (49.9−68.0) (Alpaslan and Hayta, 2006). The amino acid sequences of apricot seed kernels were Val-Leu-Tyr-Ile-Trp and Ser-Val-Pro-Tyr-Glu, respectively (Zhang et al., 2018). The oil content of the kernels varies from 27.7% to 66.7% (Alpaslan and Hayta, 2006; Halenar et al., 2017; Kopcekova et al., 2018, 2021).

Monounsaturated and polyunsaturated fatty acids, as well as minor lipid components, play an important role in human nutrition and health. Diets rich in these compounds can decrease blood pressure and total blood cholesterol levels, fight oxidative stress, and maintain body weight (Turan et al., 2007). Apricot seeds contain monounsaturated fatty acids approximately 65.3% and polysaturated fatty acids 27.1% (Kopčeková et al., 2018, 2021). The contents of unsaturated fatty acids (91.5%−91.8%) and saturated fatty acids (5.9%−8.3%) (Alpaslan and Hayta, 2006; Kopcekova et al., 2018, 2021) have been reported, as well as neutral lipids (95.7%−95.2%), glycolipids (1.3%−1.8%), and phospholipids (2.0%) (Alpaslan and Hayta, 2006). Oleic acid (64.5%−70.70%) was the predominant fatty acid found in the oils, followed by linoleic (22.41%−27.1%), palmitic (3.14%−4.6%), stearic (1.2%−1.4%), linolenic (0.90%−27.1%), and palmitoleic (0.70%−0.8%) acid (Stryjecka et al., 2019; Kopcekova et al., 2018, 2021). The content of α-, γ-, and δ-tocopherols in the oils from the five apricot cultivars was 19.6—40.0, 315.4—502.3, and 28.3—58.5 mg/kg, respectively (Stryjecka et al., 2019).

Furthermore, apricot kernel oil contains some biologically active substances, such as β -carotene (61.05 mg/g), tocopherols (50.76 mg/100 g), phenolic compounds, campesterol (11.8 mg/100 g), stigmasterol (9.8 mg/100 g), sitosterol (177.0 mg/100 g), provitamin A (Zhou et al., 2016; Matthaus et al., 2016) and gallic acid (30.1 mg/g), ferulic acid (2.9 mg/g), cinnamic acid (0.9 mg/g), genistein (5.6 mg/g) and rutin (11.3 mg/g) (Kopčeková et al., 2021). Rudzińska et al. (2017) were investigated the profile of sterols and squalene content in oils recovered from the kernels of 15 apricot (*Prunus armeniaca* L.) varieties. Nine sterols (campesterol, β-sitosterol, Δ5-avenasterol, 24-methylene-cycloartanol, cholesterol, gramisterol, Δ7-stigmasterol, Δ7-avenasterol and citrostadienol) were identified in apricot kernel oils. The β-sitosterol was the predominant sterol in each cultivar and consisted of 76%—86% of the total detected sterols. The content of total sterols and squalene were significantly affected by the variety and ranged between 215.7—973.6 and 12.6—43.9 mg/100 g of oil, respectively (Rudzińska et al., 2017).

The total sugar content ranges 4.10%—7.76% (Alpaslan and Hayta, 2006; Kopcekova et al., 2018, 2021). Invert sugar content was 5.86% (Alpaslan and Hayta, 2006). The kernels contain thiamine, riboflavin, niacin, vitamin C, α -tocopherol, and δ -tocopherol (Alpaslan and Hayta, 2006; Zhang et al., 2011). The mineral content ranges of apricot kernel (mg/100 g dry matter) are as follows: Na: 35.2—64.2 mg; K: 473—592.5; Ca: 1.8—177.4; Mg: 113—290; P: 470; Fe: 2.14—2.82; Zn: 2.33—5.9; Mn: 0.48—0.6 mg; Ni: 0.14 mg and Co: 0.002 mg (Alpaslan and Hayta, 2006; Halenar et al., 2017; Kopcekova et al., 2018, 2021). Depending on the variety, apricot kernels contain the cyanogenic glycoside amygdalin (also known as laetrile or vitamin B17). Amygdalin (D-mandelonitrile-β-D-gentiobioside) is one of the main bioactive substances abundantly present in the kernels of the various species of *Rosaceae* family such in the bitter seeds of apricot, apple, almond, peaches, cherries, plums, grains, millets, sprouts, and nuts (Lv et al., 2017; Thodberg et al., 2018; Tanwar et al., 2018; Saleem et al., 2018; Ayaz et al., 2020). Approximately 50 g/kg (3%—5%) amygdalin is found in bitter almond (Lee et al., 2013) and 5%—6% in apricot kernels (Halenar et al., 2017; Kopcekova et al., 2018, 2021). In contrast, the amygdalin content of seeds from apples (*Malus domestica*) range from 1 to 4 g/kg (Bolarinwa et al., 2015).

3.1.3 Physiological and therapeutic actions

The apricot seed consumption could be useful for treatment of various diseases. On the other hand, the use of apricot seeds for human nutrition is limited because of their content of the toxic, cyanogenic glycoside amygdalin.

84 Contaminants and Plants Action on Female Reproduction

A short-term consumption of apricot seeds does not represent risk for human (Tusimova et al., 2017) and animal health (Kovacikova et al., 2019). The study on rabbits has shown that short-term application of oral consumption of crushed bitter apricot seeds (*Prunus armeniaca* L.) did not represent a risk for biochemical, hematological parameters (Kovacikova et al., 2019), as well as it did not affect endocrine profile (Halenar et al., 2017). On the other hand, a long-term consumption of apricot seeds might negatively affect the liver microscopic structure (Kolesárová et al., 2020).

Pharmacological studies demonstrated cardioprotective (Zhang et al., 2011), antiinflammatory (Minaiyan et al., 2014; Chang et al., 2005), anti-asthmatic (Do et al., 2006; Badr and Tawfik, 2010), analgesic, antimutagenic (Lee et al., 2014), antitussive (Minaiyan et al., 2014), antioxidant (Korekar et al., 2011; Gomaa et al., 2013; Minaiyan et al., 2014; Lee et al., 2014; Gomaa et al., 2013, 2013; Durmaz and Alpaslan, 2007; Chang et al., 2005), anticancer (Gomaa et al., 2013; Chen et al. 2013, 2020; Yang et al., 2014; Hu et al., 2015; Yamshanov et al., 2016; Qin et al., 2019; Cassiem et al., 2019; Ayaz et al., 2020; Aamazadeh et al., 2020), and antimicrobial (Lee et al., 2014; Gomaa et al., 2013) effects of apricot or almond seeds. Apricot kernel has been traditionally used in gastric inflammations, dermatitis and also as a carminative agent (Minaiyan et al., 2014). The detailed description of apricot seeds is presented below.

In Chinese traditional medicine bitter apricot or almond seeds have long been used for the treatment of cure respiratory disorders (coughing, wheeze, asthma, emphysema, and bronchitis) and skin diseases (acne vulgaris, dandruff, and furuncle) (Lee et al., 2014; Geng et al., 2016). Semen *Armeniacae amarum* has long been used to control allergic asthma in Korean traditional medicine (Do et al., 2006).

Traditionally, in oriental medicine, apricot seed (Semen *Armeniacae amarum*) is used to treat skin diseases, such as furuncle, acne vulgaris, dandruff. Moreover, amygdalin from *Armeniacae semen* can induce apoptosis of skin cells (Lee et al., 2014). However, the underlying mechanism of action has not been systematically elucidated (Li et al., 2016). Li et al. (2016) studied the antiproliferative effect of bitter apricot essential oil on cultured immortalized human keratinocytes. Bitter apricot essential oil could efficiently inhibit proliferation and induce apoptosis of immortalized human keratinocytes.

However, very few studies have evaluated the effects of apricot seed consumption on human lipid profile and other risk factors of cardiovascular diseases so far. Jiagang et al. (2011) first examined its therapeutic effect in atherosclerosis. Triglycerides and total cholesterol levels in mice was decreased by amygdalin treatment (1 mg/kg). Meanwhile, amygdalin treatment also decreased the low-density lipoprotein (LDL) cholesterol but did not induce any decline in high-density lipoprotein (HDL) cholesterol. Our previous studies suggest that daily consumption of apricot seeds may affect plasma lipid profile (Kopcekova et al., 2018; Kopčeková et al., 2021). Consumption of bitter apricot seeds at 60 mg/kg body weight also caused significant reductions in triglycerides and LDL cholesterol levels in healthy volunteers for 6 and 12 weeks (Kopcekova et al., 2018). The blood levels of total cholesterol, HDL cholesterol did not change, however, the level of LDL cholesterol decreased significantly after 42 days in women of reproductive age. On the other hand, there was increase of total cholesterol and triglycerides after 21 days (Kopčeková et al., 2021). In another *in vivo* study in chickens, an aqueous extract of *Prunus armeniaca* seeds in drinking water at 200 mg/kg body weight (along with basal diet for 6 weeks) reduced the level of plasma cholesterol and triglyceride (Kalia et al., 2017). Zibaeenezhad et al. (2017) hypothesized that *Amygdalus scoparia* kernel oil supplementation may have a positive effect in reducing serum triglyceride level in patients with dyslipidemia without significant effect on serum cholesterol levels. The study of Lv et al. (2017) demonstrated that amygdalin inhibited the progression of atherosclerosis and also markedly alleviated hypercholesterolemia and hypertriglyceridemia in the mice and affected the extent of atherosclerotic lesions. Apricot kernel oil has potent cardioprotective effects and could be developed as a nutriment for the treatment and prevention of myocardial infarcts (Zhang et al., 2011).

The antioxidant, anticancer effects on a human hepatoma and antibacterial properties of the bitter apricot *Armeniaca sibirica* (L.) kernel skins polyphenol extracts were further characterized *in vitro*. Apricot kernels inhibited the growth of both tumors murine mammary adenocarcinoma and lymphosarcoma (Yamshanov et al., 2016). The antiproliferative effect of apricot and peach kernel extracts on human colon cancer cells *in vitro* was confirmed (Cassiem et al., 2019). Bitter apricot ethanolic extract induces apoptosis human pancreatic cancer cells (Aamazadeh et al., 2020). Apricot seeds have been used also for the treatment of lung or liver cancer in Chinese or Pakistani traditional medicine (Shen and Wen, 2018; Ramzan et al., 2017; Hu et al., 2015). The apricot kernel skins extract inhibit proliferation and induce apoptosis of liver carcinoma cells HepG2 (Qin et al., 2019). Also, from the study of Ayaz et al. (2020) results that natural sources of amygdalin, i.e., bitter kernels could be used in a specified amount to treat various types of cancer in human. Antiproliferative effect on human hepatocellular carcinoma cells HepG2 in the kernels of 19 apricot cultivars was found (Chen et al., 2020). This exposed that some unique phytochemicals in apricot kernels could be responsible for the inhibition of HepG2 tumor cells (Chen et al., 2020).

86 Contaminants and Plants Action on Female Reproduction

There is an evidence of apricot seeds or amygdalin effect on male reproductive functions in animals (Kolesar et al., 2018; Halenár et al., 2016a; Duracka et al., 2016). A 28-day period of daily consumption of apricot seeds (at doses 60 and 300 mg/kg b.w.) does not induce any negative change in rabbit spermatozoa *in vivo*. The study has shown that oral consumption of apricot seeds causes no significant impact on motility parameters in comparison to intramuscular amygdalin application, which resulted in a significant time- and dose-dependent decrease of sperm motility as well as progressive motility (Kolesar et al., 2018). On the other hand, long-term administration of amygdalin at high doses can increase protein oxidation and lipid peroxidation as a result of its ability to compromise oxidative balance in the testicular tissue, as well as increase the reactive oxygen species (ROS) production, which can lead to cellular oxidative stress (Duracka et al., 2016).

The gastroprotective effect of apricot kernel oil on ethanol induced gastric ulcer in rats apricot kernel oil protects rat gastric mucosa against ethanol induced injury by its antiinflammatory, antioxidative and antiapoptotic effects, and might be useful for reducing the severity of gastric ulcers (Karaboğa et al., 2018).

Study of Kovacova et al. (2019) was designed to reveal whether long-term peroral consumption of bitter apricot seeds induces changes in bone microstructure of rabbits. Long-term administration of bitter apricot seeds did not affect total body weight, femoral weight and femoral length of rabbits. Similarly, microcomputed tomography (3D analysis) did not demonstrate a significant impact of chronic amygdalin toxicity on all measured parameters of cortical and trabecular bone tissues. On the contrary, histological (2D) analysis of cortical bone tissue revealed a reduced density of secondary osteons. All these negative changes are consistent with a different vascularization and biomechanical properties of the cortical bone.

These results demonstrated the potential of apricot seeds including amygdalin in the control and treatment of civilization diseases and support the potential of the clinical application. However, amygdalin as a therapeutic agent is not used in more countries owing to insufficient clinical verification of its therapeutic efficacy and adverse side-effects. Therefore, the character of apricot seeds effect remains controversial.

3.1.4 Mechanisms of action

One of the main bioactive substances abundantly present in the kernels of the various species of *Rosaceae* family, such in the bitter apricot seeds, is amygdalin (D-mandelonitrile-β-D-gentiobioside) (Lv et al., 2017; Thodberg et al., 2018; Tanwar et al., 2018; Saleem et al., 2018; Ayaz et al., 2020). Mechanism of the amygdalin action is described in detail (the specific chapter). However, the underlying mechanism of action has not been completely elucidated (Li et al., 2016).

The antioxidant activity of extracts of bitter and sweet apricot kernels (*Prunus armeniaca* L.) *in vitro* was described, whereas the highest percent inhibition of lipid peroxidation (69%) and total phenolic content were detected in the extract of sweet kernels (Yiğit et al., 2009). The serum creatine kinase and aspartate aminotransferase activities demonstrated beneficial effects of apricot kernel oil in rat model *in vivo* (Zhang et al., 2011). Myocardial catalase, superoxide dismutase (SOD), glutathione peroxidase (GPx), and constitutive nitric oxide synthase activities, as well as nitric oxide (NO) concentrations, were all increased, whereas malondialdehyde content and inducible nitric oxide synthase were decreased in apricot kernel oil-treated rats (Zhang et al., 2011).

The essential oil acts also as a potent *in vitro* antiproliferative agent for cultured human epidermal keratinocytes—through both death receptor and mitochondrial pathways and can represent an interesting aid in the treatment of psoriasis, one of the most common human diseases of the skin (Lee et al., 2014).

Bitter apricot ethanolic extract could mediate apoptosis induction in human pancreatic cancer through a mitochondria dependent pathway. These findings suggest that bitter apricot ethanolic extract functions as a potent proapoptotic factor for human pancreatic cancer cells without a significant effect on normal epithelial cells (Aamazadeh et al., 2020).

The findings of Chang et al. (2005) indicated that the seed extract of *Prunus armeniaca* was able to suppress synthesis of prostaglandin E2 and nitric oxide production via inhibiting the lipopolysaccharide -stimulated enhancement of cyclooxygenase-2 and inducible nitric oxide synthase mRNA expression in mouse BV2 microglial cells (Chang et al., 2005). Besides, reduction of c-Fos and inflammatory cytokines such as tumor necrosis factor α (TNF-α) and interleukin-1 β (IL-1β) was also observed after amygdalin administration (Hwang et al., 2008).

3.1.5 Effects on female reproductive processes

There is an evidence of apricot seed effect on female reproductive functions in animals (Halenar et al. 2014, 2015, 2016, 2016b, 2017; Kolesar et al., 2015; Dogru et al., 2017; Kopčeková et al., 2021). Previous *in vitro* and *in vivo* studies by our group also examined the effects of amygdalin as bioactive component of bitter apricot seeds on female reproductive functions.

3.1.5.1 Effect on reproductive hormones

Endocrine regulators were evaluated after intramuscular application of amygdalin and oral application of apricot seeds in rabbits focused on ovarian functions (Halenar et al., 2017; Michalcová et al., 2016). Findings indicate that selected doses of apricot seeds (60, 300, 420 mg/kg of body weight) did not affect the plasma levels of luteinizing hormone (LH) and prolactin (PRL).

88 Contaminants and Plants Action on Female Reproduction

However, plasma level of follicle stimulating hormone (FSH) was decreased by apricot seeds at the highest dose 420 mg/kg of body weight. These results indicate that the natural substances present in apricot seeds may affect ovarian gonadotropin release and the gonadotropin-dependent reproductive processes (Michalcová et al., 2016). On the other hand, intramuscular application of amygdalin and oral application of apricot seeds did not affect the plasma steroid levels of progesterone, 17β-estradiol, and testosterone in experimental rabbit model (Halenar et al., 2017). Short-term intake of amygdalin at the recommended doses did not present risk for animal (Halenar et al., 2017). Moreover, a daily consumption of apricot seeds may down-regulates release of some reproductive hormones in women of reproductive age: their plasma level of prolactin and testosterone increased and 17β-estradiol decreased after apricot seeds consumption (Kopčeková et al., 2021).

3.1.5.2 Effect on ovarian follicular cell functions

In vitro studies demonstrated the dose-dependent modulatory amygdalin on secretion activity of ovarian cells and process of steroidogenesis (Halenar et al., 2014, 2015a,b, 2016b). *In vitro* amygdalin application caused a dose-dependent stimulation of 17β-estradiol but not of progesterone release by porcine ovarian granulosa cells (Halenar et al., 2015a,b). Amygdalin application positively affected ovarian cell viability and stimulated testosterone release by porcine ovarian granulosa cells (Kolesar et al., 2015; Halenár et al., 2016). Amygdalin induces apoptosis in human cervical cancer cell line (Chen et al., 2013).

Taken together, the aforementioned findings indicate the influence of apricot seeds on physiological processes including female reproduction—ovarian cell viability and release of reproductive hormones which can be involved in control of ovarian folliculogenesis and fecundity. Such action however requires validation by further experiments.

3.1.6 Mechanisms of action on female reproductive processes

There is not an evidence of mechanism of apricot seed action on female reproductive processes. There is an evidence of the evaluation of **endocrine regulators** after oral application of apricot seeds using young (Michalcová et al., 2016) and adult female rabbits (Halenar et al., 2017) as a biological model and human (Kopčeková et al., 2021). Findings indicate that selected doses of apricot seeds did not affect the plasma levels of LH and PRL in young animals. On the other hand, plasma level of FSH was decreased by apricot seeds. These results indicate that the natural substances present in apricot seeds may be involved in control of reproductive processes of young rabbits via changes in some reproductive hormones (Michalcová et al., 2016).

Furthermore, plasma levels of steroid (progesterone, 17β-estradiol, testosterone), thyroid (triiodothyronine—T3, thyroxine—T4, thyroid-stimulating hormone—TSH), as well as anterior pituitary (PRL, LH) hormones in adult female rabbits were not changed by apricot seed treatment (Halenár et al., 2017).

Mechanisms of amygdalin action as compound of apricot seeds on female reproductive processes is described in Chapter 4.1.6. Amygdalin stimulates **apoptotic process** by upregulating expression of Bax (proapoptotic protein) and caspase-3 and downregulating expression of B-cell lymphoma 2 (Bcl-2) (antiapoptotic protein). It also promotes arrest of **cell cycle** in G0/G1 phase and decrease number of cells entering S and G2/M phases. Thus, it is proposed to enhance deceleration of cell cycle by blocking cell proliferation and growth (Saleem et al., 2018). Similarly, previous studies suggest that amygdalin regulates apoptosis-related proteins and signaling molecules and increased the expression level of proapoptotic protein Bax and decreased that of antiapoptotic Bcl-2. (Lee and Moon, 2016). β-Glucosidase (β-Glu) can accelerate the hydrolysis of amygdalin into hydrogen cyanide, which can effectively kill tumor cells by inhibiting cytochrome C oxidase in mitochondria, resulting in a significant increase in the cell mortality rate (Blaheta et al., 2016).

The effects of the apricot seeds and amygdalin on healthy female reproductive functions and ovarian cancer cells suggest several extra- and intracellular mechanisms of apricot substances: through inhibition of release of some peptide and steroid hormones, blockage of the cell cycle and induction of apoptosis through increased expression of the proapoptotic protein Bax, and decreased expression of antiapoptotic Bcl-2.

3.1.7 Application in reproductive biology and medicine

The ability of both apricot seed and amygdalin to suppress reproductive functions of healthy rabbits indicate that their consumption could be harmful for reproductive processes. Amygdalin as a therapeutic agent is not used in more countries owing to insufficient clinical verification of its therapeutic efficacy and adverse side-effects. Therefore, the character of apricot seeds effect remains controversial. To protect reproductive health, it is very important to study effect of natural substances in apricot seeds and to know the right harmless dose. Nevertheless, their application for suppression of these processes in some conditions (for example, for synchronization of the cell cycles or for prevention of ovarian hyper stimulation) cannot be excluded.

The down-regulation of reproductive hormones secretion, ovarian cell proliferation and promotion of their apoptosis induced by apricot seeds indicate that they could be promising for treatment of ovarian cancer. However, future studies are required to identify the biological activity of different apricot seed constituents, to understand the mechanisms by which apricot seeds affect reproductive system, their efficient, but safe doses, and character of their action on animal and human organism, reproductive processes and cancerogenesis.

3.1.8 Conclusions and possible direction of future studies

Due to the significant amounts of nutrients such as oils, proteins, amino acids, carbohydrates, soluble sugars, minerals, vitamins, and fibers in the apricot, apricot seed exhibits a wide range of biological activity including action on reproductive system.

The available data suggest that apricot seed and it constituents could downregulate some female reproductive functions (suppress the release of reproductive hormones, cell proliferation, and viability of ovarian cells and to promote their apoptosis).

Despite some contradictions in the available data about benefit and toxic effects of apricot seeds, they indicate applicability of amygdalin in low doses as a promising tool for regulation of various reproductive and nonreproductive processes and treatment of their disorders, primarily in cancer in phytotherapy, animal production, medicine, and biotechnology. Moreover, further research is required to understand the apricot seed constituents responsible for particular effects, the character and mechanisms of their functions and their applicability in various animal species and humans.

References

Aamazadeh F, Ostadrahimi A, Saadat YA, Barar J. Bitter apricot ethanolic extract induces apoptosis through increasing expression of Bax/Bcl-2 ratio and caspase-3 in PANC-1 pancreatic cancer cells. Mol Biol Rep 2020;47(3):1895−904. https://doi.org/10.1007/s11033-020-05286-w.

Alpaslan M, Hayta M. Apricot kernel: physical and chemical properties. J Am Oil Chem Soc 2006;83(5):469−71.

Ayaz Z, Zainab B, Khan S, Abbasi AM, Elshikh MS, Munir A, Al-Ghamdi AA, Alajmi AH, Alsubaie QD, Mustafa AEMA. In silico authentication of amygdalin as a potent anticancer compound in the bitter kernels of family Rosaceae. Saudi J Biol Sci 2020;27(9):2444−51. https://doi.org/10.1016/j.sjbs.2020.06.041.

Badr JM, Tawfik MK. Analytical and pharmacological investigation of amygdalin in *Prunus armeniaca* L. kernels. J Pharm Res 2010;3(9). ISSN: 0974-6943:2134−7.

Bolarinwa IF, Orfila C, Morgan MRA. Amygdalin content of seeds, kernels and food products commercially-available in the UK. Food Chem 2014;152:133−9. https://doi.org/10.1016/j.foodchem.2013.11.002.

Bolarinwa IF, Orfila C, Morgan MR. Determination of amygdalin in apple seeds, fresh apples and processed apple juices. Food Chem 2015;170:437−42. https://doi.org/10.1016/j.foodchem.2014.08.083.

Cassiem W, de Kock M. The anti-proliferative effect of apricot and peach kernel extracts on human colon cancer cells in vitro. BMC Compl Alternative Med 2019;19(1):32. https://doi.org/10.1186/s12906-019-2437-4.

Chang HK, Yang HY, Lee TH, Shin MC, Lee MH, Shin MS, Kim CJ, Kim OJ, Hong SP, Cho S. Armeniacae semen extract suppresses lipopolysaccharide-induced expressions of cycloosygenase-2 and inducible nitric oxide synthase in mouse BV2 microglial cells. Biol Pharm Bull 2005;28(3):449—54. https://doi.org/10.1248/bpb.28.449.

Chen Y, Ma J, Wang F, Hu J, Cui A, Wei C, Yang Q, Li F. Amygdalin induces apoptosis in human cervical cancer cell line HeLa cells. Immunopharmacol Immunotoxicol 2013;35(1):43—51. https://doi.org/10.3109/08923973.2012.738688.

Chen Y, Al-Ghamdi AA, Elshikh MS, Shah MH, Al-Dosary MA, Abbasi AM. Phytochemical profiling, antioxidant and HepG2 cancer cells' antiproliferation potential in the kernels of apricot cultivars. Saudi J Biol Sci 2020;27(1):163—72. https://doi.org/10.1016/j.sjbs.2019.06.013.

Dhen N, Rejeb IB, Boukhris H, Gargouri M. Physicochemical and sensory properties of wheat-Apricot kernels composite bread. LWT Food Sci Technol 2018;95:262—7. https://doi.org/10.1016/j.lwt.2018.04.068.

Dogru YH, Kunt Isguder C, Arici A, Zeki Ozsoy A, Bahri Delibas I, Cakmak B. Effect of amygdalin on the treatment and recurrence of endometriosis in an experimental rat study. Period Biol 2017;119:173—80. https://doi.org/10.18054/pb.v119i3.4767.

Do JS, Hwang JK, Seo HJ, Woo WH, Nam SY. Antiasthmatic activity and selective inhibition of type 2 helper T cell response by aqueous extract of semen armeniacae amarum. Immunopharmacol Immunotoxicol 2006;28:213—25. https://doi.org/10.1080/08923970600815253.

Duracka M, Tvrda E, Halenar M, Zbynovska K, Kolesar E, Lukac N, Kolesarova A. The impact of amygdalin on the oxidative profile of rabbit testicular tissue. Proc Int Confer MendelNet 2016;23:770—5.

Durmaz G, Alpaslan M. Antioxidant properties of roasted apricot (*Prunus armeniaca* L.) kernel. Food Chem 2007;100:1177—81.

Geng H, Yu X, Lu A, Cao H, Zhou B, Zhou L, Zhao Z. Extraction, chemical composition, and antifungal activity of essential oil of bitter almond. Int J Mol Sci 2016;17(9):1421—35. https://doi.org/10.3390/ijms17091421.

Gezer I, Haciseferogullari H, Ozcan MM, Arslan D, Asma BM, Unver A. Physico-chemical properties of apricot (*Prunus armeniaca* L.) kernels. South-West J Hortic Biol Environ 2011;2(1):1—13.

Gomaa EZ. In vitro antioxidant, antimicrobial, and antitumor activities of bitter almond and sweet apricot (*Prunus armeniaca* L.) kernels. Food Sci Biotechnol 2013;22(2):455—63. https://doi.org/10.1007/s10068-013-0101-1.

Halenár M, Medveďová M, Maruniaková N, Packová D, Kolesárová A. Dose-response of porcine ovarian granulosa cells to amygdalin treatment combined with deoxynivalenol. J Microbiol Biotechnol Food Sci 2014;3(2):77—9.

Halenar M, Medvedova M, Maruniakova N, Kolesarova A. Assessment of a potential preventive ability of amygdalin in mycotoxin-induced ovarian toxicity. J Environ Sci Health 2015a;50:411—6. https://doi.org/10.1080/03601234.2015.1011956.

92 Contaminants and Plants Action on Female Reproduction

Halenar M, Medveďová M, Maruniaková N, Kolesárová A. Ovarian hormone production affected by amygdalin addition in vitro. J Microbiol Biotechnol Food Sci 2015b;4(2):19−22. https://doi.org/10.15414/jmbfs.2015.4.special2.19-22.

Halenar M, Tvrda E, Slanina T, Ondruska L, Kolesar E, Massanyi P, Kolesarova A. In vitro effects of amygdalin in the functional competence of rabbit spermatozoa. Int J Biol Biomol Agric Food Biotechnol Eng 2016a;10:664−8.

Halenar M, Tušimová E, Nynca A, Sadowska A, Ciereszko R, Kolesárová A. Stimulatory effect of amygdalin on the viability and steroid hormone secretion by porcine ovarian granulosa cells in vitro. J Microbiol Biotechnol Food Sci 2016b;5(1):44−6. https://doi.org/10.15414/jmbfs.2016.5.special1.44-46.

Halenar M, Chrastinova M, Ondruska L, Jurcik R, Zbynovska K, Tusimova E, Kovacik A, Kolesarova A. The evaluation of endocrine regulators after intramuscular and oral application of cyanogenic glycoside amygdalin in rabbits. Biologia 2017;72:448−74. https://doi.org/10.1515/biolog-2017-0044.

Hu B, Wang SS, Du Q. Traditional Chinese medicine for prevention and treatment of hepatocarcinoma: from bench to bedside. World J Hepatol 2015;7(9):1209−32. https://doi.org/10.4254/wjh.v7.i9.1209.

Hwang HJ, Kim P, Kim CJ, Lee HJ, Shim I, Yin CS, Yang Y, Hahm DH. Antinociceptive effect of amygdalin isolated from *Prunus armeniaca* on formalin-induced pain in rats. Biol Pharm Bull 2008;31(8):1559−64. https://doi.org/10.1248/bpb.31.1559.

Jiagang D, Li C, Wang H, Hao E, Du Z, Bao C, Lv J, Wang Y. Amygdalin mediates relieved atherosclerosis in apolipoprotein E deficient mice through the induction of regulatory T cells. Biochem Biophys Res Commun 2011;411(3):523-529. https://doi.org/10.1016/j.bbrc.2011.06.162.

Kalia S, Bharti VK, Giri A, Kumar B. Effect of *Prunus armeniaca* seed extraction health, survivability, antioxidant, blood biochemical and immune status of broiler chickens at high altitude cold desert. J Adv Res 2017;8(6):677−86. https://doi.org/10.1016/j.jare.2017.08.005.

Karaboğa I, Ovalı MA, Yılmaz A, Alpaslan M. Gastroprotective effect of apricot kernel oil in ethanol-induced gastric mucosal injury in rats. Biotech Histochem 2018;93(8):601−7. https://doi.org/10.1080/10520295.2018.1511064.

Kolesar E, Halenar M, Kolesarova A, Massanyi P. Natural plant toxicant − cyanogenic glycoside amygdalin: characteristic, metabolism and the effect on animal reproduction. J Microbiol Biotechnol Food Sci 2015;4:49−50. https://doi.org/10.15414/jmbfs.2015.4.special2.49-50.

Kolesar E, Tvrda E, Halenar M, Schneidgenova M, Chrastinova L, Ondruska L, Jurcik R, Kovacik A, Kovacikova E, Massanyi P, Kolesarova A. Assessment of rabbit spermatozoa characteristics after amygdalin and apricot seeds exposure in vivo. Toxicol Rep 2018;5:679−86. https://doi.org/10.1016/j.toxrep.2018.05.015.

Kolesárová A, Pivko J, Halenár M, Zbyňovská K, Chrastinová Ľ, Ondruška Ľ, Jurčík R, Kopčeková J, Valuch J, Kolesárová A. Effect of apricot seeds on renal structure of rabbits. Potravinarstvo Slovak J Food Sci 2017;11(1):309−14. https://doi.org/10.5219/751.

Kolesárová A, Džurňáková V, Michalcová K, Baldovská S, Chrastinová Ľ, Ondruška Ľ, Jurčík R, Tokárová K, Kováčiková E, Kováčik A, et al. The effect of apricot seeds on microscopic structure of rabbit liver. J Microb Biotech Food Sci 2020;10:321−4. https://doi.org/10.15414/jmbfs.2020.10.2.321-324.

Kopcekova J, Kolesarova A, Kovacik A, Kovacikova E, Gazarova M, Chlebo P, Valuch J, Kolesarova A. Influence of long-term consumption of bitter apricot seeds on risk factors for cardiovascular diseases. J Environ Sci Health 2018;53:298−303. https://doi.org/10.1080/03601234.2017.1421841.

Kopčeková J, Kováčiková E, Kováčik A, Kolesárová A, Mrázová J, Chlebo P, Kolesárová A. Consumption of bitter apricot seeds affects lipid and endocrine profile in women. J Environ Sci Health B 2021;56(4):378−86. https://doi.org/10.1080/03601234.2021.1890513.

Korekar G, Stobdan T, Arora R, Yadav A, Singh S. Antioxidant capacity and phenolics content of apricot (*Prunus armeniaca* L.) kernel as a function of genotype. Plant Foods Hum Nutr 2011;66(4):376−83. https://doi.org/10.1007/s11130-011-0246-0.

Kovacikova E, Kovacik A, Halenar M, Tokarova K, Chrastinova L, Ondruska L, Jurcik R, Kolesar E, Valuch J, Kolesarova A. Potential toxicity of cyanogenic glycoside amygdalin and bitter apricot seed in rabbits—health status evaluation. J Anim Physiol Anim Nutr 2019;103:695−703. https://doi.org/10.1111/jpn.13055.

Kovacova V, Sarocka A, Omelka R, Bauerova M, Grosskopf B, Formicki G, Kolesarova A, Martiniakova M. Subacute exposure to amygdalin influences compact bone remodeling of rabbits. Physiol Pharmacol 2019;70(4). https://doi.org/10.26402/jpp.2019.4.15.

Lee J, Zhang G, Wood E, Castillo C, Mitchell AE. Quantification of amygdalin in nonbitter, semibitter, and bitter almonds (*Prunus dulcis*) by UHPLC-(ESI)QqQ MS/MS. J Agric Food Chem 2013;61(32):7754−9. https://doi.org/10.1021/jf402295u.

Lee HH, Ahn JH, Kwon AR, Lee ES, Kwak JH, Min YH. Chemical composition and antimicrobial activity of the essential oil of apricot seed. Phytother Res 2014;28(12):1867−72. https://doi.org/10.1002/ptr.5219.

Lee HM, Moon A. Amygdalin regulates apoptosis and adhesion in Hs578T triple-negative breast cancer cells. Biomol Ther (Seoul) 2016;24(1):62−6. https://doi.org/10.4062/biomolther.2015.172.

Li K, Yang W, Li Z, Li J, Zhang P, Xiao T. Bitter apricot essential oil induces apoptosis of human HaCaT keratinocytes. Int Immunopharm 2016;34:189−98. https://doi.org/10.1016/j.intimp.2016.02.019.

Lim T. *Prunus armeniaca*, edible medicinal and non-medicinal plants, vol. 4. New York: Springer; 2012. p. 442−50.

Lv J, Xiong W, Lei T, Wang H, Sun M, Hao E, Wang Y. Amygdalin ameliorates the progression of atherosclerosis in LDL receptor-deficient mice. Mol Med Rep 2017;16(6):8171−9. https://doi.org/10.3892/mmr.2017.7609.

Matthaus B, Özcan MM, Al Juhaimi F. Fatty acid composition and tocopherol content of the kernel oil from apricot varieties (Hasanbey, Hacihaliloglu, Kabaasi and Soganci) collected at different harvest times. Eur Food Res Technol 2016;242(2):221−6. https://doi.org/10.1007/s00217-015-2533-8.

Minaiyan M, Ghannadi A, Asadi M, Etemad M, Mahzouni P. Anti-inflammatory effect of *Prunus armeniaca* L. (Apricot) extracts ameliorates TNBS-induced ulcerative colitis in rats. Res Pharm Sci 2014;9(4):225−31.

Michalcová K, Halenár M, Tusimová E, Kováčik A, Chrastinová L, Ondruska L, Jurčík R, Kolesár E, Kolesárová A. Influence of apricot kernels on blood plasma levels of selected anterior pituitary hormones in male and female rabbits in vivo. Anim Sci Biotechnol 2016;49:109−14.

Ozturk F, Gul M, Ates B, Ozturk IC, Cetin A, Vardi N, Otlu A, Yilmaz I. Protective effect of apricot (*Prunus armeniaca* L.) on hepatic steatosis and damage induced by carbon tetrachloride in wistar rats. Br J Nutr 2009;102(12):1767−75. https://doi.org/10.1017/S0007114509991322.

Qin F, Yaoa L, Lua C, Lia C, Zhoua Y, Sua C, Chena B, Shena Y. Phenolic composition, antioxidant and antibacterial properties, and in vitro anti-HepG2 cell activities of wild apricot

94 Contaminants and Plants Action on Female Reproduction

(*Armeniaca sibirica* L. Lam) kernel skins. Food Chem Toxicol 2019;129:354−64. https://doi.org/10.1016/j.fct.2019.05.007.

Rai I, Bachheti RK, Saini CK. A review on phytochemical, biological screening and importance of Wild Apricot (*Prunus armeniaca* L.). Orient Pharm Exp Med 2016;16:1−15. https://doi.org/10.1007/s13596-015-0215-5.

Ramzan S, Soelberg J, Jäger AK, Cantarero-Arévalo L. Traditional medicine among people of Pakistani descent in the capital region of Copenhagen. J Ethnopharmacol 2017;196:267−80. https://doi.org/10.1016/j.jep.2016.11.048.

Rudzińska M, Górnaś P, Raczyk M, Soliven A. Sterols and squalene in apricot (*Prunus armeniaca* L.) kernel oils: the variety as a key factor. Nat Prod Res 2017;31(1):84−8. https://doi.org/10.1080/14786419.2015.1135146.

Ruiz D, Egea J, Gil MI, Barberan FAT. Characterization and quantitation of phenolic compounds in new apricot (*Prunus armeniaca* L.) varieties. J Agric Food Chem 2005;53(24):9544−52. https://doi.org/10.1021/jf051539p.

Saleem M, Asif J, Asif M, Saleem U. Amygdalin, from apricot kernels, induces apoptosis and causes cell cycle arrest in cancer cells: an updated review. Anticancer Agents Med Chem 2018;18(12):1650−5. https://doi.org/10.2174/1871520618666180105161136.

Shen HS, Wen SH. Effect of early use of Chinese herbal products on mortality rate in patients with lung cancer. J Ethnopharmacol 2018;211:1−8. https://doi.org/10.1016/j.jep.2017.09.025.

Stryjecka M, Kiełtyka-Dadasiewicz A, Michalak M, Rachoń L, Głowacka A. Chemical composition and antioxidant properties of oils from the seeds of five apricot (*Prunus armeniaca* L.) cultivars. J Oleo Sci 2019;68(8):729−38. https://doi.org/10.5650/jos.ess19121.

Tanwar B, Modgil R, Goyal A. Antinutritional factors and hypocholesterolemic effect of wild apricot kernel (*Prunus armeniaca* L.) as affected by detoxification. Food Funct 2018;9(4):2121−35. https://doi.org/10.1039/c8fo00044a.

Tanwar B, Modgil R, Goyal A. Effect of detoxification on biological quality of wild apricot (*Prunus armeniaca* L.) kernel. J Sci Food Agric 2019;99(2):517−28. https://doi.org/10.1002/jsfa.9209.

Thodberg S, Del Cueto J, Mazzeo R, Pavan S, Lotti C, Dicenta F, Jakobsen Neilson EH, Møller BL, Sánchez-Pérez R. Elucidation of the amygdalin pathway reveals the metabolic basis of bitter and sweet almonds (*Prunus dulcis*). Plant Physiol 2018;178(3):1096−111. https://doi.org/10.1104/pp.18.00922.

Turan S, Topcu A, Karabulut I, Vural H, Hayaloglu AA. Fatty acid, triacylglycerol, phytosterol, and tocopherol variations in kernel oil of Malatya apricots from Turkey. J Agric Food Chem 2007;55(26):10787−94. https://doi.org/10.1021/jf071801p.

Tušimová E, Zbyňovská K, Kováčik A, Michalcová K, Halenár M, Kolesárová A, Kopčeková J, Valuch J, Kolesárová A. Human urine alterations caused by apricot seeds consumption. Adv Res Life Sci 2017;1(1):68−74. https://doi.org/10.1515/arls-2017-0012.

Yamshanov VA, Kovan'ko EG, Pustovalov YI. Effects of amygdaline from apricot kernel on transplanted tumors in mice. Bull Exp Biol Med 2016;160(5):712−4. https://doi.org/10.1007/s10517-016-3257-x.

Yang C, Li X, Rong J. Amygdalin isolated from Semen Persicae (Tao Ren) extracts induces the expression of follistatin in HepG2 and C2C12 cell lines. Chin Med 2014;9:28. https://doi.org/10.1186/1749-8546-9-23.

Yiğit D, Yiğit N, Mavi A. Antioxidant and antimicrobial activities of bitter and sweet apricot (*Prunus armeniaca* L.) kernels. Braz J Med Biol Res 2009;42:346−52. https://doi.org/10.1590/S0100-879X2009000400006.

Zhang J, Gu HD, Zhang L, Tian ZJ, Zhang ZQ, Shi XC, Ma WH. Protective effects of apricot kernel oil on myocardium against ischemia-reperfusion injury in rats. Food Chem Toxicol 2011;49:3136−41. https://doi.org/10.1016/j.fct.2011.08.015.

Zhang H, Xue J, Zhao H, Zhao X, Xue H, Sun Y, Xue W. Isolation and structural characterization of antioxidant peptides from degreased apricot seed kernels. J AOAC Int 2018;101(5):1661−3. https://doi.org/10.5740/jaoacint.17-0465.

Zhou B, Wang Y, Kang J, Zhong J, Prenzler PD. The quality and volatile-profile changes of Longwangmo apricot (*Prunus armeniaca* L.) kernel oil prepared by different oil-producing processes. Eur J Lipid Sci Technol 2016;118:236−43. https://doi.org/10.1002/ejlt.201400545.

Zibaeenezhad MJ, Shahamat M, Mosavat SH, Attar A, Bahramali E. Effect of Amygdalus scoparia kernel oil consumption on lipid profile of the patients with dyslipidemia: a randomized, open-label controlled clinical trial. Oncotarget 2017;8(45):79636−41. https://doi.org/10.18632/oncotarget.18956.

96 Contaminants and Plants Action on Female Reproduction

Chapter 3.2

Black elder (*Sambucus nigra* L.)

3.2.1 Introduction

Black elderberry (*Sambucus nigra* L., black elder) is an extremely accessible and abundant plant native to the northern hemisphere. Their seeds are spread rapidly by birds and other animals to colonize forest edges and disturbed areas, leading to being nowadays diffused in different habitat as subtropical regions of Asia, North Africa and North America (Finn et al., 2008; Senica et al., 2016; Młynarczyk et al., 2018). Since the beginning of 1980s the black elder has been planted and commercialized in some countries in Europe, the USA, Canada, New Zealand, and Chile (Finn et al., 2008). In Europe, elderberries are intensively used for centuries both in food industry to produce pies, jellies, jams, ice creams, yogurts, and different alcoholic beverages (Senica et al., 2016), and in folk medicine, used in the treatment of many diseases and ailments thanks for their antioxidant, anticarcinogenic, immune-stimulating, antiallergic, antiviral, and antibacterial properties (Oniszczuk et al., 2016). These aspects of elderberry species recently have received significant attention, especially for antioxidant capacity, as function compounds in food application such as natural conservations or food supplements (Marisa Ribeiro et al., 2019). Due to its health-promoting and sensory properties, elderberry is used primarily in food and pharmaceutical industry (Młynarczyk et al., 2018). In this last years, there is an increasing interest for this wild growing plant species, due to their simple cultivation, high availability and rich in bioactive compounds (Lorenzo et al., 2018; Milena et al., 2019). Additionally to health implications, there is a growing interest in the use of new natural additives in food industry (Domínguez et al., 2020; Lorenzo et al., 2018; Munekata et al., 2017). To this regard, replacing synthetic additives by natural bioactive compounds extracted from plants was an important strategy for food manufacturers (Lorenzo et al., 2018; Munekata et al., 2017).

3.2.2 Provenance and properties

Black elderberry (*Sambucus nigra* L.) belongs to the *Adoxaceae* family and its common names are elder, elderberry, black elder, European elder, European elderberry and European black elderberry (ITIS, 2016). *Sambucus nigra* is a small tree or shrub, 1−8 m tall having a strong odor. The bark is brownish in color, with longitudinal fractures and deep grooves. The leaves are opposite, imparipennate, with 5−7 elliptic−lanceolate, dentate leaflets.

The inflorescence is an umbel with many milky-white flowers. The fruit is a shiny black−purple, subspheric drupe (Pignatti, 1982). The plant is found in woods, clearings, and hedges from sea level to mountainous elevations (1400 m a.s.l.). All parts of this plant (flower, bark, leaf, and fruits) are the rich sources of dietary phytochemicals, such as carbohydrates, lipids, terpenoids, flavonoids, phenolic acids, alkaloids, and so forth (Finn et al., 2008; Agalar, 2019).

Sambucus nigra L. is widely distributed throughout Europe, and its berries are an exceptional source of compounds of great nutritional and technological value. From a nutritional point of view, the elderberries contain a high content of water, total soluble solids and protein. The fat level is 0.35% (of fresh berry), although this fat has an exceptional fatty acid profile. The content of the essential fatty acids (linoleic and α-linolenic) are very high (approximately 39% each) and the polyunsaturated fatty acids represent 78% of the total fatty acids. The content of phenolic compounds, anthocyanins and the antioxidant capacity of the extracts obtained reveal that these extracts contain a large number of bioactive compounds, which gives them a high antioxidant capacity. Regarding phenolic compounds, elderberries were abundant in flavonoids (rutin and quercetin) and phenolic acids (gallic acid and gentisic acid) (Domínguez et a., 2020). The most prominent compounds present in *Sambucus nigra* fruits are anthocyanins. Additionally, other types of polyphenols like flavonol glycosides and flavonol esters are present in elderberries. Other known metabolites include lectins, cyanogenic glycosides, essential oil, fatty acids, organic acids, carbohydrates, vitamins, and minerals (Vlachojannis et al., 2010; Młynarczyk et al., 2018). However, the bark, leaves, seeds, and raw or unripe fruit contain the cyanogenic glycoside sambunigrin, which is potentially toxic (Ulbricht et al., 2014; Buhrmester et al., 2000).

Elderberries are rich in vitamins, organic acids, and a good source of major and trace elements. Vitamin B2 (65 mg), vitamin B6 (0.25 mg), vitamin C (18−26 mg), folic acid (17 mg), biotin (1.8 mg), β-carotene (0.36 mg), pantothenic acid (0.18 mg), nicotinamide (1.48 mg), potassium (288−305 mg), phosphor (49−57 mg), pectin (0.16%), and glucose and fructose (7.5%) are all contained in 100 g of fresh berries (Diviš et al., 2015). Elderberries are rich in anthocyanins, the majority of which includes cyanidin-3-glucoside and cyanidin-3-sambubioside, which are found in elderberry juice and polar extracts (Vlachojannis et al., 2015; Wu et al., 2004). In addition, elderberries are a rich source of flavanols, phenolic acids, and procyanidins. They also contain terpenes and lectins, free and conjugated forms of amino acids, proteins, fatty acids, and fiber (Glensk et al., 2014; Krüger et al., 2015; Salvador et al., 2015; Tejero et al., 2015). The characteristic aroma of elderberries is a result of (E)-β damascenone, dihydroedulan, ethyl-9-decenoate, 2-phenyl ethanol, phenylacetaldehyde, and nonanal. Alcohols, esters, and aldehydes are frequently identified volatile groups in elderberries. Elderflowers are particularly rich in flavonoids (up to 3%), such as kaempferol, astragalin, quercetin, quercetin-3-O-glucoside, rutin, isoquercitrin, and hyperoside (Krauze-Baranowska et al., 2009). Other major secondary metabolites comprise approximately 1% triterpenes (as α- and β-amyrin, ursolic acid, and oleanolic acid) and about 1% sterols (β-sitosterol, campesterol, and stigmasterol). In addition, pectins, tannins, and phenolic acids are found in the flowers (Ho et al., 2016, 2017). Elderflowers have a strong, flowery, pleasant odor, and an essential oil yield from the flowers of 0.03%–0.14%. The aroma composition of elderflowers includes aldehydes, ketones, alcohols, esters, oxides, terpenes, and free fatty acids (Jørgensen et al., 2000).

3.2.3 Physiological and therapeutic actions

Elderberry is widely used in domestic medicine through its pharmacological properties (Torabian et al., 2018) and currently present one of the most-used medicinal plants worldwide (Porter and Bode, 2017). In folk medicine it is used in the treatment of many diseases and ailments thanks for their antioxidant, anticarcinogenic, immune-stimulating, antiallergic, antiviral, antibacterial (Młynarczyk et al., 2018; Oniszczuk et al., 2016), antidepressant, hypoglycemic properties, and to reduce body fat and lipid concentration (Młynarczyk et al., 2018). What is more, scientific evidence implied that elderberry flowers can be used both in the prevention and therapy of various diseases, due to the expressed antiinflammatory (Olejnik et al., 2015), antioxidant (Dawidowicz et al., 2006), antimicrobial (Arjoon et al., 2012) and anticancer activity (Chen et al., 2013; Araújo et al., 2013; Schröder et al., 2016; Pereira et al., 2020). Elderberry products have been marked as antioxidant (Denev et al., 2010; Sun-Waterhouse et al., 2013; Topolska et al., 2015), antiviral (Roschek et al., 2009; Kinoshita et al., 2012; Porter and Bode, 2017), immunomodulatory (Waknine-Grinberg et al., 2009; Frøkiæ et al., 2012), antiinflammatory (Burns et al., 2010; Olejnik et al., 2015), and antimicrobial (Arjoon et al., 2012). There are some reports on the anticonvulsant (Ataee et al., 2016) and antidepressant activities (Mahmoudi et al., 2014) of elderberry extracts.

Elderberry and elderflower constituents may possess inflammatory modulating activity, which increases their nutritional value (Ho et al., 2017). Elderflowers have been used in traditional medicine for treatment of inflammation, skin disorders, diuretic, colds, fever and other respiratory disturbances (Blumenthal et al., 2000; Weiss and Fintelmann, 2000; EMA, 2008). These aspects of elderberry species recently have received significant attention, especially for antioxidant capacity, as function compounds in food application such as natural conservations or food supplements (Marisa Ribeiro et al., 2019). The strong antioxidant capacity of elderberries is related to phenolic compounds content, the most abundant (and more actives) were rutin and quercetin (regarding flavonols) (Iacopini et al., 2008) and gallic acids (regarding phenolic acids). Many studies confirm the strong relationship between the phenolic content of a fruit and its radical scavenging capacity (Bahorun et al., 2004). Isoflavone rutin (the most abundant polyphenol in elderflower) inhibits the viability of human neuroblastoma cells (Chen et al., 2013). Similarly, a concentration-dependent cytotoxic action of rutin was also noted on leukemic cells (Araújo et al., 2013). A cytotoxic effect of rutin was also noted on human estrogenic breast cancer cell line, and osteosarcoma cell lines. However, the viability of these cells was intact (Abdel-Naim et al., 2018). Although most studies have focused on the antiviral properties of *Sambucus nigra* L. particularly owing to the presence of beneficial polyphenolic compounds (Mlynarczyk et al., 2018), the potential use of elderflower extract in breast cancer prevention and/or treatment has previously been advocated (Schröder et al., 2016).

Elderberry extract has the potential to ameliorate skin photoaging and inflammation (Lin et al., 2019). Elderberry extract inhibited the infectious bronchitis virus at an early point during replication (Chen et al., 2014). Consumption of elderberry extract has also been suggested for people with diabetic osteoporosis for improving lipid profile and reducing atherogenic risk and hyperglycemia (Badescu et al., 2012). *In vitro* studies suggested that elderflower extracts stimulate insulin-dependent glucose uptake (Bhattacharya et al., 2013; Christensen et al., 2010; Gray et al., 2000). Elderberries' curative capacity is due to the presence of high amounts of polyphenolic compounds, primarily flavonols, phenolic acids and anthocyanins. Indeed, these compounds are well known free radical scavengers and are able to protect the human body against oxidative stress and lipid peroxidation processes (Duymuş et al., 2014; Moldovan et al., 2016; Silva et al., 2017).

3.2.4 Mechanisms of action

Elderberry extract showed good efficiency on scavenging free radicals and dose-dependently reduced reactive oxygen species (ROS) generation. Elderberry extract notably decreased UVB-induced matrix metalloproteinase-1 (MMP-1) expression and inflammatory cytokine secretion through the inhibition of mitogen-activated protein kinases/activator protein 1 (MAPK/AP-1) and nuclear factor-κB (NF-κB) signaling pathways, blocking extracellular matrix (ECM) degradation and inflammation in UVB-irradiated human skin keratinocytes (HaCaTs). In addition, Elderberry extract improved nuclear factor E2-related factor 2/heme oxygenase-1 (Nrf2/HO-1) signaling to increase oxidative defense capacity, and enhanced transforming growth factor β (TGF-β) signaling activation to promote procollagen type I synthesis, relieving UVB-induced skin cell damage. These results indicated that elderberry extract has the potential to ameliorate UVB-induced skin photoaging and inflammation (Lin et al., 2019).

Elderberry, a rich source of flavonoids, has been shown to exert protective effects against influenza virus by stimulating the immune system of the host through enhancing the production of cytokines (Kinoshita et al., 2012).

Additionally, acid polysaccharides in elderberry such as pectins may contribute to boosting the immune function (Ho et al., 2015; Kinoshita et al., 2012), possibly through macrophage stimulation (Barsett et al., 2012). Direct effects of elderberry extract by blocking viral glycoproteins as well as indirect effects by increased expression of interleukins IL-6, IL-8, and tumor necrosis factor (TNF) were confirmed (Torabian et al., 2019). The action of elderberry is both direct—suppressing viral entry, affecting the postinfection phase, and viral transmission from cell to cell, and indirect—by modulating the release of cytokines such as IL-6, IL-8, and TNF. These data support the use of *Sambucus nigra* berries as nutraceutical ingredients for the management of influenza infection (Torabian et al., 2019).

The effect of elderflower extracts on hormone production and proliferation of breast cancer cells was confirmed. A significant estrogen receptor α (ERα) downregulation and the progesterone receptor (PR) upregulation were observed. In addition, the production of estradiol was upregulated in breast cancer cells in a concentration dependent manner. The downregulating effect of elderflower extracts on ERα expression and the upregulation of the PR expression in breast cancer cells are promising results (Schröder et al., 2016).

Most of the data suggest that the nonreproductive effects of *Sambucus nigra* L. are due to the activation of immune, antiinflammatory processes (cytokine production, etc.) and regulation of hormones and their receptors in cancer cells.

3.2.5 Effects on female reproductive processes

3.2.5.1 Effect on ovarian cell functions

Our recent *in vitro* study demonstrated the promising stimulatory effect of elderberry and elderflower extract on the viability of human ovarian granulosa cells and the secretion of steroid hormones (Baldovska et al., not published data). The results of the screening of polyphenols showed that elderflowers present a rich source of polyphenols, and the most prominent compounds present in both, elderflower and elderberry extract was flavonoid rutin (Baldovska et al., not published data). In this *in vitro* study was indicated an increase of the number of viable ovarian cells after the addition of the elderflower and elderberry extract. Moreover, the secretion of progesterone and 17β-estradiol by the cells was increased after supplementation by either elderflower extract or elderberry extract. The results of tis study demonstrates the benefits of elderberry extracts using different parts (flower and fruit) in up-regulation of biosynthesis of ovarian steroid hormones *in vitro*. The data suggest that elderberry may be a source of phytoestrogens (Kolesarova et al., not published data). The most abundant polyphenol in elderflower is isoflavone rutin (Chen et al., 2013). In healthy cultured porcine ovarian granulosa cells, rutin administration was able to reduce the 17β-estradiol output, without affecting progesterone release (Sirotkin et al., 2021). Rutin administration reduced the viability of porcine ovarian granulosa cells (Sirotkin et al., 2021). Previous study of Chowdhury et al. (2017) suggested that *Sambucus nigra* agglutinin can reduce viability and mitochondrial activity and induce apoptosis of cultured ovarian cancer cells.

The present observations suggest the stimulatory influence of black elder substances on viability and steroidogenesis of healthy ovarian cells and its ability to promote apoptosis and reduce viability of cultured ovarian cancer cells.

3.2.5.2 Effect on embryo

Schröder et al. (2016) studied the effect of elderflower extracts on production of estradiol and ERα and on proliferation of trophoblast tumor cell lines JEG-3 and BeWo. Elderflower extracts inhibited estradiol production in JEG-3 cells and prompted in BeWo cells. Furthermore, it upregulated accumulation of ERα in JEG-3 cell lines. Elderflower extracts contained a substantial amount of lignans, which could be responsible for its effect. No significant influence of elderflower extracts on trophoblast cell proliferation was noted.

This observation demonstrate that black elder extract can affect embryonal trophoblast and its estradiol-estradiol receptor system, which could be responsible for embryo development and maintenance gravidity. Furthermore, it indicates black elder influence on trophoblast cancerogenesis.

3.2.6 Mechanisms of action on female reproductive processes

The intracellular mechanisms of *Sambucus nigra* agglutinin on ovarian cancer cells were investigated by Chowdhury et al. (2017). They demonstrated that the suppressive action of this black elder constituent on ovarian cancer cells are mediated by a cascade of events:

(1) *Sambucus nigra* agglutinin activates the signaling pathways of Akt and the extracellular signal-regulated kinase 1/2 (ERK1/2), which promotes dephosphorylation of dynamin-related protein-1 (Drp-1),

(2) Upon its translocation to the mitochondrial fission loci, Drp-1 indice fragmentation of mitochondrial membrane,

(3) Mitochondrial outer membrane permeabilization results generation of ROS and cytochrome-c release into the cytosol—the signs of mitochondrial apoptosis.

(4) These changes may result cell cycle arrest before G2/M phase.

The results of this study demonstrated that *Sambucus nigra* agglutinin can suppress ovarian cancer cell functions via down-regulation of Akt- and ERK1/2-dependent intracellular signaling pathways—promoters of cell proliferation and by induction of cytoplasmic/mitochondrial apoptosis.

The mechanisms mediating stimulatory action black elder extract on steroidogeneis (Baldovska et al., 2021) remain to be detected. Nevertheless, this extract promoted estradiol release, which is considered as promoter of ovarian cell proliferation and viability and ovarian folliculogenesis (Sirotkin, 2014). Therefore, the stimulatory action of black elder extract on healthy ovarian cells may be mediated by estrogen.

Estrogens and estrogen receptors could mediate also black elder molecules on trophoblast tumor cells (Schröder et al., 2016).

3.2.7 Application in reproductive biology and medicine

There is a few of evidence that elderberry can be involved to regulation of female reproductive processes. Elderberry extracts application using different parts (flower and fruit) affected ovarian steroidogenesis. Furthermore, Baldovska et al. (2021) showed that elderberry may be a source of phytoestrogens. As a promoter of steroidogenesis and regulators of steroid hormones receptors, black elderberry could be activator of steroid hormones-dependent reproductive events and related processes. It is not to be excluded that it can support ovarian folliculogenesis, oogenesis, fecundity, embryogenesis, and gravidity to relied on the age-related reproductive insufficiency and menopause-related problems.

On the other hand, the ability of black elder substances to affect ovarian and trophoblast tumor cells indicates potential ability of these substances to affect cancerogenesis, although the first studies failed to find strong evidence for it.

3.2.8 Conclusions and possible direction of future studies

The reproductive effects of black elder is very poorly investigated. Nevertheless, the few performed studies demonstrate the stimulatory influence of black elder substances on viability and steroidogenesis of healthy ovarian cells and its ability to promote apoptosis and reduce viability of cultured ovarian cancer cells. Furthermore, they demonstrate that black elder extract can affect embryonal trophoblast and its estradiol receptor system, which could be responsible for embryo development and maintenance gravidity. Furthermore, it indicates black elder influence on ovarian and trophoblast cancerogenesis. Effects of black elder could be mediated by intracellular signaling pathways regulating cell proliferation and cytoplasmic/mitochondrial apoptosis and viability, as well as by estrogen and estrogen receptors.

The current state-of-art induces more queries concerning black elder action on female reproduction that the answers. The biologically active constituents of *Sambucus nigra* remain to be identified. The action of black elder extract has not been studied in *in vivo* experiments. The performed studies were performed mainly on cancer cells, whist the number of such studies and their results don't enable to generate any definitive conclusion concerning curative value of this plant. Therefore, although black elder influence on female reproductive processes has been demonstrated, its action and application require further studies.

References

Abdel-Naim AB, Alghamdi AA, Algandaby MM, Al-Abbasi FA, Al-Abd AM, Eid BG, Abdallah HM, El-Halawany AM. Rutin isolated from chrozophora tinctoria enhances bone cell proliferation and ossification markers. Oxid Med Cell Longev 2018;2018:5106469. https://doi.org/10.1155/2018/5106469.

Agalar H. Elderberry (*Sambucus nigra* L.). Nonvit Nonmin Nutri Suppl 2019:211—5.

Araújo KC, de M B Costa EM, Pazini F, Valadares MC, de Oliveira V. Bioconversion of quercetin and rutin and the cytotoxicity activities of the transformed products. Food Chem Toxicol 2013;51:93—6. https://doi.org/10.1016/j.fct.2012.09.015.

Arjoon AV, Saylor CV, May M. In vitro efficacy of antimicrobial extracts against the atypical ruminant pathogen Mycoplasma mycoides subsp. capri. BMC Complement Altern 2012;169.

Ataee R, Falahati A, Ebrahimzadeh MA, Shokrzadeh M. Anticonvulsant activities of *Sambucus nigra*. Eur Rev Med Pharmacol Sci 2016;20:3123—6.

Badescu L, Badulescu O, Badescu M, Ciociou M. Mechanism by *Sambucus nigra* extract improves bone mineral density in experimental diabetes. Evid Based Complement Altern Med 2012. https://doi.org/10.1155/2012/848269.

Bahorun T, Luximon-Ramma A, Crozier A, Aruoma OI. Total phenol, flavonoid, proanthocyanidin and vitamin C levels and antioxidant activities of Mauritian vegetables. J Sci Food Agric 2004;84(12):1553—61. https://doi.org/10.1002/jsfa.1820.

Barsett H, Aslaksen TH, Gildhyal P, Michaelsen TE, Paulsen BS. Comparison of carbohydrate structures and immunomodulating properties of extracts from berries and flowers of *Sambucus nigra* L. Eur J Med Plants 2012;2(3):216—29.

Bhattacharya S, Christensen KB, Olsen LC, Christensen LP, Grevsen K, Færgeman NJ, Kristiansen K, Young JF, Oksbjerg N. Bioactive components from flowers of *Sambucus nigra* L. increase glucose uptake in primary porcine myotube cultures and reduce fat accumulation in *Caenorhabditis elegans*. J Agric Food Chem 2013;61(46):11033—40. https://doi.org/10.1021/jf402838a.

Blumenthal M, Goldberg A, Brinckmann J. Herbal medicine: Expanded Commission E Monographs. Austin: American Botanical Council; 2000.

Buhrmester RA, Ebinger1a JE, Seigler DS. Sambunigrin and cyanogenic variability in populations of *Sambucus canadensis* L. (Caprifoliaceae). Biochem Systemat Ecol 2000;28(7):689—95. https://doi.org/10.1016/s0305-1978(99)00105-2.

Burns JJ, Zhao L, Taylor EW, Spelman K. The influence of traditional herbal formulas on cytokine activity. Toxicology 2010;278:140—59.

104 Contaminants and Plants Action on Female Reproduction

Chen H, Miao Q, Geng M, Liu J, Hu Y, Tian L, Pan J, Yang Y. Anti-tumor effect of rutin on human neuroblastoma cell lines through inducing G2/M cell cycle arrest and promoting apoptosis. Sci World J 2013;2013:269165. https://doi.org/10.1155/2013/269165.

Chen C, Zuckerman DM, Brantley S, Sharp M, Childress K, Hoiczyk E, Pendleton FAR. *Sambucus nigra* extracts inhibit infectious bronchitis virus at an early point during replication. BMC Vet Res 2014;10(24). https://doi.org/10.1186/1746-6148-10-24.

Chowdhury S, Ray U, Chatterjee BP, Roy SS. Targeted apoptosis in ovarian cancer cells through mitochondrial dysfunction in response to *Sambucus nigra* agglutinin. Cell Death Dis 2017;8:e2762. https://doi.org/10.1038/cddis.2017.77.

Christensen KB, Petersen RK, Kristiansen K, Christensen LP. Identification of bioactive compounds from flowers of black elder (*Sambucus nigra* L.) that activate the human peroxisome proliferator-activated receptor (PPAR) γ. Phytother Res 2010;24:129−32.

Dawidowicz AL, Wianowska D, Baraniak B. The antioxidant properties of alcoholic extracts from *Sambucus nigra* L. (antioxidant properties of extracts). LWT Food Sci Technol 2006;39:308−15.

Denev P, Ciz M, Ambrozova G, Lojek A, Yanakieva I, Kratchanova M. Solid-phase extraction of berries' anthocyanins and evaluation of their antioxidative properties. Food Chem 2010;123:1055−61.

Diviš P, Pořízka J, Vespalcová M, Matějíček A, Kaplan J. Elemental composition of fruits from different black elder (*Sambucus nigra* L.) cultivars grown in the Czech Republic. J Elem 2015;20(3):549−57.

Domínguez R, Zhang L, Rocchetti G, Lucini L, Pateiro M, Munekata PES, Lorenzo José M. Elderberry (*Sambucus nigra* L.) as potential source of antioxidants. Characterization, optimization of extraction parameters and bioactive properties. Food Chem 2020;330:127266. https://doi.org/10.1016/j.foodchem.2020.127266.

Duymuş HG, Göger F, Başer KHC. In vitro antioxidant properties and anthocyanin compositions of elderberry extracts. Food Chem 2014;155:112−9. https://doi.org/10.1016/j.foodchem.2014.01.028.

EMA. Assessment report for the development of community monographs and for inclusion of herbal substance(s). Preparation(s) or combinations thereof in the list: *Sambucus nigra* L, London flows. 2008. EMEA/HMPC/283170/2007.

Finn CE, Thomas AL, Byers PL, Serçe S. Evaluation of American (*Sambucus canadensis*) and European (*S. nigra*) elderberry genotypes grown in diverse environments and implications for cultivar development. Hortscience 2008;43(5):1385−91.

Frøkiær H, Henningsen L, Metzdorff SB, Weiss G, Roller M, Flanagan J, Fromentin E, Ibarra A. Astragalus root and elderberry fruit extracts enhance the IFN-β stimulatory effects of *Lactobacillus acidophilus* in murine-derived dendritic cells. PLoS One 2012;7(10). https://doi.org/10.1371/journal.pone.0047878.

Glensk M, Glinski JA, Wlodarczyk M, Stefanowicz P. Determination of ursolic and oleanolic acid in *Sambuci fructus*. Chem Biodivers 2014;11:1939−44.

Gray AM, Abdel-Wahab YHA, Flatt PR. The traditional plant treatment, *Sambucus nigra* (elder) exhibits insulin-like and insulin-releasing actions in vitro. J Nutr 2000;130(1):15−20.

Ho GTT, Ahmed A, Zou YF, Aslaksen T, Wangensteen H, Barsett H. Structure−activity relationship of immunomodulating pectins from elderberries. Carbohydr Polym 2015;125:314−22. https://doi.org/10.1016/j.carbpol.2015.02.

Ho GTT, Zhou YF, Wangensteen H, Barsett H. RG-I regions from elderflower pectins substituted on GalA are strong immunomodulators. Int J Biol Macromol 2016;92:731—8.

Ho GTT, Wangensteen H, Barsett H. Elderberry and elderflower extracts, phenolic compounds, and metabolites and their effect on complement, RAW 264.7 macrophages and dendritic cells. Int J Mol Sci 2017a;18(3):E584. https://doi.org/10.3390/ijms18030584.

Ho GTT, Kase ET, Wangensteen H, Barsett H. Effect of phenolic compounds from elderflowers on glucose- and fatty acid uptake in human myotubes and HepG2-cells. Molecules 2017b;22(90). https://doi.org/10.3390/molecules22010090.

Iacopini P, Baldi M, Storchi P, Sebastiani L. Catechin, epicatechin, quercetin, rutin and resveratrol in red grape: content, in vitro antioxidant activity and interactions. J Food Compos Anal 2008;21(8):589—98. https://doi.org/10.1016/j.jfca.2008.03.011.

ITIS. *Sambucus* L. taxonomic serial no.: 35315. Integrated taxonomic information system taxonomy and nomenclature. 2016. Accessed Google Scholar, http://www.itis.gov/servlet/SingleRpt/SingleRpt?search_topic=TSN&search_value=35315.

Jørgensen U, Hansen M, Christensen LP, Jensen K, Kaack K. Olfactory and quantitative analysis of aroma compounds in elder flower (*Sambucus nigra* L.) drink processed from five cultivars. J Agric Food Chem 2000;48:2376—83.

Kinoshita E, Hayashi K, Katayama H, Hayashi T, Obata A. Anti-influenza virus effects of elderberry juice and its fractions. Biosci Biotechnol Biochem 2012;76(9):1633—8. https://doi.org/10.1271/bbb.120112.

Krauze-Baranowska M, Malinowska I, Głód D, Majdan M, Wilczańska A. UTLC of flavonols in *Sambucus nigra* flowers. J Planar Chromat 2009;22(5):385—7.

Krüger S, Mirgos M, Morlock GE. Effect-directed analysis of fresh and dried elderberry (*Sambucus nigra* L.) via hyphenated planar chromatography. J Chromatogr 2015;1426:209—19.

Lin P, Hwang E, Ngo HTT, Seo SA, Yi TH. *Sambucus nigra* L. ameliorates UVB-induced photoaging and inflammatory response in human skin keratinocytes. Cytotechnology 2019;71(5):1003—17. https://doi.org/10.1007/s10616-019-00342-1.

Lorenzo JMM, Pateiro M, Domínguez R, Barba FJ, Putnik P, Kovačević DB, Franco D. Berries extracts as natural antioxidants in meat products: a review. Food Res Int 2018;106:1095—104. https://doi.org/10.1016/j.foodres.2017.12.005.

Mahmoudi M, Ebrahimzadeh MA, Dooshan A, Arimi A, Ghasemi N, Fathiazad F. Antidepressant activities of Sambucus ebulus and *Sambucus nigra*. Eur Rev Med Pharmacol Sci 2014;18:3350—3.

Marisa Ribeiro A, Estevinho BN, Rocha F. Microencapsulation of polyphenols - the specific case of the microencapsulation of *Sambucus nigra* L. extracts - a review. Trends Food Sci Technol 2019:1—14. https://doi.org/10.1016/j.tifs.2019.03.011.

Milena V, Tatjana M, Gökhan Z, Ivana B, Aleksandra C, Mohammad MF, Marija R. Advantages of contemporary extraction techniques for the extraction of bioactive constituents from black elderberry (*Sambucus nigra* L.) flowers. Ind Crop Prod 2019;136:93—101. https://doi.org/10.1016/j.indcrop.2019.04.058.

Młynarczyk K, Walkowiak-Tomczak D, Łysiak GP. Bioactive properties of *Sambucus nigra* L. as a functional ingredient for food and pharmaceutical industry. J Funct Foods 2018;40:377—90. https://doi.org/10.1016/j.jff.2017.11.025.

Moldovan B, David L, Achim M, Clichici S, Filip GA. A green approach to phytomediated synthesis of silver nanoparticles using *Sambucus nigra* L. fruits extract and their antioxidant activity. J Mol Liq 2016;221:271—8. https://doi.org/10.1016/j.molliq.2016.06.003.

106 Contaminants and Plants Action on Female Reproduction

Munekata PES, Domínguez R, Campagnol PCB, Franco D, Trindade MA, Lorenzo JM. Effect of natural antioxidants on physicochemical properties and lipid stability of pork liver pâté manufactured with healthy oils during refrigerated storage. J Food Sci Technol 2017;4(13):4324−34. https://doi.org/10.1007/s13197-017-2903-2.

Olejnik A, Kowalska K, Olkowicz M, Rychlik J, Juzwa W, Myszka K, Dembczyński R, Białas W. Anti-inflammatory effects of gastrointestinal digested *Sambucus nigra* L. fruit extract analyzed in co-cultured intestinal epithelial cells and lipopolysaccharide- stimulated macrophages. J Funct Foods 2015;19:649−60.

Oniszczuk A, Olech M, Oniszczuk T, Wojtunik-Kulesza K, Wójtowicz A. Extraction methods, LC-ESI-MS/MS analysis of phenolic compounds and antiradicalproperties of functional food enriched with elderberryflowers or fruits. Arabi J Chem 2016;12:4719−30. https://doi.org/10.1016/j.arabjc.2016.09.003.

Pereira DI, Amparo TR, Almeida TC, Costa FSF, Brandão GC, Santos ODHD, da Silva GN, Bianco de Souza GH. Cytotoxic activity of butanolic extract from *Sambucus nigra* L. flowers in natura and vehiculated in micelles in bladder cancer cells and fibroblasts. Nat Prod Res 2020;25:1−9. https://doi.org/10.1080/14786419.2020.1851220.

Pignatti S. Flora d'Italia, vol. 2. Bologna: Edagricole; 1982. p. 639.

Porter RS, Bode RF. A review of the antiviral properties of black elder (*Sambucus nigra* L.) products. Phytother Res 2017;31(4):533−54. https://doi.org/10.1002/ptr.5782.

Roschek B, Fink RC, McMichael MD, Li D, Alberte RS. Elderberry flavonoids bind to and prevent H1N1 infection in vitro. Phytochemistry 2009;70(10):1255−61. https://doi.org/10.1016/j.phytochem.2009.06.003.

Salvador AC, Rocha SM, Silvestre AJD. Lipophilic phytochemicals from elderberries (*Sambucus nigra* L.): influence of ripening, cultivar and season. Ind Crop Prod 2015;71:15−23.

Schröder L, Richter DU, Piechulla B, Chrobak M, Kuhn C, Schulze S, Abarzua S, Jeschke U, Weissenbacher T. Effects of phytoestrogen extracts isolated from elder flower on hormone production and receptor expression of trophoblast tumor cells JEG-3 and BeWo, as well as MCF7 breast cancer cells. Nutrients 2016;8(10):616. https://doi.org/10.3390/nu8100616.

Senica M, Stampar F, Veberic R, Mikulic-Petkovsek M. Processed elderberry (*Sambucus nigra* L.) products: a beneficial or harmful food alternative? LWT - Food Sci Technol 2016;72:182−8. https://doi.org/10.1016/j.lwt.2016.04.056.

Silva P, Ferreira S, Nunes FM. Elderberry (*Sambucus nigra* L.) by-products a source of anthocyanins and antioxidant polyphenols. Ind Crop Prod 2017;95:227−34. https://doi.org/10.1016/j.indcrop.2016.10.018.

Sirotkin A, Záhoranska Z, Tarko A, Fabova Z, Alwasel S, Halim Harrath A. Plant polyphenols can directly affect ovarian cell functions and modify toluene effects. J Anim Physiol Anim Nutr 2021;105(1):80−9. https://doi.org/10.1111/jpn.13461.

Sun-Waterhouse D, Jin D, Waterhouse GIN. Effect of adding elderberry juice concentrate on the quality attributes, polyphenol contents and antioxidant activity of three fibre-enriched pastas. Food Res Int 2013;54:781−9.

Tejero J, Jimez P, Quinto EJ, Cordoba-Diaz D, Garrosa M, Cordoba-Diaz M, Gayoso MJ, Girbes T. Elderberries: a source of ribosome inactivating proteins with lectin activity. Molecules 2015;20:2364−87.

Topolska D, Valachova K, Rapta P, Silhar S, Panghyova E, Horvath A, Soltes L. Antioxidative properties of *Sambucus nigra* extracts. Chem Paper 2015;69(9):1202−10.

Torabian G, Bahramian B, Zambon A, Spilimbergo S, Adil Q, Schindeler A, Valtchev P, Dehghani F. A hybrid process for increasing the shelf life of elderberry juice. J Supercrit Fluids 2018;140:406—14. https://doi.org/10.1016/j.supflu.2018.07.023.

Torabian G, Valtchev P, Adil Q, Dehghani F. Anti-influenza activity of elderberry (*Sambucus nigra*). J Funct Foods 2019;54:353—60. https://doi.org/10.1016/j.jff.2019.01.031.

Ulbricht C, Basch E, Cheung L, Goldberg H, Hammerness P, Isaac R, Khalsa KPS, Romm A, Rychlik I, Varghese M, Weissner W, Windsor RC, Wortley J. An evidence-based systematic review of elderberry and elderflower (*Sambucus nigra*) by the Natural Standard Research Collaboration. J Diet Suppl 2014;11(1):80—120.

Vlachojannis JE, Cameron M, Chrubasik S. A systematic review on the Sambuci fructus effect and efficacy profiles. Phytother Res 2010;24:1—8.

Vlachojannis C, Zmmermann BF, Chrubasik-Hausmann S. Quantification of anthocyanins in elderberry and chokeberry dietary supplements. Phytother Res 2015;29:561—5.

Waknine-Grinberg JH, El-On J, Barak V, Barenholz Y, Golenser J. The immunomodulatory effect of Sambucol on leishmanial and malarial infections. Planta Med 2009;75:581—6.

Weiss R, Fintelmann V. Herbal medicine. Stuttgart: Thieme; 2000.

Wu X, Gu L, Prior RL, Mckay S. Characterization of anthocyanins and proanthocyanidins in some cultivars of Ribes, Aronia, and Sambucus and their antioxidant capacity. J Agric Food Chem 2004;52:7846—56.

Chapter 3.3

Buckthorn (*Hippophae rhamnoides* L.)

3.3.1 Introduction

Sea buckthorn (*Hippophae rhamnoides* L., *Elaeagnaceae*) is an economically and ecologically important medicinal plant comprising species which are winter hardy, dioecious, wind-pollinated multipurpose shrubs bearing yellow or orange berries with nitrogen-fixing ability. It grows widely in cold regions of Indian Himalayas, China, Russia, Europe and many other countries. It is commonly known as "cold desert gold" due to its high potential as a bio-resource for land reclamation, reducing soil erosion and its multifarious uses (Kalia et al., 2011). Every part of the plant is used as medicine, nutritional supplement, fuel and fence, and therefore sea buckthorn is popularly known as Gold Mine, Wonder Plant, or Golden Bush. This plant contains a series of chemical compounds, possessing various biological as well as therapeutic activities (Hakeem et al., 2018). Sea buckthorn has been used for centuries as a medicine and nutritional supplement in Asia and Europe (Wen et al., 2020). However, its fruits have been used as a raw material for foods and medicines for decades in various regions of the world including China, Russia, North America, and Europe. Berry products of sea buckthorn are nowadays becoming popular foods in the United States, Canada, Finland, Germany, and some other European countries. However, the commercial cultivation and exploitation of sea buckthorn berries using its secondary compounds such as flavonoids, vitamins, and carotenes for high quality food products or even to produce basic products such as juices is differently developed in Europe, Asia, and Russia. The medicinal and/or nutritional components of berries will provide very cheap raw material for national and international pharmaceutical industries, benefiting humanity worldwide. That is why, sea buckthorn is one among the research and development subjects of the pharmaceutical industries all over the world (Hakeem et al., 2018).

Hippophae rhamnoides has great economic value due to various uses of different parts of the plan for example pharmaceutical and cosmetic ingredients, source of energy, soil enhancer, and also rich nutritious value (Gray et al., 2010). As botanicals and dietary supplements are used increasingly in many countries, the issue of safety is particularly critical for regulation of food products containing these substances. However, data regarding to the safety assessment of the plant and its extracts is still rare (Wen et al., 2020).

Food/medicinal herbs and their influence **Chapter | 3 109**

In the present evidence-based study, we have tried to summarize the current knowledge concerning the properties and other health benefits of sea buckthorn with emphasis on its potential mechanism of action on female reproductive processes at various regulatory levels.

3.3.2 Provenance and properties

Hippophae rhamnoides L. (family—*Elaeagnaceae*, common name—Sea buckthorn) (Swenson and Bartish, 2002; Pundir et al., 2021) is a flowering shrub native to cold temperate regions of Eurasia (Pundir et al., 2021). There are seven known species in this genus, but *Hippophae rhamnoides* L. subsp. *rhamnoides* has the widest distribution (Swenson and Bartish, 2002).

The whole plant (fruit, leaf, and tree) is known to be a great source of a broad range of active biological substances (Suryakumar and Gupta, 2011). Berries are the most prominent feature of the plant (Pundir et al., 2021) and are orange-yellow to red color (Suryakumar and Gupta 2011). Berries are orange-yellow to red color (Suryakumar and Gupta, 2011) and together with leaves are plentiful in a diversity of vitamins and other biologically active substances phenols, carotenoids (lycopene, carotene, lutein, and zeaxanthin), flavonoids (isorhamnetin, quercetin, glycosides, and kaempferol), tocopherols, sterols (Sajfrtová et al., 2010; Enkhtaivan et al., 2017; Pundir et al., 2021), polyunsaturated fatty acids, minerals, vitamins, Omega 3, 6, 9 and rarest Omega 7 and about 190 bioactive compounds (Pundir et al., 2021), and have potent medicinal properties and high nutrients (Sajfrtová et al., 2010; Enkhtaivan et al., 2017). Oil extracted from sea buckthorn seed is the only natural oil that provides a 1:1 ratio of omega-3 and omega-6 fatty acids (linolenic and linoleic acids), respectively. The major constituent of phytosterols in the sea buckthorn oil has been identified as β-sitosterol (Sajfrtová et al., 2010).

The phytochemical composition of the berries of four different subspecies of sea buckthorn *Hippophae rhamnoides* L. subsp. *yunnanensis* (Yunnanensis), *Hippophae rhamnoides* L. subsp. *mongolica* (Mongolica), *Hippophae rhamnoides* L. subsp. *turkestanica* (Turkestanica) and *Hippophae rhamnoides* L. subsp. *sinensis* (Sinensis) was studied. Sinensis demonstrated the highest total phenolic content and corresponding total antioxidant activity, while the greatest cellular antioxidant and antiproliferative properties were observed in *H. rhamnoides* L. subsp. *yunnanensis* (Guo et al., 2017). The sea buckthorn has a total polyphenol content of approximately 46 mg gallic acid (GA) equivalent/g dried extract, total flavonoid content of approximately 23 mg quercetin equivalent/g dried extract (Zhamanbayeva et al., 2016). Berries also contain tocopherols, tocotrienols, and essential polyunsaturated fatty acids, while leaves are a good source of polyphenols (Yang and Kallio, 2002). Recently, a number of bioactive compounds have been isolated from the

110 Contaminants and Plants Action on Female Reproduction

berries of sea buckthorn such as oleanolic acid, 19-alpha-hydroxyursolic acid, dulcioic acid, ursolic acid, 5-hydroxymethyl-2-furancarbox-aldehyde, octacosanoic acid, palmitic acid, hippophae cerebroside, and 1-O-hexadecanolenin (Zheng et al., 2019). Oleanolic acid and hippocorosolate, new corosolic ester derivative isolated from sea buckthorn berries, were responsible for anticancer activity (Ali et al., 2019). However, the bioactive composition of berries depends on maturity, species, fruit size, climate, geographic location, and also methods of extraction (Leskinen et al., 2010).

The polyphenolic compounds in the leaves are represented by flavonols, leucoanthocyanidins, epicatechin, gallocatechin, bepigallocatechin, and gallic acid (Ozturk et al., 2018) and others. Some of bioactive phenolic constituents, such as quercetin-3-O-galactoside, quercetin-3-O-glucoside, kaempferol, and isorhamnetin (Upadhyay, 2010), and flavonol glycosides (diglycosides and triglycosides) (Enkhtaivan et al., 2017) were quantified in leaf extracts. Six compounds kaempferol-3-O- β-α-(6″-O-coumaryl) glycoside, 1-feruloyl-β-α-glucopyranoside, isorhamnetin-3-O-glucoside, quercetin-3-O-β-α-glucopyranoside, quercetin-3-O-β-α-glucopyranosyl-7-O-α-L-rhamnopyranoside, and isorhamnetin-3-O-rutinoside from sea buckthorn leaf extract were isolated, too (Kim et al., 2017). Tannin fraction has been isolated from leaves and the principal components are hydrolysable gallo- and ellagi-tannins of monomeric type: strictinin, isostrictinin, casuarinin, and casuarictin (Shipulina et al., 2006).

Isorhamnetin isolated from the refuse of berries of sea buckthorn that has been pressed for other products has shown significant antioxidant activity. Zeaxanthin and beta-cryptoxanthin esters from sea buckthorn berries can be used as food additives, cosmetic ingredients, or nutraceuticals, too (Pintea et al., 2005).

3.3.3 Physiological and therapeutic actions

In recent years, many types of research demonstrated many positive effects of sea buckthorn on human health. Sea buckthorn is believed to be helpful in preventing and healing diseases such as viral infections and cancer (Enkhtaivan et al., 2017) due to its antioxidant (Kashyap et al., 2020), anti-inflammatory (Kwon et al., 2011), antiviral (Jain et al., 2008), antimicrobial (Chauhan et al., 2007; Upadhyay et al., 2010), and antibacterial (Kumar et al., 2013; Widén et al., 2015) properties. Among many others, it was also demonstrated its cardioprotective, antiantherogenic, hepatoprotective effect (Wang et al., 2018; Kashyap et al., 2020), dermatological (Pundir et al., 2021), antiproliferative (Zhamanbayeva et al., 2016) and also anticancer activity against colon and liver (Gray et al., 2010), lung (Enkhtaivan et al., 2017; Ali et al., 2019), cervical, ovarian (Enkhtaivan et al., 2017), and breast cancer cells (Ali et al., 2019). Phenolic compounds such as proanthocyanidins, curcumin, and resveratrol, have been found to offer significant benefits in cancer chemoprevention and radiotherapy (Ko et al., 2017). Similarly, the inhibition of breast cancer growth by kaempferol was reported (Yang et al., 2019). It is well-known that higher dietary intake of phenolic compounds especially flavonoids and procyanidins are associated with a lower risk of cancer (Kristo et al., 2016).

3.3.4 Mechanisms of action

Several principal extra- and intracellular mechanisms of sea buckthorn action and its compounds on the cells have been proposed, although the evidence for each of this mechanism is limited, and the interrelationships between these mechanisms are poorly investigated yet.

A new strategy for designing chemical agents as modulators of the nuclear factor erythroid 2—related factor 2 (Nrf2) dependent pathways to target the immune response is suggested (Saha et al., 2020). The activation of Nrf-2/ heme oxygenase 1 (HO-1)- superoxide dismutase 2 (SOD2) signaling pathway is associated with the protective effect of polysaccharide from sea buckthorn berry (Wang et al., 2018). Inflammation is a key driver in many pathological conditions such as allergy, cancer, and many others. Nrf2 plays a pivotal role controlling the expression of antioxidant genes that ultimately exert antiinflammatory functions. Nrf2 is proved to contribute to the regulation of the HO-1 axis, which is a potent antiinflammatory target (Saha et al., 2020). Recent studies showed a connection between the Nrf2/antioxidant response element system and the expression of inflammatory mediators, nuclear factor kappa B (NF-κB) pathway and macrophage metabolism (Saha et al., 2020). Recently, the polysaccharide was found to have hepatoprotective activity against acetaminophen (APAP)-induced hepatotoxicity. Sea buckthorn polysaccharide decreased the level of enzymes such as alanine aminotransferase (ALT) and aspartate aminotransferase (AST) in mice (Wang et al., 2018). It also increased glutathione (GSH) and plasma glutathione peroxidase (GSH-Px) level, and the expression of nitric oxide (NO) and the inducible nitric oxide synthase (iNOS) was reduced with diminished liver injuries. The increased expression of Bcl-2/ Bax was revealed and APAP-induced c-Jun N-terminal kinase (JNK) phosphorylation was suppressed. Furthermore, expression of Nrf-2 was also increased which targets the HO-1 gene in APAP mice (Wang et al., 2018). Antitumor activity of sea buckthorn can be attributed to antioxidant compounds, particularly phenolic compounds such as flavonoids kaempferol, isorhamnetin, and quercetin. These flavonoids protect cells from oxidative damage that can lead to cancer and genetic mutations (Olas et al., 2018).

Isorhamnetin is one of the most important active ingredients in the fruits of *Hippophae rhamnoides* L., which possesses extensive pharmacological activities. The related mechanisms involve the regulation of phosphatidylinositol 3-kinase (PI3K)/Akt/ protein kinase B (PKB), nuclear factor-κB (NF-κB), mitogen-activated protein kinase (MAPK) and other signaling pathways as well as the expression of related cytokines and kinases. Isorhamnetin has a high value of development and application. However, the investigations on its mechanism of action are limited and lack of detailed scientific validation (Gong et al., 2020). The mechanism of isorhamnetin (3′-O-methylated metabolite of quercetin) action may be the apoptosis of cells induced by the downregulation of oncogenes and up-regulation of apoptotic genes. Isorhamnetin suppresses the proliferation of human colorectal cancer cells, induces cell cycle arrest at the G2/M phase, and suppresses cell proliferation by inhibiting the PI3K-Akt-mTOR pathway (phosphatidylinositol 3-kinase—protein kinase B—the mammalian target of rapamycin). In addition, isorhamnetin reduces the phosphorylation levels of Akt, phosph-p70S6 kinase, and phosph-4E-BP1 proteins and enhances the expression of cyclin B1 protein (Li et al., 2014).

Further, isorhamnetin treatment also significantly reduces the levels of aspartate aminotransferase, alanine amino transferase and blood urea nitrogen, suggesting that it can improve liver and kidney function in infected mice. Docking studies reveal that isorhamnetin binds deep in the hydrophobic binding pocket of myeloid differentiation 2 (MD-2) via extensive hydrophobic interactions and hydrogen bonding with Tyr102, preventing the toll-like receptor 4 (TLR4)/MD-2 dimerization. Notably, binding and secreted alkaline phosphatase reporter gene assays show that isorhamnetin can interact directly with the TLR4/MD-2 complex, thus inhibiting the TLR4 cascade, which eventually causes systemic inflammation, resulting in death due to cytokine storms. We therefore presume that isorhamnetin could be a suitable therapeutic candidate to treat bacterial sepsis (Chauhan et al., 2019)

Recently, Masoodi et al. (2020) suggested that sea buckthorn inhibits cellular proliferation and decreases the expression of a specific antigen in prostate cancers. Androgen receptor is a ligand-dependent transcription factor that governs the expression of androgen-responsive genes. If somehow androgen receptor is retained in the cytoplasm and its shuttling into the nucleus is inhibited it cannot activate the target androgen responsive genes and prevent prostate cancer progression. Leaf extract was applied on prostate cancer cells to determine the inhibition of cellular proliferation and inhibition of growth on prostate cancer cells. The extract caused a dose-dependent decline in the colony-forming ability of prostate cancer cells, inhibited cell regrowth and induces cancer cell death (Masoodi et al., 2020). Sea buckthorn leaf extract may inhibit the rapid proliferation of rat glioma cells, probably by inducing the early events of cell apoptosis. The reduction of glioma cell proliferation and viability following administration of the sea buckthorn extract was accompanied by a decrease in the production of reactive oxygen species (ROS), which are critical mediators for the proliferation of tumor cells. Sea buckthorn extract not only upregulated the expression of the proapoptotic protein Bcl-2-associated X (Bax), but also promoted its localization, accumulation, and translocation in the nucleus (Kim et al., 2017).

However, quercetin-induced increase in the levels of cytochrome c together with the activation of caspase-3 and caspase-9 leads to apoptosis in cancer cells, as shown in human retinoblastoma cells (Liu, 2017). Also, procyanidins were found to induce cell apoptosis in a dose-dependent manner. These procyanidins can induce human breast cancer cell apoptosis by inhibiting intracellular fatty acid synthase (FAS) activity. FAS is a key enzyme for de novo long-chain fatty acid biosynthesis, high levels of which are found in cancer cells. This inhibition was dose-dependent. And cell growth was suppressed by treatment with sea buckthorn procyanidins (Wang et al., 2014).

On the other hand, *Hippophae rhamnoides* treatment protected spermatogenesis by enhancing the spermatogonial proliferation, enhancing the stem cell survival and reducing sperm abnormalities. The presence of polyphenolic flavonoids and tannins in the extract and the radical scavenging activity might be responsible for the protective action of *Hippophae rhamnoides* (Goel et al., 2006).

3.3.5 Effects on female reproductive processes

3.3.5.1 Effect on ovarian follicular cell functions

The available scientific databases don't contain publications concerning buckthorn extract on ovarian functions. Nevertheless, the action of several biological active buckthorn constituents on female reproductive processes have been reported. There is a direct action of quercetin on basic ovarian cell functions (proliferation, apoptosis, and hormones release) (Sirotkin et al., 2019a,b) which can be species-specific (Sirotkin et al., 2019a) (see also chapter concerning quercetin in this book). Quercetin inhibits also human metastatic ovarian cancer cell growth and modulates components of the intrinsic apoptotic pathway (Bhat et al., 2014; Teekaraman et al., 2019).

Other biologically active constituents of buckhorn have also shown beneficial effects on ovarian cancer cells. Apigenin, myricetin, and luteolin were found to induce apoptosis, suppress cell growth, inhibited cell invasion, and arrested the ovarian cancer cell cycle. The authors further proposed apigenin, myricetin, and luteolin as promising candidates for ovarian cancer prevention and adjuvant therapy (Tavsan et al., 2019). Apigenin action has been described in details in the corresponding chapter.

Kaempherol is the next sea buckthorn molecule, which can suppress ovarian cancerogenesis. At least it induces apoptosis and blocked cell cycle in cultured ovarian cancer cells (Choi and Kim, 2008; Imran et al., 2019; El-Kott et al., 2020).

Therefore, buckthorn constituents could affect proliferation, apoptosis and hormones release of healthy ovarian cells, as well as to induce ovarian cancer cell apoptosis, suppress cell growth, invasion, and to arrest the cell cycle.

3.3.5.2 Effect on vagina and uterus

Sea buckthorn traditionally have been used for the treating gynecological disorders including uterus inflammation and endometriosis. Its oil was able to suppress the signs of endometriosis and uterus inflammation. These effects could possibly be attributed to its carotenoid, sterol, and hypericin contents (Ilhan et al., 2016).

Furthermore, it can be useful for prevention of vaginal complications during menopause, which is associated with vaginal atrophy and the thinning and drying of vaginal mucosa. Application of sea buckthorn oil to menopausal women improved the integrity of vaginal epithelium and tended to improve vaginal health index. It was considered as an alternative to estrogen substitution for vaginal health of postmenopausal women (Larmo et al., 2014). Furthermore, a new vaginal gel composed of sea buckthorn oil, aloe vera, 18β-glycyrrhetic acid, hyaluronic acid, and glycogen has been recently registered. It seems to be a valid choice as a single, local agent for relieving vulvovaginal atrophy symptoms (vaginal dryness, itching, and burning sensations) and improving sexual function in postmenopausal women (De Seta et al., 2021).

Sea buckthorn contains a large amount of vitamins C and E. The intake of vitamins C and E by infertile or subfertile women subjected to controlled ovarian stimulation could improve their uterine characteristics, endometrial thickness, and endometrial blood flow. Furthermore, they postulated the antioxidant and anticoagulant roles of vitamins C and E in improving fertility (Nasri et al., 2007).

Therefore, the available evidence demonstrates a therapeutic action of sea buckthorn and its compounds on treatment of gynecological disorders including uterus inflammation, endometriosis and relieving vulvovaginal atrophy symptoms in postmenopausal women.

3.3.6 Mechanisms of action on female reproductive processes

The evidence about the mechanisms of *Hippophae rhamnoides* L. action on female reproductive processes are insufficient.

A mixture of sea buckthorn oil regresses induced rat endometriosis. This treatment reduced level of inflammatory **cytokines** (markers and promoters of inflammatory processes) and **vascular endothelial growth factor (VEGF)** (marker and promoter of angiogenesis) (Ilhan et al., 2016). Therefore, cytokines and VEGF could be extracellular mediators of curative action of sea buckthorn on endometriosis.

Furthermore, Imran et al. (2019) predicted kaempferol-induced inhibition of tumor growth and angiogenesis by decreasing the VEGF expression via downregulation of hypoxia-inducible factor 1α (HIF-1α), a regulator of VEGF expression.

A mechanism of flavonoids actions in ovarian cells including intracellular regulators of **proliferation** and **apoptosis** are described in special chapters. For example, guercetin increases the expression of caspase-3 which leads to DNA fragmentation and apoptosis. Also, it has been shown that quercetin can downregulate antiapoptotic proteins and upregulate proapoptotic members in various cancer cell lines (Bhat et al., 2014; Teekaraman et al., 2019). Kampherol induced expression of morphological signs of apoptosis (membrane blebbing) and accumulation of intracellular markers and promoters of apoptosis (caspases 3, 8, and 9, and Bax) as well as decreased the expression of antiapoptotic B-cell lymphoma 2 (Bcl-2). Furthermore, kaempferol induced blockage of cells cycle at the G0/G1 checkpoint accompanied by the suppression of cyclin B1 and cyclin dependent kinase 2 (Cdc2) expression (Choi and Kim, 2008). Imran et al. (2019) suggested that kaempferol can induce apoptosis and cell cycle arrest at the G2/M phase via upregulation of checkpoint kinase 2/cell division cycle 25C/cyclin-dependent kinase 2 (Chk2/Cdc25C/Cdc2), receptors DR5 and DR4, c-Jun N-terminal kinase (JNK), C/EBP homologous protein (CHOP), p38, p21, the extracellular signal-regulated kinase 1/2 (ERK1/2) proteins, caspase-3, -7, -8, Bad, Bax, and p53 proteins.

El-Kott et al. (2020) reported that inhibition of ovarian cancer by kaempferol was associated with **endoplasmic reticulum stress,** glucose-regulated protein 78 (GRP78), the RNA-like endoplasmic reticulum kinase (PERK), ATF6, IRE-1, LC3II, and beclin 1 suggesting the activation of cytotoxic **autophagy**. In addition, kaempferol increased the sensitivity of ovarian cancer cells to cisplatin by decreasing the protein levels of **phosphorylated Akt (p-Akt).**

Therefore, flavonoids of *Hippophae rhamnoides* L. can suppress ovarian cancer cells via downregulating VEGF, antiapoptotic proteins, upregulate proapoptotic proteins, suppress cell cycle at various checkpoints, p-Akt, as well as to induce endoplasmic reticulum stress and autophagy.

3.3.7 Potential for application in reproductive biology and medicine

Sea buckthorn attracts special attention, and it is one of the most interesting plants—functional food and drug for animal and human use affecting health and medicine against several female reproductive disorders. Although sea buckhorn molecule quercetin action on healthy ovarian cells has been reported, it remains unknown, whether sea buckthorn extract or its constituents could be useful to influence healthy female reproductive processes.

Much more evidence is obtained for the application of sea buckthorn and its constituents to prevent and to treat ovarian cancer. Furthermore, the applicability of buckthorn and its compounds for treatment of gynecological disorders including uterus inflammation, endometriosis, and relieving vulvovaginal atrophy symptoms in postmenopausal women has been demonstrated.

116 Contaminants and Plants Action on Female Reproduction

No sea buckthorn berry oil treatment-related maternal toxicity or embryo toxicity was observed. The available evidence suggests that sea buckthorn products can be useful as functional food dietary supplement and medicine (Wen et al., 2020).

3.3.8 Conclusions and possible direction of future studies

The analysis of the available literature demonstrates that buckthorn constituents could affect proliferation, apoptosis and hormones release of healthy ovarian cells, as well as to suppress ovarian cancer (induce ovarian cancer cell apoptosis, suppress cell growth, invasion, and to arrest the cell cycle). These effects can be mediated by downregulating VEGF, antiapoptotic proteins, upregulating proapoptotic proteins, suppression cell cycle at various checkpoints, p-Akt, as well as to induction of endoplasmic reticulum stress and autophagy.

Moreover, sea buckthorn and its compounds can treat gynecological disorders including uterus inflammation, endometriosis and relieving vulvovaginal atrophy symptoms in postmenopausal women targeting inflammatory cytokines and VEGF.

Nevertheless, many aspects of sea buckhorn action and application are still waiting for their understanding. The effects of sea buckhorn extract on female reproductive processes, and the role of its particular constituents are studied insufficiently. The performed studies of this plant were focused on its therapeutic action and application, but the action of buckhorn extract and of some its key molecules on healthy female reproductive system remained practically unknown. The optimal forms of delivery of biologically active molecules of sea buckhorn to organisms are to be found. The medicinal potency of this plant and its molecules should be verified not only in *in vitro* and animal experiments, but also in clinical studies. The few reported studies could be only first steps toward understanding biological and therapeutic effects of this plant and its constituents, which requires more profound investigations.

References

Ali I, Zahra NA, Reazuddin SH, Mudassar BHA. A new potent anti-cancer corosolic ester identified from the super miracle plant *Hippophae rhamnoides* (sea buckthorn). Biochem Mod Appl 2019;2:24−9. https://doi.org/10.33805/2638-7735.119.

Bhat FA, Sharmila G, Balakrishnan S, Singh PR, Srinivasan N, Arunakaran J. Epidermal growth factor-induced prostate cancer (PC3) cell survival and proliferation is inhibited by quercetin, a plant flavonoid through apoptotic machinery. Biomed Prevent Nutr 2014;4(4):459−68. https://doi.org/10.1016/j.bionut.2014.07.003.

Chauhan AS, Negi PS, Ramteke RS. Antioxidant and antibacterial activities of aqueous extract of Seabuckthorn (*Hippophae rhamnoides*) seeds. Fitoterapia 2007;78(7-8):590−2. https://doi.org/10.1016/j.fitote.2007.06.004.

Chauhan AK, Kim J, Lee Y, Balasubramanian PK, Kim Y. Isorhamnetin has potential for the treatment of *Escherichia coli*-induced sepsis. Molecules 2019;24(21):3984. https://doi.org/10.3390/molecules24213984.

Choi EJ, Kim T. Equol induced apoptosis via cell cycle arrest in human breast cancer MDA-MB-453 but not MCF-7 cells. Mol Med Rep 2008;1(2):239–44. https://doi.org/10.3892/mmr.1.2.239.

De Seta F, Caruso S, Di Lorenzo G, Romano F, Mirandola M, Nappi RE. Efficacy and safety of a new vaginal gel for the treatment of symptoms associated with vulvovaginal atrophy in postmenopausal women: a double-blind randomized placebo-controlled study. Maturitas 2021;147:34–40. https://doi.org/10.1016/j.maturitas.2021.03.002. Epub 2021 Mar 4.

El-Kott AF, Shati AA, Al-Kahtani MA, Alharbi SA. Kaempferol induces cell death in A2780 ovarian cancer cells and increases their sensitivity to cisplatin by activation of cytotoxic endoplasmic reticulum-mediated autophagy and inhibition of protein kinase B. Folia Biol 2020;66(1):36–46.

Enkhtaivan G, John KM, Pandurangan M, Hur JH, Leutou AS, Kim DH. Extreme effects of Seabuckthorn extracts on influenza viruses and human cancer cells and correlation between flavonol glycosides and biological activities of extracts. Saudi J Biol Sci 2017;24(7):1646–56. https://doi.org/10.1016/j.sjbs.2016.01.004.

Goel HC, Samanta N, Kannan K, Kumar IP, Bala M. Protection of spermatogenesis in mice against gamma ray induced damage by *Hippophae rhamnoides*. Andrologia 2006;38(6):199–207. https://doi.org/10.1111/j.1439-0272.2006.00740.x.

Gong G, Guan YY, Zhang ZL, Rahman K, Wang SJ, Zhou S, Luan X, Zhang H. Isorhamnetin: A review of pharmacological effects. Biomed Pharmacother 2020;128:110301. https://doi.org/10.1016/j.biopha.2020.110301.

Gouvinhas I, Pinto R, Santos R, Saavedra MJ, Barros AI. Enhanced phytochemical composition and biological activities of grape (*Vitis vinifera* L.) stems growing in low altitude regions. Sci Hortic 2020;265:109248.

Grey C, Widén C, Adlercreutz P, Rumpunen K, Duan RD. Antiproliferative effects of sea buckthorn (*Hippophae rhamnoides* L.) extracts on human colon and liver cancer cell lines. Food Chem 2010;120(4):1004–10. https://doi.org/10.1016/j.foodchem.2009.11.039.

Guo R, Guo X, Li T, Fu X, Liu RH. Comparative assessment of phytochemical profiles, antioxidant and antiproliferative activities of Sea buckthorn (*Hippophaë rhamnoides* L.) berries. Food Chem 2017;221:997–1003. https://doi.org/10.1016/j.foodchem.2016.11.063.

Hakeem KR, Ozturk M, Altay V, Letchamo W. An alternative potential natural genetic resource: sea buckthorn [*Elaeagnus rhamnoides* (syn.: *Hippophae rhamnoides*)]. Glob Perspect Underutil Crops 2018:25–82. https://doi.org/10.1007/978-3-319-77776-4_2.

İlhan M, Süntar İ, Demirel MA, Yeşilada E, Keleş H, Küpeli Akkol E. A mixture of St. John's wort and sea buckthorn oils regresses endometriotic implants and affects the levels of inflammatory mediators in peritoneal fluid of the rat: a surgically induced endometriosis model. Taiwan J Obstet Gynecol 2016;55(6):786–90. https://doi.org/10.1016/j.tjog.2015.01.006.

Imran M, Salehi B, Sharifi-Rad J, Aslam Gondal T, Saeed F, Imran A, Shahbaz M, Fokou PVT, Arshad MU, Khan H, Guerreiro GS, Martins N, Estevinho LM. Kaempferol: a key emphasis to its anticancer potential. Molecules 2019;24(12):2277. https://doi.org/10.3390/molecules24122277.

Jain M, Ganju L, Katiyal A, Padwad Y, Mishra KP, Chanda S, Karana D, Yogendra KMS, Sawhney RC. Effect of *Hippophae rhamnoides* leaf extract against Dengue virus infection in human blood-derived macrophages. Phytomedicine 2008;15(10):793–9. https://doi.org/10.1016/j.phymed.2008.04.017.

118 Contaminants and Plants Action on Female Reproduction

Kalia RK, Singh R, Rai MK, Mishra GP, Singh SR, Dhawan AK. Biotechnological interventions in sea buckthorn (Hippophae L.): current status and future prospects. Trees 2011;25:559−75. https://doi.org/10.1007/s00468-011-0543-0.

Kashyap P, Riar CS, Jindal N. Sea Buckthorn. Antioxidants in fruits: properties and health benefits. 2020. p. 201−25. https://doi.org/10.1007/978-981-15-7285-2_11.

Kim SJ, Hwang E, Yi SS, Song KD, Lee HK, Heo TH, Park SK, Jung YJ, Jun HS. Sea buckthorn leaf extract inhibits glioma cell growth by reducing reactive oxygen species and promoting apoptosis. Appl Biochem Biotechnol 2017;182(4):1663−74. https://doi.org/10.1007/s12010-017-2425-4.

Ko JH, Sethi G, Um JY, Shanmugam MK, Arfuso F, Kumar AP, Bishayee A, Ahn KS. The role of resveratrol in cancer therapy. Int J Mol Sci 2017;18(12):2589. https://doi.org/10.3390/ijms18122589.

Kristo AS, Klimis-Zacas D, Sikalidis AK. Protective role of dietary berries in cancer. Antioxidants 2016;5(4):37. https://doi.org/10.3390/antiox5040037.

Kumar MY, Tirpude RJ, Maheshwari DT, Bansal A, Misra K. Antioxidant and antimicrobial properties of phenolic rich fraction of Sea buckthorn (Hippophae rhamnoides L.) leaves in vitro. Food Chem 2013;141(4):3443−50. https://doi.org/10.1016/j.foodchem.2013.06.057.

Kwon DJ, Bae YS, Ju SM, Goh AR, Choi SY, Park J. Casuarinin suppresses TNF-α-induced ICAM-1 expression via blockade of NF-κB activation in HaCaT cells. Biochem Biophys Res Commun 2011;409(4):780−5. https://doi.org/10.1016/j.bbrc.2011.05.088.

Larmo PS, Yang B, Hyssälä J, Kallio HP, Erkkola R. Effects of sea buckthorn oil intake on vaginal atrophy in postmenopausal women: a randomized, double-blind, placebo-controlled study. Maturitas 2014;79(3):316−21. https://doi.org/10.1016/j.maturitas.2014.07.010.

Leskinen HM, Suomela JP, Yang B, Kallio HP. Regioisomer compositions of vaccenic and oleic acid containing triacylglycerols in sea buckthorn (Hippophae rhamnoides) pulp oils: influence of origin and weather conditions. J Agric Food Chem 2010;58(1):537−45. https://doi.org/10.1021/jf902679v.

Li C, Yang X, Chen C, Cai S, Hu J. Isorhamnetin suppresses colon cancer cell growth through the PI3K-Akt-mTOR pathway. Mol Med Rep 2014;9(3):935−40. https://doi.org/10.3892/mmr.2014.1886.

Liu H, Zhou M. Antitumor effect of Quercetin on Y79 retinoblastoma cells via activation of JNK and p38 MAPK pathways. BMC Complement Altern Med 2017;17(1):1−8. https://doi.org/10.1186/s12906-017-2023-6.

Masoodi KZ, Wani W, Dar ZA, Mansoor S, Anam-ul-Haq S, Farooq I, Hussain K, Wani AS, Nehvi AF, Ahmed N. Sea buckthorn (Hippophae rhamnoides L.) inhibits cellular proliferation, wound healing and decreases expression of prostate specific antigen in prostate cancer cells in vitro. J Funct Foods 2020;73:104102. https://doi.org/10.1016/j.jff.2020.104102.

Nasir SA. Improved endometrial thickness and vascularity following vitamins E and C administration in infertile women undergoing controlled ovarian stimulation. Al-Qadisiyah Med J 2017;13(23):174−9.

Olas B, Skalski B, Ulanowska K. The anticancer activity of sea buckthorn [Elaeagnus rhamnoides (L.) A. Nelson]. Front Pharmacol 2018;9:232. https://doi.org/10.3389/fphar.2018.00232.

Ozturk M, Hakeem KR, Ashraf M, Ahmad MSA. Global perspectives on underutilized crops [online]. Switzerland: Springer International Publishing; 2018, ISBN 978-3-319-77775-7. p. 53. https://doi.org/10.1007/978-3-319-77776-4.

Pintea A, Varga A, Stepnowski P, Socaciu C, Culea M, Diehl HA. Chromatographic analysis of carotenol fatty acid esters in Physalis alkekengi and Hippophae rhamnoides. Phytochem Anal 2005;16(3):188−95. https://doi.org/10.1002/pca.844.

Pundir S, Garg P, Dviwedi A, Ali A, Kapoor VK, Kapoor D, Kulshrestha S, Lal UR, Negi P. Ethnomedicinal uses, phytochemistry and dermatological effects of *Hippophae rhamnoides* L.: a review. J Ethnopharmacol 2021;266:113434. https://doi.org/10.1016/j.jep.2020.113434. Epub 2020 Oct 2.

Saha S, Buttari B, Panieri E, Profumo E, Saso L. An overview of Nrf2 signaling pathway and its role in inflammation. Molecules 2020;25(22):5474. https://doi.org/10.3390/molecules25225474.

Sajfrtová M, Ličková I, Wimmerová M, Sovová H, Wimmer Z. β-Sitosterol: supercritical carbon dioxide extraction from sea buckthorn (*Hippophae rhamnoides* L.) seeds. Int J Mol Sci 2010;11(4):1842—50. https://doi.org/10.3390/ijms11041842.

Shipulina LD, Tolkachev ON, Krepkova LV, Bortnikova VV, Shkarenkov AA. Anti-viral, antimicrobial and toxicological studies on seabuckthorn (*Hippophae rhamnoides* L.). Seabuckthorn (Hippophae L.). A multipurpose wonder plant. Biochem Pharmacol 2006;2:471—83. ISBN:817035415.

Sirotkin AV, Hrabovszká S, Štochmaľová A, Grossmann R, Alwasel S, Halim Harrath A. Effect of quercetin on ovarian cells of pigs and cattle. Anim Reprod Sci 2019a;205:44—51. https://doi.org/10.1016/j.anireprosci.2019a.04.002.

Sirotkin AV, Štochmaľová A, Alexa R, Kádasi A, Bauer M, Grossmann R, Alrezaki A, Alwasel S, Harrath AH. Quercetin directly inhibits basal ovarian cell functions and their response to the stimulatory action of FSH. Eur J Pharmacol 2019b;860:172560. https://doi.org/10.1016/j.ejphar.2019.172560.

Suryakumar G, Gupta A. Medicinal and therapeutic potential of Sea buckthorn (*Hippophae rhamnoides* L.). J Ethnopharmacol 2011;138(2):268—78. https://doi.org/10.1016/j.jep.2011.09.024.

Swenson U, Bartish IV. Taxonomic synopsis of hippophae (Elaeagnaceae). Nord J Bot 2002;22(3):369—74. https://doi.org/10.1111/j.1756-1051.2002.tb01386.x.

Tavsan Z, Kayali HA. Flavonoids showed anticancer effects on the ovarian cancer cells: involvement of reactive oxygen species, apoptosis, cell cycle and invasion. Biomed Pharmacother 2019;116:109004. https://doi.org/10.1016/j.biopha.2019.109004.

Teekaraman D, Elayapillai SP, Viswanathan MP, Jagadeesan A. Quercetin inhibits human metastatic ovarian cancer cell growth and modulates components of the intrinsic apoptotic pathway in PA-1 cell line. Chem Biol Interact 2019;300:91—100. https://doi.org/10.1016/j.cbi.2019.01.008.

Upadhyay NK, Kumar MY, Gupta A. Antioxidant, cytoprotective and antibacterial effects of Sea buckthorn (*Hippophae rhamnoides* L.) leaves. Food Chem Toxicol 2010;48(12):3443—8. https://doi.org/10.1016/j.fct.2010.09.019.

Wang Y, Nie F, Ouyang J, Wang X, Ma X. Inhibitory effects of sea buckthorn procyanidins on fatty acid synthase and MDA-MB-231 cells. Tumor Biol 2014;35(10):9563—9. https://doi.org/10.1007/s13277-014-2233-1.

Wang X, Liu J, Zhang X, Zhao S, Zou K, Xie J, Wang X, Liu C, Wang J, Wang Y. Sea buckthorn berry polysaccharide extracts protect against acetaminophen induced hepatotoxicity in mice via activating the Nrf-2/HO-1-SOD-2 signaling pathway. Phytomedicine 2018;38:90—7. https://doi.org/10.1016/j.phymed.2017.11.007.

Wen P, Zhao P, Qin G, Tang S, Li B, Zhang J, Peng L. Genotoxicity and teratogenicity of seabuckthorn (*Hippophae rhamnoides* L.) berry oil. Drug Chem Toxicol 2020;43(4):391—7. https://doi.org/10.1080/01480545.2018.1497047.

Widén C, Renvert S, Persson GR. Antibacterial activity of berry juices, an in vitro study. Acta Odontol Scand 2015;73(7):539—43. https://doi.org/10.3109/00016357.2014.887773.

Yang B, Kallio H. Supercritical CO2-extracted sea buckthorn (*Hippophae rhamnoides*) oils as new food ingredients for cardiovascular health. Proc Health Ingred Euro 2002;17:19.

Yang S, Si L, Jia Y, Jian W, Yu Q, Wang M, Lin R. Kaempferol exerts anti-proliferative effects on human ovarian cancer cells by inducing apoptosis, G0/G1 cell cycle arrest and modulation of MEK/ERK and STAT3 pathways. J Buon 2019;24(3). ISSN: 2241-6293:975−81.

Zhamanbayeva GT, Aralbayeva AN, Murzakhmetova MK, Tuleukhanov ST, Danilenko M. Cooperative antiproliferative and differentiation-enhancing activity of medicinal plant extracts in acute myeloid leukemia cells. Biomed Pharmacother 2016;82:80−9. https://doi.org/10.1016/j.biopha.2016.04.062.

Zheng WH, Bai HY, Han S, Bao F, Zhang KY, Sun LL, Du H, Yang ZG. Analysis on the constituents of branches, berries, and leaves of *Hippophae rhamnoides* L. by UHPLC-ESI-QTOF-MS and their anti-inflammatory activities. Nat Product Commun 2019;14:8. https://doi.org/10.1177/1934578X19871404.

Chapter 3.4

Buckwheat (*Fagopyrum tataricum*, L., *Fagopyrum esculentum* Moench)

3.4.1 Introduction

One of the promising functional food and medicinal plant is buckwheat. Some publications summarized the information concerning chemical or medicinal characteristics of their plants (Giménez-Bastida and Zieliński, 2015; Jing et al., 2016; Kreft, 2016). On the other hand, they did not reflect the recent publications in this field, some buckwheat constituents, as well as some important aspects of buckwheat action, for example its effect on reproductive processes. The present chapter is the first review concerning physiological and reproductive effects of buckwheat, which summarizes the available data concerning mechanisms of action of this plant on various targets, as well as outlines the direction of the further studies of this functional food plant.

3.4.2 Provenance and properties

Buckwheat is an ancient pseudocereal crop under Polygonaceae family and genus *Fagopyrum*. The genus *Fagopyrum* (Polygonaceae), currently comprising 15 species of plants, includes three important buckwheat species: *Fagopyrum esculentum* (*F. esculentum*) *Moench*. (common buckwheat), *Fagopyrum tataricum* (*F. tataricum*) (L.) *Gaertn*. (tartary buckwheat), and *Fagopyrum dibotrys* (*F. dibotrys*) (*D. Don*) *Hara*. (perennial buckwheat), which have been well explored due to their long tradition of both edible and medicinal use (Jing et al., 2016).

Buckwheats have numerous ecological adaptabilities, so they can be cultivated in high altitude regions with low rainfall and temperature. The most popular common and tartary buckwheats are originated from mountainous provinces of southern China, but now it is broadly cultivated in Asia, Europe, and the Americas. The world-leading buckwheat producing countries are China, Russia, France, Ukraine, Poland, the United States, and Kazakhstan

(Huda et al., 2021). It is generally used as a cereal after removal of the husk (cereal grain) or milled for flour. Buckwheat flour is generally processed into other foods such as noodles, confectionery, bread, and the products of fermentation such as vinegar, and alcoholic spirits. The leaves are also used as leafy vegetables and can be dried to make teas and powder. Furthermore, buckwheat plants are cultivated for landscaping and for honey production (Giménez-Bastida et al., 2015; Suzuki et al., 2020).

Buckwheat has high nutritional value: it contains several times more proteins, fat, fiber, and vitamins than other popular grains, such as rice, wheat, maize, and oat (Francis Raguindin et al., 2021; Huda et al., 2021). Buckwheat proteins are of high quality with a balanced essential amino acid composition characterized by abundant amounts of sulfur-rich amino acids (Martínez-Villaluenga et al., 2020). The protein quality of buckwheat seeds varies between the tartary and common buckwheat types, but both are gluten-free and contain considerable amount of indispensable amino acids (Jin et al., 2020a,b).

Buckwheat possesses high nutritional and health-promoting value due to its constituents. 178 bioactive compounds have been detected in buckwheats (Francis Raguindin et al., 2021). There are fatty acids, polysaccharides, proteins, and amino acids, iminosugars, dietary fiber, fagopyrins, resistant starch, vitamins, as well as triterpenoid saponins, flavinoids (i.e., rutin, quercetin, kaempherol) flavones (luteolin, vitexin, orientin, metabolites of quercetin), flavanones (hesperitins, naringenins), flavanols (catechins), anthocyanins (cyanidins) fagopyrins, isoflavones (genistein and others), stilbene resveratrol and many others. Buckwheat is also a good source of minerals (calcium, iron, and zinc) (Luthar et al., 2020; Martínez-Villaluenga et al., 2020; Huda et al., 2021).

In damaged or milled grain under wet conditions, most of the rutin is degraded to quercetin by rutin-degrading enzymes (e.g., rutinosidase). From tartary buckwheat varieties with low rutinosidase activity it is possible to prepare foods with high levels of rutin, with the initial levels preserved in the grain (Luthar et al., 2020).

3.4.3 Physiological and therapeutical actions

Various buckwheat compounds have antioxidant, antitumor, antiinflammatory, hepatoprotective, antibacterial, antifungal, antiviral, antiulcer, antifatigue, hypolipidemic, prebiotic, immunomodulatory, neuroprotective, cardioprotective, hypotensic, antidiabetic, antiatherosclerotic, antineoplastic, antithrombotic, and antiaging activities. They decrease blood glucose and cholesterol level indicating the applicability of buckwheat and its constituents for treatment of metabolic syndrome and related disorders (Giménez-Bastida et al., 2015; Kreft et al., 2016; Lu et al., 2017; Kawai et al., 2018; Huda et al., 2021). Buckwheat constituents can suppress adipocyte differentiation and fat accumulation, and therefore obesity and obesity-related disorders (Park et al., 2016). They have protective action against ethanol-induced liver injury (Jin et al., 2020a,b). In addition, buckwheat does not contain substantial amount of gluten, therefore it can prevent metabolic disorders induced by gluten (Jin et al., 2020a,b; Martínez-Villaluenga et al., 2020). Buckwheat constituents are prebiotics supporting and normalizing good microflora (Giménez-Bastida et al., 2015; Peng et al., 2019; Jin et al., 2020a,b; Huda et al., 2021).

On the other hand, it is important to note that these data concerning the health-promoting effect of buckwheat compounds were obtained during *in vitro* experiments or few experiments performed on laboratory animals. Human studies showed only that buckwheat can be a good basis for functional food to mitigate the manifestation of gluten-related diseases such as celiac disease, nonceliac sensitivity, and wheat allergy. Furthermore, the few clinical trials demonstrated that both acute and chronic administration of buckwheat to diabetic subjects modulated metabolic and cardiovascular markers (Martínez-Villaluenga et al., 2020).

3.4.4 Mechanisms of action

Due to a number of biologically active constituents with various properties and targets, the mechanisms of buckwheat action is not easy to formulate.

The drug target prediction, network analysis, and molecular docking simulation enabled to predict 97 putative target molecules, which can mediate buckwheat action on type II diabetes, hypertension and hyperlipidemia (Lu et al., 2017). The experiments indicated involvement of some of these predicted targets in mediating buckwheat action on these and other illnesses.

For example, the ability of buckwheat to suppress fat storage can be explained by its inhibitory action on lipid accumulation, triglyceride content, and glycerol-3-phosphate dehydrogenase activity during adipocyte differentiation. This action was associated with down-regulation of the mRNA levels of genes involved in fatty acid synthesis, such as peroxisome proliferator-activated receptor-γ (PPAR-γ), CCAAT/enhancer binding protein-α (CEBP-α), adipocyte protein 2 (aP2), acetyl-CoA carboxylase (ACC), fatty acid synthase (FAS), and stearoylcoenzyme A desaturase-1 (SCD-1) (Lee et al., 2017).

The anti-oxidant properties of 11 kinds of buckwheat substances, especially of rutin and quercetin (Huda et al., 2021) indicate that their ability to neutralize reactive oxygen species and to prevent the oxidative stress-induced

124 Contaminants and Plants Action on Female Reproduction

DNA damage and apoptosis can explain their antitumor, antiinflammatory, antidiabetic, antiaging, neuro-, hepatic-, and cardioprotective properties (Kreft, 2016; Zhang et al., 2018; Huda et al., 2021).

The antioxidant properties of buckwheat molecules and their ability to suppress obesity and obesity-related disorders can be due to antiinflammatory action of buckwheat. At least, buckwheat extract can suppress adipocyte inflammation reducing the mRNA levels of inflammatory mediators such as tumor necrosis factor-α (TNF-α), interleukin-6 (IL-6), monocyte chemo-attractant protein 1 (MCP-1), inducible nitric oxide synthase (iNOS) and nitric oxide (NO) production (Lee et al., 2017). The antiinflammatory action of buckwheat molecule quercetin-3-O-glucuronide can be explained by its accumulation and metabolization by macrophages and resulted suppression of their actions (Kawai, 2018).

The prebiotic properties of buckwheat polysaccharides, fiber and phenols, especially rutin, quercetin could be responsible for some of its effect on lipid and carbohydrate metabolism and immune system regulated by gut microbiota (Giménez-Bastida et al., 2015; Peng et al., 2019; Jin et al., 2020a,b; Huda et al., 2021). Buckwheat polypeptide glutaredoxin can promote lifespan via up-regulation of both antioxidant enzymes and of heat shock transcription factor (Li et al., 2018). The buckwheat extract and trypsin inhibitor isolated from buckwheat can suppress cancerogenesis via up-regulation of cancer cell nuclear apoptosis associated with DNA fragmentation and of cytoplasmic apoptosis associated with mitochondria dysfunctions manifested by release of cytochrome C from the mitochondria to the cytosol and activation of cyto-plasmic caspase-3, -8, and -9 (Bai et al., 2015; Wang et al., 2015), proapoptotic transcription factor p53 (Peng et al., 2015) and apoptosis regulators bcl-2 and FAS (Guo et al., 2010). The other buckwheat molecules, flavonoids, have similar mechanism of action: they induce leukemia cell apoptosis via release of cytochrome C from mitochondria to the cytosol, as well as via upregulation of proapoptotic Fas expression on the cell surface, through a caspase-3-dependent mechanism in cytoplasm and via inactivation of nuclear tran-scription factor NF-kappaB (Ren et al., 2003). Furthermore, D-chiro-inositol, and other bioactive compound of tartary buckwheat, can prevent cardiovas-cular diseases and type II diabetes via prevention of mitochondrial dysfunc-tions as well (Zhang et al., 2018). In addition, it can prevent these illnesses via mitigation of endoplasmic reticulum stress, an inactivation of inflammation-associated interleukin 6 and Jun n-terminal kinase, which are considered triggers of endothelial dysfunction and lesion (Zhang et al., 2018).

In adipocytes, buckwheat extract reduced cyclin-dependent kinase 2 and cyclin expression and increased p21 and p27 expression, thus causing cell cycle arrest at the G1/S phase (Hong et al., 2017). The arrest of the cancer cell cycle at G(0) phase and prevention of transition from G(0)/G(1) phase to S phase induced also the anticancer protein isolated from buckwheat (Guo et al., 2010).

Finally, buckwheat-containing food can prevent adipocyte differentiation and fat storage via inhibition of the expression of adipogenic transcription factor, peroxisome proliferator-activated receptor γ, and AMP-activated protein kinase (Park et al., 2016).

These observations demonstrate the action of numerous buckwheats bioactive molecules via multiple intracellular signaling pathways. In some cases, the same intracellular pathway could mediate action of several buckwheat molecules on different processes. On the other hand, one buckwheat molecule can use different mediators of its action.

3.4.5 Effects on female reproductive processes

3.4.5.1 Effect of on ovarian and reproductive state

The accessible databases do not contain any reports concerning buckwheat influence on ovarian, reproductive state, reproductive cycle, fecundity, and other reproductive characteristics *in vivo*.

3.4.5.2 Effect on ovarian cell functions

It can be postulated that fecundity depends on ovarian follicle development, which is in turn determined by follicular cell viability. The viability of cell population in turn depends on the proliferation: apoptosis rate. The available observations concerning character buckwheat on these processes in ovarian cells are contradictory. In some experiments, buckwheat extract reduced viability and accumulation of marker and promoter of proliferation PCNA, not influencing cytoplasmic/mitochondrial apoptosis marker bax (Sirotkin et al., 2020c) in cultured porcine ovarian granulosa cells. In other experiments, this extract did not influence viability and this apoptosis marker, but PCNA accumulation was promoted (Sirotkin et al., 2020b). In other similar experiments, buckwheat extract enhanced cell viability and accumulation of bax not influencing PCNA (Sirotkin et al., 2020a).

Despite the variations in the observed effects of buckwheat, the performed *in vitro* experiments demonstrated its direct action on ovarian cell proliferation, apoptosis, and viability.

3.4.5.3 Effect on reproductive hormones

Inconclusive are also the reports concerning influence of buckwheat extract on steroid hormones' release by cultured porcine ovarian granulosa cells. In some experiments, the addition of buckwheat extract increased the release of progesterone, but not of estradiol (Sirotkin et al., 2020c). Other experiments did not show buckwheat influence on progesterone release, but it reduced estradiol output (Sirotkin et al., 2020b). In next experiments, buckwheat extract affected neither progesterone nor estradiol release, but inhibited testosterone output (Sirotkin et al., 2020a).

126 Contaminants and Plants Action on Female Reproduction

Therefore, the performed studies on cultured porcine ovarian granulosa cells showed that buckwheat extract can directly influence basic ovarian cell functions—proliferation, apoptosis, viability, and steroidogenesis, although the character of this influence varied among the experiments. Such variability in response of cultured porcine ovarian cells to buckwheat treatment could be due to variations in the initial state of ovarian cells used in experiments.

The functional interrelationships between buckwheat effects and their mechanisms remain to be elucidated. Nevertheless, it might be proposed that buckwheat could directly affect ovarian cells and therefore female reproductive processes via changes in release of steroid hormones—the known regulators of ovarian cell proliferation, apoptosis, and ovarian foliculogenesis and fecundity (Sirotkin, 2014)

3.4.5.4 Effect on ovarian cell response to environmental contaminants

The publications cited above demonstrated that buckwheat extract can not only affect basic ovarian cell functions, but also prevent the adverse influence of some environmental contaminants on these cells. Xylene-suppressed viability, accumulation of PCNA, bax and release of progesterone and estradiol by cultured porcine ovarian granulosa cells. Addition of buckwheat extract prevented the suppressive influence of xylene on four from five measured ovarian cell parameters (viability, PCNA, bax, and estradiol output) (Sirotkin et al., 2020b). Buckwheat was able to modify the influence of another oil-related environmental contaminant—benzene on cultured porcine granulosa cells. Benzene reduced cell viability, as well as P and E release, but not PCNA and bax accumulation. Buckwheat addition induced the stimulatory influence of benzene on accumulation of proliferation marker PCNA (Sirotkin et al., 2020a). Finally, buckwheat modified the influence of copper nanoparticles supported in titania on these cells. These nanoparticles $CuNPs/TiO_2$ increased cell viability, proliferation, apoptosis, and testosterone but not progesterone release, and reduced the 17β-estradiol output. Addition of buckwheat extract to culture medium mitigated the nanoparticles' effects on cell viability, PCNA and estradiol release (Sirotkin et al., 2020c).

Taken together, the available reports demonstrate the ability of buckwheat to directly affect ovarian cells, to change their basic functions (proliferation, apoptosis, viability, steroidogenesis), as well as to mitigate or prevent the influence of some environmental contaminants (xylene, benzene, and copper nanoparticles) on these functions.

3.4.6 Mechanisms of action on female reproductive processes

The action of buckwheat on **ovarian hormones**, as well as on **proliferation** and **apoptosis** of ovarian cells could indicate that hormones and intracellular regulators of ovarian cell proliferation, apoptosis, and viability could be mediators of buckwheat action on female reproductive processes. Neverthe-less, the direct evidence for the mediatory role of these signaling molecules are absent now. It is not to be excluded that buckwheat can influence reproductive processes through the same mediators as on nonreproductive events listed above. Nevertheless, mechanisms of action of buckwheat and its molecules on female reproductive processes remain to be elucidated yet.

3.4.7 Application in reproductive biology and medicine

If the suppressive action of buckwheat on ovarian functions occurs not only *in vitro*, but also *in vivo*, the potential adverse influence of this plant on female reproduction is to be taken into account before medical application of buckwheat or its constituents or before overconsumption of buckwheat containing food.

On the other hand, the ability of buckwheat to prevent the influence of various harmful environmental contaminants on ovarian cells suggests its potential applicability as a novel natural protector of female reproductive system from contaminant action.

These applications of buckwheat and its molecules remain rather specu-lative as of yet; they could be validated by the corresponding *in vivo* studies.

3.4.8 Conclusions and possible direction of future studies

The available literature demonstrates the high nutritional value of buckwheat, as well as the high contents and number of regulatory molecules in this functional food plant. These molecules can, via multiple signaling pathways, affect a wide spectrum of physiological processes and illnesses, which sug-gests a therapeutic value of buckwheat substances. Furthermore, recent reports demonstrate ability of buckwheat extract to affect directly basic ovarian cell functions (proliferation, apoptosis, viability, steroidogenesis).

On the other hand, the major data concerning therapeutic effects of buckwheat and its molecules were obtained in animal or *in vitro* experiments, while no clinical trials have been performed yet. Understanding the hierar-chical functional interrelationships between the physiological effects and mediators of buckwheat action requires further studies. No studies concerning buckwheat influence on female reproductive processes *in vivo* has been re-ported yet. Influence of buckwheat on ovarian cells was demonstrated only in *in vitro* experiment on one model (cultured porcine granulosa cells), while the obtained results are variable, sometimes contradictory and therefore incon-clusive. Therefore, understanding the character and applicability of buckwheat influence on female reproductive processes requires further studies.

The buckwheat application could be promoted also by higher output of its biological active constituents. The genomic analysis and selection of this plant, as well as the improvement of processing of buckwheat-based food can increase the content of desirable molecules (for example, of proteins, rutin, and quercetin; Jin et al., 2020a,b; Luthar et al., 2020; Rodriquex et al., 2020) in this food, to improve their bioavailability (Kawai, 2018) and its physiological

128 Contaminants and Plants Action on Female Reproduction

and medicinal benefits. Isolation (Li et al., 2017; Jin et al., 2020a,b) or production of recombinant bioactive buckwheat molecules (Li et al., 2018) could provide new drugs for prevention and treatment of various disorders.

References

Bai CZ, Feng ML, Hao XL, Zhao ZJ, Li YY, Wang ZH. Anti-tumoral effects of a trypsin inhibitor derived from buckwheat in vitro and in vivo. Mol Med Rep 2015;12(2):1777−82. https://doi.org/10.3892/mmr.2015.3649.

Baldovska S, Roychoudhury S, Bandik M, et al. Ovarian steroid hormone secretion by human granulosa cells after supplementation of sambucus nigra L. extract. Physiological Research 2021. PMID: 34505534.

Francis Raguindin P, Adam Itodo O, Stoyanov J, Dejanovic GM, Gamba M, Asllanaj E, Minder B, Bussler W, Metzger B, Muka T, Glisic M, Kern H. A systematic review of phytochemicals in oat and buckwheat. Food Chem 2021;338:127982. https://doi.org/10.1016/j.foodchem.2020.127982.

Giménez-Bastida JA, Zieliński H. Buckwheat as a functional food and its effects on health. J Agric Food Chem 2015;63(36):7896−913. https://doi.org/10.1021/acs.jafc.5b02498.

Guo X, Zhu K, Zhang H, Yao H. Anti-tumor activity of a novel protein obtained from tartary buckwheat. Int J Mol Sci 2010;11(12):5201−11. https://doi.org/10.3390/ijms11125201.

Hong H, Park J, Lumbera WL, Hwang SG. Monascus ruber-fermented buckwheat (red yeast buckwheat) suppresses adipogenesis in 3T3-L1 cells. J Med Food 2017;20(4):352−9. https://doi.org/10.1089/jmf.2016.3761.

Huda MN, Lu S, Jahan T, Ding M, Jha R, Zhang K, Zhang W, Georgiev MI, Park SU, Zhou M. Treasure from garden: bioactive compounds of buckwheat. Food Chem 2021;335:127653. https://doi.org/10.1016/j.foodchem.2020.127653.

Jin HR, Lee S, Choi SJ. Pharmacokinetics and protective effects of tartary buckwheat flour extracts against ethanol-induced liver injury in rats. Antioxidants 2020a;9(10):913. https://doi.org/10.3390/antiox9100913.

Jin J, Ohanenye IC, Udenigwe CC. Buckwheat proteins: functionality, safety, bioactivity, and prospects as alternative plant-based proteins in the food industry. Crit Rev Food Sci Nutr 2020:1−13. https://doi.org/10.1080/10408398.2020.1847027.

Jing R, Li HQ, Hu CL, Jiang YP, Qin LP, Zheng CJ. Phytochemical and pharmacological profiles of three Fagopyrum buckwheats. Int J Mol Sci 2016;17(4):589. https://doi.org/10.3390/ijms17040589.

Kawai Y. Understanding metabolic conversions and molecular actions of flavonoids in vivo:toward new strategies for effective utilization of natural polyphenols in human health. J Med Invest 2018;65(3.4):162−5. https://doi.org/10.2152/jmi.65.162.

Kreft M. Buckwheat phenolic metabolites in health and disease. Nutr Res Rev 2016;29(1):30−9. https://doi.org/10.1017/S0954422415000190.

Lee W, Bae JS. Anti-inflammatory effects of aspalathin and nothofagin from rooibos (aspalathus linearis) In vitro and In vivo. Inflammation 2015;38(4):1502−16. https://doi.org/10.1007/s10753-015-0125-1.

Lee MS, Shin Y, Jung S, Kim SY, Jo YH, Kim CT, Yun MK, Lee SJ, Sohn J, Yu HJ, Kim Y. The inhibitory effect of tartary buckwheat extracts on adipogenesis and inflammatory response. Molecules 2017;22(7):1160. https://doi.org/10.3390/molecules22071160.

Li F, Zhang X, Li Y, Lu K, Yin R, Ming J. Phenolics extracted from tartary (*Fagopyrum tartaricum* L. Gaerth) buckwheat bran exhibit antioxidant activity, and an antiproliferative effect on

Food/medicinal herbs and their influence Chapter | 3 **129**

human breast cancer MDA-MB-231 cells through the p38/MAP kinase pathway. Food Funct 2017;8(1):177—88. https://doi.org/10.1039/c6fo01230b.

Li F, Ma X, Cui X, Li J, Wang Z. Recombinant buckwheat glutaredoxin intake increases lifespan and stress resistance via hsf-1 upregulation in *Caenorhabditis elegans*. Exp Gerontol 2018;104:86—97. https://doi.org/10.1016/j.exger.2018.01.028.

Lu CL, Zheng Q, Shen Q, Song C, Zhang ZM. Uncovering the relationship and mechanisms of Tartary buckwheat (*Fagopyrum tataricum*) and Type II diabetes, hypertension, and hyperlipidemia using a network pharmacology approach. PeerJ 2017;5:e4042. https://doi.org/10.7717/peerj.4042.

Luthar Z, Germ M, Likar M, Golob A, Vogel-Mikuš K, Pongrac P, Kušar A, Pravst I, Kreft I. Breeding buckwheat for increased levels of rutin, quercetin and other bioactive compounds with potential antiviral effects. Plants 2020;9(12):1638. https://doi.org/10.3390/plants9121638.

Martínez-Villaluenga C, Peñas E, Hernández-Ledesma B. Pseudocereal grains: nutritional value, health benefits and current applications for the development of gluten-free foods. Food Chem Toxicol 2020;137:111178. https://doi.org/10.1016/j.fct.2020.111178.

Park GS, Jeon YM, Kim JH, Park SK, Lee MY. In vitro studies on anti-obesity activity of Korean Memilmuk through AMPK activation. J Environ Biol 2016;37(1):1—5.

Peng W, Hu C, Shu Z, Han T, Qin L, Zheng C. Antitumor activity of tatariside F isolated from roots of *Fagopyrum tataricum* (L.) Gaertn against H22 hepatocellular carcinoma via up-regulation of p53. Phytomedicine 2015;22(7-8):730—6. https://doi.org/10.1016/j.phymed.2015.05.003.

Peng L, Zhang Q, Zhang Y, Yao Z, Song P, Wei L, Zhao G, Yan Z. Effect of tartary buckwheat, rutin, and quercetin on lipid metabolism in rats during high dietary fat intake. Food Sci Nutr 2019;8(1):199—213. https://doi.org/10.1002/fsn3.1291.

Ren W, Qiao Z, Wang H, Zhu L, Zhang L, Lu Y, Zhang Z, Wang Z. Molecular basis of Fas and cytochrome c pathways of apoptosis induced by tartary buckwheat flavonoid in HL-60 cells. Methods Find Exp Clin Pharmacol 2003;25(6):431—6. https://doi.org/10.1358/mf.2003.25.6.769647.

Rodríguez JP, Rahman H, Thushar S, Singh RK. Healthy and resilient cereals and pseudo-cereals for marginal agriculture: molecular advances for improving nutrient bioavailability. Front Genet 2020;11:49. https://doi.org/10.3389/fgene.2020.00049.

Sirotkin AV, Macejková M, Tarko A, Fabova Z, Alrezaki A, Alwasel S, Harrath AH. Effects of benzene on gilts ovarian cell functions alone and in combination with buckwheat, rooibos, and vitex. Environ Sci Pollut Res Int 2020. https://doi.org/10.1007/s11356-020-10739-7.

Sirotkin AV, Macejková M, Tarko A, Fabova Z, Alwasel S, Harrath AH. Buckwheat, rooibos, and vitex extracts can mitigate adverse effects of xylene on ovarian cells in vitro. Environ Sci Pollut Res Int 2020. https://doi.org/10.1007/s11356-020-11082-7.

Sirotkin AV, Radosová M, Tarko A, Fabova Z, Martín-García I, Alonso F. Abatement of the stimulatory effect of copper nanoparticles supported on titania on ovarian cell functions by some plants and phytochemicals. Nanomaterials 2020c;10(9):1859. https://doi.org/10.3390/nano10091859.

Sirotkin AV. Regulators of ovarian functions. New York: Nova Publishers Inc.; 2014. 194 pp.

Suzuki T, Noda T, Morishita T, Ishiguro K, Otsuka S, Brunori A. Present status and future perspectives of breeding for buckwheat quality. Breed Sci 2020;70(1):48—66. https://doi.org/10.1270/jsbbs.19018.

Wang Z, Li S, Ren R, Li J, Cui X. Recombinant buckwheat trypsin inhibitor induces mitophagy by directly targeting mitochondria and causes mitochondrial dysfunction in hep G2 cells. J Agric Food Chem 2015;63(35):7795—804. https://doi.org/10.1021/acs.jafc.5b02644.

Zhang B, Gao C, Li Y, Wang M. D-chiro-inositol enriched *Fagopyrum tataricum* (L.) Gaench extract alleviates mitochondrial malfunction and inhibits ER stress/JNK associated inflammation in the endothelium. J Ethnopharmacol 2018;214:83—9. https://doi.org/10.1016/j.jep.2017.12.002.

130 Contaminants and Plants Action on Female Reproduction

Chapter 3.5

Curcuma/turmeric (*Curcuma longa* L., *Curcuma zedoaria* (Christm.) Roscoe)

3.5.1 Introduction

Curcuma/turmeric is currently a popular cooking additive and functional food for prevention of various disorders and illnesses. Its health-promoting effects have been described in several reviews (Lestari and Indrayanto, 2014; Kunnumakkara et al., 2017; Dosoky and, Setzer, 2018). On the other hand, only one publication (Mohebbati et al., 2017) reviewed curcuma action on reproductive processes, while it was focused on cancerogenesis, but not on curcuma action on healthy reproductive system. The present publication is the first attempting to review the current available knowledge concerning the action of curcuma and its constituents on healthy female reproductive system, as well as of the data concerning their ability to prevent and to treat reproductive disorders.

3.5.2 Provenance and properties

The genus *Curcuma* L. (Zingiberaceae) represents a group of perennial rhizomatous herbs native to tropical and subtropical regions. Curcuma is extensively cultivated in tropical and subtropical regions of Asia, Australia, and South America. There are approximately 93–100 Curcuma species. The genus is best known for being an essential source of coloring and flavoring agents in the Asian cuisines, traditional medicines, spices, dyes, perfumes, cosmetics, and ornamental plants. Curcumas are a part of curry powder and other spices currently considered as functional food. *C. longa* (Indian turmeric) and *C. zedoaria* (zedoary) are the most extensively studied species of Curcuma due to their high commercial value. Turmeric is native to Southeast Asia but is cultivated extensively worldwide now. Zedoary is native to northeast India and Indonesia, but widely cultivated in subtropical regions around the world. The rhizome is the most commonly used part of these plants.

Turmeric rhizome typically contains carbohydrates (69.4%), protein (6.3%), fat (5.1%), and minerals (3.5%). The main active components of the rhizome are the nonvolatile curcuminoids and the volatile oil (Dosoky and Setzer, 2018). A major biologically active component of *Curcuma longa* L. are curcumoids curcumin, which is chemically known as diferuloylmethane (the chemical name 1, 7-bis (4-hydroxy-3-methoxyphenyl)-1, 6-eptadiene-3, 5-dione, Fig. 3.1) and its polyphenolic derivatives demethoxycurcumin and bisdemethoxycurcumin (Lestari and Indrayanto, 2014). The major components of curcuma oils are sesquiterpenoids and monoterpenoids (Dosoky and Setzer, 2018).

3.5.3 Physiological and therapeutic actions

The Curcuma species are used medicinally in Bangladesh, Malaysia, India, Nepal, and Thailand to treat pneumonia, bronchial complaints, leucorrhoea, diarrhea, dysentery, infectious wounds or abscesses, and insect bites. In traditional medicine, turmeric is extensively used as a carminative, digestive aid, stomachic, appetizer, anthelmintic, tonic and laxative. It is also used for treating fever, gastritis, dysentery, infections, chest congestion, cough, hypercholesterolemia, hypertension, rheumatoid arthritis, jaundice, liver and gall bladder problems, urinary tract infections, skin diseases, diabetic wounds, and menstrual discomfort (Dosoky and Setzer, 2018).

Polyphenol curcumin has antioxidant, antiinflammatory, antimicrobial, antiangiogenic, and anticancer properties. Numerous studies have indicated that curcumin is a highly potent antimicrobial agent and has been shown to be active against various chronic diseases including various types of cancers, diabetes, obesity, cardiovascular, pulmonary, neurological and autoimmune diseases (Vallianou et al., 2015; Nayak et al., 2016; Kunnumakkara et al., 2017). Furthermore, this compound has also been shown to be synergistic with other nutraceuticals such as resveratrol, piperine, catechins, quercetin and genistein (Kunnumakkara et al., 2017).

The essential oil of Curcuma species possesses a wide variety of pharmacological properties, including antiinflammatory, anticancerous, antiproliferative, hypocholesterolemic, antidiabetic, antihepatotoxic, antidiarrheal, carminative, diuretic, antirheumatic, hypotensive, antioxidant, antimicrobial,

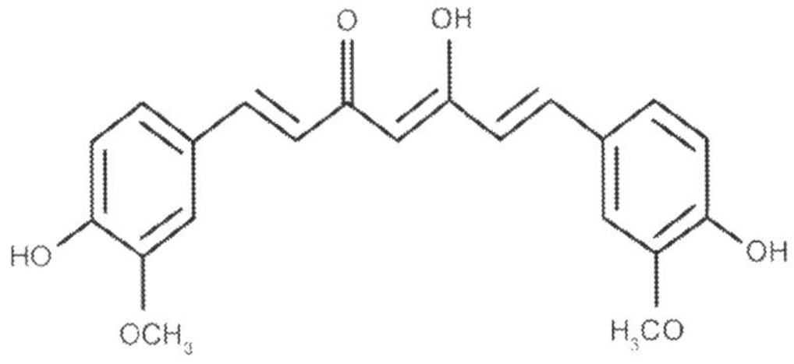

FIGURE 3.1 The chemical structure of curcumin. *From www.reserachgate.com.*

132 Contaminants and Plants Action on Female Reproduction

antiviral, insecticidal, larvicidal, antivenomous, antithrombotic, antityrosinase, and cyclooxygenase-1 (COX-1) inhibitory activities. Curcuma oils are also known to enhance immune function, promote blood circulation, accelerate toxin elimination, and stimulate digestion (Dosoky and Setzer, 2018).

Curcuma longa extracts possess renal protective, hepatoprotective, cardioprotective, antidiabetic, neuroprotective and gonadoprotective effects (Kunnumakkara et al., 2017; Mohebbati et al., 2017).

3.5.4 Mechanisms of action

Common molecular targets of curcumin include transcription factors, inflammatory mediators, protein kinases and enzymes like protein reductases and histone acetyltransferase. It can interact with a huge number of different signaling and regulator proteins such as nuclear factor E2-related factor 2 (Nrf2), β-catenin, NF-κB, p38 MAPK, DNA (cytosine-5)-methyltransferase-1, COX-2, 5-lipoxygenase, PGE2, FOXO3, inducible NOS, ROS, cyclin D1, VEGF, glutathione, cytosolic PLA2, p-Tau (p-τ) and TNF-α. For example, curcumin has been shown to exhibit antiinflammatory activity through the suppression of numerous cell signaling pathways including NF-κB, STAT3, Nrf2, ROS and COX-2 (Kunnumakkara et al., 2017). In addition, curcumin can suppress cancerogenesis (Liu et al., 2019) and other illnesses (Momtazi et al., 2016) targeting microRNAs and long noncoding RNAs. The physiological action of turmeric oils could be explained by their ability to inhibit reactive oxygen species, and COX-2 and other signaling molecules too (Dosoky and Setzer, 2018). This ability of turmeric molecules to modulate multiple cell signaling pathways linked to different chronic diseases, which strongly suggests that it is a potent multitargeted polyphenol (Kunnumakkara et al., 2017).

3.5.5 Effect on female reproductive processes

3.5.5.1 Effect on ovarian and reproductive state

The dietary administration of curcumin does not affect the ovarian weight or litter size of rats (Inano et al., 2000), but an artificial pegylated curcumin analogue reportedly suppresses rat ovarian folliculogenesis, sexual maturation, and fecundity (Murphy et al., 2012). On the other hand, curcumin promoted ovarian growth, folliculogenesis (Azami et al., 2020) and mitigated the age-dependent suppression of ovarian functions in mice (Tiwari-Pandey and Ram Sairam, 2009), D-galactose —induced ovarian failure in mice (Yan et al., 2018) and cyclophosphamide-induced ovarian failure in rat (Melekoglu et al., 2018). Feeding of rabbits with turmeric powder failed to affect their ovarian length and weight but did increase the number of primary follicles, diameter of primary, secondary and tertiary follicles. It did not affect conception or kindling rate, but increased the number of liveborn and weaned pups and decreased the number of stillborn pups (Sirotkin et al., 2018).

Therefore, curcumin can both up- and down-regulated puberty, ovarian folliculogenesis and fecundity. The character of its effect depends on animal species and experimental model.

3.5.5.2 Effect on ovarian cell functions

One *in vitro* study demonstrated the inhibitory action of curcumin on basic healthy ovarian cell functions—proliferation, viability, and apoptosis. In experiments of Kádasi et al. (2017), curcumin addition reduced proliferation, viability and promoted apoptosis in cultured porcine ovarian granulosa cells. Numerous studies on ovarian cancer cells showed curcumin's ability to suppress cell cycle and to promote their apoptosis (Inano et al., 2000; Shehzad et al., 2010; Watson et al., 2010; Rath et al., 2013; Vallianou et al., 2015; Seo et al., 2016; Bondì et al., 2017; Dwivedi et al., 2018; Hatasmpour et al., 2018; Duse, 2019; Liu et al., 2019; Fatemi et al., 2020)

The *in vivo* studies performed on mices demonstrated the opposite, stimulatory action of curcumin or its analogue on ovarian functions—it promoted proliferation and reduced apoptosis in murine ovarian cells (Voznesens'ka et al., 2010; Aktas et al., 2012; Yan et al., 2018; Fatemi Abhari et al., 2020). Moreover, curcumin supported murine ovarian oogenesis (Voznesens'ka et al., 2010; Alekseyeva et al., 2011; Azami et al., 2020).

3.5.5.3 Effect on reproductive hormones

In vivo experiments demonstrated the ability of curcumin to decrease concentration of follicle-stimulating hormone and luteinizing hormone in rat serum (Melekoglu et al., 2018; Azami et al., 2020). Surprisingly, decrease in gonadotropin release was associated with increase in serum level of estradiol and anti-Mullerian hormone (Melekoglu et al., 2018; Azami et al., 2020), which should be up-regulated by gonadotropins (Sirotkin, 2014). Feeding with curcumin increased also ovarian production of progesterone, but decreased ovarian release of testosterone and leptin in rabbits. Moreover, feeding with curcumin altered the response of ovarian cells to LH: LH decreased the leptin output of control rabbits but increased that of rabbits fed with curcumin (Sirotkin et al., 2018).

In vitro experiments also showed the stimulatory influence of curcumin on progesterone and testosterone release by cultured porcine ovarian granulosa cells (Kádasi et al., 2017), but its inhibitory action on progesterone release by cultured rat luteal cells (Purwaningsih et al., 2012).

134 Contaminants and Plants Action on Female Reproduction

Some experiments (Inano et al., 2000), however did not show curcumin influence on FSH, progesterone and estradiol in rat plasma. Moreover, they demonstrated increased levels of LH and decreased levels of prolactin in rats fed with curcumin.

Thus, the performed studies demonstrated curcumin influence on (1) release of gonadotropins by pituitary, (2) release of steroid and peptide hormones by ovarian cells, and (3) response of ovarian cells to gonadotropins.

3.5.5.4 Mechanisms of action on female reproductive processes

It can be easily noted that curcumin effects observed in *in vivo* and *in vitro* studies are different, sometimes contradictory. This fact indicates action of curcumin directly on ovarian cells and on their upstream regulators. The best-known endocrine and paracrine/autocrine regulators of reproductive and other physiological actions are hormones. It is generally accepted that fecundity depends on the growth, development, and atresia of ovarian follicles, and these processes are regulated by gonadotropins (LH and FSH), ovarian steroids (progestogens, androgens, and estrogens), and peptide hormones (Sirotkin, 2014). The previous subchapter demonstrated that curcumin can influence **pituitary and ovarian hormones**, which can mediate curcumin action on reproductive processes.

Curcumin is a phytoestrogen (Bachmeier et al., 2010) and can, therefore, interact with the endocrine system, affect the hypothalamo-hypophysial-ovarian axis, and treat its disorders (Sirotkin and Harrath, 2014). For example, in mouse ovaries, curcumin has been shown to downregulate androgen receptors and upregulate 3-beta-hydroxysteroid dehydrogenase (Tiwari-Pandey and Ram Sairam, 2009), although it had no effect on aromatase (Valentine et al., 2006). Furthermore, curcumin mitigated the effect of dehydroepiandrosterone on mouse ovarian cell apoptosis (Fatemi Abhari et al., 2020). Therefore, curcumin action on the ovary could be mediated by its action on ovarian **steroid hormones receptors**.

There are indications that curcumin action on the ovary (at least the ovary suffering from polycystic ovarian syndrome) can be mediated by its suppressive action on **inflammation** (Mohebbati et al., 2017) **and inflammation-related cytokines** C-reactive protein, interleukin-6 and tumor necrosis factor-α (Mohammadi et al., 2017), **growth factors GDF-9, BMP-15** (Azami et al., 2020; Heshmati et al., 2020) and **vascular endothelial growth factor** (Mohebbati et al., 2017).

Furthermore, the action of curcumin on intracellular **regulators of cell cycle and apoptosis** in healthy (Voznesens'ka et al., 2010; Aktas et al., 2012; Kádasi et al., 2017; Yan et al., 2018; Fatemi Abhari et al., 2020) and cancer (Watson et al., 2010; Seo et al., 2016; Mohebbati et al., 2017; Duse et al., 2019; Hatamipour et al., 2018; Liu et al., 2019) cells suggests that these regulators could be mediators of curcumin action on ovarian functions.

Viability, proliferation and apoptosis of ovarian cells should affect their hormone release mentioned earlier, although the opposite effect of steroid and peptide hormones on ovarian cell proliferation and apoptosis has been documented too (Sirotkin, 2014; Fatemi Abhari et al., 2020).

Curcumin action on ovarian cell proliferation and apoptosis can be mediated by its action on **transcription factor p53**, which stops cell cycle, promotes DNA reparation, and induces apoptosis in defect or cancer ovarian cells (Mohebbati et al., 2017). On the other hand, there is evidence that curcumin-induced apoptosis in ovarian carcinoma cells is p53-independent and involves activation of proliferation promoter p38 **mitogen-activated protein kinase** and downregulation of antiapoptotic peptides **Bcl-2** and **survivin** expression and **Akt signaling** (Watson et al., 2010). The involvement of mitogen-activated protein kinase in mediating curcumin action has been demonstrated for healthy ovarian cells too (Purwaningsih et al., 2012).

The results of other studies suggested that curcumin action on ovarian cancer cells or ovarian cells suffered from polycystic ovarian syndrome could be mediated by **transcription factor STAT, matrix metalloproteinase-9** (Mohebbati et al., 2017), sarco/endoplasmic reticulum **Ca2 + ATPase** (Seo et al., 2016), **5′ adenosine monophosphate-activated protein kinase, nuclear factor-κB, heat shock protein 70** (Voznesens'ka et al., 2010; Hatamipour et al., 2018), **Nrf2/HO-1 and PI3K/Akt** (Yan et al., 2018), **AKT/mTOR/p70S6K** (Liu et al., 2019), **Wnt/β-catenin signaling pathways** (Hatamipour et al., 2018) and **enzymes SIRT-1 and SIRT-3** (Heshmati et al., 2020).

The antioxidant curcumin can bind **reactive oxygen species**, and, therefore, prevent peroxidation and degradation of nucleic acids, peptides and lipids and the consequent apoptosis, aging or malignant transformation of healthy, but not of cancer cells (Mohebbati et al., 2017; Azami et al., 2020; Heshmati et al., 2020).

Finally, recently the involvement of some **microRNAs** and **long non-coding RNAs** in regulating various types of cancer cells and curcumin action on these cells have been indicated (Hatamipour et al., 2018; Liu et al., 2019).

3.5.6 Application in reproductive biology and medicine

The available reports concerning curcumin effect on reproductive processes allow to outline possible areas of its application.

The stimulatory action of curcumin on ovarian functions and fecundity in laboratory rodents (Voznesens'ka et al., 2010; Aktas et al., 2012; Murphy et al., 2012; Yan et al., 2018; Azami et al., 2020; Fatemi Abhari et al., 2020) and rabbits (Sirotkin, 2018) *in vivo* indicates its potential applicability as natural biostimulator of reproduction in large farm animal production. Furthermore, its ability to improve oocyte maturation, quality, and

136 Contaminants and Plants Action on Female Reproduction

developmental capacity (Voznesens'ka et al., 2010; Alekseyeva et al., 2011; Aktas et al., 2012; Azami et al., 2020) suggests that its addition could be beneficial for *in vitro* oocyte maturation, fertilization, and embryo production.

Furthermore, a number of evidences suggest its applicability for prevention, mitigation and maybe even treatment of some age-dependent reproductive disorders including late sexual maturation (Tiwari-Pandey and Ram Sairam, 2009; Azami et al., 2020), reproductive aging and ovarian insufficiency and failure (Tiwari-Pandey et al., 2009; Voznesens'ka et al., 2010; Alekseyeva et al., 2011; Melekoglu et al., 2018; Yan et al., 2018). Moreover, its potential to prevent, mitigate and treat signs of ovarian cancer (Inano et al., 2000; Shehzad et al., 2010; Watson et al., 2010; Rath et al., 2013; Vallianou et al., 2015; Seo et al., 2016; Bondì et al., 2017; Dwivedi et al., 2018; Hatasmpour et al., 2018; Duse, 2019; Liu et al., 2019; Fatemi et al., 2020), polycystic ovarian syndrome (Mohammadi et al., 2017; Fatemi Abhari et al., 2020; Heshmati et al., 2020), and adverse effects of immunological shock (Voznesens'ka et al., 2010), hypoxia (Sak et al., 2013), ionizing irradiation (Aktas et al., 2012), ischemia (Eser et al., 2015), oxidative stress (Qin et al., 2015; Heshmati et al., 2020) and mytotoxins (Qin et al., 2015) on ovarian function suggests large therapeutic potential of curcumin and its molecules.

It is however necessary to note that wider application of curcumin is limited by its relatively low solubility, bioavailability, bio-absorption (Lestari and Indrayanto, 2014; Rath et al., 2013; Arozal et al., 2019) and stability in light and in organism (Duse et al., 2019). To improve these physicochemical and therapeutic characteristics, several approaches have been tested:

- synthetic curcumin analogues with enhanced bio-absorption and higher antioxidant and anticancer activity (Rath et al., 2013),
- application of curcumin nanoparticles with higher absorption, transport through cell membranes and accumulation in the ovary (Arozal et al., 2019) or with increased stability and proapoptotic effect on cancer cells (Duse et al., 2019),
- combination of chemical modification of curcumin and generation of nanoparticles from such modified curcumin (Dwivedi et al., 2018), and
- application of lipid (Nayak et al., 2016; Bondì et al., 2017) or metal (Fatemi et al., 2020) nanoparticles as carriers for curcumin transport.

The chemical modifications of curcumin and its carriers could expand the possibilities and efficiency of its application in animal production, biotechnology, human and veterinary medicine.

3.5.7 Conclusion and possible directions of further studies

Rhizomes of curcumas contain various biologically active substances, the best-known of which is polyphenol curcumin with a wide array of biological and medical effects. Its influence on female reproduction (puberty, reproductive aging, ovarian folliculo- and oogenesis and fecundity) has been well documented. Curcumin can affect these processes via changes in release and reception of pituitary and ovarian hormones, growth factors and cytokines, response of ovarian cells to these substances and external environmental factors, regulators of cell proliferation and apoptosis, oxidative processes and numerous intracellular signaling pathways. These effects suggest the applicability of curcumin for stimulation of female reproductive processes *in vivo* and *in vitro*, as well as for prevention, mitigation, and treatment of various reproductive disorders from ovarian insufficiency and infertility to polycystic ovarian syndrome and ovarian cancer.

On the other hand, numerous aspects of curcumas' influence and application in control of female reproduction require further studies. Previous investigations were focused on curcumin, while reproductive effects of other constituents of curcumas including curcumin derivatives and curcuma oils remain not studied as of yet. Hierarchical interrelationships between numerous mechanisms of curcumin action on reproduction require further elucidation. It could help to understand the causes of different actions of curcumin *in vivo* and *in vitro*, on healthy and on cancer cells. The available information concerning effect and potential applicability of curcumin was obtained mainly during animal and *in vitro* experiments, while the clinical studies remain rare and insufficient. Increased bioavailability and efficiency of curcumin requires further efforts of specialists in chemistry and medicine. Nevertheless, available information demonstrates that curcumin can be a promising cheap, accessible and efficient natural regulator, protector, and medicine for animal and human female reproductive processes.

References

Aktas C, Kanter M, Kocak Z. Antiapoptotic and proliferative activity of curcumin on ovarian follicles in mice exposed to whole body ionizing radiation. Toxicol Ind Health 2012;28(9):852−63. https://doi.org/10.1177/0748233711425080.

Alekseyeva IN, Makogon NV, Bryzgina TM, Voznesenskaya TY, Sukhina VS. Effects of NF-κB blocker curcumin on oogenesis and immunocompetent organ cells in immune ovarian injury in mice. Bull Exp Biol Med 2011;151(4):432−5. https://doi.org/10.1007/s10517-011-1349-1.

Arozal W, Ramadanty WT, Louisa M, Satyana RPU, Hartono G, Fatrin S, Purbadi S, Estuningtyas A, Instiaty I. Pharmacokinetic profile of curcumin and nanocurcumin in plasma, ovary, and other tissues. Drug Res 2019;69(10):559−64. https://doi.org/10.1055/a-0863-4355.

Azami SH, Nazarian H, Abdollahifar MA, Eini F, Farsani MA, Novin MG. The antioxidant curcumin postpones ovarian aging in young and middle-aged mice. Reprod Fertil Dev 2020;32(3):292−303. https://doi.org/10.1071/RD18472.

Bachmeier BE, Mirisola V, Romeo F, Generoso L, Esposito A, Dell'eva R, Blengio F, Killian PH, Albini A, Pfeffer U. Reference profile correlation reveals estrogen-like transcriptional activity of Curcumin. Cell Physiol Biochem 2010;26(3):471−82. https://doi.org/10.1159/000320570.

138 Contaminants and Plants Action on Female Reproduction

Bondì ML, Emma MR, Botto C, Augello G, Azzolina A, DI Gaudio F, Craparo EF, Cavallaro G, Bachvarov D, Cervello M. Biocompatible lipid nanoparticles as carriers to improve curcumin efficacy in ovarian cancer treatment. J Agric Food Chem 2017. https://doi.org/10.1021/acs.jafc.6b04409.

Chang VC, Cotterchio M, Boucher BA, Jenkins DJA, Mirea L, McCann SE, Thompson LU. Effect of dietary flaxseed intake on circulating sex hormone levels among postmenopausal women: a randomized controlled intervention trial. Nutr Cancer 2019;71(3):385−98. https://doi.org/10.1080/01635581.2018.1516789.

Dosoky NS, Setzer WN. Chemical composition and biological activities of essential oils of curcuma species. Nutrients 2018;10(9):1196. https://doi.org/10.3390/nu10091196.

Duse L, Agel MR, Pinnapireddy SR, Schäfer J, Selo MA, Ehrhardt C, Bakowsky U. Photodynamic therapy of ovarian carcinoma cells with curcumin-loaded biodegradable polymeric nanoparticles. Pharmaceutics 2019;11(6):282. https://doi.org/10.3390/pharmaceutics11060282.

Dwivedi P, Yuan S, Han S, Mangrio FA, Zhu Z, Lei F, Ming Z, Cheng L, Liu Z, Si T, Xu RX. Core-shell microencapsulation of curcumin in PLGA microparticles: programmed for application in ovarian cancer therapy. Artif Cells Nanomed Biotechnol 2018;46(suppl 3):S481−91. https://doi.org/10.1080/21691401.2018.1499664.

Eser A, Hizli D, Haltas H, Namuslu M, Kosus A, Kosus N, Kafali H. Effects of curcumin on ovarian ischemia-reperfusion injury in a rat model. Biomed Rep 2015;3(6):807−13. https://doi.org/10.3892/br.2015.515.

Fatemi Abhari SM, Khanbabaei R, Hayati Roodbari N, Parivar K, Yaghmaei P. Curcumin-loaded super-paramagnetic iron oxide nanoparticle affects on apoptotic factors expression and histological changes in a prepubertal mouse model of polycystic ovary syndrome-induced by dehydroepiandrosterone - a molecular and stereological study. Life Sci 2020;249:117515. https://doi.org/10.1016/j.lfs.2020.117515.

Hatamipour M, Ramezani M, Tabassi SAS, Johnston TP, Ramezani M, Sahebkar A. Demethoxycurcumin: a naturally occurring curcumin analogue with antitumor properties. J Cell Physiol 2018;233(12):9247−60. https://doi.org/10.1002/jcp.27029.

Heshmati J, Golab F, Morvaridzadeh M, Potter E, Akbari-Fakhrabadi M, Farsi F, Tanbakooei S, Shidfar F. The effects of curcumin supplementation on oxidative stress, Sirtuin-1 and peroxisome proliferator activated receptor γ coactivator 1α gene expression in polycystic ovarian syndrome (PCOS) patients: a randomized placebo-controlled clinical trial. Diabetes Metab Syndr 2020;14(2):77−82. https://doi.org/10.1016/j.dsx.2020.01.002.

Inano H, Onoda M, Inafuku N, Kubota M, Kamada Y, Osawa T, Kobayashi H, Wakabayashi K. Potent preventive action of curcumin on radiation-induced initiation of mammary tumorigenesis in rats. Carcinogenesis 2000;21:1835−41.

Kádasi A, Maruniaková N, Štochmaĺová A, Bauer M, Grossmann R, Harrath AH, Kolesárová A, Sirotkin AV. Direct effect of curcumin on porcine ovarian cell functions. Anim Reprod Sci 2017;182:77−83. https://doi.org/10.1016/j.anireprosci.2017.05.001.

Kunnumakkara AB, Bordoloi D, Padmavathi G, Monisha J, Roy NK, Prasad S, Aggarwal BB. Curcumin, the golden nutraceutical: multitargeting for multiple chronic diseases. Br J Pharmacol 2017;174(11):1325−48. https://doi.org/10.1111/bph.13621.

Lestari ML, Indrayanto G. Curcumin. Profiles Drug Subst Excipients Relat Methodol 2014;39:113−204. https://doi.org/10.1016/B978-0-12-800173-8.00003-9.

Liu Y, Sun H, Makabel B, Cui Q, Li J, Su C, Ashby Jr CR, Chen Z, Zhang J. The targeting of non-coding RNAs by curcumin: facts and hopes for cancer therapy (review). Oncol Rep 2019;42(1):20−34. https://doi.org/10.3892/or.2019.7148.

Melekoglu R, Ciftci O, Eraslan S, Cetin A, Basak N. Beneficial effects of curcumin and capsaicin on cyclophosphamide-induced premature ovarian failure in a rat model. J Ovarian Res 2018;11(1):33. https://doi.org/10.1186/s13048-018-0409-9.

Mohammadi S, Kayedpoor P, Karimzadeh-Bardei L, Nabiuni M. The effect of curcumin on TNF-α, IL-6 and CRP expression in a model of polycystic ovary syndrome as an inflammation state. J Reproduction Infertil 2017;18(4):352–60.

Mohebbati R, Anaeigoudari A, Khazdair MR. The effects of *Curcuma longa* and curcumin on reproductive systems. Endocr Regul 2017;51(4):220–8. https://doi.org/10.1515/enr-2017-0024.

Momtazi AA, Derosa G, Maffioli P, Banach M, Sahebkar A. Role of microRNAs in the therapeutic effects of curcumin in non-cancer diseases. Mol Diagn Ther 2016;20(4):335–45. https://doi.org/10.1007/s40291-016-0202-7.

Murphy CJ, Tang H, Van Kirk EA, Shen Y, Murdoch WJ. Reproductive effects of a pegylated curcumin. Reprod Toxicol 2012;34(1):120–4. https://doi.org/10.1016/j.reprotox.2012.04.005.

Nayak AP, Mills T, Norton I. Lipid based nanosystems for curcumin: past, present and future. Curr Pharmaceut Des 2016;22(27):4247–56. https://doi.org/10.2174/1381612822666160614083412.

Purwaningsih E, Soejono SK, Dasuki D, Meiyanto E. Curcumin inhibits luteal cell steroidogenesis by suppression of extracellular signal regulated kinase. Univ Med 2012;31:73–80.

Qin X, Cao M, Lai F, Yang F, Ge W, Zhang X, Cheng S, Sun X, Qin G, Shen W, Li L. Oxidative stress induced by zearalenone in porcine granulosa cells and its rescue by curcumin in vitro. PLoS One 2015;10(6):e0127551. https://doi.org/10.1371/journal.pone.0127551.

Rath KS, McCann GA, Cohn DE, Rivera BK, Kuppusamy P, Selvendiran K. Safe and targeted anticancer therapy for ovarian cancer using a novel class of curcumin analogs. J Ovarian Res 2013;6(1):35. https://doi.org/10.1186/1757-2215-6-35.

Sak ME, Soydinc HE, Sak S, Evsen MS, Alabalik U, Akdemir F, Gul T. The protective effect of curcumin on ischemia-reperfusion injury in rat ovary. Int J Surg 2013;11(9):967–70. https://doi.org/10.1016/j.ijsu.2013.06.007.

Seo JA, Kim B, Dhanasekaran DN, Tsang BK, Song YS. Curcumin induces apoptosis by inhibiting sarco/endoplasmic reticulum Ca2+ ATPase activity in ovarian cancer cells. Cancer Lett 2016;371(1):30–7. https://doi.org/10.1016/j.canlet.2015.11.021.

Shehzad A, Wahid F, Lee YS. Curcumin in cancer chemoprevention: molecular targets, pharmacokinetics, bioavailability, and clinical trials. Arch Pharm 2010;343:489–99.

Sirotkin AV, Harrath AH. Phytoestrogens and their effects. Eur J Pharmacol 2014;741:230–6. https://doi.org/10.1016/j.ejphar.2014.07.057.

Sirotkin AV, Kadasi A, Stochmalova A, Balazi A, Földesiová M, Makovicky P, Chrenek P, Harrath AH. Effect of turmeric on the viability, ovarian folliculogenesis, fecundity, ovarian hormones and response to luteinizing hormone of rabbits. Animal 2018;12(6):1242–9. https://doi.org/10.1017/S175173111700235X.

Sirotkin AV. Regulators of ovarian functions. Hauppauge, NY, USA: Nova Science Publishers, Inc.; 2014. 194 pp.

Tiwari-Pandey R, Ram Sairam M. Modulation of ovarian structure and abdominal obesity in curcumin- and flutamide-treated aging FSH-R haploinsufficient mice. Reprod Sci 2009;16(6):539–50. https://doi.org/10.1177/1933719109332822.

Valentine SP, Le Nedelec MJ, Menzies AR, Scandlyn MJ, Goodin MG, Rosengren RJ. Curcumin modulates drug metabolizing enzymes in the female Swiss Webster mouse. Life Sci 2006;78(20):2391–8. https://doi.org/10.1016/j.lfs.2005.09.017.

Vallianou NG, Evangelopoulos A, Schizas N, Kazazis C. Potential anticancer properties and mechanisms of action of curcumin. Anticancer Res 2015;35(2):645–51.

Voznesens'ka TI, Bryzhina TM, Sukhina VS, Makohon NV, Aleksieieva IM. Effect of NF-kappaB activation inhibitor curcumin on the oogenesis and follicular cell death in immune ovarian failure in mice. Fiziol Zh 2010;56(4):96−101.

Watson JL, Greenshields A, Hill R, Hilchie A, Lee PW, Giacomantonio CA, Hoskin DW. Curcumin-induced apoptosis in ovarian carcinoma cells is p53-independent and involves p38 mitogen-activated protein kinase activation and downregulation of Bcl-2 and survivin expression and Akt signaling. Mol Carcinog 2010;49(1):13−24. https://doi.org/10.1002/mc.20571.

Yan Z, Dai Y, Fu H, Zheng Y, Bao D, Yin Y, Chen Q, Nie X, Hao Q, Hou D, Cui Y. Curcumin exerts a protective effect against premature ovarian failure in mice. J Mol Endocrinol 2018;60(3):261−71. https://doi.org/10.1530/JME-17-0214.

Chapter 3.6

Flaxseed (*Linum usitatissimum* L.)

3.6.1 Introduction

Flaxseed (*Linum usitatissimum* L.) contains a number of biologically active substances, which defines its nutritional, physiological and therapeutic value. The physiological and medicinal properties of this plant have been described in a number of reviews (Singh et al., 2011; Martinchik et al., 2012; Parikh et al., 2018; Shayan et al., 2020). On the other hand, no special reviews of available information concerning flaxseed action on reproductive processes have been published yet. The present chapter is the first attempting to review the current available knowledge concerning the action of flaxseed on healthy female reproductive system, as well as of the data concerning their ability to prevent and to treat reproductive disorders.

3.6.2 Provenance and properties

Flaxseed or linseed (*Linum usitatissimum* L.) is a plant from the Linaceae family. The earliest evidence of flaxseed products comes from approximately the 7th century BCE from Iran. Cultivated flaxseed reached Europe approximately 4000 years later (discovery in the Switzerland). The wild predecessor of flaxseed remains unknown. Cultivated flaxseed can be divided from the economic standpoint: flax for fiber/linen, for oil (as component of human and animal food and of fuel), or for oil and linen. In addition, in the recent decades, flaxseed started being used for phytoremedication of soils contaminated by heavy metals (Prasad and Dhar, 2016; Parikh et al., 2018; Saleem et al., 2020).

Seed of this plant contains 30%−45% fat (drying oil whose main components are unsaturated acids − linoleic and linolenic acids), 20%−30% protein, 8% water, and up to 28% soluble and insoluble fiber. In seed coat, the soluble fiber creates gel, which can swell to multiple times its size. The seed contains 1% cyanide glycosides, of which the best-known are lignans as well as high content of E vitamin (Singh et al., 2011; Martinchik et al., 2012; Prasad and Dhar, 2016; Parikh et al., 2018; Shayan et al., 2020).

Flaxseed is the richest source of the lignan secoisolariciresinol diglucoside-a compound found in the outer layers of flaxseed. In rumen it is conversed to the enterolignans enterodiol and enterolactone, while the important role of rumen microbiota in such conversion has been demonstrated (Brito and Zang, 2018).

The available data concerning bioavailability and metabolic conversion of flaxseed oils are contradictory: several animal studies did not show an increase in plasma omega-3 (*n*-3) fatty acids level after flaxseed consumption (Lane et al., 2021). Dietary flaxseed did not affect the concentration of its components—lignin enterolactone (Zachut, 2015), β-hydroxybutyrate and nonesterified fatty acids (Jahani-Moghadam et al., 2015) in cow plasma. In rats, dietary flaxseed oil did not increase, but even reduced blood cholesterol, triglycerides and LDL (Komal et al., 2020). On the other hand, feeding of gilts with flaxseed increased levels of alpha-linolenic acid, timodonic and cervonic acids and decreased concentrations of myristic, palmitic and palmitoleic acids in gilt plasma (Vlčková et al., 2018). Moreover, dietary flaxseed increased the level of its enterolignans in women plasma (Chang et al., 2019). Feeding of cows with flaxseed α-linolenic acid resulted accumulation of this acid and its metabolites in cow milk (Petit et al., 2004; Brito and Zang, 2018), plasma (Zachut, 2015), ovarian follicular fluid and oocyte-cumulus complexes (Zachut et al., 2010; Moallem et al., 2013; Zachut, 2015). Metabolites of lignans, enterolactones can also be accumulated in milk and together with milk can enter animal and human organism (Brito and Zang, 2018). Therefore, the dietary flaxseed molecules can enter general circulation and accumulate in female reproductive organs.

3.6.3 Physiological action

The performed *in vitro*, animal and clinical studies (see reviews of Rhee and Brunt, 2011, Singh et al., 2011, Akhtar et al., 2013, Piermartiri et al., 2015; Ren et al., 2016, Brito and Zang, 2018; Parikh et al., 2018; Mali et al., 2019; Shayan et al., 2020) report a number of physiological and therapeutic effects of flaxseed. Antioxidant and antithrombotic effects enable to reduce blood level of overall cholesterol, LDL cholesterol and triacylglycerols, to prevent metabolic (type II diabetes, obesity) and cardiovascular (hypertension, arteriosclerosis and ischemia) diseases and to improve memory. Flaxseed can prevent or treat tumor development and chemical intoxication including intoxication induced by neurotoxic warfare organophosphate nerve agents. Flaxseed is a stimulator of immunity, but it has antiinflammatory effect and can prevent autoimmune disorders (psoriasis, systemic lupus erythematosus, asthma, rheumatoid arthritis, etc.). Flaxseed can improve health and hygiene of colon and maintenance of gastrointestinal microbiota. For the valuable properties of its gel, flaxseed is used as a treatment against constipation and to stimulate bowel activity, treat stomach irritation, stomach ulcers and to prevent stomach cancer, for bronchitis and inflammations of urinal tract.

Flaxseed oil is applied externally in dermatology. It is very well applicable in treatment of skin diseases, burns, and as regenerative cosmetic preparation for regular skin treatment. Isolated esters of fatty acids are applied in medicinal cosmetology as ingredients in regenerative preparations.

Flaxseed constituents help maintain healthy hair and skin—reduce redness and flaking, treat acne and eczema. Flaxseed can serve as a substitute for flour containing gluten for those, who suffers from celiac disease.

Well-known are the positive effect of flaxseed on general and fat metabolism. Studies on rats showed that flaxseed meal decreased the levels of fatty acids in their blood, the size of their adipose cells, but not the feed intake or body weight (Ribeiro et al., 2016). In humans, flaxseed consumption mitigated inflammation, insensitivity to glucose and insulin (Rhee and Brunt, 2011) and reduced digestibility of lipids (Kristensen et al., 2008). On the other hand, clinical studies did not determine its effect on body weight and obesity of humans (Kristensen et al., 2008; Wu et al., 2010; Rhee and Brunt, 2011).

Taken together, the available data demonstrate positive effect of flaxseed on a wide plethora of physiological processes, dysfunctions, and illnesses.

No adverse side-effects were determined in consumption of flaxseed products at therapeutic doses. However, when large doses are ingested, the possible intoxication by hydrogen cyanide, which the seeds release during milling cannot be excluded. Long-term use of flaxseed at therapeutic doses is not recommended without medical supervision in these cases: intestinal blockage, acute appendicitis, pancreatitis, peritonitis, and acute painful hernia. Pregnant women should opt for flaxseed oil rather than milled flaxseed—the oil does not contain lignans with phytoestrogenic effect, which could trigger complications in pregnancy. These possible dangers, however, have not been validated by experiments, and flaxseed products can be considered safe. US Food and Drug Administration (FDA) considers flaxseed products safe for nursing women and nutrition of children (Drugs and Lactation Database, 2006).

3.6.4 Mechanisms of action

3.6.4.1 Flaxseed constituents responsible for its physiological effects

The comparison of physiological effects of flaxseed and its constituents listed above enabled to identify flaxseed molecules responsible for its physiological and therapeutic effects (see reviews Rhee and Brunt, 2011, Singh et al., 2011, Akhtar et al., 2013, Ren et al., 2016, Parikh et al., 2018; Brito and Zang, 2018; Mali et al., 2019; Pal et al., 2019; Shayan et al., 2020). These authors reported that cardiovascular diseases (arteriosclerosis and ischemic disease, hypertension, hypercholisteronemia) can be prevented by flaxseed alpha-linoleic fatty acid, lignans, phenolic acids, E vitamin, folic acid, which possess antioxidant, anticholesterol, and antithrombotic properties. Flaxseed has antitumor properties due to the presence of alpha-linoleic fatty acid, lignans, gamma tocopherol (E vitamin), folic acid (B9 vitamin), magnesium, phenolic acid, and flavonoids with antioxidative and antiproliferative properties. The flaxseed

144 Contaminants and Plants Action on Female Reproduction

alpha-linoleic acid, docosahexaenoic acid, and 2-methoxyestradiol, steroid metabolite up-regulated by flaxseed, suppress angiogenesis during tumor development. The antitype II diabetic effect of flaxseed could be due to its alpha-linoleic acid, lignans, gamma tocopherol (E vitamin), lignans and fiber. Alpha-linoleic acid, lignans, and gamma tocopherol (E vitamin) can be responsible also for immune-stimulatory action of flaxseed. Alpha-linoleic fatty acid could be beneficial also for treatment of autoimmune disorders (psoriasis, systemic lupus erythematosus, asthma, rheumatoid arthritis, etc.). Flaxseed fiber, E vitamin, lignans, vitamins, and minerals can promote gastrointestinal tract activity and maintenance of gastrointestinal microbiota, which in turn is important for fat metabolism and immune processes. Positive effect of flaxseed on memory can be explained by the presence of omega-3 fatty acids. These acids and lignans suppress also inflammatory processes. The ability of flaxseed to bind and to prevents chemical toxins can be due to presence of omega-3 fatty acids, fiber, and lignans.

3.6.4.2 Mediators of flaxseed effects

The intracellular mediators of flaxseed action on some physiological processes and illnesses have been described in a series of special reviews (Piermartiri et al., 2015; DeLuca et al., 2018; Parikh et al., 2018; Yadav et al., 2018; McGrowder et al., 2020; Teodor et al., 2020). The antihypertensive, anti-atherosclerotic, antiplatelet aggregation, and cardioprotective effects of flaxseed proteins and other molecules could be due to its ability to suppress inflammatory processes via downregulation of proinflammatory transcription factor NF-κB (Parikh et al., 2018). The involvement of intracellular promoters of cell proliferation and apoptosis in control of these processes and cancerogenesis, which are changed under influence of flaxseed molecules, is also documented. Flaxseed can promote vascular hyperemia of various organs via promotion of generation of vasodilatory nitric oxide. In addition, flaxseed α-linolenic acid can inhibit soluble epoxide hydrolase, which in turn can promote vasodilation and induce production of inflammatory and cytotoxic oxylipins, metabolites of w-3 (n-3) polyunsaturated fatty acids (Parikh et al., 2018). Flaxseed enterolactones have high antioxidant properties (ability to neutralize reactive oxygen species directly or via up-regulation of antioxidant enzymes (Brito and Zang, 2018). Due to antioxidant (Brito and Zang, 2018) and antiestrogenic (Dikshit et al., 2016; Domínguez-López et al., 2020) properties, flaxseed and enterolactones can suppress the development of cancer in various organs suppressing mutagenesis and promoters of cell proliferation, survival, angiogenesis, inflammation and metastasis (Mali et al., 2019; McGrowder et al., 2020; Teodor et al., 2020). The neuroprotective action of flaxseed α-linolenic acid can be explained by activation of endogenous neuroprotective and neurorestorative pathways in brain via up-regulation of the transcription factor NF-κB, brain-derived neurotrophic factor and its receptor (Piermartiri et al., 2015).

There is evidence that flaxseed can exert its physiological actions not only through intracellular, but also extracellular regulators—hormones, growth factors, and cytokines. For example, flaxseed proteins could suppress inflammatory processes via downregulation of cytokines interleukines and tumor necrosis factor and to affect angiogenesis through changes in vascular endothelial growth factor (Parikh et al., 2018). Furthermore, flaxseed lignans possess high phytoestrogen properties. They could either up- and down-regulate estrogen receptors alpha (Dikshit et al., 2016; Domínguez-López et al., 2020) and decrease the expression of estrogen receptors alpha (Dikshit et al., 2016) and increase production of 2-methoxyestradiol and to affect estrogen metabolism (Pal et al., 2019). These actions on steroid hormones production and reception could explain their ability to affect development of estrogen-dependent breast cancer or estrogen-dependent glucose metabolism and cardiovascular risk in postmenopausal individuals (Domínguez-López.et l., 2020). Finally, flaxseed can downregulate the IGF/insulin signaling pathway (Dikshit et al., 2016), which is involved in control of metabolic, proliferative and reparative processes.

These examples illustrate the multiple molecules and pathways mediating influence of flaxseed and its constituents on various targets and defining the curative effect of flaxseed on some disorders.

3.6.5 Effects on female reproductive processes

3.6.5.1 Effect on ovarian and reproductive state

The experiments performed *in vivo* on laboratory and farm animals demonstrated that flaxseed and its constituents can affect ovarian growth, follicle development and resulting puberty and reproductive cycles.

Experiments on rats demonstrated both stimulatory and inhibitory action of flaxseed on their reproductive processes. In experiments of Tou et al. (1999), flaxseed given at higher dose (10%) caused earlier puberty onset, higher relative ovarian weight and lengthened estrous cycles (Tou et al., 1999). Moreover, their offspring, which received secoisolariciresinol diglycoside with mother milk, had increased uterine and ovarian relative weights, earlier puberty, lengthened estrous cycle, and persistent estrus. In similar experiments of Jelodar et al. (2018) and Mehraban et al. (2020), treatment of rats with flaxseed or flaxseed and spearmint mixture resulted in an increase in number of primary, preantral and antral ovarian follicles, decrease in the number of cystic follicles, increased the thickness of granulosa cell layer and decreased the

146 Contaminants and Plants Action on Female Reproduction

thickness of theca layer in rats with polycystic ovarian syndrome. Leal Soares et al. (2010) did not observe any influence of dietary flaxseed on rat estrous cycle length and regularity. Moreover, in experiments of Tou et al. (1999), the exposure of rats to diet containing 5% flaxseed or its lignan precursor, secoisolariciresinol diglycoside reduced the ovarian weight, delayed puberty and prolonged diestrous phase of the estrous cycle demonstrating the inhibitory influence of flaxseed given at this dose and its molecule on rat reproduction. In experiments of Pourjafari et al. (2021) feeding with flaxseed decreased the number of rat ovarian follicles.

In the experiments of Pourjafari et al. (2019), feeding of mice with flaxseed reduced the number of healthy ovarian follicles and increased the number of atretic follicles, i.e., in mice flaxseed promoted atresia of ovarian follicles.

In cows, the dietary flaxseed oil increased the number of ovarian follicles (Moallem et al., 2013). Ulfina et al. (2015) also reported the stimulatory action of flaxseed on bovine ovarian folliculogenesis: it increased the size of dominant ovarian follicle and corpus luteum, promoted uterine involution during postpartum period and shortened the time up to entrance to the next reproductive cycle. The shortening of postpartum period was reported also by Jahani-Moghadam et al. (2015). They observed the ability of flaxseed to decrease the incidence of cystic follicles, but no changes in general pregnancy rate. Furthermore, Ambrose et al. (2006) reported the ability of diet enriched in flaxseed alpha-linolenic acid to reduce incidence of pregnancy losses in cows. All these publications demonstrated the stimulatory action of flaxseed on bovine female reproductive processes. On the other hand, some authors (Petit et al., 2004; Hutchinson et al., 2012) did not observe flaxseed on bovine ovarian folliculogenesis. In the experiments of Zachut et al. (2011) feeding of cows with flaxseed oil prolonged their estrous cycle.

Numerous animal and *in vitro* studies (Eilati et al., 2013; Dikshit et al., 2015, 2016, 2017; Mali et al., 2019; Pal et al., 2019, 2020) demonstrated that flaxseed consumption suppress ovarian cancer development.

There are practically no reports concerning flaxseed influence on female reproductive organs besides ovary. Nevertheless, Sacco et al. (2012) observed the histomorphological changes in uteri of rats fed with flaxseed similar to changes induced by estrogen and indicating uteral activation and growth.

Therefore, the performed animal and *in vitro* studies demonstrated both stimulatory and inhibitory action of flaxseed and its molecule on ovarian folliculogenesis and cyclicity. The character of this effect depended on the model object/species and flaxseed dose tested. The influence of flaxseed on other female reproductive organs is possible, but it has not been adequately verified.

3.6.5.2 Effect on oocytes and embryos

The ability of flaxseed to affect ovarian folliculogenesis, which defines oocyte maturation and further development indicates a possible flaxseed influence on oo- and embryogenesis. Indeed, feeding of cows with flaxseed oil increased the number of recovered oocytes, their ability to cleavage and to reach the blastocyst stage after *in vitro* fertilization (Zachut et al., 2010; Moallem et al., 2013). Diet containing flaxseed at moderate (4% or 8%), but not at high (12%) dose increased the number and morphological quality of goat embryos (Dutra et al., 2019).

These observations suggest the beneficial influence of flaxseed on farm animal oo- and embryogenesis.

3.6.5.3 Effect on reproductive hormones

Some animal *in vivo* and *in vitro* studies demonstrated the ability of flaxseed to affect production, metabolism, and reception of steroid hormones in ovarian cells.

Treatment of rats with flaxseed increased the plasma level of progesterone and decreased the level of testosterone and sometimes estradiol, but not of dehydroepiandrosterone (Jelodar et al., 2018; Komal et al., 2020; Mehraban et al., 2020). In experiments of Tou et al. (1999), the consumption of flaxseed increased estradiol levels in rat plasma.

In mice, dietary flaxseed resulted in a decrease in estradiol in their blood plasma (Pourjafari et al., 2019).

In chicken, flaxseed diet affected estradiol metabolism by increasing CYP1A1 expression with a corresponding increase in the oncoprotective estradiol metabolite, 2-methoxyestradiol (Dikshit et al., 2015, 2016, 2017). Furthermore, it decreased the expression of ovarian alpha estrogen receptors (Dikshit et al., 2015, 2016).

As concerns cows, Ulfina et al. (2015) and Jahani-Moghadam et al. (2015) reported increased, while Hutchinson et al. (2012) reduced progesterone level in plasma of cows fed with flaxseed. Hutchinson et al. (2012) did not observe flaxseed influence on bovine plasma estradiol level. Nevertheless, Zachut et al. (2011) reported that the dietary flaxseed tended to reduce estradiol levels in ovarian follicles. The later studies of these author however showed an increase in follicular estradiol level in cows fed with flaxseed (Zachut, 2015).

Flaxseed diet did not affect steroid hormones' levels in porcine plasma (Vlčková et al., 2018).

In women, eating of flaxseed increased serum 2-hydroxyestrone level (Chang et al., 2019). Treatment of patients suffering from polycystic ovarian syndrome decreased total and free testosterone level in their plasma (Nowak et al., 2007; Ebrahimi et al., 2017). On the other hand, Mirmasoumi et al. (2018) did not find any changes in plasma steroid hormones level in patients treated with flaxseed against polycystic ovarian syndrome.

Besides steroid hormones, flaxseed appears to affect peptide hormones. Flaxseed diet decreased LH level in rat (Pourjafari et al., 2019) and bovine (Komal et al., 2020) plasma, and reduced prolactin concentration in women's blood (Chang et al., 2019). Dietary flaxseed was able to reduce IGF-I level in pig plasma (Vlčková et al., 2018), insulin level in rat (Komal et al., 2020) and to increase insulin concentration in women's (Nowak et al., 2007) plasma. On the contrary, subsequent experiments (Ebrahimi et al., 2017; Mirmasoumi et al., 2018; Haidari et al., 2020) showed a decrease in plasma insulin level in women after flaxseed consumption. In rats, flaxseed suppressed production of anti-Mullerian hormone by ovarian cells (Pourjafari et al., 2021). Flaxseed reduced leptin level in women's plasma (Haidari et al., 2020) and leptin release by cultured porcine ovarian granulosa cells (Štochmal'ová et al., 2019).

Moreover, flaxseed diet reduced the expression of prostaglandin E2 and COX-2, the enzyme responsible for its synthesis in chicken plasma (Eilati et al., 2013; Dikshit et al., 2015). On the other hand, in cows, dietary flaxseed did not affect the endometrial expression of genes involved in prostaglandin synthesis (Hutchinson et al., 2012).

Therefore, a number of evidence demonstrates the ability of flaxseed to act on production, release and reception of hormones and growth factors produced by pituitary, ovary and uterus, which are involved in endocrine and paracrine/autocrine regulation of female reproductive processes in various species.

3.6.6 Mechanisms of action on female reproductive processes

3.6.6.1 Flaxseed constituents responsible for its effects on female reproductive processes

The similarity in inhibitory effects of whole flaxseed and of its lignan precursor, secoisolariciresinol diglycoside on rat ovarian growth, puberty and estrous cycle (Tou et al., 1999) indicates that flaxseed influence on these processes can be due to presence of secoisolariciresinol diglycoside. Similarly, pregnancy in cows was affected not only by whole flaxseed, but also by its alpha-linolenic acid (Ambrose et al., 2006).

Animal and *in vitro* studies (Eilati et al., 2013; Dikshit et al., 2016, 2017; Pal et al., 2019) demonstrated that flaxseed can suppress ovarian cancer due to the presence of alpha-linoleic acid metabolite docosahexaenoic acid and estrogen metabolite 2-methoxyestradiol. The antiovarian cancer and antimetastatic effect has also another flaxseed alpha-linolic acid metabolite—lignin enterolactone (Mali et al., 2019).

These observations suggest that flaxseed influence on healthy and cancer ovarian cell functions are determined by presence of its lignans, alpha-linolenic acid, and their products.

3.6.6.2 Mediators of flaxseed effects on female reproductive processes

The numerous target processes and molecules, which can be affected by flaxseed and its molecules have been outlined above. Principally, all of them could be involved in mediating the influence of flaxseed and its molecules on female reproductive processes too. Nevertheless, only few of they were indicated in relation to flaxseed action on female reproduction, while the substantial part of the available knowledge concerned flaxseed action not on healthy, but on ovarian cancer cells.

Estrogens are important regulators of female reproduction (ovarian folliculogenesis, oogenesis, embryo implantation in uterus, and gravidity/pregnancy; Sirotkin, 2014). Previously we have mentioned the ability of flaxseed to affect estradiol release (Tou et al., 1999; Zachut et al. 2011, 2015; Dikshit et al., 2015, 2016, 2017; Pourjafari et al., 2019), estradiol metabolism (Dikshit et al., 2015, 2016; Chang et al., 2019), expression of estrogen receptors (Dikshit et al., 2015, 2016), and similarity of its action to action of estrogen (Tou et al., 1998). These data indicate that flaxseed molecules with phytoestrogen activity can affect estrogen-dependent reproductive processes and ovarian carcinogenesis via estrogen receptors.

The ability of flaxseed to affect synthesis of **prostaglandin E2** mentioned earlier (Eilati et al., 2013; Dikshit et al., 2015) indicates the involvement of this prostaglandin in mediating flaxseed effects too.

Furthermore, flaxseed decreased the expression of markers of **IGF/insulin signaling pathways** (IRS1, IGFBP4, IGFBP5), which are known regulators of healthy ovarian cells (Sirotkin, 2014) and their malignant transformation (Dikshit et al., 2016).

The involvement of steroid hormones, **gonadotropins, leptin,** insulin, IGF-I, **anti-Mullerian hormone,** and prostaglandins in control of female reproductive processes is well documented (Sirotkin, 2014). The ability of flaxseed to affect these molecules suggests that these hormones could be extracellular mediators of flaxseed action on female reproduction.

In addition, flaxseed oils are rich in calories, affecting general metabolic state, which in turn via metabolic hormones affects reproductive processes (Roa and Tena-Sempere, 2014; Evans and Anderson, 2017) and their dysfunctions (Romualdi et al., 2018). These facts indicate that flaxseed can affect reproduction via changes in **general metabolic state**.

The involvement of several intracellular signaling mechanisms in mediating flaxseed action on female reproduction might be proposed too. The key mediators of flaxseed action could be regulators of ovarian cell **proliferation and apoptosis**. Addition of flaxseed extract decreased proliferation (accumulation of proliferation marker PCNA) and increased cytoplasmic

150 Contaminants and Plants Action on Female Reproduction

apoptosis marker (accumulation of Bax) in cultured ovarian granulosa cells (Štochmal'ová et al., 2019). Similar effect on these markers in ovarian cells was observed in pigs fed with flaxseed diet *in vivo* (Vlčková et al., 2018). On the other hand, *in vivo* experiments on mice did not show substantial influence of dietary flaxseed on ovarian cell apoptosis and the expression of antiapoptotic molecule bcl-2 (Pourjafari et al., 2019). *In vivo* experiments of Dixit et al. (2016, 2017) revealed the ability of flaxseed diet to induce the apoptosis of chicken ovarian cancer cells. The reported action of flaxseed on intracellular regulators of proliferation and apoptosis in healthy (Vlčková et al., 2018; Štochmal'ová et al., 2019) and cancer (Dixit et al.,2016, 2017; Pal et al., 2019, 2020; Pourjafari et al., 2019) ovarian cells suggests that its influence on ovarian state and cell turnover could be mediated by its influence on ovarian cell proliferation: apoptosis rate.

Furthermore, flaxseed decreased the expression of **AKT** and altered expression of proinflammatory **transcription factor NF-κB**. All these intracellular pathways could be involved in mediating flaxseed action on ovarian cell proliferation, apoptosis, angiogenesis and tumorgenesis (Dikshit et al., 2016).

The flaxseed-derived molecules, 2-methoxyestradiol and docosahexaenoic acid can suppress tumor development via promotion of malignant cell apoptosis and suppression of tumor angiogenesis via down-regulation of **p38-MAPK intracellular signaling pathway** (Pal et al., 2019, 2020). The proapoptotic action of 2-methoxyestradiol can be partially mediated also by **protein kinase Cδ** (Pal et al., 2020).

These reports demonstrate multiple extra- and intracellular pathways mediating action of flaxseed and its constituents on healthy ovarian functions and their dysfunctions.

Other mediators of flaxseed action on female reproductive processes and their dysfunctions could be proposed. For example, several flaxseed molecules have **antioxidant properties**, which can be responsible for some flaxseed physiological and curative effects on nonreproductive cells. It is highly probable that flaxseed antioxidants could influence ovarian functions and that they could explain the preventive and therapeutic action of flaxseed on ovarian cancer and other reproductive disorders. Nevertheless, the involvement of antioxidants in mediating flaxseed action on female reproductive processes seems to not have been examined yet.

3.6.7 Application in reproductive biology and medicine

The results of some basic studies performed on laboratory and farm animals and on cell cultures suggest the applicability of flaxseed for improvement of farm animal reproductive efficiency. As it was described above, feeding of cows and goats with flaxseed can promote the development of their ovarian follicles (Moallem et al., 2013; Ulfina et al., 2015), increase the number and quality of their oocytes and embryos (Zachut et al., 2010; Moallem et al., 2013; Dutra et al., 2019; Pal et al., 2019).

The numerous demonstrations of flaxseed effect on steroid and peptide hormones (see above), which define puberty (Blaustein et al., 2016), ovarian follicular reserve, premature ovarian failure (Lew, 2019) and symptoms of menopause (Minkin, 2019; Paciuc, 2020). Furthermore, the flaxseed is able to affect rat puberty (Tou et al., 1998, 1999). These facts indicate that flaxseed could be potentially applicable for control of puberty, reproductive age and menopause and for prevention and treatment of their disorders. Nevertheless, this hypothesis requires preclinical and clinical validation.

Furthermore, animal and *in vitro* studies indicated applicability of flaxseed and its constituents for prevention and treatment of polycystic ovarian syndrome (Nowak et al., 2007; Jelodar et al., 2018; Komal et al., 2020; Mehraban et al., 2020) and of ovarian cancer (Eilati et al., 2013; Dikshit et al., 2015, 2016, 2017; Mali et al., 2019; Pal et al., 2019, 2020), although these indices have not been confirmed by the corresponding clinical studies.

Therefore, the available preclinical and animal studies indicate the potential practical applicability of flaxseed for improvement of farm animal reproduction, ovarian cancer, polycystic ovarian syndrome and maybe of other reproductive dysfunctions.

3.6.8 Conclusions and possible direction of future studies

Flaxseed contains a number of biological active molecules, which, acting through multiple signaling pathways, can determine numerous physiological, protective and therapeutic effects of flaxseed. The available publications demonstrate the action of flaxseed and its constituents on female reproductive system - ovarian growth, follicle development, the resulting puberty and reproductive cycles, ovarian cell proliferation and apoptosis, oo- and embryogenesis, hormonal regulators of reproductive processes and their dysfunctions. These effects can be determined by flaxseed lignans, alpha-linolenic acid and their products. Their action can be mediated by changes in general metabolism, metabolic and reproductive hormones, their binding proteins, receptors and several intracellular signaling pathways including protein kinases, transcription factors regulating cell proliferation, apoptosis, angiogenesis and malignant transformation. Flaxseed and its active molecules are found potentially useful for improvement of farm animal reproductive efficiency and in treatment of polycystic ovarian syndrome and ovarian cancer.

On the other hand, a number of important details of flaxseed action on female reproduction and other processes require further elucidation. The flaxseed constituents and mediators of their action responsible for some effects remain to be determined. The published data concerning character of flaxseed action on many female reproductive processes in various species and even in one model are contradictory. The main data concerning flaxseed effects are obtained in laboratory animal and *in vitro* studies, sometimes on models with dysfunctions (cancer or polycystic ovarian syndrome), while the reports of clinical studies are rare and sometimes inconsistent and inconclusive.

152 Contaminants and Plants Action on Female Reproduction

Therefore, the available information suggests substantial physiological and therapeutic potential of flaxseed, but also the need for more profound basic and applied studies of its physiological, protective, and therapeutic effects.

References

Akhtar S, Ismail T, Riaz M. Flaxseed - a miraculous defense against some critical maladies. Pak J Pharm Sci 2013;26(1):199—208.

Ambrose DJ, Kastelic JP, Corbett R, Pitney PA, Petit HV, Small JA, Zalkovic P. Lower pregnancy losses in lactating dairy cows fed a diet enriched in alpha-linolenic acid. J Dairy Sci 2006;89(8):3066—74. https://doi.org/10.3168/jds. S0022-0302(06)72581-4.

Blaustein JD, Ismail N, Holder MK. Review: puberty as a time of remodeling the adult response to ovarian hormones. J Steroid Biochem Mol Biol 2016;160:2—8. https://doi.org/10.1016/j.jsbmb.2015.05.007.

Brito AF, Zang Y. A review of lignan metabolism, milk enterolactone concentration, and anti-oxidant status of dairy cows fed flaxseed. Molecules 2018;24(1):41. https://doi.org/10.3390/molecules24010041.

DeLuca JAA, Garcia-Villatoro EL, Allred CD. Flaxseed bioactive compounds and colorectal cancer prevention. Curr Oncol Rep 2018;20(8):59. https://doi.org/10.1007/s11912-018-0704-z.

Dikshit A, Gomes Filho MA, Eilati E, McGee S, Small C, Gao C, Klug T, Hales DB. Flaxseed reduces the pro-carcinogenic micro-environment in the ovaries of normal hens by altering the PG and oestrogen pathways in a dose-dependent manner. Br J Nutr 2015;113(9):1384—95. https://doi.org/10.1017/S000711451500029X.

Dikshit A, Gao C, Small C, Hales K, Hales DB. Flaxseed and its components differentially affect estrogen targets in pre-neoplastic hen ovaries. J Steroid Biochem Mol Biol 2016;159:73—85. https://doi.org/10.1016/j.jsbmb.2016.02.028.

Dikshit A, Hales K, Hales DB. Whole flaxseed diet alters estrogen metabolism to promote 2-methoxtestradiol-induced apoptosis in hen ovarian cancer. J Nutr Biochem 2017;42:117—25. https://doi.org/10.1016/j.jnutbio.2017.01.002.

Domínguez-López I, Yago-Aragón M, Salas-Huetos A, Tresserra-Rimbau A, Hurtado-Barroso S. Effects of dietary phytoestrogens on hormones throughout a human lifespan: a review. Nutrients 2020;12(8):2456. https://doi.org/10.3390/nu12082456.

Drugs and lactation database (LactMed) [internet]. Bethesda (MD): National Library of Medicine (US); 2006. Available from: http://www.ncbi.nlm.nih.gov/books/NBK501895/.

Dutra PA, Pinto LFB, Cardoso Neto BM, Gobikrushanth M, Barbosa AM, Barbosa LP. Flaxseed improves embryo production in Boer goats. Theriogenology 2019;127:26—31. https://doi.org/10.1016/j.theriogenology.2018.12.038.

Ebrahimi FA, Samimi M, Foroozanfard F, Jamilian M, Akbari H, Rahmani E, Ahmadi S, Taghizadeh M, Memarzadeh MR, Asemi Z. The effects of omega-3 fatty acids and vitamin E co-supplementation on indices of insulin resistance and hormonal parameters in patients with polycystic ovary syndrome: a randomized, double-blind, placebo-controlled trial. Exp Clin Endocrinol Diabetes 2017;125(6):353—9. https://doi.org/10.1055/s-0042-117773.

Eilati E, Bahr JM, Hales DB. Long term consumption of flaxseed enriched diet decreased ovarian cancer incidence and prostaglandin E$_2$in hens. Gynecol Oncol 2013;130(3):620—8. https://doi.org/10.1016/j.ygyno.2013.05.018.

Evans MC, Anderson GM. Neuroendocrine integration of nutritional signals on reproduction. J Mol Endocrinol 2017;58(2):R107—28. https://doi.org/10.1530/JME-16-0212.

Haidari F, Banaei-Jahromi N, Zakerkish M, Ahmadi K. The effects of flaxseed supplementation on metabolic status in women with polycystic ovary syndrome: a randomized open-labeled controlled clinical trial. Nutr J 2020;19(1):8. https://doi.org/10.1186/s12937-020-0524-5.

Hutchinson IA, Hennessy AA, Waters SM, Dewhurst RJ, Evans AC, Lonergan P, Butler ST. Effect of supplementation with different fat sources on the mechanisms involved in reproductive performance in lactating dairy cattle. Theriogenology 2012;78(1):12−27. https://doi.org/10.1016/j.theriogenology.2011.12.031.

Jahani-Moghadam M, Mahjoubi E, Dirandeh E. Effect of linseed feeding on blood metabolites, incidence of cystic follicles, and productive and reproductive performance in fresh Holstein dairy cows. J Dairy Sci 2015;98(3):1828−35. https://doi.org/10.3168/jds.2014-8789.

Jelodar G, Masoomi S, Rahmanifar F. Hydroalcoholic extract of flaxseed improves polycystic ovary syndrome in a rat model. Iran J Basic Med Sci 2018;21(6):645−50. https://doi.org/10.22038/IJBMS.2018.25778.6349.

Komal F, Khan MK, Imran M, Ahmad MH, Anwar H, Ashfaq UA, Ahmad N, Masroor A, Ahmad RS, Nadeem M, Nisa MU. Impact of different omega-3 fatty acid sources on lipid, hormonal, blood glucose, weight gain and histopathological damages profile in PCOS rat model. J Transl Med 2020;18(1):349. https://doi.org/10.1186/s12967-020-02519-1.

Kristensen M, Damgaard TW, Sørensen AD, Raben A, Lindeløv TS, Thomsen AD, Bjergegaard C, Sørensen H, Astrup A, Tetens I. Whole flaxseeds but not sunflower seeds in rye bread reduce apparent digestibility of fat in healthy volunteers. Eur J Clin Nutr 2008;62(8):961−7.

Lane KE, Wilson M, Hellon TG, Davies IG. Bioavailability and conversion of plant based sources of omega-3 fatty acids - a scoping review to update supplementation options for vegetarians and vegans. Crit Rev Food Sci Nutr 2021:1−16. https://doi.org/10.1080/10408398.2021.1880364.

Leal Soares L, Ferreira Medeiros de França Cardozo L, Andrade Troina A, De Fonte Ramos C, Guzmán-Silva MA, Teles Boaventura G. Influence of flaxseed during lactation on the reproductive system of female Wistar rats. Nutr Hosp 2010;25(3):437−42.

Lew R. Natural history of ovarian function including assessment of ovarian reserve and premature ovarian failure. Best Pract Res Clin Obstet Gynaecol 2019;55:2−13. https://doi.org/10.1016/j.bpobgyn.2018.05.005.

Mali AV, Padhye SB, Anant S, Hegde MV, Kadam SS. Anticancer and antimetastatic potential of enterolactone: clinical, preclinical and mechanistic perspectives. Eur J Pharmacol 2019;852:107−24. https://doi.org/10.1016/j.ejphar.2019.02.022.

Martinchik AN, Baturin AK, Zubtsov VV, Molofeev V. Nutritional value and functional properties of flaxseed. Vopr Pitan 2012;81(3):4−10.

McGrowder DA, Miller FG, Nwokocha CR, Anderson MS, Wilson-Clarke C, Vaz K, Anderson-Jackson L, Brown J. Medicinal herbs used in traditional management of breast cancer: mechanisms of action. Medicines (Basel) 2020;7(8):47. https://doi.org/10.3390/medicines7080047.

Mehraban M, Jelodar G, Rahmanifar F. A combination of spearmint and flaxseed extract improved endocrine and histomorphology of ovary in experimental PCOS. J Ovarian Res 2020;13(1):32. https://doi.org/10.1186/s13048-020-00633-8.

Minkin MJ. Menopause: hormones, lifestyle, and optimizing aging. Obstet Gynecol Clin N Am 2019;46(3):501−14. https://doi.org/10.1016/j.ogc.2019.04.008.

Mirmasoumi G, Fazilati M, Foroozanfard F, Vahedpoor Z, Mahmoodi S, Taghizadeh M, Esfeh NK, Mohseni M, Karbassizadeh H, Asemi Z. The effects of flaxseed oil omega-3 fatty acids supplementation on metabolic status of patients with polycystic ovary syndrome: a randomized, double-blind, placebo-controlled trial. Exp Clin Endocrinol Diabetes 2018;126(4):222−8. https://doi.org/10.1055/s-0043-119751.

Moallem U, Shafran A, Zachut M, Dekel I, Portnick Y, Arieli A. Dietary α-linolenic acid from flaxseed oil improved folliculogenesis and IVF performance in dairy cows, similar to eicosapentaenoic and docosahexaenoic acids from fish oil. Reproduction 2013;146(6):603−14. https://doi.org/10.1530/REP-13-0244.

154 Contaminants and Plants Action on Female Reproduction

Nowak DA, Snyder DC, Brown AJ, Demark-Wahnefried W. The effect of flaxseed supplementation on hormonal levels associated with polycystic ovarian syndrome: a case study. Curr Top Nutraceutical Res 2007;5(4):177−81.

Paciuc J. Hormone therapy in menopause. Adv Exp Med Biol 2020;1242:89−120. https://doi.org/10.1007/978-3-030-38474-6_6.

Pal P, Hales K, Petrik J, Hales DB. Pro-apoptotic and anti-angiogenic actions of 2-methoxyestradiol and docosahexaenoic acid, the biologically derived active compounds from flaxseed diet, in preventing ovarian cancer. J Ovarian Res 2019;12(1):49. https://doi.org/10.1186/s13048-019-0523-3.

Pal P, Hales K, Hales DB. The pro-apoptotic actions of 2-methoxyestradiol against ovarian cancer involve catalytic activation of PKCδ signaling. Oncotarget 2020;11(40):3646−59. https://doi.org/10.18632/oncotarget.27760.

Parikh M, Netticadan T, Pierce GN. Flaxseed: its bioactive components and their cardiovascular benefits. Am J Physiol Heart Circ Physiol 2018;314(2):H146−59.

Petit HV, Germiquet C, Lebel D. Effect of feeding whole, unprocessed sunflower seeds and flaxseed on milk production, milk composition, and prostaglandin secretion in dairy cows. J Dairy Sci 2004;87(11):3889−98. https://doi.org/10.3168/jds.S0022-0302(04)73528-6.

Piermartiri T, Pan H, Figueiredo TH, Marini AM. α-linolenic acid, a nutraceutical with pleiotropic properties that targets endogenous neuroprotective pathways to protect against organophosphate nerve agent-induced neuropathology. Molecules 2015;20(11):20355−80. https://doi.org/10.3390/molecules201119698.

Pourjafari F, Haghpanah T, Sharififar F, Nematollahi-Mahani SN, Afgar A, Asadi Karam G, Ezzatabadipour M. Protective effects of hydro-alcoholic extract of *foeniculum vulgare* and *linum usitatissimum* on ovarian follicle reserve in the first-generation mouse pups. Heliyon 2019;5(10):e02540. https://doi.org/10.1016/j.heliyon.2019.e02540.

Pourjafari F, Haghpanah T, Sharififar F, Nematollahi-Mahani SN, Afgar A, Ezzatabadipour M. Evaluation of expression and serum concentration of anti-Mullerian hormone as a follicle growth marker following consumption of fennel and flaxseed extract in first-generation mice pups. BMC Complement Med Ther 2021;21(1):90. https://doi.org/10.1186/s12906-021-03267-5.

Prasad K, Dhar A. Flaxseed and diabetes. Curr Pharm Des 2016;22(2):141−4. https://doi.org/10.2174/1381612822666151112151230.

Ren GY, Chen CY, Chen GC, Chen WG, Pan A, Pan CW, Zhang YH, Qin LQ, Chen LH. Effect of flaxseed intervention on inflammatory marker C-reactive protein: a systematic review and meta-analysis of randomized controlled trials. Nutrients 2016;8(3):136.

Rhee Y, Brunt A. Flaxseed supplementation improved insulin resistance in obese glucose intolerant people: a randomized crossover design. Nutr J 2011;10:44. https://doi.org/10.1186/1475-2891-10-44.

Ribeiro DC, Pereira AD, da Silva PC, dos Santos Ade S, de Santana FC, Boueri BF, Pessanha CR, de Abreu MD, Mancini-Filho J, da Silva EM, do Nascimento-Saba CC, da Costa CA, Boaventura GT. Flaxseed flour (*Linum usitatissimum*) consumption improves bone quality and decreases the adipocyte area of lactating rats in the post-weaning period. Int J Food Sci Nutr 2016;67(1):29−34.

Roa J, Tena-Sempere M. Connecting metabolism and reproduction: roles of central energy sensors and key molecular mediators. Mol Cell Endocrinol 2014;397(1-2):4−14. https://doi.org/10.1016/j.mce.2014.09.027.

Romualdi D, Immediata V, De Cicco S, Tagliaferri V, Lanzone A. Neuroendocrine regulation of food intake in polycystic ovary syndrome. Reprod Sci 2018;25(5):644−53. https://doi.org/10.1177/1933719117728803.

Sacco SM, Jiang JM, Thompson LU, Ward WE. Flaxseed does not enhance the estrogenic effect of low-dose estrogen therapy on markers of uterine health in ovariectomized rats. J Med Food 2012;15(9):846−50. https://doi.org/10.1089/jmf.2011.0314.

Saleem MH, Ali S, Hussain S, Kamran M, Chattha MS, Ahmad S, Aqeel M, Rizwan M, Aljarba NH, Alkahtani S, Abdel-Daim MM. Flax (*Linum usitatissimum* L.): a potential candidate for phytoremediation? Biological and economical points of view. Plants 2020;9(4):496. https://doi.org/10.3390/plants9040496.

Shayan M, Kamalian S, Sahebkar A, Tayarani-Najaran Z. Flaxseed for health and disease: review of clinical trials. Comb Chem High Throughput Screen 2020;23(8):699−722. https://doi.org/10.2174/1386207323666200521121708.

Singh KK, Mridula D, Rehal J, Barnwal P. Flaxseed: a potential source of food, feed and fiber. Crit Rev Food Sci Nutr 2011;51(3):210−22.

Štochmaľová A, Harrath AH, Alwasel S, Sirotkin AV. Direct inhibitory effect of flaxseed on porcine ovarian granulosa cell functions. Appl Physiol Nutr Metabol 2019;44(5):507−11. https://doi.org/10.1139/apnm-2018-0547.

Teodor ED, Moroeanu V, Radu GL. Lignans from medicinal plants and their anticancer effect. Mini Rev Med Chem 2020;20(12):1083−90. https://doi.org/10.2174/1389557520666200212110513.

Tou JC, Chen J, Thompson LU. Flaxseed and its lignan precursor, secoisolariciresinol diglycoside, affect pregnancy outcome and reproductive development in rats. J Nutr 1998;128(11):1861−8. https://doi.org/10.1093/jn/128.11.1861.

Tou JC, Chen J, Thompson LU. Dose, timing, and duration of flaxseed exposure affect reproductive indices and sex hormone levels in rats. J Toxicol Environ Health 1999;56(8):555−70. https://doi.org/10.1080/00984109909350177.

Ulfina GG, Kimothi SP, Oberoi PS, Baithalu RK, Kumaresan A, Mohanty TK, Imtiwati P, Dang AK. Modulation of post-partum reproductive performance in dairy cows through supplementation of long- or short-chain fatty acids during transition period. J Anim Physiol Anim Nutr 2015;99(6):1056−64. https://doi.org/10.1111/jpn.12304.

Vlčková R, Andrejčáková Z, Sopková D, Hertelyová Z, Kozioł K, Koziorowski M, Gancarčíková S. Supplemental flaxseed modulates ovarian functions of weanling gilts via the action of selected fatty acids. Anim Reprod Sci 2018;193:171−81. https://doi.org/10.1016/j.anireprosci.2018.04.066.

Wu H, Pan A, Yu Z, Qi Q, Lu L, Zhang G, Yu D, Zong G, Zhou Y, Chen X, Tang L, Feng Y, Zhou H, Chen X, Li H, Demark-Wahnefried W, Hu FB, Lin X. Lifestyle counseling and supplementation with flaxseed or walnuts influence the management of metabolic syndrome. J Nutr 2010;140(11):1937−42. https://doi.org/10.3945/jn.110.126300.

Yadav RK, Singh M, Roy S, Ansari MN, Saeedan AS, Kaithwas G. Modulation of oxidative stress response by flaxseed oil: Role of lipid peroxidation and underlying mechanisms. Prostaglandins Other Lipid Mediat 2018;135:21−6. https://doi.org/10.1016/j.prostaglandins.2018.02.003.

Zachut M, Dekel I, Lehrer H, Arieli A, Arav A, Livshitz L, Yakoby S, Moallem U. Effects of dietary fats differing in n-6:n-3 ratio fed to high-yielding dairy cows on fatty acid composition of ovarian compartments, follicular status, and oocyte quality. J Dairy Sci 2010;93(2):529−45. https://doi.org/10.3168/jds.2009-2167.

Zachut M, Arieli A, Moallem U. Incorporation of dietary n-3 fatty acids into ovarian compartments in dairy cows and the effects on hormonal and behavioral patterns around estrus. Reproduction 2011;141(6):833−40. https://doi.org/10.1530/REP-10-0518.

Zachut M. Short communication: concentrations of the mammalian lignan enterolactone in pre-ovulatory follicles and the correlation with intrafollicular estradiol in dairy cows fed extruded flaxseed. J Dairy Sci 2015;98(12):8814−7. https://doi.org/10.3168/jds.2015-9699.

156 Contaminants and Plants Action on Female Reproduction

Chapter 3.7

Ginkgo (*Ginkgo biloba,* L.)

3.7.1 Introduction

The medicinal properties of *Ginkgo biloba* L. (Anonimous, 1998; Diamond and Baily, 2013; Drug and Lactation Dayabase, 2018; Achete de Souza et al., 2020; Nguyen and Alzahram, 2021) and its constituents (Ude et al., 2013; Fang et al., 2020; Nguyen and Alzahram, 2021) are well known. Nevertheless, the influence of these plant substances on female reproductive processes has not been summarized yet. This chapter reviews provenance, biologically active constituents of *Ginkgo biloba,* L., its general health effects, the currently available knowledge concerning both inhibitory and stimulatory actions of ginkgo and its constituents on female reproductive processes, as well as possible extra- and intracellular mediators of their effects.

3.7.2 Provenance and properties

Ginkgo biloba L. (Ginkgo, Maidenhair Tree) (family: Ginkgoaceae) is believed to be the eldest plant growing on the Earth, its origin dating 250−260 million years back. Given that it does not show any botanical or taxonomic relatedness with other plant species currently inhabiting our planet, it is called a living fossil plant. This tree is a native from Japan, Korea, and China; however, it has been cultivated around the world as an ornamental tree. The ginkgo tree is a deciduous plant with green leaves turning golden in autumn and the ginkgo seeds are contained in ginkgo fruits born on the female trees. The nutlike gametophytes inside the seeds and ginkgo leaves have been used in traditional medicine (see below) and in cooking of some traditional Chinese and Japanese dishes (Anonimous, 1998; Mei et al., 2017).

Among roughly 50 compounds have been isolated from the plant so far, those most relevant for its biological activity are the following two groups of substances: flavonoids and terpene trilactones. The composition of the plant is predominated by flavonoids, among which the most represented are flavonol mono-, di- and triglycosides, which have an ester bond with the

Food/medicinal herbs and their influence Chapter | 3 **157**

p-coumarin acid. Aglyconic molecular parts of typical ginkgo flavone glycosides are quercetin, kaempferol and isorhamnetin. The terpene trilactone group is represented by diterpene lactone ginkgolides A, B, C, J, K, L and M, and sesquiterpene lactone bilobalide. Ginkgolides and bilobalide are unique terpenoid components found solely in the *Ginkgo biloba* L. tree (Drug and Lactation Database, 2018; Feng et al., 2019; Budeč et al., 2019; Eisvand et al., 2020; Martiniakova et al., 2020).

Commercial extracts of the dried leaves of Ginkgo biloba for therapeutical use (EGb761 and LI1370) are standardized in a multi-step procedure designed to concentrate the desired active principles from the plant. These extracts contain approximately 24% flavone glycosides (quercetin, kaempferol, and isorhamnetin) and 6% terpene lactones (2.8%−3.4% ginkgolides A, B, and C, and 2.6%−3.2% bilobalide). Other constituents include proanthocyanadins, glucose, rhamnose, organic acids (hydroxykinurenic, kynurenic, protocatechic, vanillic, shikimic), D-glucaric acid and ginkgolic acid, and related alkylphenols (Anonimous, 1998). The quality and safety of consumed ginkgo products must be oriented on a defined quantity of terpene trilactones, flavonoids, and ginkgolic acid (Ude et al., 2013).

3.7.3 Physiological and therapeutic actions

The eldest data on therapeutic virtues of the *Ginkgo biloba* L. and its medicinal use can be found in the roughly 5000-year old Chinese Pharmacopeia. The leaves, seeds, and fruit are used in traditional Chinese medicine for cough, asthma, enuresis, pyogenic skin infections, and intestinal tract worm infections (Anonimous, 1998; Eimazoudy and Attia, 2012; Mei et al., 2017).

Modern scientific studies demonstrated the antioxidant, antiinflammatory, antiapoptotic, and antigenotoxic effects of ginkgo (Budeč et al., 2019; Lejri et al., 2019; Omidkhoda et al., 2019). These properties of ginkgo define its applicability for prevention and treatment of a number of disorders. Animal and clinical studies indicated that ginkgo extract may be efficacious in the treatment of a wide array of conditions associated with physical and mental deterioration.

The best-known physiological and therapeutical effect of ginkgo is neuroprotective and curative action against Alzheimer's disease/senile dementia. Ginkgo extracts appear to be capable of stabilizing and, in some cases, improving the cognitive performance and the social functioning of patients with dementia (Ude et al., 2013; Liu et al., 2019; Singh et al., 2019; Achete de Souza et al., 2020; Nguyen and Alzahrani, 2021). The next positive action of ginkgo on brain includes treatment of cerebral vascular insufficiency and impaired cerebral performance—visual field disturbances associated with chronic lack of bloodflow, oculomotor and complex choice reaction, vigilance and reaction times, intermittent claudication, low memory and mental performance, and dizziness, as well as such psychiatric disorders like depression,

158 Contaminants and Plants Action on Female Reproduction

schizophrenia, multiinfarct dementia, tardive dyskinesia, and caregiver distress (Anonimous, 1998; Diamond et al., 2013; Tian et al., 2017; Achete de Souza et al., 2020; Nguyen and Alzahrani, 2021). Ginkgo could be useful for treatment of cerebral injury following ischemic strokes (Feng et al., 2019; Achete de Souza et al., 2020). It could be helpful also in cases of vertigo/equilibrium Disorders (Nguyen and Alzahrani, 2021).

Ginkgo can be helpful in treatment of cardiovascular diseases. In animal models, ginkgo can have positive effect on myocardial injury (Achete de Souza et al., 2020). Treatment with Ginkgo biloba extract lowers fibrinogen levels and decreases plasma viscosity (Anonimous, 1998; Chen et al., 2019). In humans, the positive effect of ginkgo extract on ischemic colitis was shown (Achete de Souza et al., 2020). Treatment using ginkgo extract induced an improvement in neurological function among individuals with acute ischemic stroke, but not in their mortality or cerebrovascular bleeding (Feng et al., 2019; Chong et al., 2020; Nguyen and Alzahrani, 2021).

Ginkgo biloba can be applicable in case of metabolic syndrome, including obesity, high blood pressure, dyslipidemia, and hyperglycemia (Chen et al., 2019; Achete de Souza et al., 2020; Eisvand et al., 2020) and diabetes mellitus. Although human clinical trials have not been conducted, in animal experimental models, ginkgo biloba extract appears to improve insulin signaling and some complications associated with diabetes (Anonimous, 1998; Hirata et al., 2015). Animal studies demonstrated the ability of ginkgo extract to reduce food consumption, body weight, activation and inflammation of adipocytes and obesity (Hirata et al., 2015; Banin et al., 2017).

There is evidence for therapeutic action of ginkgo against other illnesses including liver fibrosis associated with chronic hepatitis (Anonimous, 1998) and macular degeneration and glaucoma (Anonimous, 1998; Evans, 2013; Nguyen and Alzahrani, 2021).

Ginkgo can weaken some symptoms of premenstrual syndrome. Ginkgo extract was effective in treatment of memory impairment, loss of sustained attention, congestive (particularly breast symptoms) and neuropsychological symptoms of this syndrome, and in the alleviation of idiopathic cyclic edema (Anonimous, 1998; Elsabagh et al., 2005; Kargozar et al., 2017; Nguyen and Alzahrani, 2021).

Animal and human studies demonstrated the antitumor activity of ginkgo polysaccharides (Jiao et al., 2016).

Animal studies suggested the antiinflammatory and anelgesic effects of ginkgo extract (Abdel-Salam et al., 2004).

Ginkgo can promote gastric acid secretion in rats (Abdel-Salam et al., 2004).

Ginkgo can affect bone remodeling, osteoclastogenesis, osteoblastogenesis, bone cell activity, death, and oxidative stress and prevent osteoporosis (Martiniakova et al., 2020).

Ginkgo can be a natural protector against natural toxins, chemical toxicities, and radiation (Abdel-Salam et al., 2004; de Souza Predes et al., 2011; Omidkhoda et al., 2019).

The results of some early studies indicated beneficial effect of ginkgo on impotence in men and women's sexual desire (Anonimous, 1998; McKay, 2004), but these effects have not been confirmed by subsequent clinical experiments, or positive effects of ginkgo-containing mixture could be explained by presence of components other than ginkgo (Nguyen and Alzahrani, 2021).

Early studies have shown contradictory results in the treatment of tinnitus, which might be due to the diverse etiology of this condition (Anonimous, 1998). Later studies, however, did not reveal the positive influence of ginkgo on this dysfunction (Nguyen and Alzahrani, 2021).

Ginkgo did not affect altitude sickness (Nguyen and Alzahrani, 2021).

In general, ginkgo is well tolerated. The most dangerous adverse effect of ginkgo consumption is its cancerogenic effect, which has been detected in rodent and human studies (Mei et al., 2017). In addition, the uncommon negative side effect like gastrointestinal disturbances (nausea, vomiting, increased salivation, loss of appetite), headaches, dizziness, tinnitus, peripheral visual shimmering, allergic and hypersensitivity reactions, such as skin rash, increased bleeding after surgery, lowering of seizure threshold have been reported to occur in some individuals (Anonimous, 1998; Kargozar et al., 2017; Drug and Lactation Database, 2018). In addition, there is some weak scientific evidence from animal and *in vitro* studies that ginkgo leaf has antiplatelet activity, which may be of concern during labor as ginkgo use could prolong bleeding time. Low-level evidence based on expert opinion shows that ginkgo leaf may be an emmenagogue and have hormonal (endocrine disrupter) properties (Dugoua et al., 2006).

Raw ginkgo seeds contain potentially toxic cyanogenic glycosides and should not be used; roasted seeds do not carry this risk (Drug and Lactation Database, 2018). Because of their toxic potential the amount of ginkgolic acid in consumed ginkgo products should be less than 5 ppm (Ude et al., 2013).

Besides its possible toxic side-effects, ginkgo can affect the level and effect of other drugs in organism. For example, it decreases the plasma concentrations of omeprazole, ritonavir and tolbutamide. Clinical cases indicate interactions of ginkgo with monoamine oxidase inhibitors, antiepileptics, alprazolam, aspirin (acetylsalicylic acid), diuretics, haloperidol, ibuprofen, risperidone, rofecoxib, trazodone, nifedipine and warfarin (Izzo et al., 2009; Diamond et al., 2013).

3.7.4 Mechanisms of action

3.7.4.1 Ginkgo constituents responsible for particular effects

Comparison of physiological and therapeutic effects of ginkgo extract and of its single constituents enables to identify the ginkgo molecules responsible for

160 Contaminants and Plants Action on Female Reproduction

particular ginkgo actions. For example, several ginkgo phytoestrogens are antioxidants, i.e., have the capacity of binding oxygen free radicals, which induce oxidative stress, which in turn can induce mitochondrial dysfunctions, development of cancer, inflammation and obesity (Diamond et al., 2013; Budeč et al., 2019; Lejri et al., 2019; Singh et al., 2019; Achete de Souza et al., 2020). Quercetin is considered also as an anti-cholesterol and anti-low density lipoprotein and antihistaminic drug, which can prevent obesity and inflammatory response respectively. Kaempferol aids in the prevention of atherosclerosis owing to the inhibition of low density lipoproteins. It also inhibits platelet aggregation and development of malignant cells. Terpene trilactones are believed to be capable of inhibiting the platelet activation factor that induces platelet aggregation, as well as the development of edemas, smooth muscle contractions, hypertension, bronchial constriction, increased microvascular permeability, increased hematocrit and lysosomal enzyme secretion (see Diamond et al., 2013; Budeč et al., 2019 for review).

Effect of ginkgo extract on bone and osteoporosis are defined by the presence of phytoestrogens (e.g., kaempferol, quercetin), terpenoids and ginkgolic acid (Martiniakova et al., 2020).

The ginkgolides (especially A, B and C), bilobalide, quercetin and kaempferol have demonstrated neuroprotective properties toward cerebral injury following ischemic strokes mentioned above (Feng et al., 2019; Achete de Souza et al., 2020). Moreover, ginkgolide A is a protective compound against toxicities, which had similar effects to those of total ginkgo extract (Omidkhoda et al., 2019). Animal studies demonstrated that the ginkgolides A, B, and C, bilobalide, quercetin, and kaempferol can be responsible also for ginkgo action on myocardial injury, renal and liver damage. In humans, these molecules have therapeutic action on diabetes, metabolic syndrome, and ischemic colitis (Achete de Souza et al., 2020).

Ginkgo polysaccharides have also remarkable biological activities, including antioxidant, antiviral, antitumor, antiinflammation, immunomodulatory, and hepatoprotection effects (Fang et al., 2020).

3.7.4.2 Mediators of ginkgo and its constituents' effects

Studies using cell and animal models enabled to outline not only ginkgo constituents responsible for particular procersses, but also extra- and intracellular signaling pathways mediating effect of ginkgo on various processes.

Hormones, cytokines and growth factors could be important extracellular mediators of ginkgo action on hormone-dependet physiological processes and illnesses. For example, animal and *in vitro* studies demonstrated ability of ginkgo extract to affect prostata estrogen and androgen receptors (Real et al., 2015) and on plasma insulin and muscle insulin receptor level (Banin et al., 2014). The antiinfammatory action of ginkgo molecules are due to their ability to inhibit the expression of proinflammatory cytokines, such as IL-1, IL-6, and

TNF-α (Achete de Souza et al., 2020). Ginkgo's role includes a regulator of membrane stabilizer, and vasodilator. In the arterial endothelium, ginkgo biloba extract triggers the release of endogenous relaxing factors, such as endothelium-derived relaxing factor and prostacyclin (Nguyen and Alzahrani, 2021). Ginkgo constituents' terpene lactones are potent antagonists of the platelet-activating factor (Nguyen and Alzahrani, 2021).

Nervous system mechanisms could be involved in mediating ginkgo action on some processes as well. At least, ginkgo extract appears to have inhibitory effects on rat brain monoamine oxidase by inhibiting the uptake of serotonin and dopamine (Nguyen and Alzahrani, 2021). The ability of ginkgo to affect food consumption and obesity can be mediated by up-regulation of the brain serotoninergic mechanisms regulating appetite (Banin et al., 2017). It is hypothesised that ginkgo can prevent inhibitory action of serotonin on erectile functions (McKay, 2004) (see also below).

A number of intracellular signaling pathways are involved in mediating the action of ginkgo and its constituents on particular processes and illnesses. For example, *in vitro* studies of Koltermann et al. (2009) showed that the anti-angiogenic action of EGb 761 *in vitro* (impairment proliferation, migration and tube formation of endothelial cells) can be mediated by inhibition of the Raf-MEK-ERK pathway. Animal *in vivo* studies of Banin et al. (2014) demonstrated that this ginkgo extract can prevent obesity through changes in muscle levels of insulin receptor substrate 1, protein tyrosine phosphatase 1B, protein kinase B (Akt), as well as Akt phosphorylation. *In vitro* animal and clinical studies indicated that ginkgo polysaccharides could down-regulate reactive oxygen species-induced intracellular pathways (e.g., mitochondrial autophagy, MAPK and JNK) and transcription factor-related pathways (e.g., NF-kB and HIF) which are in turn involved in inflammatory and death receptor pathways. Some of the polysaccharides may also down-regulate malignant transformation of cells via an action on intracellular regulators of tumorigenic pathways Wnt and p53 (Jiao et al., 2016).

Ginkgo effect can be mediated also by inhibition of some esterases. Ginkgo extract EGb761 can inhibit activity of anticholinesterase, hence increasing cholinergic transmission in the brain (Nguyen and Alzahrani, 2021). The protective effects of ginkgo and its ginkgolide A against some specific toxic agents can be mediated by acetylcholine esterase inhibition in the aluminum neurotoxicity or membrane-bond phosphodiesterase activation in the triethyltin toxicity (Omidkhoda et al., 2019).

Such ginkgo constituents as glyceollins and quercetin can affect the expression of some microRNAs affecting apoptosis, cancerogenesis and cancer cell metastasis suggesting that ginkgo can affect physiological functions and dysfunctions via RNA interference/microRNAs (Ahmed et al., 2020).

162 Contaminants and Plants Action on Female Reproduction

Finally, ginkgo effect can be mediated by gas nitric oxide. Ginkgo can strengthen the activity of nitric oxide synthase, presumably bypassing serotonin's ability to block nitric oxide production. Increasing nitric oxide bioavailability may improve overall vascular perfusion and the resulted vascular health, erectile functions, and maybe other age-related chronic diseases (McKay, 2004). Under tissue-damaging inflammatory conditions such as ischemia, ginkgo can promote nitric oxide production and exert vasorelaxation properties (Nguyen and Alzahrani, 2021).

Therefore, the performed studies detected a number of extra- and intracellular signaling pathways (including hormones, cytokines, growth factors, neuromediators, their receptors, postreceptory signaling molecules, esterases, miRNAs and nitric oxide) mediating the action of ginkgo and its molecules on various physiological processes and their disfunctions.

3.7.5 Effects on female reproductive processes

3.7.5.1 Effect on ovarian and reproductive state

The *in vivo* experiments on rats demonstrated the reproductive toxicity of ginkgo extract EGb 761: treatments with this extract disrupted estrous cycles, reduced ovarian follicle counts, embryo resorption and implantation indexes and viability (Eimazoudy and Attia, 2012).

Other *in vivo* experiments, however, showed that ginkgo flavonoids, amifostine, or leuprorelin can increase rat ovarian weight (Chang et al., 2014).

Ginkgo can prevent and treat some illnesses related to female reproduction. The epidemiological studies demonstrated that ginkgo can reduce risk of ovarian cancer (Ye et al., 2007; Jiang et al., 2011). Moreover, ginkgo can be applicable in cancer treatment for impairment of reproductive toxicity of anticancer treatment: its extract mitigated the ovarian toxicity of platinum (Chang et al., 2014). Ginkgo extract also reduced the adverse effects of ischemia/reperfusion injury (torsion/de-torsion) on rat ovarian follicular degeneration, vascular congestion, edema, hemorrhage, and leukocyte infiltration (Yildirim et al., 2018). The efficiency of ginkgo for treatment of menopausal syndrome has been mentioned above (Anonimous, 1998; Elsabagh et al., 2005; Kargozar et al., 2017; Nguyen and Alzahrani, 2021), but it remains to be established whether this action is directly linked with ginkgo action on women reproductive system.

The plant preparation containing extract of Ginkgo biloba promoted sexual desire in menopausal women (Palacios et al., 2019), but it remains unclear whether this effect was induced just by ginkgo.

Taken together, the current literature reports that ginkgo can both up- and down-regulate female reproductive processes and treat their dysfunctions.

3.7.5.2 Effect on ovarian cell functions

In vitro experiments demonstrated the inhibitory action of ginkgo extract on healthy and cancer ovarian cells. This extract was able to inhibit proliferation (down-regulates PCNA, cyclin B1 and their mRNAs), to promote apoptosis (accumulation of bax) in cultured porcine ovarian granulosa cells (Štochmaľová et al., 2018). *In vitro* studies demonstrated the ability of ginkgo extract to suppress proliferation and to promote apoptosis of ovarian cancer cells, too (Ye et al., 2007; Jiang et al., 2011, 2014). It is noteworthy that cytotoxicity of ginkgo extract for epithelial ovarian cancer cells was less than that for healthy cells (Jiang et al., 2014). Ginkgolide B promoted apoptosis in mouse embryos *in vivo* and *in vitro* (Shiao et al., 2009).

On the other hand, *in vivo* experiments on rats showed that ginkgo flavonoids amifostine or leuprorelin suppressed ovarian cell apoptosis and probably induced ovarian cell proliferation (Chang et al., 2014).

Therefore, ginkgo and its constituents can both up- and down-regulate cell proliferation and apoptosis in female reproductive system.

3.7.5.3 Effect on oocytes and embryos

In vivo and *in vitro* studies on mice oocytes and embryos demonstrated ability of ginkgolide B to suppress oocyte maturation, fertilization, *in vivo*, and *in vitro* embryonic development to decrease embryo cell number, viability, and implantation rate (Chan et al., 2006; Shiao et al., 2009).

3.7.5.4 Effect on reproductive hormones

Ginkgo constituents have phytoestrogen properties, and therefore they should presumably affect reproductive steroid hormones and/or their receptors (Dugoua et al., 2006). Ex-vivo experiments on rats showed that ginkgo extract EGb 761 can affect biotransformation of androstendione as well as metabolism of endogenous steroids. This extract failed to affect the urinary steroid profile in humans (Chatterjee et al., 2005). Nevertheless, treatment of menopausal women with plant preparation containing extract of *Ginkgo biloba* L. increased testosterone and decreased sex hormone-binding globulin level in plasma (Palacios et al., 2019). However, it remained unknown, whether this effect was due to ginkgo or other components of the plant mixture. The *in vitro* experiments demonstrated the direct action of ginkgo extract on ovarian hormone production: it suppressed leptin and progesterone release by cultured porcine ovarian granulosa cells (Štochmaľová et al., 2018).

These observations demonstrate the influence of ginkgo extract on production and metabolism of ovarian hormones, although character of this influence depends on the used experimental model.

3.7.6 Mechanisms of action on female reproductive processes

There are some indications that ginkgo can affect reproduction through **central mechanisms**. At least, ginkgo can affect mental functions (Ude et al., 2013; Liu et al., 2019; Singh et al., 2019; Achete de Souza et al., 2020; Nguyen and Alzahrani, 2021) and prevent some symptoms of menopause via improvement to the central serotonin transmission (Banin et al., 2017). Furthermore, plant mixture containing ginkgo promoted sexual behavior regulated by brain centers (Palacios et al., 2019).

The ability of ginkgo to affect ovarian **hormones** *in vivo* (Palacios et al., 2019) and *in vitro* (Štochmaľová et al., 2018) suggests that the known hormonal regulators of female reproduction (Sirotkin, 2014) can can be extracellular mediators of ginkgo action on female reproductive processes including sexual behavior (Palacios et al., 2019).

A number of **intracellular mediators** of ginkgo and its constituents on ovarian cells can be proposed. At least, ginkgo ginkgolide B affected the expression of a 50 intracellular proteins involved in control of **cell proliferation, tumor suppression, and DNA damage repair** (Jiang et al., 2011). Ginkgo extract suppressed proliferation and promoted apoptosis of both healthy and cancer ovarian surface epithelial cell via upregulation of p21, p27, cleaved capase-3, and cleaved caspase-8 and downregulation of cyclin D1 (Jiang et al., 2014). Ginkgolide B can suppress embryo development via generation the reactive oxygen species in embryonal stem cells, which in turn induct apoptosis associated with activation of c-Jun N-terminal kinase, loss of mitochondrial membrane potential and the activation of apoptosis inductor caspase-3 (Chan et al., 2006).

The opposite, antiapoptotic action of ginkgo flavonoids amifostine can be due to increase in production of cytoplasmic cytochrome *c* and apoptotic protease activating factor-1 (Chang et al., 2014).

The diterpene lactones of ginkgo, ginkgolides A, B and C are antagonists of oocyte **GABA receptors,** which can be involved in control of oocyte development and fertilization (Huang et al., 2012).

Taken together, the available data demonstrate a wide array of extra- and intracellular mediators of ginkgo action on female reproductive processes—from hormones up to intracellular signaling substances and their receptors regulating cell cycle and apoptosis. The variability in reported character and mediators of ginkgo constituents on reporoductive processes indicates that different ginkgo molecules could influence different target cells and events via different mechanisms.

3.7.7 Application in reproductive biology and medicine

Despite well-known medicinal use of Ginkgo biloba for treatment of nonreproductive disorders, the safe and reliable application of this plant for control of reproductive processes and their disorders is limited as of yet. It is due to lack of sufficient knowledge or contradiction in the results obtained in experiments on different models concerning character of ginkgo and its constituents' effects. Both inhibitory and stimulatory action of ginkgo and its molecules on female reproductive processes have been reported. Nevertheless, both effects could be useful or should be at least taken into account before ginkgo application.

The potential reproductive toxicity (Eimazoudy and Attia, 2012) and embryotoxicity (Chan et al., 2006; Shiao et al., 2009) of ginkgo, as well as its ability to suppress oogenesis (Shiao et al., 2009) and ovarian cell proliferation, to induce their apoptosis and to inhibit the release of ovarian hormones (Štochmaľová et al., 2018) should be taken into account before ginkgo application for treatment of both nonreproductive and reproductive disorders. On the other hand, these antireproductive effects indicate potential usefulness of ginkgo for synchronisation of animal reproductive cycles or as a natural contraceptive.

Ginkgo can suppress the functions not only of healthy, but also of cancer ovarian cells, while cancer cells look more sensitive to toxic effects of ginkgo (Ye et al., 2007; Jiang et al., 2011). This characteristic makes ginkgo a potential drug for treatment of ovarian cancer. It could be used in combination with other anticancer drugs because ginkgo can not only suppress cancer cells, but also to protect healthy ovarian cells from the toxic influence of anticancer treatment (platinum: Chang et al., 2014). The protective effect of ginkgo against other dysfunctions (ischemia: Yildirim et al., 2018) could be possible too.

Several studies demonstrated the opposite, stimulatory action of ginkgo on female reproductive processes—downregulation of ovarian cell apoptosis and promotion of ovarian growth (Chang et al., 2014), suppression of sex hormone-binding globulin release, promotion of testosterone production and sexual behavior (Palacios et al., 2019). Such positive effects of ginkgo could be useful for stimulation of animal and human reproductive processes and for treatment of their dysfunctions, for example, of menopausal syndrome (Banin et al., 2017).

3.7.8 Conclusions and possible direction of future studies

Ginkgo biloba L. contains a number of biological active molecules which, via numerous signaling pathways, can influence a number of physiological processes and illnesses. Ginkgo and its constituents can influence (mainly downregulate) female reproductive functions, as well as mitigate the reproductive

166 Contaminants and Plants Action on Female Reproduction

disorders via several extra- and intracellular mediators. Therefore, it could be a promising drug for prevention and treatment of some reproductive disorders (including ovarian cancer).

Nevertheless, the available data concerning reproductive effects of ginkgo remain insufficient and contradictory—character of the reported influence on female reproductive processes highly depends on the used experimental model. The majority of data concerns the whole ginkgo extract, but not its constituents, which can possess different and even opposite properties. The data concerning medicinal effects of ginkgo were obtained predominantly in *in vitro* and animals, but not in human experiments. This imperfect knowledge limit wide application of *Ginkgo biloba* L. and its molecules. Therefore, the available information concerning ginkgo suggests both its biological and therapeutic potential and the necessity of more profound studies of its biological and medicinal effects.

References

Abdel-Salam OM, Baiuomy AR, El-batran S, Arbid MS. Evaluation of the anti-inflammatory, antinociceptive and gastric effects of Ginkgo biloba in the rat. Pharmacol Res 2004;49(2):133—42. https://doi.org/10.1016/j.phrs.2003.08.004.

Achete de Souza G, de Marqui SV, Matias JN, Guiguer EL, Barbalho SM. Effects of ginkgo biloba on diseases related to oxidative stress. Planta Med 2020;86(6):376—86. https://doi.org/10.1055/a-1109-3405.

Ahmed F, Ijaz B, Ahmad Z, Farooq N, Sarwar MB, Husnain T. Modification of miRNA Expression through plant extracts and compounds against breast cancer: mechanism and translational significance. Phytomedicine 2020;68:153168. https://doi.org/10.1016/j.phymed.2020.153168.

Anonimous. Ginkgo biloba. Altern Med Rev 1998;3(1):54—7.

Banin RM, Hirata BK, Andrade IS, Zemdegs JC, Clemente AP, Dornellas AP, Boldarine VT, Estadella D, Albuquerque KT, Oyama LM, Ribeiro EB, Telles MM. Beneficial effects of Ginkgo biloba extract on insulin signaling cascade, dyslipidemia, and body adiposity of diet-induced obese rats. Braz J Med Biol Res 2014;47(9):780—8. https://doi.org/10.1590/1414-431x20142983.

Banin RM, de Andrade IS, Cerutti SM, Oyama LM, Telles MM, Ribeiro EB. *Ginkgo biloba* extract (GbE) stimulates the hypothalamic serotonergic system and attenuates obesity in ovariectomized rats. Front Pharmacol 2017;8:605. https://doi.org/10.3389/fphar.2017.00605.

Budeč M, Bošnir J, Racz A, Lasić D, Brkić D, Mosović Ćuić A, Kuharić Ž, Jurak G, Barušić L. Verification of authenticity of *Ginkgo biloba* L. leaf extract and its products present on the croatian market by analysis of quantity and ratio of ginkgo flavone glycosides (quercetin, kaempferol and isorhamnetin) to terpene trilactones to the effect of unmasking counterfeit drugs endangering patient health. Acta Clin Croat 2019;58(4):672—92. https://doi.org/10.20471/acc.2019.58.04.15.

Chan WH. Ginkgolide B induces apoptosis and developmental injury in mouse embryonic stem cells and blastocysts. Hum Reprod 2006;21(11):2985—95. https://doi.org/10.1093/humrep/del255.

Chang Z, Wang HL, Du H. Protective effect of Ginkgo flavonoids, amifostine, and leuprorelin against platinum-induced ovarian impairment in rats. Genet Mol Res 2014;13(3):5276—84. https://doi.org/10.4238/2014.July.24.6.

Food/medicinal herbs and their influence Chapter | 3 **167**

Chatterjee SS, Doelman CJ, Nöldner M, Biber A, Koch E. Influence of the Ginkgo extract EGb 761 on rat liver cytochrome P450 and steroid metabolism and excretion in rats and man. J Pharm Pharmacol 2005;57(5):641−50. https://doi.org/10.1211/0022357056046.

Chen H, Zhou C, Yu M, Feng S, Ma Y, Liu Z, Zhang J, Ding T, Li B, Wang X. The effect of ginkgo biloba dropping pills on hemorheology and blood lipid: a systematic review of randomized trials. Evid Based Complement Alternat Med 2019;2019:2609625. https://doi.org/10.1155/2019/2609625.

Chong PZ, Ng HY, Tai JT, Lee SWH. Efficacy and safety of *Ginkgo biloba* in patients with acute ischemic stroke: a systematic review and meta-analysis. Am J Chin Med 2020;48(3):513−34. https://doi.org/10.1142/S0192415X20500263.

de Souza Predes F, Monteiro JC, Matta SL, Garcia MC, Dolder H. Testicular histomorphometry and ultrastructure of rats treated with cadmium and *Ginkgo biloba*. Biol Trace Elem Res 2011;140(3):330−41. https://doi.org/10.1007/s12011-010-8702-5.

Diamond BJ, Bailey MR. *Ginkgo biloba*: indications, mechanisms, and safety. Psychiatr Clin 2013;36(1):73−83. https://doi.org/10.1016/j.psc.2012.12.006.

Drugs and Lactation Database (LactMed) [Internet]. Ginkgo. Bethesda (MD): National Library of Medicine (US); 2018. 30000868.

Dugoua JJ, Mills E, Perri D, Koren G. Safety and efficacy of ginkgo (Ginkgo biloba) during pregnancy and lactation. Can J Clin Pharmacol 2006;13(3):e277−84.

Eimazoudy RH, Attia AA. Efficacy of Ginkgo biloba on vaginal estrous and ovarian histological alterations for evaluating anti-implantation and abortifacient potentials in albino female mice. Birth Defects Res B Dev Reprod Toxicol 2012;95(6):444−59. https://doi.org/10.1002/bdrb.21032.

Eisvand F, Razavi BM, Hosseinzadeh H. The effects of Ginkgo biloba on metabolic syndrome: a review. Phytother Res 2020;34(8):1798−811. https://doi.org/10.1002/ptr.6646.

Elsabagh S, Hartley DE, File SE. Limited cognitive benefits in Stage +2 postmenopausal women after 6 weeks of treatment with Ginkgo biloba. J Psychopharmacol 2005;19(2):173−81. https://doi.org/10.1177/0269881105049038.

Evans JR. Ginkgo biloba extract for age-related macular degeneration. Cochrane Database Syst Rev 2013;2013(1):CD001775. https://doi.org/10.1002/14651858.CD001775.pub2.

Fang J, Wang Z, Wang P, Wang M. Extraction, structure and bioactivities of the polysaccharides from Ginkgo biloba: a review. Int J Biol Macromol 2020;162:1897−905. https://doi.org/10.1016/j.ijbiomac.2020.08.141.

Feng Z, Sun Q, Chen W, Bai Y, Hu D, Xie X. The neuroprotective mechanisms of ginkgolides and bilobalide in cerebral ischemic injury: a literature review. Mol Med 2019;25(1):57. https://doi.org/10.1186/s10020-019-0125-y.

Hirata BK, Banin RM, Dornellas AP, de Andrade IS, Zemdegs JC, Caperuto LC, Oyama LM, Ribeiro EB, Telles MM. Ginkgo biloba extract improves insulin signaling and attenuates inflammation in retroperitoneal adipose tissue depot of obese rats. Mediat Inflamm 2015;2015:419106. https://doi.org/10.1155/2015/419106.

Huang SH, Lewis TM, Lummis SC, Thompson AJ, Chebib M, Johnston GA, Duke RK. Mixed antagonistic effects of the ginkgolides at recombinant human ρ1 GABAC receptors. Neuropharmacology 2012;63(6):1127−39. https://doi.org/10.1016/j.neuropharm.2012.06.067.

Izzo AA, Ernst E. Interactions between herbal medicines and prescribed drugs: an updated systematic review. Drugs 2009;69(13):1777−98. https://doi.org/10.2165/11317010-000000000-00000.

Jiang W, Qiu W, Wang Y, Cong Q, Edwards D, Ye B, Xu C. Ginkgo may prevent genetic-associated ovarian cancer risk: multiple biomarkers and anticancer pathways induced by ginkgolide B in BRCA1-mutant ovarian epithelial cells. Eur J Canc Prev 2011;20(6):508−17. https://doi.org/10.1097/CEJ.0b013e328348fbb7.

Jiang W, Cong Q, Wang Y, Ye B, Xu C. Ginkgo may sensitize ovarian cancer cells to cisplatin: antiproliferative and apoptosis-inducing effects of ginkgolide B on ovarian cancer cells. Integr Canc Ther 2014;13(3):NP10−N17. https://doi.org/10.1177/1534735411433833.

Jiao R, Liu Y, Gao H, Xiao J, So KF. The anti-oxidant and antitumor properties of plant polysaccharides. Am J Chin Med 2016;44(3):463−88. https://doi.org/10.1142/S0192415X16500269.

Kargozar R, Azizi H, Salari R. A review of effective herbal medicines in controlling menopausal symptoms. Electron Phys 2017;9(11):5826−33. https://doi.org/10.19082/5826.

Koltermann A, Liebl J, Fürst R, Ammer H, Vollmar AM, Zahler S. Ginkgo biloba extract EGb 761 exerts anti-angiogenic effects via activation of tyrosine phosphatases. J Cell Mol Med 2009;13(8B):2122−30. https://doi.org/10.1111/j.1582-4934.2008.00561.x.

Lejri I, Agapouda A, Grimm A, Eckert A. Mitochondria- and oxidative stress-targeting substances in cognitive decline-related disorders: from molecular mechanisms to clinical evidence. Oxid Med Cell Longev 2019;2019:9695412. https://doi.org/10.1155/2019/9695412.

Martiniakova M, Babikova M, Omelka R. Pharmacological agents and natural compounds: available treatments for osteoporosis. J Physiol Pharmacol 2020;71(3). https://doi.org/10.26402/jpp.2020.3.01.

McKay D. Nutrients and botanicals for erectile dysfunction: examining the evidence. Alternative Med Rev 2004;9(1):4−16.

Mei N, Guo X, Ren Z, Kobayashi D, Wada K, Guo L. Review of Ginkgo biloba-induced toxicity, from experimental studies to human case reports. J Environ Sci Health C Environ Carcinog Ecotoxicol Rev 2017;35(1):1−28. https://doi.org/10.1080/10590501.2016.1278298.

Nguyen T, Alzahrani T. Ginkgo biloba. In: StatPearls [Internet]. Treasure Island (FL: StatPearls Publishing; 2021. 31082068.

Omidkhoda SF, Razavi BM, Hosseinzadeh H. Protective effects of *Ginkgo biloba* L. against natural toxins, chemical toxicities, and radiation: a comprehensive review. Phytother Res 2019;33(11):2821−40. https://doi.org/10.1002/ptr.6469.

Palacios S, Soler E, Ramírez M, Lilue M, Khorsandi D, Losa F. Effect of a multi-ingredient based food supplement on sexual function in women with low sexual desire. BMC Womens Health 2019;19(1):58. https://doi.org/10.1186/s12905-019-0755-9.

Real M, Molina-Molina JM, Jimenez J, Diéguez HR, Fernández MF, Olea N. Assessment of hormone-like activities in *Ginkgo biloba*, *Elettaria cardamomum* and *Plantago ovata* extracts using in vitro receptor-specific bioassays. Food Addit Contam Part A Chem Anal Control Expo Risk Assess 2015;32(9):1531−41. https://doi.org/10.1080/19440049.2015.1071922.

Shiao NH, Chan WH. Injury effects of ginkgolide B on maturation of mouse oocytes, fertilization, and fetal development in vitro and in vivo. Toxicol Lett 2009;188(1):63−9. https://doi.org/10.1016/j.toxlet.2009.03.004.

Singh SK, Srivastav S, Castellani RJ, Plascencia-Villa G, Perry G. Neuroprotective and antioxidant effect of ginkgo biloba extract against AD and other neurological disorders. Neurotherapeutics 2019;16(3):666−74. https://doi.org/10.1007/s13311-019-00767-8.

Sirotkin AV. Regulators of ovarian functions. 2nd ed. New York: Nova Science Publishers, Inc.; 2014, ISBN 978-1-62948-574-4. 194 pp.

Štochmaľová A, Kádasi A, Alexa R, Bauer M, Harrath AH, Sirotkin AV. Direct effect of polyphenol-rich plants, rooibos and ginkgo, on porcine ovarian cell functions. J Anim Physiol Anim Nutr 2018;102(2):e550−7. https://doi.org/10.1111/jpn.12795.

Tian J, Liu Y, Chen K. Ginkgo biloba extract in vascular protection: molecular mechanisms and clinical applications. Curr Vasc Pharmacol 2017;15(6):532−48. https://doi.org/10.2174/1570161115666170713095545.

Tricco AC, Lillie E, Zarin W, O'Brien KK, Colquhoun H, Levac D, Moher D, Peters MDJ, Horsley T, Weeks L, Hempel S, Akl EA, Chang C, McGowan J, Stewart L, Hartling L, Aldcroft A, Wilson MG, Garritty C, Lewin S, Godfrey CM, Macdonald MT, Langlois EV, Soares-Weiser K, Moriarty J, Clifford T, Tunçalp Ö, Straus SE. PRISMA extension for scoping reviews (PRISMA-ScR): checklist and explanation. Ann Intern Med 2018;169(7):467−73. https://doi.org/10.7326/M18-0850.

Ude C, Schubert-Zsilavecz M, Wurglics M. Ginkgo biloba extracts: a review of the pharmacokinetics of the active ingredients. Clin Pharmacokinet 2013;52(9):727−49. https://doi.org/10.1007/s40262 013 0074-5.

Ye B, Aponte M, Dai Y, Li L, Ho MC, Vitonis A, Edwards D, Huang TN, Cramer DW. Ginkgo biloba and ovarian cancer prevention: epidemiological and biological evidence. Cancer Lett 2007;251(1):43−52. https://doi.org/10.1016/j.canlet.2006.10.025.

Yildirim N, Simsek D, Kose S, Yildirim AGS, Guven C, Yigitturk G, Erbas O. The protective effect of Gingko biloba in a rat model of ovarian ischemia/reperfusion injury: improvement in histological and biochemical parameters. Adv Clin Exp Med 2018;27(5):591−7. https://doi.org/10.17219/acem/68896.

Chapter 3.8

Grape (*Vitis vinifera* L.)

3.8.1 Introduction

Vitis vinifera (grape) is the most important fruit crops (Fabres et al., 2017) and one of the "most-produced fruit" in the world (Tabeshpour et al., 2018) and. It is popularly consumed for fresh consumption and wine and spirit production (Fabres et al., 2017) but also as processed products and derivatives, including dried fruits, jellies, juices, but about 70% of its crop intended especially for wine production (Liang et al., 2014; Brenes et al., 2016; Cadiz-Gurrea et al., 2017; Lopes de Menezes et al., 2018).

Grape it is a matrix rich in bioactive compounds and its production generates large quantities of by-products, such as grape stems, which, to date, present low commercial value (Leal et al., 2020). However, grape pomace, consisting of peel, seed, stem, and pulp, is discarded during grape processing, including juice extraction and winemaking (Averilla et al., 2019). Wine pomace, also called grape pomace are an important byproduct in the grape processing process (Rombaut et al., 2014; de Souza et al., 2020). This byproduct can be considered as a valuable source of health-beneficial compounds such as phenolic compounds including resveratrol (Tabeshpour et al., 2018) and procyanidin (Zhao et al., 2013) that could add value to products in the food industries (Amorim et al., 2019), cosmetics (Glampedaki et al., 2014), and pharmaceutical (Nassiri-Asl and Hosseinzadeh, 2016).

The current knowledge concerning its physiological and therapeutic effects are summarized in several recent reviews (Teixeira et al., 2014a, Nassiri-Asl and Hosseinzadeh, 2016; Lim and Song, 2017; Huang et al., 2018; Gouvinhas et al., 2020), but the available data concerning its action on reproductive processes has not been sufficient reviewed yet.

The present paper aimed to make a snapshot of the available information concerning characteristics of *Vitis vinifera* and its biological active constituents present in grape, metabolism, general physiological, and therapeutic effects of grape. The main purpose of this review is however to describe, to summarize and to analyze the current knowledge concerning grape effects on female reproductive processes and their dysfunctions, which have not been reviewed yet, but which could be useful for better understanding and application of grape in reproductive biology and medicine.

3.8.2 Provenance and properties

Vitis vinifera, L. fruit (grape) contains various phenolic compounds, flavonoids, and stilbenes. The active constituents found in the fruits, seeds, stems, skin, products, and pomaces of grapes have been identified (Nassiri-Asl and Hosseinzadeh, 2016). A multibiofunctional compound found in grape skin and seeds, red and white wine is resveratrol (trans-3, 5,-4′-trihydroxystilebene) (Kuršvietienė et al., 2016; Tabeshpour et al., 2018) as stilbenoid compound with antioxidant and phytoestrogenic properties attributable to its bioactive transresveratrol content (Nashine et al., 2020) (see also the special chapter concerning resveratrol). A natural pigment in grapes is delphinidin as a member of the anthocyanidin family (Lim and Song, 2017). Another biological active constituent of grape seed is polyphenol flavonoid procyanidin, which possess antiinflammatory and antioxidant properties (Zhao et al., 2013; Zhang et al., 2019). The grape seed oils are a good source of γ-tocotrienol (499−1575 mg/kg), α-tocopherol (85.5−244 mg/kg) and α-tocotrienol (69−319 mg/kg). Concerning fatty acid profile, linoleic (C18:2cc), oleic (C18:1), palmitic (C16:0), and stearic (C18:0) acids were the predominant. Grape seed oils demonstrated to be a good source of polyunsaturated fatty acids (PUFAs) (63.64%−73.53%), whereas monounsaturated fatty acid (MUFA) and saturated fatty acid (SFA) ranged between 14.19%−21.29% and 11.64%−14.94%, respectively (Fernandes et al., 2013). Grape seed extracts contained monomeric and dimeric flavanol mono- and diglycosides. With the exception of diglycosylated flavanol dimers, these compounds were also detected in grape skin extracts. The concentration of the mono and diglycosides depended largely on both grape variety and whether they occurred in grape skins or seeds (Zerbib et al., 2018). After intake, less than 5%−10% of plant polyphenols are absorbed through the enterocytes, either by passive diffusion or by selective transportations, but only if they are in the form of aglycones or simple glycosides. These facts compromise its kinetics through the digestive tract and its availability for later metabolization through normal biological pathways, so bioavailability of polyphenols may be insufficient to induce significant biological effects in most of the cases (Gessner et al., 2017).

3.8.3 Physiological and therapeutic actions

The active constituents of different parts of *Vitis vinifera* and their pharmacological effects including skin protection, antioxidant, antibacterial, antimicrobial, anticancer, antiinflammatory, antidiabetic, metabolism- and immunity-modulatory activities as well as hepatoprotective, cardioprotective and neuroprotective effects were described (Teixeira et al., 2014a; Nassiri-Asl and Hosseinzadeh, 2016; Huang et al., 2018).

Grape seed procyanidin extracts derived from grape seeds have been reported to possess a broad spectrum of pharmacological and medicinal properties (Barbe et al., 2020). Grape seed flavonoid proanthocyanidin B2 (GSPB2) is one of the most important components of grape seed procyanidin extracts and is probably more powerful than other polyphenols. Some studies have shown that GSPB2 exhibits protective effects against stress, inflammation, and cardiovascular diseases (Yu et al., 2012; Yin et al., 2015).

The rich composition in phenolic compounds presented by grape stems can explain the high antioxidant and antimicrobial activities against, essentially, Gram-positive bacteria, and in some cases with higher efficacy than commercial antibiotics. Thus, demonstrating that this wine by-product should deserve greater attention from the pharmaceutical industries due to its excellent biological properties and characteristics not yet applied (Leal et al., 2020).

The other grape constituent, a natural pigment delphinidin possesses nutraceutical properties against various chronic diseases including cancer (Lim and Song, 2017).

Resveratrol isolated from the stems of *Vitis vinifera* possess anticancer and apoptotic effects on cell proliferation, and therefore, can be used as new approach of pharmaceutical drugs (Elgizawy et al., 2021). There is an increasing interest by the scientific community in the application of this matrix as a source of phenolic compounds, for use as food supplements and/or active ingredients for the cosmetic and pharmaceutical industries (Anastasiadi et al., 2012).

The available data demonstrate the multiple physiological and medicinal effects of Vitis vinifera. These effects can depend on biological actives substances present grape.

3.8.4 Mechanisms of action

Grape and its phenolic compounds including GSPB2, resveratrol and other bioactive constituents, resveratrol, exert their protective effects against different natural or chemical toxins which could alter physiological homeostasis through a variety of mechanisms.

Some of these mechanisms of actions include increase in superoxide dismutase (SOD), hemeoxygenase-1 (HO-1), and glutathione peroxidase (GPx) activities and reduced glutathione content and decrease in malondialdehyde (MDA) levels and activation of the nuclear factor erythroid 2-related factor 2 (Nrf2)/antioxidant-responsive element (ARE) pathway (Tabeshpour et al., 2018). Mechanism of resveratrol action is described in the special subchapter. Resveratrol can affect a wide array of cellular processes via multiple intracellular signaling pathways (see chapter 4.10.4).

3.8.5 Effects on female reproductive processes

The bioavailability of polyphenols has been confirmed in different reproductive organs including brain (hypothalamus and hypophysis) responsible for regulation of reproductive hormones, ovary, uterus placenta, and fetus (Ly et al., 2015; Woclawek-Potocka et al., 2013; Milojevic et al., 2020). This fact indicates the ability of polyphenols to pass different blood barriers of the reproductive organs and therefore to presumptively influence their physiological functions. There is a number of evidence for grape extract and its phenolic constituents action on female reproductive processes at different regulatory levels.

3.8.5.1 Effect on ovaries

Oral administration of grape seed extract was able to alter changes in progesterone release during heifers estrous cycle (Colitti et al., 2007). Addition of either grape seed extract or GSPB2 to cultured human healthy or cancer granulosa cells increased progesterone and estradiol secretion, which was associated with a higher level of the cholesterol carriers, steroidogenic acute regulatory protein (StAR) (Barbe et al., 2019). These observations demonstrate the ability of grapeseed and its constituent GSBP2 to up-regulate ovarian steroidogenesis.

Grape constituent GSPB2, when added to incubation medium, was able to promote mouse oocyte viability, maturation and developmental capacity, which was associated with improvement the state of oocyte mitochondria (Luo et al., 2020).

GSPB2 does not affect nuclear and cytoplasmic apoptosis in cultured healthy porcine ovarian granulosa cells (Zhang et al., 2019), but it is able to reduce viability of human ovarian cancer cells and their resistence to chemotherapy (Zhao et al., 2013). Delphinidin, a natural pigment in grape and other plants, is also considered to be promising drug for treatment of various chronic diseases and types of cancer including ovarian cancer cells (Lim and Song, 2017). The influence of grape seed, GSBP2 and delphinidin on cancerogenesis has been described in detail below.

Grape seed extract is a dietary supplement comprised of monomeric and oligomeric catechins and has health benefits in models of age-related reproductive insufficiency and menopause (Cutts et al., 2013). Their administration is able to prevent the destructive ovarian morphological changes induced by reproductive aging (Liu et al., 2018). This effect of grape seed extract could be due to presence of proanthocyanidin. GSPB2, an antioxidative polyphenol in grape seed, has been reported to protect against the age-dependent degenerative changes in rat ovaries (Zhang et al., 2016; Liu et al., 2018).

Proanthocyanidin was shown to be a natural protector against damage of rat ovarian tissue induced by ischemia and/or ischemia/reperfusion (Yıldırım et al., 2015; Zhang et al., 2016; Liu et al., 2018).

Short-term grape seed extract therapy provided beneficial therapeutic impacts on polycystic ovarian syndrome (PCOS) positive women's metabolic status thus, this approach could be effective in PCOS complications management (Sedighi et al., 2020).

174 Contaminants and Plants Action on Female Reproduction

Therefore, the limited number of performed studies indicated the ability of grape seed and its constituents (procyanidin B2 and maybe delphinidin) to influence ovarian steroidogenesis, oocyte maturation and developmental capacity and ovarian cell apoptosis. Furthermore, these substances could be useful for treatment of ovarian cancer, age-related reproductive insufficiency and menopausal syndrome, ovarian ischemia, PCOS.

3.8.5.1 Effect on uterus

The oral administration of grape seed extract to heifers affects the expression of a number of genes in their uterine endometrium suggesting influence of grapeseed substances on endometrial functions (Colitti et al., 2007).

Furthermore, a grapeseed constituent resveratrol can alter the response of endometrium to progesterone and estrogen during decidualization. Decidualization is the differentiation of endometrial stromal cells (ESC), which is essential to regulate trophoblast invasion and to support pregnancy establishment and progression. Decidualization follows ESC proliferation and cell cycle arrest contributes to a proper decidualization. Resveratrol could be an effective supplementation to reinforce hormone action during human ESC decidualization. This fact could facilitate the identification of new therapeutic strategies for the improvement of women's health (Mestre Citrinovitz et al., 2020).

Furthermore, the antiinflammatory effect of resveratrol can contribute to the prevention of endometriosis, this phenolic compound now being considered a new innovative drug in the prevention and treatment of this disease (Dull et al., 2019).

Grape seed proanthocyanidins could inhibit the growth of cervical cancer indicating that these proanthocyanidins may be a potential chemopreventive and/or chemotherapeutic agent against cervical cancer (Chen et al., 2014).

These reports indicate the involvement of grape seed molecules to affect state of uterine endometrium, its decidualization, as well as to prevent and to treat endometriosis and cervical cancer.

3.8.6 Mechanisms of action on female reproductive processes

The character and mechanisms of grape action on female reproductive processes are studied insufficiently, but the few available reports enable outline several possible mechanisms of their action on female reproduction.

Polyphenols may intervene in the regulation of all reproductive events through **hypothalamic neurohormones** (gonadotropin-releasing hormone, GnRH, and oxytocin), **gonadotropins** (luteinizing hormone, LH, and follicle-stimulating hormone, FSH), **steroids** (estradiol, progesterone, and testosterone), and **prostaglandins** (Hashem et al., 2020).

Polyphenols may also control **steroid hormones** function by binding or inactivating sex production enzymes, such as P450 aromatase, 5α-reductase, 17β-hydroxysteroid dehydrogenase (17β-OHDH), topoisomerases, and tyrosine kinases. In addition, polyphenols can alter sex hormone-binding globulin (SHBG) affinity, and thus levels of active steroids in blood circulation (Yousif, 2019). Addition of either grape extract or GSBP2 to cultured healthy and cancer human ovarian granulosa cells increased production of cholesterol carriers, StAR, progesterone and estradiol secretion (Barbe et al., 2019). Feeding of hens with grape extract resulted down-regulation of StAR and progesterone secretion by their granulosa cells too (Barbe et al., 2020).

The unique chemical structure of polyphenols and its similarity to the chemical structure of mammalian estrogens enables them to possess hormone-like effects through binding and activating **estrogen receptors** (ERs: ERα and ERβ), which result in estrogen-agonistic or antagonistic effects. The binding affinity of polyphenols for ERs is determined by their chemical structure, with the presence of a phenolic ring being responsible for binding to ERs, molecular weight, and optimal hydroxylation pattern (Kuiper et al., 1998; Yildiz et al., 2005). Overall, binding affinity of polyphenols is always lower than the natural ligand estradiol. Different affinities of polyphenols toward both subtypes of ERs and different distribution of ERs among reproductive tissues greatly affect the final result of exposure to polyphenols (Reed, 2016; Cipolletti et al., 2018).

Resveratrol supplementation increased the expression levels of **prolactin** and **insulin-like growth factor binding protein 1 (IGFBP1)**, indicating an enhanced *in vitro* decidualization of human ESC (Mestre Citrinovitz et al., 2020).

Oxidative stress is an important inducement in ovarian aging which results in fecundity decline in human and diverse animals (Liu et al., 2018). On the other hand, oxidative stress is a causal factor and key promoter of all kinds of reproductive disorders related to ovarian granulosa cell apoptosis that acts by dysregulating the expression of related genes (Zhang et al., 2019). Stress activates the hypothalamo—pituitary—adrenal axis leading to enhanced glucocorticoid secretion and concurrently disrupts ovarian cycle (Colitti et al., 2007). The grape seed extract (Barbe et al., 2019), grape constituents resveratrol (Mikuła-Pietrasik et al., 2015; Barbe et al., 2019), and flavonoid GSBP2 (Barbe et al., 2019; Zhang et al., 2019) can promote antioxidant enzymes and to reduce oxidative stress in healthy and cancer ovarian granulosa cells. Similarly, grape seed GSBP2 prevents ovarian aging by inhibiting oxidative stress in the hens (Liu et al., 2018). Finally, the ability of GSBP2 to improve maturation, developmental capacity, and viability of murine oocytes were associated with GSBP2-induced reduction in reactive oxygen species within the oocytes. This indicates that GSBP2 can improve oocyte quality as antioxidant and protector against oxidative stress (Luo et al., 2020).

176 Contaminants and Plants Action on Female Reproduction

Oxidative stress can destroy ovarian functions via promotion of ovarian granulosa cell **apoptosis** that acts by dysregulating the expression of related genes. Oxidative stress-induced granulosa cells apoptosis is believed to be a major cause of follicular atresia (Shen et al., 2012). GSPB2 does not affect nuclear and cytoplasmic apoptosis in cultured healthy porcine granulosa cells (Zhang et al., 2019), but it was able to reduce viability of ovarian cancer cells and their resistance to chemotherapy (Zhao et al., 2013), as well as to promote expression of markers of apoptosis in human cancer granulosa cells (Barbe et al., 2019). Grape seed GSPB2 could not only promote apoptosis in ovarian cancer cells, but also to inhibit the growth of cervical cancer by inducing apoptosis through the mitochondrial pathway. This provides evidence indicating that grape seed proanthocyanidins may be a potential chemopreventive and/or chemotherapeutic agent for cervical cancer (Chen et al., 2014).

Some miRNAs have been reported to interfere with and modulate granulosa cells apoptosis signaling (Worku et al., 2017). GSPB2 can suppress ovarian cell apoptosis, while **microRNA let-7a** can be involved in mediating this effect. At least, GSPB2 treatment protected cultured porcine granulosa cells from a hydrogen peroxide (H_2O_2)-induced apoptosis and increased let-7a expression in these cells. Furthermore, let-7a overexpression markedly increased cell viability and inhibited H_2O_2-induced granulosa cell apoptosis. Finally, the results indicate that GSPB2 inhibits H_2O_2-induced apoptosis of GCs, possibly through the upregulation of let-7a (Zhang et al., 2019).

Grape molecules can affect ovarian cell functions and dysfunctions also through changes in their **proliferation.** GSPB2 could maintain the homeostasis between cell apoptosis and cell proliferation in the natural aging ovaries (Liu et al., 2018). Both grape seed and GSBP2 delayed in G1 to S phase cell cycle progression and inhibited the expression of its intracellular regulators (a mitogen-activated protein kinase (MAPK), cyclin D2, Akt phosphorylation) and promoted proliferation inhibitors (p21, p27 cyclin-dependent kinase) in cultured human healthy (Barbe et al., 2019) and cancer ovarian granulosa cells (Zhao et al., 2013; Barbe et al., 2019; Homayoun et al., 2020). An other grape constituent, delphinidin, was able to down-regulate the expression of intracellular prompters of proliferation (phosphorylated protein kinase B, Akt), ribosomal protein S6 kinase β-1, P70S6K, ribosomal protein S (S6), extracellular signal-regulated kinase (ERK1/2 and p38) in cultured ovarian cancer cells. Furthermore, the combination of certain pharmacological inhibitors, including phosphoinositide 3-kinase (PI3K), ERK1/2 and delphinidin significantly reduced the proliferation of the cells and the phosphorylation of each of those target proteins (Lim and Song, 2017). These results indicate that delphinidin inhibits the proliferation of cells through inactivation of PI3K/Akt and ERK1/2 mitogen-activated protein kinase signaling cascades, and that this cell signaling pathway may be a pivotal therapeutic target for the prevention of epithelial ovarian cancer, including paclitaxel-resistant ovarian cancer (Lim and Song, 2017).

Oxidative stress can induce also **inflammatory processes** which in turn promote numerous reproductive disorders including disorders of ovarian folliculogenesis and ovarian cycle, oocyte maturation, release of reproductive hormones, polycystic ovarian syndrome (Snider and Wood, 2019), endometriosis (Wang et al., 2020) and cancer (Predescu et al., 2019). Grape seed extract reduced the expression of proinflammatory interleukines in rats with polycystic ovarian syndrome (Salmabadi et al., 2017). Resveratrol was able to prevent ovarian cell damage induced by ionizing radiation (Said et al., 2016) and endometriosis (Dull et al., 2019) via down-regulating of synthesis of prostaglandins, interleukines and transcription factor nuclear factor-κB (NF-κB)—inductor of inflammation. Similarly, GSBP2 was able to protect ovarian cancer cells from chemotherapeutical agents by down-regulating NF-κB (Zhao et al., 2013).

Resveratrol, a silencing information regulator 1 (SIRT1) activator, counteract the inflammatory signaling associated with radiotherapy-induced premature ovarian failure. Resveratrol-activated SIRT1 expression is associated with inhibition of poly(ADP-ribose) polymerase (PARP-1) and the NF-κB expression-mediated inflammatory cytokines. Resveratrol restored ovarian function through increasing anti-Mullerian hormone (AMH) levels, and diminishing ovarian inflammation, predominantly via upregulation of PPAR-c and SIRT1 expression leading to inhibition of NF-κB provoked inflammatory cytokines (Said et al., 2016). Resveratrol action on inflammatory processes and regulators could be mediated by inhibition of prostaglandin (Dull et al., 2019).

These reports demonstrate that grape seed extract and int constituents, proanthocyanidin B2 and probably delphinidin can affect female reproductive functions and ovarian cancer via reproductive hormones (GnRH, gonadotropins, prolactin, prostaglandin, insulin-like growth factor binding protein), steroid hormones receptors, intracellular regulators of oxidative stress and subsequent inflammation, apoptosis, and proliferation.

3.8.7 Application in reproductive biology and medicine

The ability of grape molecules to protect reproductive system from oxidative stress and inflammation, as well as their ability to improve oocyte maturation, viability and developmental capacity indicate that grape seed extract or its molecules could be promising as biostimulators of farm animal and human reproduction in animal production, biotechnology, and assisted reproduction. Furthermore, its action on endometrium could be useful for assurance of healthy gravidity, embryogenesis and labor.

Moreover, a number of evidence demonstrate applicability of grape seed extract and its constituents for prevention and treatment of age-related menopausal reproductive insufficiency and complications, gynecological cancer and endometriosis. Moreover, their phytoestrogenic properties might be potentially useful for prevention and treatment of other disorders induced by

178 Contaminants and Plants Action on Female Reproduction

estrogen deficit (menopausal syndrome, osteoporosis) as alternative to hormonal treatment. To out knowledge, such application of grape seed has not been examined yet.

3.8.8 Conclusions and possible direction of future studies

Therefore, the limited number of performed studies indicated the ability of grape seed and its constituents (GSPB2 and maybe delphinidin) to influence ovarian steroidogenesis, oocyte maturation and developmental capacity and ovarian cell apoptosis. Furthermore, these substances could be useful for treatment of ovarian cancer, age-related reproductive insufficiency and menopausal syndrome, ovarian ischemia, PCOS. In addition, grape seed molecules can affect state of uterine endometrium, its decidualization, and they can be potentially useful to prevent and to treat endometriosis and cervical cancer.

Grape seed extract and int constituents, proanthocyanidin B2 and probably delphinidin can affect female reproductive functions disfunctions via changes in reproductive hormones (GnRH, gonadotropins, prolactin, prostaglandin, and insulin-like growth factor binding protein), steroid hormones receptors, intracellular regulators of oxidative stress and subsequent inflammation, apoptosis, and proliferation.

The potential usefulness of these molecules for improvement reproductive processes and for treatment of gynecological cancer and endometriosis, age-related reproductive insufficiency, and menopause-related complications is proposed.

Nevertheless, a number of aspects of grape seed effects on female reproductive processes and their disfunctions remains not investigated yet. Sometimes, the constituent responsible for particular effect of grape seed remain not identified. The action of all grape seed constituents on reproductive processes have not been tested yet. The main studies were performed on *in vitro* or animal model, while clinical studies are missing. The action of grape seed or its constituents on ovarian folliculogenesis, ovulation, and fecundity have not been properly examined. Therefore, the lack of knowledge limites understanding and application of grape seed, despite its large biological and therapeutic potential.

References

Amorim FL, de Cerqueira Silva MB, Cirqueira MG, Oliveira RS, Machado BAS, Gomes RG, de Souza CO, Druzian JI, de Souza Ferreira E, Umsza-Guez MA. Grape peel (Syrah var.) jam as a polyphenol-enriched functional food ingredient. Food Sci Nutr 2019;7(5):1584—94. https://doi.org/10.1002/fsn3.981.

Anastasiadi M, Pratsinis H, Kletsas D, Skaltsounis AL, Haroutounian SA. Grape stem extracts: polyphenolic content and assessment of their in vitro antioxidant properties. LWT - Food Sci Technol 2012;48(2):c316—322. https://doi.org/10.1016/j.lwt.2012.04.006.

Averilla JN, Oh J, Kim HJ, Kim JS, Kim JS. Potential health benefits of phenolic compounds in grape processing by-products. Biol Med 2019;28:1607–15. https://doi.org/10.1007/s10068-019-00628-2.

Barbe A, Ramé C, Mellouk N, Estienne A, Bongrani A, Brossaud A, Riva A, Guérif F, Froment P, Dupont J. Effects of grape seed extract and proanthocyanidin B2 on in vitro proliferation, viability, steroidogenesis, oxidative stress, and cell signaling in human granulosa cells. Int J Mol Sci 2019;20(17):4215. https://doi.org/10.3390/ijms20174215.

Barbe A, Mellouk N, Ramé C, Grandhaye J, Anger K, Chahnamian M, Ganier P, Brionne A, Riva A, Froment P, Dupont J. A grape seed extract maternal dietary supplementation improves egg quality and reduces ovarian steroidogenesis without affecting fertility parameters in reproductive hens. PloS one 2020;15(5):e0233169. https://doi.org/10.1371/journal.pone.0233169.

Blaheta RA, Nelson K, Haferkamp A, Juengel E. Amygdalin, quackery or cure? Phytomedicine 2016;23(4):367–76. https://doi.org/10.1016/j.phymed.2016.02.004.

Brenes A, Viveros A, Chamorro S, Arija I. Use of polyphenol-rich grape by-products in monogastric nutrition. A review. Anim Feed Sci Technol 2016;211:1–17. https://doi.org/10.1016/j.anifeedsci.2015.09.016.

Cadiz-Gurrea MDLL, Borras-Linares I, Lozano-Sanchez J, Joven J, Fernandez-Arroyo S, Segura-Carretero A. Cocoa and grape seed by products as a source of antioxidant and anti-inflammatory proanthocyanidins. Int J Mol Sci 2017;18(2):376. https://doi.org/10.3390/ijms18020376.

Cipolletti M, Solar Fernandez V, Montalesi E, Marino M, Fiocchetti M. Beyond the antioxidant activity of dietary polyphenols in cancer: the modulation of estrogen receptors (ers) signaling. Int J Mol Sci 2018;19(9):2624. https://doi.org/10.3390/ijms19092624.

Chen Q, Liu XF, Zheng PS. Grape seed proanthocyanidins (GSPs) inhibit the growth of cervical cancer by inducing apoptosis mediated by the mitochondrial pathway. PLoS One 2014;9(9):e107045. https://doi.org/10.1371/journal.pone.0107045. eCollection 2014.

Colitti M, Sgorlon S, Stradaioli G, Farinacci M, Gabai G, Stefanon B. Grape polyphenols affect mRNA expression of PGHS-2, TIS11b and FOXO3 in endometrium of heifers under ACTH-induced stress. Theriogenology 2007;68(7):1022–30. https://doi.org/10.1016/j.theriogenology.2007.07.018. Epub 2007.

Cutts JK, Peavy TR, Moore DR, Prasain J, Barnes S, Kim H. Ovariectomy lowers urine levels of unconjugated (+)-catechin, (-)-epicatechin, and their methylated metabolites in rats fed grape seed extract. Horm Mol Biol Clin Invest 2013;16(3):129–38. https://doi.org/10.1515/hmbci-2013-0044.

De Souza RC, Machado BAS, Barreto GA, Leal IL, Anjos JPD, Umsza-Guez MA. Effect of experimental parameters on the extraction of grape seed oil obtained by low pressure and supercritical fluid extraction. Molecules 2020;25(7):1634. https://doi.org/10.3390/molecules25071634.

Dull AM, Moga MA, Dimienescu OG, Sechel G, Burtea V, Anastasiu CV. Therapeutic approaches of resveratrol on endometriosis via anti-inflammatory and anti-angiogenic pathways. Molecules 2019;24(4):667. https://doi.org/10.3390/molecules24040667.

Elgizawy HA, Ali AA, Hussein MA. Resveratrol: isolation, and its nanostructured, inhibits cell proliferation, induces cell apoptosis in certain human cell lines carcinoma and exerts protective effect against paraquat-induced hepatotoxicity. J Med Food 2021;24(1):89–100. https://doi.org/10.1089/jmf.2019.0286.

180 Contaminants and Plants Action on Female Reproduction

Fabres PJ, Collins C, Cavagnaro TR, Rodríguez López CM. A concise review on multi-omics data integration for terroir analysis in Vitis vinifera. Front Plant Sci 2017;20(8):1065. https://doi.org/10.3389/fpls.2017.01065.

Fernandes L, Casal S, Cruz R, Pereira JA, Ramalhosa E. Seed oils of ten traditional Portuguese grape varieties with interesting chemical and antioxidant properties. Food Res Int 2013;50(1):161−6. https://doi.org/10.1016/j.foodres.2012.09.039.

Gessner DK, Ringseis R, Eder K. Potential of plant polyphenols to combat oxidative stress and inflammatory processes in farm animals. J Anim Physiol Anim Nutr 2017;101:605−28. https://doi.org/10.1111/jpn.12579.

Glampedaki P, Dutschk V. Stability studies of cosmetic emulsions prepared from natural products such as wine, grape seed oil and mastic resin. Colloids Surf A Physicochem Eng Asp 2014;460:306−11. https://doi.org/10.1016/j.colsurfa.2014.02.048.

Hashem NM, Gonzalez-Bulnes A, Simal-Gandara J. Polyphenols in farm animals: source of reproductive gain or waste? Antioxidants 2020;9(10):1023. https://doi.org/10.3390/antiox9101023.

Homayoun M, Ghasemnezhad Targhi R, Soleimani M. Anti-proliferative and anti-apoptotic effects of grape seed extract on chemo-resistant OVCAR-3 ovarian cancer cells. Res Pharm Sci 2020;15(4):390−400. https://doi.org/10.4103/1735-5362.293517.

Huang Q, Liu X, Zhao G, Hu T, Wang Y. Potential and challenges of tannins as an alternative to in-feed antibiotics for farm animal production. Anim Nutr 2018;4(2):137−50. https://doi.org/10.1016/j.aninu.2017.09.004.

Kuiper GG, Lemmen JG, Carlsson B, Corton JC, Safe SH, van der Saag PT, van der Burg B, Gustafsson JA. Interaction of estrogenic chemicals and phytoestrogens with estrogen receptor beta. Endocrinology 1998;139(10):4252−63. https://doi.org/10.1210/endo.139.10.6216.

Kuršvietienė L, Stanevičienė I, Mongirdienė A, Bernatonienė J. Multiplicity of effects and health benefits of resveratrol. Medicina 2016;52(3):148−55. https://doi.org/10.1016/j.medici.2016.03.003.

Leal C, Santos RA, Pinto R, Queiroz M, Rodrigues M, José Saavedra M, Barros A, Gouvinhas I, Saudi J. Recovery of bioactive compounds from white grape (Vitis vinifera L.) stems as potential antimicrobial agents for human health. Biol Sci 2020;27(4):1009−15. https://doi.org/10.1016/j.sjbs.2020.02.013.

Liang Z, Cheng L, Zhong GY, Liu RH. Antioxidant and antiproliferative activities of twenty-four Vitis vinifera grapes. PLoS One 2014;9(8):e105146. https://doi.org/10.1371/journal.pone.0105146.

Lim W, Song G. Inhibitory effects of delphinidin on the proliferation of ovarian cancer cells via PI3K/AKT and ERK 1/2 MAPK signal transduction. Oncol Lett 2017;14(1):810−8. https://doi.org/10.3892/ol.2017.6232.

Liu X, Lin X, Mi Y, Li J, Zhang C. Grape seed proanthocyanidin extract prevents ovarian aging by inhibiting oxidative stress in the hens. Oxid Med Cell Longev 2018;2018:16. https://doi.org/10.1155/2018/9390810.9390810z.

Lopes de Menezes M, Johann G, Diório A, Pereira NC, da Silva EA. Phenomenological determination of mass transfer parameters of oil extraction from grape biomass waste. J Clean Prod 2018;176:130−9. https://doi.org/10.1016/j.jclepro.2017.12.128.

Loren DJ, Seeram NP, Schulman RN, Holtzman DM. Maternal dietary supplementation with pomegranate juice Is neuroprotective in an animal model of neonatal hypoxic-ischemic brain injury. Pediatric Research 2005;57(6):858−64. https://doi.org/10.1203/01.pdr.0000157722.078.

Luo Y, Zhuan Q, Li J, Du X, Huang Z, Hou Y, Fu X. Procyanidin B2 improves oocyte maturation and subsequent development in type 1 diabetic mice by promoting mitochondrial function. Reprod Sci 2020;27(12):2211−22. https://doi.org/10.1007/s43032-020-00241-3.

Ly C, Yockell-Lelievre J, Ferraro ZM, Arnason JT, Ferrier J, Gruslin A. The effects of dietary polyphenols on reproductive health and early development. Hum Reprod Update 2015;21(2):228−48. https://doi.org/10.1093/humupd/dmu058.

Mestre Citrinovitz AC, Langer L, Strowitzki T, Germeyer A. Resveratrol enhances decidualization of human endometrial stromal cells. Reproduction 2020;159(4):453−63. https://doi.org/10.1530/REP-19-0425.

Mikuła-Pietrasik J, Sosińska P, Murias M, Wierzchowski M, Brewińska-Olchowik M, Piwocka K, Szpurek D, Książek K. High potency of a novel resveratrol derivative, 3,3',4,4'-Tetrahydroxy-trans-stilbene, against ovarian cancer is associated with an oxidative stress-mediated imbalance between DNA damage accumulation and repair. Oxid Med Cell Longev 2015;2015:135691. https://doi.org/10.1155/2015/135691.

Milojevic V, Sinz S, Kreuzer M, Chiumia D, Marquardt S, Giller K. Partitioning of fatty acids into tissues and fluids from reproductive organs of ewes as affected by dietary phenolic extracts. Theriogenology 2020;144:174−84. https://doi.org/10.1016/j.theriogenology.2020.01.012.

Nashine S, Nesburn AB, Kuppermann BD, Kenney MC. Role of resveratrol in transmitochondrial AMD RPE cells. Nutrients 2020;12(1):159. https://doi.org/10.3390/nu12010159.

Nassiri-Asl M, Hosseinzadeh H. Review of the pharmacological effects of *Vitis vinifera* (grape) and its bioactive constituents: an update. Phytother Res 2016;30(9):1392−403. https://doi.org/10.1002/ptr.5644.

Predescu DV, Crețoiu SM, Crețoiu D, Pavelescu LA, Suciu N, Radu BM, Voinea SC. G protein-coupled receptors (GPCRs)-Mediated calcium signaling in ovarian cancer: focus on GPCRs activated by neurotransmitters and inflammation-associated molecules. Int J Mol Sci 2019;20(22):5568. https://doi.org/10.3390/ijms20225568.

Reed K. Fertility of herbivores consuming phytoestrogen-containing medicago and trifolium species. Agriculture 2016;6(3):35. https://doi.org/10.3390/agriculture6030035.

Rombaut N, Savoire R, Thomasset B, Bélliard T, Castello J, Van Hecke É, Lanoisellé JL. Grape seed oil extraction: interest of supercritical fluid extraction and gas-assisted mechanical extraction for enhancing polyphenol co-extraction in oil. Compt Rendus Chem 2014;17(3):284−92. https://doi.org/10.1016/j.crci.2013.11.014.

Said RS, El-Demerdash E, Nada AS, Kamal MM. Resveratrol inhibits inflammatory signaling implicated in ionizing radiation-induced premature ovarian failure through antagonistic crosstalk between silencing information regulator 1 (SIRT1) and poly(ADP-ribose) polymerase 1 (PARP-1). Biochem Pharmacol 2016;103:140−50. https://doi.org/10.1016/j.bcp.2016.01.019. PMID: 26827941.

Salmabadi Z, Mohseni Kouchesfahani H, Parivar K, Karimzadeh L. Effect of grape seed extract on lipid profile and expression of interleukin-6 in polycystic ovarian syndrome wistar rat model. Int J Fertil Steril 2017;11(3):176−83. https://doi.org/10.22074/ijfs.2017.5007.

Sedighi P, Helli B, Sharhani A, Vatanpur A. Effects of grape seed extract supplementation on fasting blood glucose, insulin resistance, and lipid profile in women with polycystic ovary syndrome. Anat Sci 2020;17(2):73−82.

Shen M, Lin F, Zhang J, Tang Y, Chen WK, Liu H. Involvement of the up-regulated FoxO1 expression in follicular granulosa cell apoptosis induced by oxidative stress. J Biol Chem 2012;287(31):25727−40. https://doi.org/10.1074/jbc.M112.349902.

Snider AP, Wood JR. Obesity induces ovarian inflammation and reduces oocyte quality. Reproduction 2019;158(3):R79−90. https://doi.org/10.1530/REP-18-0583.

182 Contaminants and Plants Action on Female Reproduction

Tabeshpour J, Mehri S, Behbahani FS, Hosseinzadeh H. Protective effects of *Vitis vinifera* (grapes) and one of its biologically active constituents, resveratrol, against natural and chemical toxicities: a comprehensive review. Phytother Res 2018;32(11):2164−90. https://doi.org/10.1002/ptr.6168.

Teixeira A, Baenas N, Dominguez-Perles R, Barros A, Rosa E, Moreno DA, Garcia-Viguera C. Natural bioactive compounds from winery by-products as health promoters: a review. Int J Mol Sci 2014;15(9):15638−78. https://doi.org/10.3390/ijms150915638.

Wang Y, Nicholes K, Shih IM. The origin and pathogenesis of endometriosis. Annu Rev Pathol 2020;15:71−95. https://doi.org/10.1146/annurev-pathmechdis-012419-032654.

Woclawek-Potocka I, Mannelli C, Boruszewska D, Kowalczyk-Zieba I, Wasniewski T, Skarzynski DJ. Diverse effects of phytoestrogens on the reproductive performance: cow as a model. Internet J Endocrinol 2013;2013:650984. https://doi.org/10.1155/2013/650984.

Worku T, Rehman ZU, Talpur HS, et al. MicroRNAs: new insight in modulating follicular atresia: a review. Int J Mol Sci 2017;18(2):333. https://doi.org/10.3390/ijms18020333.

Yan Z, Dai Y, Fu H, et al. Curcumin exerts a protective effect against premature ovarian failure in mice. J Mol Endocrinol 2018;60(3):261−71. https://doi.org/10.1530/JME-17-0214.

Yamasaki M, Kitagawa T, Koyanagi N, Chujo H, Maeda H, Kohno-Murase J, Imamura J, Tachibana H, Yamada K. Dietary effect of pomegranate seed oil on immune function and lipid metabolism in mice. Nutrition 2006;22(1):54−9. https://doi.org/10.1016/j.nut.2005.03.009.

Yıldırım S, Topaloğlu N, Tekin M, Küçük A, Erdem H, Erbaş M, Yıldırım A. Protective role of Proanthocyanidin in experimental ovarian torsion. Med J Islam Repub Iran 2015;29:185.

Yildiz HB, Kiralp S, Toppare L, Yagci Y. Immobilization of tyrosinase in poly (ethyleneoxide) electrodes and determination of phenolics in red wines. React Funct Polym 2005;63(2):155−61. https://doi.org/10.1016/j.reactfunctpolym.2005.02.016.

Yin W, Li B, Li X, et al. Critical role of prohibitin in endothelial cell apoptosis caused by glycated low-density lipoproteins and protective effects of grape seed procyanidin B2. J Cardiovasc Pharmacol 2015;65(1):13−21. https://doi.org/10.1097/fjc.0000000000000157.

Yousif ANE. Effect of Flaxseed on some hormonal profile and genomic DNA concentration in Karadi lambs. IOP Conf Ser Earth Environ Sci 2019;388:012035. https://doi.org/10.1088/1755-1315/388/1/012035.

Yu F, Li BY, Li XL, et al. Proteomic analysis of aorta and protective effects of grape seed procyanidin B2 in db/db mice reveal a critical role of milk fat globule epidermal growth factor-8 in diabetic arterial damage. PLoS One 2012;7(12):e52541. https://doi.org/10.1371/journal.pone.0052541.e52541.

Zerbib M, Cazals G, Enjalbal C, Saucier C. Identification and quantification of flavanol glycosides in Vitis vinifera grape seeds and skins during ripening. Molecules 2018;23(11):2745. https://doi.org/10.3390/molecules23112745.

Zhang JQ, Gao BW, Wang J, et al. Critical role of FoxO1 in granulosa cell apoptosis caused by oxidative stress and protective effects of grape seed procyanidin B2. Oxid Med Cell Longev 2016;2016:16. https://doi.org/10.1155/2016/6147345.6147345.

Zhang JQ, Wang XW, Chen JF, Ren QL, Wang J, Gao BW, Shi ZH, Zhang ZJ, Bai XX, Xing BS. Grape seed procyanidin B2 protects porcine ovarian granulosa cells against oxidative stress-induced apoptosis by upregulating let-7a expression. Oxid Med Cell Longev 2019;2019:1076512. https://doi.org/10.1155/2019/1076512.

Zhao BX, Sun YB, Wang SQ, Duan L, Huo QL, Ren F, Li GF. Grape seed procyanidin reversal of p-glycoprotein associated multi-drug resistance via down-regulation of NF-κB and MAPK/ERK mediated YB-1 activity in A2780/T cells. PLoS One 2013;8(8):e71071. https://doi.org/10.1371/journal.pone.0071071.

Chapter 3.9

Pomegranate (*Punica granatum* L.)

3.9.1 Introduction

Nowadays, pomegranate has gained widespread interest as a functional food and nutraceutical source. It is generally considered safe to consume the fresh fruit and juice of pomegranates. The pomegranate fruit can be divided into three parts, including the seeds (about 3% of the whole fruit weight), the arils account for about 45%–52% of the weight of the whole fruit and have been valued for pomegranate juice (about 30% of the fruit weight), and the peels, which also include the interior network of membranes and present almost 55% of whole fruit weight (Lansky and Newman, 2007; Viuda-Martos et al., 2010). The aim of the present chapter was to discuss the scientific evidence that suggests that pomegranate and its phytonutrients possesses a diverse array of biological actions and may have a potential to be helpful in the regulation of reproductive functions, prevention of various diseases, including cancer therapy.

3.9.2 Provenance and properties

Pomegranate plant, a deciduous shrub, native to Iran and Mediterranean region, belongs to family *Punicaceae*, genus *Punica*, with one predominant species called *Punica granatum*, which has been cultivated and consumed since 3000 BCE. Nowadays, multidimensional beneficial applications of pomegranate fruit are reported (Zarfeshany et al., 2014; Shaygannia et al., 2016; Saeed et al., 2018; Kandylis et al., 2020). The metabolites of various part of pomegranate fruit and plant include polyphenols (ellagitannins, ellagic acid, gallic acid, punicalagin, etc.), flavonoids (luteolin, kaempferol, etc.), and anthocyanins (delphinidin, cyanidin, pelargonidin), which are responsible for red color of edible parts of pomegranate. Moreover, pomegranate fruit and content alkaloids (pelletierine, isopelletierine, methylpelletierine, pseudopelletierine, etc.), sugars (glucose, fructose, sucrose, maltose), organic acids (fumaric acid, oxalic acid, succinic acid, citric acid, malic acid, tartaric acid, etc.), fatty acids (linolenic acid, linoleic acid, oleic acid, palmitic acid, stearic acid, palmitoleic acid, arachidonic acid, lauric acid, etc), and vitamins mainly C, B1, and B2. Pomegranate peel contains minerals such as calcium, phosphorus, magnesium, and potassium. In addition, biologically active substances found in pomegranate peel include ellagic acid, gallic acid, chlorogenic acid, cinnamic acid, hydroxy benzoic acid, caffeic acid, ferulic acid, coumaric acid,

184 Contaminants and Plants Action on Female Reproduction

pelletierine, isopelletierine, punicalagin, punicalin, quercetin, and catechin (Van Elswijk et al., 2004). Hydrolyzable tannins such as ellagitannins, gallotannins, gallagyl esters, hydroxybenzoic acids and hydroxycinnamic acids are the main class of polyphenolic compounds identified in pomegranate juice. Punicalagin (2,3-hexahydroxydiphenoyl-4,6-gallagylglucoside) and its isomers present the major ellagitannin in pomegranate. Pomegranate juice also contains punicalin α and β (Gil et al., 2000). The pomegranate seed oil has recently received more attention and can be used in providing necessary fatty acids (Melgarejo et al., 2000). Additionally, it contents steroidal estrogens (γ-tocopherol, 17α-estradiol, stigmasterol, β-estriol sitosterol, testosterone) and nonsteroidal compounds (compestrol, coumestrol) (Van Elswijk et al., 2004).

3.9.2.1 Bioavailability, metabolism and pharmacokinetics

Phytonutrients from pomegranate is becoming an increasingly popular dietary supplement. The antioxidant activity of pomegranate has been attributed to its phenolic content, specifically punicalagins, punicalins, gallagic acid, and ellagic acid, which are metabolized during digestion to ellagic acid and urolithins (Johanningsmeier and Harris, 2011). Ellagitannins have been claimed to possess stable structural properties under physiological conditions in the stomach (Usta et al., 2013). On the other hand, Larrosa et al. (2010) have reported that ellagic acid is methyl conjugated by the action of catechol O-methyltransferase quickly following its absorption in digestive tract. The pharmacokinetic profile showed that only a part of this absorption takes place in the stomach (Lei et al., 2003). The metabolism of ellagic acid proceeds by conversion to the most abundant metabolite detected up to date via a two-step reaction—dimethyl-ellagic acid-glucuronide (Whitley et al., 2003). Bioavailability and metabolism of ellagitannins and ellagic acid have been assessed and urolithin A and B. Ellagic acid is poorly absorbed in the stomach and small intestine and mainly metabolized by gut bacteria in the intestinal lumen to produce urolithins. Microbial metabolism starts in the small intestine, the first metabolites produced retain four phenolic hydroxyls and further metabolized along the intestinal tract to remove hydroxyl units leading to urolithin A and B in the distal parts of the colon (Devipriya et al., 2007). Study in humans has demonstrated the rapid absorption and plasma clearance of ellagitannins. The maximum ellagic acid blood concentration was 31.9 ng/mL 1 h after 180 mL of pomegranate juice consumption, with rapid plasma clearance by 4 hours postingestion (Seeram et al., 2004). Another study suggested an explanation of the benefits of long-term pomegranate administration and confirmed urolithin metabolites excreted in the urine can persist for 48 h after pomegranate juice ingestion (Seeram et al., 2006). In another study, it was stated that the maximum concentration of ellagic acid in plasma was 213 ng/mL, approximately 1 h after oral administration of 0.8 g/kg of pomegranate leaf extract (Lei et al., 2003). In addition, it was suggested that the persistence of urolithin A and B in the urine may be responsible for pomegranate's long-term antioxidant effects (Johanningsmeier and Harris., 2011). In a 13-day clinical trial, three metabolites were detected in the plasma—urolithin A, urolithin B, and an unidentified minor metabolite after 5-days consumption of 1 L of pomegranate juice containing 4.37 g/L punicalagins and 0.49 g/L anthocyanins. Urinalysis revealed six metabolites, whereas significant variability of observed urinary metabolite concentrations may be attributed to differences in colonic microflora (Cerda et al., 2004). Current research demonstrated a significant increase (31.8%) in plasma antioxidant capacity 30 min after pomegranate extract (containing 330.4 mg punicalagins and 21.6 mg ellagic acid) administration (Mertens-Talcott et al., 2006).

3.9.3 Physiological and therapeutic actions

Several studies have confirmed beneficial properties of *Punica granatum* products or their phenolic compounds including antioxidant effects (Rosenblat et al., 2006; Guo et al., 2008; Gil et al., 2000; Sudheesh et al., 2005; Johanningsmeier and Harris, 2011), antiinflammatory activities (Hora et al., 2003; Seeram et al., 2005; Lee et al., 2010; Motaal et al., 2011; Zarfeshany et al., 2014; Zhao et al., 2016; Reis et al., 2016; Shah et al., 2016; Danesi and Ferguson, 2017; Sahebkar et al., 2017), neuroprotective (Amri et al., 2017), antidiabetic (Katz et al., 2007; Banihani et al., 2013), antiatherosclerotic, antihypertensive, cardioprotective (Rock et al., 2008; Asgary et al., 2014, 2017; Reis et al., 2016; Sahebkar et al., 2017), antimicrobial (Choi et al., 2011), antibacterial (Viladomiu et al., 2013), antifungal (Pirzadeh et al., 2020), antiproliferative, apoptotic (Kim et al., 2002; Mehta et al., 2004; Jeune et al., 2005; Seeram et al., 2005; Jeune et al., 2005, 2005; Modaeinama et al., 2015), anticancer (Malik et al., 2005), chemopreventive and therapeutic actions on oral, colon (Adams et al., 2006; Seeram et al., 2007), lung (Khan et al., 2007; Zahin et al., 2014; Modaeinama et al., 2015), prostate (Hong et al., 2008; Malik et al., 2005; Paller et al., 2017), ovarian (Modaeinama et al., 2015; Baldovská et al., 2019, 2020) and breast (Kim et al., 2002; Mehta et al., 2004; Jeune et al., 2005; Mandal et al., 2015; Modaeinama et al., 2015) cancer and regulatory functions in reproductive processes (Türk et al., 2008; Packova et al., 2015; Packova and Kolesarova, 2016). Unlike *in vitro* studies, clinical research in patients with various types of cancer is still sparse (Paller et al., 2017).

Pomegranate fruit extract may have strong potential as a chemopreventive or chemotherapeutic agent in preventing of human lung cancer (Khan et al., 2007). Another study reported that pomegranate peel and seed extracts possess antimicrobial activity due to their rich content of phenolic compounds (Choi et al., 2011). The protective effects of pomegranate seed oil and pomegranate extracts on skin cells may be beneficial for both cancer prevention and reduction of photoaging (Afaq et al., 2005). Amri et al. (2017) ascribed protective benefits of pomegranate seed oil, leaves, juice and peel on brain oxidative stress and lipid profile. The strong antioxidant activity is attributed to the presence of high polyphenolic content including punicalagins, punicalins, anthocyanins, unique fatty acids, gallagic and ellagic acid (Johanningsmeier and Harris, 2011). Ellagitannins and ellagic acid as active agents induce vasorelaxation, oxygen-free radical scavenging, hypolipidemic, antiinflammatory, and anticarcinogenic activities (Usta et al., 2013). Moreover, anthocyanins and fatty acid profile of the pomegranate seed oil may also play a role in pomegranate's health effects (Johanningsmeier and Harris., 2011). Several pomegranate-based antiinflammatory products have been patented (Hora et al., 2003; Motaal et al., 2011). There are studies on pomegranate product usage in relation to chronic pulmonary disease (Cerdá et al., 2006), Alzheimer's disease (Hartman et al., 2006), immune function (Yamasaki et al., 2006), neonatal neuroprotection (Loren et al., 2005), menopause (Newton et al., 2006), male infertility (Türk et al., 2008), and erectile dysfunction (Forest et al., 2007). Currently, Asgary et al. (2017) demonstrated that pomegranate juice possesses strong antihypertensive, antioxidant, and anti-atherosclerotic properties. Oleanolic, ursolic and gallic acids present in pomegranate juice are responsible for antidiabetic properties (Katz et al., 2007). Pomegranate juice consumption may increase epididymal sperm concentration, sperm motility, spermatogenic cell density, diameter of seminiferous tubules and germinal cell layer thickness, as well as decrease the number of abnormal sperm compared to control rat models (Türk et al., 2008). The antidiabetic activity of pomegranate flower (Li et al., 2008), and antilipidemic effect (Huang et al., 2005) was demonstrated.

Combination of pomegranate's phytocompounds exerts synergistic effects that are markedly higher than the effect of single molecules. Future research should reveal the nature of these interactions as well as the mechanisms underlying these activities (Vučić et al., 2019). Taken together, the available data demonstrate the antioxidant, antiinflamentary, anticancer and other beneficial properties of pomegranate and its related products such as pomegranate juice, seed oil, pomegranate leaves, flowers and peel.

3.9.4 Mechanisms of action

The main mechanism of pomegranate action and its compounds includes reduction of reactive oxygen species (ROS) and prevention the oxidative stress and inflammatory processes:

(1) decrease of production of ROS (Park et al., 2016), hydroperoxides concentration (Sudheesh et al., 2005), oxidative damage and of the inflammatory response (Zhao et al., 2016), decrease of oxidative stress, and cellular lipid peroxide content (Rosenblat et al., 2006),

Food/medicinal herbs and their influence Chapter | 3 **187**

(2) increase of antioxidant capacity of plasma (Guo et al., 2008; Türk et al., 2008), and sperm (Türk et al., 2008), activation the antioxidant enzymes decrease of malondialdehyde (MDA) and hydroperoxides concentrations, and enhancement of catalase, superoxide dismutase (SOD), glutathione peroxidase (GPx), and glutathione reductase activities in the liver (Sudheesh et al., 2005), SOD and GPx (Amri et al., 2017), and reduced glutathione (Rosenblat et al., 2006) in plasma and induction of the expression of nuclear factor-erythroid 2-related factor-2 (Nrf2), which regulates the expression of endogenous antioxidant enzymes (Reis et al., 2016),

(3) decrease of plasma carbonyl content as a biomarker for risk of various inflammatory diseases (Guo et al., 2008),

(4) reduction of the production of proinflammatory signaling molecules: tumor necrosis factor (TNF)-α, nuclear factor kappa B (NF-κB), transducer and activator of transcription (STAT3), interleukin (IL)1β, monocyte chemoattractant protein-1 (MCP-1), and intercellular adhesion molecule 1 (ICAM-1) (Dreiseitel et al., 2009; Hontecillas et al., 2009; Speciale et al., 2010; Park et al., 2016; Shah et al., 2016). Reduction of cyclooxigenase-2 (COX-2) expression via the inhibition of phosphatidylinositide 3-kinases (PI3K) and protein kinase B or Akt, both necessary for NF-κB activation (Sheu et al., 2005; Larrosa et al., 2010; Shah et al., 2016). Decrease of NF-κB and the nuclear factor erythroid 2 -related factor 2 (Nrf2) molecular pathways, decrease the expression of COX-2 via NF-κB and mitogen-activated protein kinase (MAPK) pathways, reducing the production of proinflammatory prostaglandins, inhibition of phosphatidylinositide 3-kinases (PI3K), protein kinase B or Akt, or NF-κB (Larrosa et al., 2010; Mandal et al., 2017). Downregulate inflammation in mucosal immune and epithelial cells via peroxisome proliferator-activated receptor (PPAR)γ and PPARδ dependent mechanism, and activate PPARγ-responsive genes expression—CD36, fatty acid-binding protein 4 (FABP4), and an insulin-regulated glucose transporter (GLUT4) (Viladomiu et al., 2013; Hontecillas et al., 2009). inhibition of transcription factor AP-1 (activation protein 1) via MAPK-induced phosphorylation of extracellular signal-regulated protein kinase (ERK)1/2, c-Jun N-terminal protein kinase (JNK)1,2,3, and p38, as well as reduce TNF-α, inducible nitric oxide synthase (iNOS), metalloproteinases (MMP), IL-6, and IL-1β (Viladomiu et al., 2013; Vučić et al., 2019). Downregulate proinflammatory and proangiogenic iNOS mediators and modulation of signaling pathways p38 MAPK (Shah et al., 2016),

(5) decrease of fasting glucose concentrations, upregulate PPAR-α and -γ responsive genes (Hontecillas et al., 2009),

188 Contaminants and Plants Action on Female Reproduction

(6) the inhibition of cholinesterase, decrease of MDA and protein carbonylation levels (Amri et al., 2017),

(7) inhibition of a gene coding an enzyme to break down warfarin in the body (CYP2CP) and increase of bioavailability of tolbutamide (substrate for CYP2CP), inhibition of cytochrome P450-3A (CYP3A)-mediated carbamazepine metabolism (Nagata et al., 2007; Hidaka et al., 2005), the decrease of total CYP450 level (Faria et al., 2007).

The next mechanism of pomegranate action and its compounds concerns its action on regulators of **steroidogenesis**:

(1) inhibition of 17β-estradiol catalyzed by 17β-hydroxysteroid enzyme, which can prevent proliferation of breast cancer cells and inhibit apoptosis by inactivation of the mentioned enzyme (Kim et al., 2002),

(2) inhibition of enzymes involved in breast carcinogenesis, such as aromatase, which converts androgen to estrogen, and 17β-hydroxysteroid dehydrogenase, which is involved in estrogen biosynthesis (Kim et al., 2002),

(3) concurrent disruption of estrogen receptors (ER) and Wnt/β-catenin signaling pathways due to decrease of expression of ER-α and ER-β, and β-catenin in breast cancer (Mandal et al., 2015),

(4) reduction of the expression of genes involved in androgen biosynthesis, such as 3β-hydroxysteroid dehydrogenase type 2, steroid 5α reductase type 1, and androgen receptor in prostate cancer (Hong et al., 2008).

Furthermore, pomegranate and its compounds can affect cells (at least cancer cells) via suppression of their **proliferation**:

(1) inhibition of proliferation of lung (Zahin et al., 2014; Modaeinama et al., 2015), and colon carcinoma cells (Cho et al., 2015),

(2) a selective cytotoxic and antiproliferative effect in human colon carcinoma, breast and prostate cancer without any toxic impact on the viability of normal human lung fibroblast cells (Losso et al., 2004),

(3) arrest the cell growth in G0/G1 phase of the cell cycle showing reduction in the protein expressions of cyclins D1, D2, E, declined cyclin-dependent kinase (CDK) − cdk2, cdk4, and cdk6 expression and activation of WAF1/p21 and KIP1/p27, decrease of phosphorylation of MAPK proteins and phosphorylation of Akt, as well as inhibit phosphoinositide 3-kinases (PI3K), NF-κB, nuclear protein Ki-67, and proliferating cell nuclear antigen (PCNA) to reduce the cancer cell proliferation in the lung (Khan et al., 2007),

(4) inhibition of promoter of proliferation MAPK (Viladomiu et al., 2013; Shah et al., 2016; Vučić et al., 2019).

The next mechanism of pomegranate action and its compounds concerns its action on regulators and development of **apoptosis**:

(1) induction of apoptosis via modulation of proteins regulating apoptosis (Malik et al., 2005),

(2) induction of apoptosis through modulation of Bcl-2 proteins (Larrosa et al., 2006), increases p21 and p27, and downregulates cyclins and cyclin kinases network, which leads to a cell cycle arrest in cancer cells (Faria and Calhau, 2011), and activation of caspase 3 and 9 (Larrosa et al., 2006),

(3) induction of cell cycle arrest at the G2/M phase, as well as activation of CDKN1A expression, and activation of caspases 3, 8, and 9, induction of both the extrinsic (death receptor-mediated; where caspase 8 is activated) and intrinsic (mitochondrial mediated; where caspase 9 is activated) apoptotic pathways in colon cancer (Cho et al., 2015),

(4) induction of apoptosis in human colon carcinoma, breast and prostate cancer without any toxic impact on the viability of normal human lung fibroblast cells (Losso et al., 2004),

(5) suppression of TNFα-induced COX-2 protein expression and Akt activation, needed for NF-κB DNA binding in human colon cancer cell (Adams et al., 2006).

In summary, phytonutrients present in pomegranate can modulate various molecular signaling pathways of inflammation, steroidogenesis, proliferation and apoptosis by alteration of protein kinases, cellular transcription factors and other signaling proteins, resulting in beneficial effects in prevention or treatment of several diseases listed below.

3.9.5 Effects on female reproductive processes

Pomegranate fruits including their phytonutrients have beneficial impact on the female reproductive system (Packova et al., 2015; Packova and Kolesarova, 2016; Baldovská et al., 2019, 2020) by regulation of secretory activity, cell proliferation, and apoptosis and potential cancer therapy (Modaeinama et al., 2015; Baldovská et al., 2019, 2020). Little is known about the impact of pomegranate on the female reproductive system. Punicalagin, as a part of the family of ellagitannins which are the most abundant polyphenol found in pomegranate, is described in the corresponding chapter focused on its effects on female reproductive functions and their mechanisms of action.

Modulatory effect of punicalagin and pomegranate extract from nonedible parts of *Punica granatum* L. on ovarian cells was found (Baldovská et al., 2019, 2020). The 17β-estradiol but not progesterone secretion by the human ovarian granulosa cells was increased after the extract treatment (Baldovská et al., 2019). Similarly, as the previous study has shown, punicalagin increases the secretion of steroid hormone 17β-estradiol but not progesterone by human ovarian granulosa cells (Baldovská et al., 2020). The next results showed increase of the 17β-estradiol and decrease of progesterone level in human ovarian granulosa cells after pomegranate peel extract supplementation (Baldovská et al., 2020). The polyphenol punicalagin and pomegranate extract may be a potential endocrine modulator of steroidogenesis in human ovarian granulosa cells (Baldovská et al., 2019, 2020). Possible effect of ellagitannins—compounds from pomegranate on process of steroidogenesis in ovaries was evaluated. Progesterone and 17beta-estradiol (but not androstenedione and testosterone) release by rabbit ovarian fragments was affected by punicalagin addition. Punicalagin increased progesterone and decrease 17beta-estradiol release by rabbit ovarian fragments. The results suggest that punicalagin could have dose-dependent impact on secretion of steroid hormones progesterone and 17β-estradiol by rabbit ovarian fragments and it may be effector in process of ovarian steroidogenesis (Packova et al., 2015). A common reproductive, endocrine, and metabolic disease in women is polycystic ovarian syndrome (PCOS). Pomegranate juice, known as a rich source of phytochemicals with high antioxidant activity, may improve PCOS by increasing insulin sensitivity and decreasing of testosterone level but not FSH and LH (Esmaeilinezhad et al., 2019).

Pomegranate extract contains bioactive compounds including punicalagin which affect on viability of human ovarian cells in a cell- and dose-dependent manner (Baldovská et al., 2019, 2020). The pomegranate extract increases viability of human ovarian granulosa cells. Moreover, the number of viable human ovarian carcinoma cells decreased after the addition of the extract (Baldovská et al., 2019). Pomegranate peel extract decreases cell viability of human ovarian carcinoma cells in a dose-dependent manner, but there was no effect on healthy ovarian granulosa cells (Baldovská et al., 2020). On the other hand, punicalagin decreases viability of human ovarian granulosa cells at high concentrations but treatment with punicalagin does not cause any changes in the viability of human ovarian cancer cells. Pomegranate peel extract seems to be a better chemopreventive agent in comparison to pure punicalagin (Baldovská et al., 2020). Similarly, antiproliferative activity of pomegranate peel extract has been confirmed in different human carcinoma cells, including ovarian cancer cells (Modaeinama et al., 2015).

Data provides evidence that the pomegranate extract might be a promising candidate as a potential modulator of steroidogenesis and as a potential chemoprotective agent. However, further research is essential to understand the therapeutic potential of pomegranate on female reproductive system and its mechanisms of action.

3.9.6 Mechanisms of action on female reproductive processes

The ability of both pomegranate extract and punicalagin to affect ovarian cell production of **steroid hormones** (Packova et al., 2015; Esmaeilinezhad et al., 2019; Baldovská et al., 2019, 2020) which are considered as the key regulators of ovarian cell proliferation, apoptosis, ovarian folliculogenesis, ovarin cycle, and fecundity (Sirotkin, 2014) suggest that pomegranate molecules can affect these reproductive events via regulation of ovarian steroidogenesis.

The ability of pomegranate to improve organism insulin sensitivity, which is associated with suppression of PCOS signs (Esmaeilinezhad et al., 2019) could be an indication that pomegranate could affect ovarian functions also via changes in metabolism and **insulin/insulin-like growth factor—dependent signaling pathway**.

The ability of pomegranate extract to affect ovarian cell **proliferation** (Modaeinama et al., 2015) and **viability** (Baldovská et al., 2019, 2020) suggest that pomegranate can influence ovarian cell functions affecting cell cycle.

The possible mechanisms of effect of the pomegranate molecule punicalagin (regulators of inflammation, proliferation, apoptosis and steroidogenesis) are described in the specific chapter.

The functional interrelationships between the putative mediators of pomegranate action on reproductive processes require further elucidation.

3.9.7 Application in reproductive biology and medicine

Previous findings about pomegranate and punicalagin effects on physiological processes including reproduction suggest that pomegranate could have potential role in regulation of female reproductive proceses including steroidogenesis, proliferation and apoptosis. In healthy ovarian cells it can decrease oxidative stress to regulate proliferation and apoptosis and to boost healthy ovarian cell viability and steroidogenesis. This ability indicates that pomegranate extracts may be used as a source of health-promoting functional foods and biostimulator of female ovarian functions in animal production, assisted reproduction and gynecology.

On the other hand, pomegranate and punicalagin exert antiproliferative activity in cancer cells via inhibition of their proliferation and induction of apoptosis. This pomegranate molecule a promising drug for prevention of ovarian and nonovarian cancerogenesis.

However, further carefully designed confirmatory studies on the potential roles of pomegranate preparations on female reproductive system are needed to reveal the exact substances, dose-response relationship, effects, and their mechanisms of action on various healthy and ill reproductive structures.

3.9.8 Conclusions and possible direction of future studies

In comparison with other physiological processes and illnesses, the available information concerning action of pomegranate substances on female reproductive processes are limited. Nevertheless, the available data provides evidence that the pomegranate and its phytonutrients might be a promising candidate as a potential modulator of ovarian cell steroidogenesis, proliferation and viability and as a potential biostimulator of healthy ovarian cells and chemoprotective agent reducing viability of ovarian cancer cells.

Nevertheless, understanding constituents, character, mechanisms, and applicability of pomegranate for control of female reproductive processes and prevention and treatment of their disorders requires further profound studies. Although punicalagin can be responsible for pomegranate action, the effects of other pomegranate molecules, and their possible synergistic or antagonistic interrelationships remain unknown. The differences in action of extracts from different part of pomegranate suggest the differences in their chemical composition and applicability. Mechanisms of pomegranate action are very poorly investigated. The potential mediator of its action is suggested only on the basis of changes induced by pomegranate treatment, but not as a result of specific experiments with up- and down-regulation of specific signaling pathways. The pomegranate action on reproductive system was demonstrated only in *in vitro* and animal experiments, but the human studies and clinical trials have not been reported yet. Therefore, understanding and use of biological, protective and therapeutic effects of pomegranate and its constituents and require further investigations.

References

Adams LS, Seeram NP, Aggarwal BB, Takada Y, Sand D, Heber D. Pomegranate juice, total pomegranate ellagitannins, and punicalagin suppress inflammatory cell signaling in colon cancer cells. J Agric Food Chem 2006;54(3):980−5. https://doi.org/10.1021/jf052005r.

Afaq F, Saleem M, Krueger CG, Reed JD, Mukhtar H. Anthocyaninand hydrolyzable tannin-rich pomegranate fruit extract modulates MAPK and NF-kappaB pathways and inhibits skin tumorigenesis in CD-1 mice. Int J Cancer 2005;113(3):423−33. https://doi.org/10.1002/ijc.20587.

Amri Z, Ghorbel A, Turki M, Akrout FM, Ayadi F, Elfeki A, Hammami M. Effect of pomegranate extracts on brain antioxidant markers and cholinesterase activity in high fat-high fructose diet induced obesity in rat model. BMC Compl Altern Med 2017;17:339. https://doi.org/10.1186/s12906-017-1842-9.

Asgary S, Sahebkar A, Afshani MR, Keshvari M, Haghjooyjavanmard S, Rafieian-Kopaei M. Clinical evaluation of blood pressure lowering, endothelial function improving, hypolipidemic and anti-inflammatory effects of pomegranate juice in hypertensive subjects. Phytother Res 2014;28(2):193−9. https://doi.org/10.1002/ptr.4977.

Asgary S, Keshvari M, Sahebkar A, Sarrafzadegan N. Pomegranate consumption and blood pressure: a review. Curr Pharmaceut Des 2017;23(7):1042—50. https://doi.org/10.2174/1381612822666161010103339.

Baldovská S, Michalcová K, Halenár M, Carbonell-Barrachina AA, Kolesárová A. Polyphenol-rich pomegranate extract as a potential modulator of steroidogenesis in human ovarian cells. J Microbiol Biotechnol Food Sci 2019;9(6):1343—6. https://doi.org/10.15414/jmbfs.2019.8.6.1343-1346.

Baldovská S, Maruniaková N, Sláma P, Pavlík A, Kohút L, Kolesárová A. Efficacy of phytonutrients from pomegranate peel on human ovarian cells in vitro. J Microbiol Biotechnol Food Sci 2020;10(3):511—6. https://doi.org/10.15414/jmbfs.2020.10.3.511-516.

Banihani S, Swedan S, Alguraan Z. Pomegranate and type 2 diabetes. Nutr Res 2013;33(5):341—8. https://doi.org/10.1016/j.nutres.2013.03.003.

Cerda B, Espin JC, Parra S, Martinez P, Tomas-Baberan FA. The potent in vitro antioxidant ellagitannins from pomegranate juice are metabolised into bioavailable but poor antioxidant hydroxy-6H-dibenzopyran 6 one derivatives by the colonic microflora of healthy humans. Eur J Nutr 2004;43:205—20. https://doi.org/10.1007/s00394-004-0461-7.

Cerdá B, Soto C, Albaladejo MD, Martínez P, Sánchez-Gascón F, Tomás-Barberán F, Espín JC. Pomegranate juice supplementation in chronic obstructive pulmonary disease: a 5-week randomized, double-blind, placebo-controlled trial. Eur J Clin Nutr 2006;60(2):245—53. https://doi.org/10.1038/sj.ejcn.1602309. PMID: 16278692.

Choi JG, Kang O-H, Lee Y-S, Chae H-S, Oh Y-C, Brice O-O, Kim M-S, Sohn D-H, Kim H-S, park H, Shin D-W, Rho J-R. In vitro and in vivo antibacterial activity of Punica granatum peel ethanol extract against salmonella. Evid Based Compl Alter Med 2011;2011:690518. https://doi.org/10.1093/ecam/nep105.

Danesi F, Ferguson LR. Could pomegranate juice help in the control of inflammatory diseases? Nutrients 2017;9(9):958. https://doi.org/10.3390/nu9090958.

Devipriya N, Sudheer AR, Menon VP. Dose-response effect of ellagic acid on circulatory antioxidants and lipids during alcohol-induced toxicity in experimental rats. Fundam Clin Pharmacol 2007;21(6):621—30. https://doi.org/10.1111/j.1472-8206.2007.00551.x.

Dreiseitel A, Korte G, Schreier P, Oehme A, Locher S, Hajak G, Sand PG. Phospholipase A2 is inhibited by anthocyanidins. J Neural Transm 2009;116:1071—7. https://doi.org/10.1007/s00702-009-0268-z.

Esmaeilinezhad Z, Babajafari S, Sohrabi Z, Eskandari MH, Amooee S, Barati-Boldaji R. Effect of synbiotic pomegranate juice on glycemic, sex hormone profile and anthropometric indices in PCOS: a randomized, triple blind, controlled trial. Nutr Metabol Cardiovasc Dis 2019;29(2):201—8. https://doi.org/10.1016/j.numecd.2018.07.002.

Faria A, Calhau C. The bioactivity of pomegranate: impact on health and disease. Crit Rev Food Sci Nutr 2011;51:626—34. https://doi.org/10.1080/10408391003748100.

Faria A, Monteiro R, Azevedo I, Calhau C. Pomegranate juice effects on cytochrome P450S expression: in vivo studies. J Med Food 2007;10(4):643—9. https://doi.org/10.1089/jmf.2007.403.

Forest CP, Padma-Nathan H, Liker HR. Efficacy and safety of pomegranate juice on improvement of erectile dysfunction in male patients with mild to moderate erectile dysfunction: a randomized, placebo-controlled, double-blind, crossover study. Int J Impot Res 2007;19:564—7. https://doi.org/10.1038/sj.ijir.3901570.

Gil MI, Tomás-Barberán FA, Hess-Pierce B, Holcroft DM, Kader AA. Antioxidant activity of pomegranate juice and its relationship with phenolic composition and processing. J Agric Food Chem 2000;48(10):4581—9. https://doi.org/10.1021/jf000404a.

194 Contaminants and Plants Action on Female Reproduction

Guo C, Wei J, Yang J, Xu J, Pang W, Jiang Y. Pomegranate juice is potentially better than apple juice in improving antioxidant function in elderly subjects. Nutr Res 2008;28(2):72–7. https://doi.org/10.1016/j.nutres.2007.12.001.

Hartman RE, Shah A, Fagan AM, Schwetye KE, Parsadanian M, Schulman RN, Finn MB, Holtzman DM. Pomegranate juice decreases amyloid load and improves behavior in a mouse model of alzheimer's disease. Neurobiol Dis 2006;24(3):506–15. https://doi.org/10.1016/j.nbd.2006.08.006.

Hidaka M, Okumura M, Fujita K-I, Ogikubo T, Yamasaki K, Iwakiri T, Setoguchi N, Arimori K. Effects of pomegranate juice on human cytochrome p450 3A (CYP3A) and carbamazepine pharmacokinetics in rats. Drug Metab Dispos 2005;33(5):644–8. https://doi.org/10.1124/dmd.104.002824.

Hong MY, Seeram NP, Heber D. Pomegranate polyphenols downregulate expression of androgen-synthesizing genes in human prostate cancer cells overexpressing the androgen receptor. J Nutr Biochem 2008;19(12):848–55. https://doi.org/10.1016/j.jnutbio.2007.11.006.

Hontecillas R, O'Shea M, Einerhand A, Diguardo M, Bassaganya- Riera J. Activation of PPAR gamma and alpha by punicic acid ameliorates glucose tolerance and suppresses obesity-related inflammation. J Am Coll Nutr 2009;28(2):184–95. https://doi.org/10.1080/07315724.2009.10719770.

Hora JJ, Maydew ER, Lansky EP, Dwivedi C. Chemopreventive effects of pomegranate seed oil on skin tumor development in CD1 mice. J Med Food 2003;6(3):157–61. https://doi.org/10.1089/10966200360716553.

Huang TH, Peng G, Kota BP, Li GQ, Yamahara J, Roufogalis BD, Li Y. Pomegranate flower improves cardiac lipid metabolism in a diabetic rat model: role of lowering circulating lipids. Br J Pharmacol 2005;145(6):767–74. https://doi.org/10.1038/sj.bjp.0706245.

Jeune MAL, Kumi-Diaka J, Brown J. Anticancer activities of pomegranate extracts and genistein in human breast cancer cells. J Med Food 2005;8(4):469–75. https://doi.org/10.1089/jmf.2005.8.469.

Johanningsmeier SD, Harris GK. Pomegranate as a functional food and nutraceutical source. Annu Rev Food Sci Technol 2011;2:181–201. https://doi.org/10.1146/annurev-food-030810-153709.

Kandylis P, Kokkinomagoulos E. Food applications and potential health benefits of pomegranate and its derivatives. Foods 2020;9(2):122. https://doi.org/10.3390/foods9020122.

Katz SR, Newman RA, Lansky EP. Punica granatum: heuristic treatment for diabetes mellitus. J Med Food 2007;10(2):213–7. https://doi.org/10.1089/jmf.2006.290.

Khan N, Hadi N, Afaq F, Syed DN, Kweon MH, Mukhtar H. Pomegranate fruit extract inhibits prosurvival pathways in human A549 lung carcinoma cells and tumor growth in athymic nude mice. Carcinogenesis 2007;28(1):163–73. https://doi.org/10.1093/carcin/bgl145.

Kim ND, Mehta R, Yu W, Neeman I, Livney T, Amichay A, Poirier D, Nicholls P, Kirby A, Jiang W, Mansel R, Ramachandran C, Rabi T, Kaplan B, Lansky E. Chemopreventive and adjuvant therapeutic potential of pomegranate (*Punica granatum*) for human breast cancer. Breast Cancer Res Treat 2002;71(3):203–17. https://doi.org/10.1023/A:1014405730585.

Lansky EP, Newman RA. *Punica granatum* (pomegranate) and its potential for prevention and treatment of inflammation and cancer. J Ethnopharmacol 2007;109(2):177–206. https://doi.org/10.1016/j.jep.2006.09.006.

Larrosa M, Tomas-Barberan FA, Espin JC. The dietary hydrolysable tannin punicalagin releases ellagic acid that induces apoptosis in human colon adenocarcinoma Caco-2 cells by using the mitochondrial pathway. J Nutr Biochem 2006;17(9):611–25. https://doi.org/10.1016/j.jnutbio.2005.09.004.

Larrosa M, Garcia-Conesa MT, Espin JC, Tomas-Barberan FA. Ellagitannins, ellagic acid and vascular health. Mol Aspect Med 2010;31(6):513–39. https://doi.org/10.1016/j.mam.2010.09.005.

Lee CJ, Chen LG, Liang WL, Wang CC. Anti-inflammatory effects of *Punica granatum* Linne *in vitro* and *in vivo*. Food Chem 2010;118(2):315–22. https://doi.org/10.1016/j.foodchem.2009.04.123.

Lei F, Xing D-M, Xiang L, Zhao Y-N, Wang W, Zhang L-J, Du L-J. Pharmacokinetic study of ellagic acid in rat after oral administration of pomegranate leaf extract. J Chromatogr B 2003;796(1):189–94. https://doi.org/10.1016/S1570-0232(03)00610-X.

Li Y, Qi Y, Huang THW, Yamahara J, Roufogalis BD. Pomegranate flower: a unique traditional antidiabetic medicine with dual PPAR-alpha/-gamma activator properties. Diabetes Obes Metabol 2008;10(1):10–7. https://doi.org/10.1111/j.1463-1326.2007.00708.x.

Losso JN, Bansode RR, Trappey A, Bawadi HA, Truax R. *In vitro* anti-proliferative activities of ellagic acid. J Nutr Biochem 2004;15(11):672–8. https://doi.org/10.1016/j.jnutbio.2004.06.004.

Malik A, Afaq F, Sarfaraz S, Adhami VM, Syed DN, Mukhtar H. Pomegranate fruit juice for chemoprevention and chemotherapy of prostate cancer. Proc Natl Acad Sci USA 2005;102(41):14813–8. https://doi.org/10.1073/pnas.0505870102.

Mandal A, Bishayee A. Mechanism of breast cancer preventive action of pomegranate: disruption of estrogen receptor and wnt/β-catenin signaling pathways. Molecules 2015;20(12):22315–28. https://doi.org/10.3390/molecules201219853.

Mandal A, Bhatia D, Bishayee A. Anti-inflammatory mechanism involved in pomegranate-mediated prevention of breast cancer: the role of NF-κB and Nrf2 signaling pathways. Nutrients 2017;9(5):436. https://doi.org/10.3390/nu9050436.

Mehta R, Lanksy EP. Breast cancer chemopreventive properties of pomegranate (*Punica granatum*) fruit extracts in a mouse mammary organ culture. Eur J Canc Prev 2004;13(4):345–8. https://doi.org/10.1097/01.cej.0000136571.70998.5a.

Melgarejo P, Artes F. Total lipid content and fatty acid composition of oilseed from lesser known sweet pomegranate clones. J Sci Food Agric 2000;80(10):1452–4. https://doi.org/10.1002/1097-0010(200008)80:10<1452::AID-JSFA665>3.0.CO;2-L.

Mertens-Talcott SU, Jilma-Stohlawetz P, Rios J, Hingorani L, Derendorf H. Absorption, metabolism, and antioxidant effects of pomegranate (*Punica granatum* L.) polyphenols after ingestion of a standardized extract in healthy human volunteers. J Agric Food Chem 2006;54(23):8956–61. https://doi.org/10.1021/jf061674h.

Modaeinama S, Abasi M, Abbasi MM, Jahanban-Esfahlan R. Anti tumoral properties of *Punica granatum* (pomegranate) peel extract on different human cancer cells. Asian Pac J Cancer Prev 2015;16(14):5697–701. https://doi.org/10.7314/APJCP.2015.16.14.5697.

Motaal AA, Shaker S. Anticancer and antioxidant activities of standardized whole fruit, pulp, and peel extracts of Egyptian pomegranate. Open Conf Proc J 2011;2:41–5. https://doi.org/10.2174/2210289201102010041.

Nagata M, Hidaka M, Sekiya H, Kawano Y, Yamasaki K, Okumura M, Arimori K. Effects of pomegranate juice on human cytochrome P450 2C9 and tolbutamide pharmacokinetics in rats. Drug Metab Dispos 2007;35(2):302–5. https://doi.org/10.1124/dmd.106.011718.

Newton KM, Reed SD, LaCroix AZ, Grothaus LC, Ehrlich K, Guiltinan J. Treatment of vasomotor symptoms of menopause with black cohosh, multibotanicals, soy, hormone therapy, or placebo: a randomized trial. Ann Intern Med 2006;145:869–79. https://doi.org/10.7326/0003-4819-145-12-200612190-00003.

196 Contaminants and Plants Action on Female Reproduction

Packova D, Kolesarova A. Do punicalagins have possible impact on secretion of steroid hormones by porcine ovarian granulosa cells? J Microbiol Biotechnol Food Sci 2016;5:57–9. https://doi.org/10.15414/jmbfs.2016.5.special1.57-59.

Packova D, Carbonell-Barrachina AA, Kolesarova A. Ellagitannins–compounds from pomegranate as possible effector in steroidogenesis of rabbit ovaries. Physiol Res 2015;64(4):583–5. https://doi.org/10.33549/physiolres.932971.

Paller CJ, Pantuck A, Carducci MA. A review of pomegranate in prostate cancer. Prostate Cancer Prostatic Dis 2017;20(3):265–70. https://doi.org/10.1038/pcan.2017.19.

Park S, Seok JK, Kwak JY, Suh H-J, Kim YM, Boo YC. Anti-inflammatory effects of pomegranate peel extract in THP-1 cells exposed to particulate matter PM10. Evid Based Compl Alter Med 2016;2016:11. https://doi.org/10.1155/2016/6836080.

Pirzadeh M, Caporaso N, Rauf A, Shariati MA, Yessimbekov Z, Khan MU, Imran M, Mubarak MS. Pomegranate as a source of bioactive constituents: a review on their characterization, properties and applications. Crit Rev Food Sci Nutr 2020:1–18. https://doi.org/10.1080/10408398.2020.1749825.

Reis JF, Monteiro VVS, de Souza Gomes R, do Carmo MM, da Costa GV, Ribera PC, Monteiro MC. Action mechanism and cardiovascular effect of anthocyanins: a systematic review of animal and human studies. J Transl Med 2016;14:315. https://doi.org/10.1186/s12967-016-1076-5.

Rock W, Rosenblat M, Miller-Lotan R, Levy AP, Elias M, Aviram M. Consumption of wonderful variety pomegranate juice and extract by diabetic patient's increases paraoxonase 1 association with high-density lipoprotein and stimulates its catalytic activities. J Agric Food Chem 2008;56(18):8704–13. https://doi.org/10.1021/jf801756x.

Rosenblat M, Volkova N, Coleman R, Aviram M. Pomegranate byproduct administration to apolipoprotein e-deficient mice attenuates atherosclerosis development as a result of decreased macrophage oxidative stress and reduced cellular uptake of oxidized low-density lipoprotein. J Agric Food Chem 2006;54(5):1928–35. https://doi.org/10.1021/jf0528207.

Saeed M, Naveed M, BiBi J, Kamboh AA, Arain MA, Shah QA, Alagawany M, El-Hack MEA, Abdel-Latif MA, Yatoo MI, Tiwari R, Chakraborty S, Dhama K. The promising pharmacological effects and therapeutic/medicinal applications of *Punica granatum* L. (Pomegranate) as a functional food in humans and animals. Recent Pat Inflamm Allergy Drug Discov 2018;12(1):24–38. https://doi.org/10.2174/1872213X12666180221154713.

Sahebkar A, Ferri C, Giorgini P, Bo S, Nachtigal P, Grassi D. Effects of pomegranate juice on blood pressure: a systematic review and meta-analysis of randomized controlled trials. Pharmacol Res 2017;115:149–61. https://doi.org/10.1016/j.phrs.2016.11.018.

Seeram NP, Lee R, Heber D. Bioavailability of ellagic acid in human plasma after consumption of ellagitannins from pomegranate (*Punica granatum* L.) juice. Clin Chim Acta 2004;348(1-2):63–8. https://doi.org/10.1016/j.cccn.2004.04.029.

Seeram NP, Adams LS, Henning SM, Niu Y, Zhang Y, Nair MG, Heber D. In vitro antiproliferative, apoptotic and antioxidant activities of punicalagin, ellagic acid and a total pomegranate tannin extract are enhanced in combination with other polyphenols as found in pomegranate juice. J *Nutr Biochem* 2005;16(6):360–7. https://doi.org/10.1016/j.jnutbio.2005.01.006.

Seeram NP, Henning SM, Zhang Y, Suchard M, Li Z, Heber D. Pomegranate juice ellagitannin metabolites are present in human plasma and some persist in urine for up to 48 hours. J Nutr 2006;136(10):2481–5. https://doi.org/10.1093/jn/136.10.2481.

Seeram NP, Aronson WJ, Zhang Y, Henning SM, Moro A, Lee R-P, Sartippour M, Harris DM, Rettig M, Suchard MA, Pantuck AJ, Belldegrun A, Heber D. Pomegranate ellagitannin-derived metabolites inhibit prostate cancer growth and localize to the mouse prostate gland. J Agric Food Chem 2007;55(19):7732−7. https://doi.org/10.1021/jf071303g.

Shah TA, Parikh M, Patel KV, Patel KG, Joshi CG, Gandhi TR. Evaluation of the effect of *Punica granatum* juice and punicalagin on NFκB modulation in inflammatory bowel disease. Mol Cell Biochem 2016;419(1-2):65−74. https://doi.org/10.1007/s11010-016-2750-x.

Shaygannia E, Bahmani M, Zamanzad B, Rafieian-Kopaei M. A review study on *Punica granatum* L. Evid Based Complement Alternat Med 2016;21(3):221−7. https://doi.org/10.1177/2156587215598039.

Sheu ML, Chao KF, Sung YJ, Lin WW, Lin-Shiau SY, Liu SH. Activation of phosphoinositide 3-kinase in response to inflammation and nitric oxide leads to the up-regulation of cyclooxygenase-2 expression and subsequent cell proliferation in mesangial cells. Cell Signal 2005;17(8):975−84. https://doi.org/10.1016/j.cellsig.2004.11.015.

Speciale A, Canali R, Chirafisi J, Saija A, Virgili F, Cimino F. Cyanidin-3-O-glucoside protection against TNF-α-induced endothelial dysfunction: involvement of nuclear factor-κB signaling. J Agric Food Chem 2010;58(22):12048−54. https://doi.org/10.1021/jf1029515.

Sudheesh S, Vijayalakshmi NR. Flavonoids from *Punica granatum* - potential antiperoxidative agents. Fitoterapia 2005;76(2):181−6. https://doi.org/10.1016/j.fitote.2004.11.002.

Türk G, Sonmez M, Aydin M, Yuce A, Gur S, Yuksel M, Aksu EH, Aksoy H. Effects of pomegranate juice consumption on sperm quality, spermatogenic cell density, antioxidant activity, and testosterone level in male rats. Clin Nutr 2008;27(2):289−96. https://doi.org/10.1016/j.clnu.2007.12.006.

Usta C, Ozdemir S, Schiariti M, Puddu PE. The pharmacological use of ellagic acid-rich pomegranate fruit. Int J Food Sci Nutr 2013;64:907−13. https://doi.org/10.3109/09637486.2013.798268.

Van Elswijk DA, Schobel UP, Lansky EP, Irth H, van der Greef J. Rapid dereplication of estrogenic compounds in pomegranate (*Punica granatum*) using on-line biochemical detection coupled to mass spectrometry. Phytochemistry 2004;65(2):233−41. https://doi.org/10.1016/j.phytochem.2003.07.001.

Viladomiu M, Hontecillas R, Lu P, Bassaganya-Riera J. Preventive and prophylactic mechanisms of action of pomegranate bioactive constituents. Evid Based Complement Alternat Med 2013;2013:18. https://doi.org/10.1155/2013/789764.

Viuda-Martos M, Fernandez-Lopez J, Perez-Alvarez JA. Pomegranate and its many functional components as related to human health: a review. Compr Rev Food Sci Food Saf 2010;9(6):635−54. https://doi.org/10.1111/j.1541-4337.2010.00131.x.

Vučić V, Grabež M, Trchounian A, Arsić A. Composition and potential health benefits of pomegranate: a review. Curr Pharmaceut Des 2019;25(16):1817−27. https://doi.org/10.2174/1381612825666190708183941.

Whitley AC, Stoner GD, Darby MV, Walle T. Intestinal epithelial cell accumulation of the cancer preventive polyphenol ellagic acid—extensive binding to protein and DNA. Biochem Pharmacol 2003;66(6):907−15. https://doi.org/10.1016/S0006-2952(03)00413-1.

198 Contaminants and Plants Action on Female Reproduction

Zahin M, Ahmad I, Gupta RC, Aqil F. Punicalagin and ellagic acid demonstrate antimutagenic activity and inhibition of benzo[a]pyrene induced DNA adducts. BioMed Res Int 2014;2014:467465. https://doi.org/10.1155/2014/467465.

Zarfeshany A, Asgary S, Javanmard SH. Potent health effects of pomegranate. Adv Biomed Res 2014;3:100. https://doi.org/10.4103/2277-9175.129371.

Zhao F, Pang W, Zhang Z, Zhao J, Wang X, Liu Y, Wang X, Feng Z, Zhang Y, Sun W, Liu J. Pomegranate extract and exercise provide additive benefits on improvement of immune function by inhibiting inflammation and oxidative stress in high-fat-diet-induced obesity in rats. J Nutr Biochem 2016;32:20−8. https://doi.org/10.1016/j.jnutbio.2016.02.003.

Chapter 3.10

Puncture vine (*Tribulus terrestris* L.)

3.10.1 Introduction

Puncture vine (*Tribulus terrestris* L.) is a popular medicinal plant, which is considered as a traditional stimulator of masculine sexual desire. The available reviews concerning this plant don't reflect the recent information (Ukani et al., 1997; Chhatre et al., 2014; Shahid et al., 2016; Zhu et al., 2017), or they are focused on puncture vine's influence and use in man's reproduction (Neychev and Mitev, 2016; GamalEl Din, 2018; Sanagoo et al., 2019; Santos et al., 2019; Abarikwu et al., 2020). Current reviews concerning Tribulus terrestris effect on female reproduction, as well as on nonreproductive physiological and therapeutical actions of puncture vine, are practically absent in the available literature. In this review we have tried to summarize briefly the available information concerning Tribulus terrestris's provenance, constituents, properties, its action and application in the control and treatment of male and female reproductive and nonreproductive processes and their disorders.

3.10.2 Provenance and properties

The genus Tribulus, belonging to the family Zygophyllaceae, comprises about 20 species in the world, of which *Tribulus terrestris*, L. is the most common and popular as medicinal and food herb of this genus. It is an annual shrub found in Mediterranean, subtropical, and desert climate regions around the world, including India, China, southern USA, Mexico and southern Europe (Spain, Greece, Romania, Bulgaria, Hungary and southern Slovakia (Šalamon et al., 2006; Chhatre et al., 2014). It is a small prostrate, hirsute or silky hairy shrub with elliptical or oblong lanceolate leaves and yellow flowers. Its fruits are of stellate, five-cornered, and covered with speckles of yellow color. There are several oily seeds in each crocus. Root is cylindrical, branched, aromatic and sweetish astringent. Odor of fruits is also faintly aromatic and taste is slightly acrid (Ukani et al., 1997; Chhatre et al., 2014; Parham et al., 2020).

The phytochemical study of Tribulus terrestris revealed the presence of steroidal saponins, flavonoids, flavanol glycosides, alkaloids, and tannins (Chhatre et al., 2014; Saiyed et al., 2016; Shahid et al., 2016; Zhu et al., 2017). The saponins included furostanol and spirostanol saponins of tigogenin, neotigogenin, gitogenin, neogitogenin, hecogenin, neohecogenin, diosgenin, chlorogenin, ruscogenin, and sarsasapogenin. In addition, four sulfated saponins of tigogenin and diosgenin type were also isolated. Tribulus terrestrius is characterized by high concentration and number (more than 18) of flavonoids. The main flavonoids isolated from leaves and fruits of puncture vine were kaempferol, kaempferol-3-glucoside, kaempferol-3-rutinoside, and tribuloside [kaempferol-3-β-d-(6″-p-coumaroyl) glucoside], caffeoyl derivatives, quercetin glycosides, including rutin and kaempferol glycosides (quercetin 3-O-rutinoside, quercetin 3-O-glycoside and kaempferol 3-O-glycoside) (Chhatre et al., 2014; Kuchakulla et al., 2020; Verma et al., 2020). The plant also contains a mixture of B-carboline alkaloids: harmane, norharmane, tetrahydroharmane, harmine, harmaline, harmol, harmalol, ruin, and dihydroruin (Al-Bayati et al., 2008). The plant size, number of seeds and the contents of furostanol saponins are very variable in dependance on the location and conditions of plant growth (Šalamon et al., 2006). Therefore, Tribulus terrestris contains a number of phytochemicals with potential biological activity.

3.10.3 Physiological and therapeutical actions

The therapeutic features of tis medicinal plant has been described in ayurvedic medicine (Ukani et al., 1997) an Avicenna's Canon of Medicine (Kamrani et al., 2019). In the ayurvedic medicine its application is recommended for the treatment of urinary affection, urinary calculi, polyuria, dyspnoea; cough, piles dysuria, heart disease and as a gastric stimulant (Ukani et al., 1997). In native Chinese medicine, the leaves of Tribulus terrestris are used for treatment of stomach problems, bladder stones, male reproductive disorders (Kumari and Singh, 2015) and ocular diseases (Yuan et al., 2020).

Some of the Tribulus terrestris effects listed above were confirmed by modern Western medicine. The current scientifical literature describes the following physiological, protective and therapeutical activities of puncture vine (Chhatre et al., 2014; Shahid et al., 2016): diuretic, antiurolithic, immunomodulatory, antidiabetic, hypolipidemic, cardiotonic, on central nervous system, hepatoprotective, antiinflammatory, analgesic, antispasmodic, antibacterial, anthelmintic, larvicidal, and anticariogenic and anticancer activities (Chhatre et al., 2014; Parham et al., 2020; Verma et al., 2020). It is noteworthy that the toxic influence of Tribulus on cancer cells was more pronounced than that on healthy fibroblasts, suggesting the possibility of selective application of this plant against cells with malignant transformation (Neychev et al., 2007). *In vitro* studies showed the protective effect of puncture vine against ocular retina injury (Yuan et al., 2020). Tribulus phytochemicals can be promising replacement of routine antibiotica and antivirotica including anti-HIV treatment (Shaheen et al., 2019; Parham et al., 2020). Animal studies revealed its applicability for protection against bone loss induced by estrogen deficit during age-related osteoporosis (Marques et al., 2019). Clinical studies indicated its applicability for weight loss (Salgado et al., 2017), prevention and treatment of hypertenzia, coronary hearth disease, cerebral arteriosclerosis, miocardian infarction, thrombosis (Verma et al., 2020), urinary tract infections, urolithiasis, dysmenorrhea, edema, hypertension, hypercholesterolemia (Shahid et al., 2016; Zhu et al., 2017; Shaheen et al., 2019) and physical fitness and muscular performance (Al-Bayati et al., 2008; Ma et al., 2017; Wu et al., 2017; Sellami et al., 2018). On the other hand, some studies did not reveal *Tribulus terestris* action on physical performance in athletes (Pokrywka et al., 2014), sportsmen's muscle mass (Ma et al., 2017), body composition and muscular endurance in resistance-trained men (Sellami et al., 2018).

The variability in the targets of Tribulus terrestris can be due to the variability in its biological active constituents. The constituents of Tribulus terrestris, which are responsible for the particular effects of this plant, are however not fully determined. Nevertheless, there is evidence that Tribulus saponins like dioscine, diosgenin, and the protodioscin can promote libido and physical fitness, while Tribulus phytosterols, in particular beta-sistosterols can be beneficial for the prostate function, the urinary system and the cardiovascular system (Sellami et al., 2018).

Some evidence indicates the adverse effect of puncture vine overconsumption: it can induce sleeping disorder, burnout and fatigue, hypertension, high heart rate (Sellami et al., 2018), photosensitization and the resulted injury in both humans and animals (Chen et al., 2019). One case of intoxication with Tribulus has been reported (Pokrywka et al., 2014). On the other hand, clinical trials did not report serious adverse events induced in women by long Tribulus terrestris treatment (Martimbianco et al., 2020).

Therefore, Tribulus terrestris can target a number of variable physiological processes. Therefore, it can be a promising tool for treatment of a wide array of illnesses.

3.10.4 Mechanisms of action

The number of processes affected by puncture vine indicates the possible variability in its mechanisms of action, which are sometimes not properly discovered. Nevertheless, the mediators of plant action on some selected processes and illnesses have been currently outlined.

Some Tribulus effects could be mediated by changes in hormones, cytokines and growth factors release, binding and reception. For example, in rats undergoing exercise Tribulus terrestris extract promoted muscle gain and physical performance, which was associated with an increase in plasma level of both insulin-like growth factor-1 (IGF-I) and its receptor (Wu et al., 2017). In boxers, the ability of Tribulus terrestris to alleviate muscle damage and to promote anaerobic performance was associated with a decrease in plasma insulin-like growth factor binding protein-3 (IGFBP-3), but not in its target—IGF-I (Ma et al., 2017). These reports indicate that physical performance (and maybe activity, proliferation and reparation of cells other than myocytes) could be promoted by puncture vine by two ways—via reduction in

IGFBP-3, which binds/inactivates the IGF-I and via upregulation of IGF-I receptors. Under the influence of puncture vine, the increased amount of free IGF-I binds and activates increased the amounts of IGF-I receptors.

The bone-protective effect of Tribulus terrestris observed in rats is explained by its ability to increase bone mineral density. This activity may be at least partially attributable to an increase in the serum level of dehydroepiandrosterone (but not of testosterone or estradiol) and a Ca^{2+}-sparing effect (Marques et al., 2019).

The anti-inflammatory action of puncture vine can be mediated by prostaglandin. For example, addition of Tribulus terrestris to mouse macrophages inhibited their Cyclooxygenase 2 (COX-2), an enzyme promoting prostaglandin E2, which is in turn involved in promotion of inflammation. It suggests that prostaglandin E can be a mediator of antiinflammatory action of Tribulus terrestris (Hong et al., 2002).

The anti-inflammatory action of Tribulus can be mediated also by the intracellular promoter of inflammation — transcription factor NFkB. For example, Tribulus can protect on rat brain from inflammation induced by formalin and carrageenan by down-regulation of NF-κB (Ranjithkumar et al., 2019). On the contrary, derivatives of *Tribulus terrestris* alkaloids can destroy cancer cells (Jurkat E6-1) via up-regulation of this transcription factor (Basaiyye et al., 2017). Gautam and Ramanathan (2019) demonstrated that the ability to Tribulus saponins to prevent brain inflammation and to relieve pain can be attributed to attenuation of both pro-inflammatory cytokines (TNF-α, IL-1β, and IL-6) and brain neurotransmitters, glutamate, and aspartate.

Moreover, the neuroprotective action of Tribulus on rat brain was associated with activation of antiapoptotic and proproliferating MAP kinase pathway (Ranjithkumar et al., 2019; Reshma et al., 2019) and neuronal survival pathway (BDNF), by changes in apoptosis-related protein kinases JNK, GSK3β/βcatenin (Ranjithkumar et al., 2019) and activation of promoters of apoptosis caspase-3 and AIF (Ranjithkumar et al., 2019; Yuan et al., 2020).). The toxic effect of Tribulus on human fibroblasts was also associated with suppression of their proliferation, activation of their apoptosis and changes in cellular polyamines' homeostasis (Neychev et al., 2007).

A number of Tribulus phytochemicals including flavonoid, tannin, and phenolic acids have antioxidant properties, which can prevent oxidative stress, inflammation apoptosis and the development of a number of illnesses mentioned above (Parham et al., 2020). For example, Tribulus terrestris can protect ocular retina cells from oxidative stress-induced injury and apoptosis and increase their resistance and viability through activating numerous antioxidant enzymes and PI3K/Akt-Nrf2 signaling pathway (Yuan et al., 2020).

Studies on rats indicated that the antihepertensive effect of puncture vine can be due to its action on membrane hyperpolarization and relaxation of

Food/medicinal herbs and their influence **Chapter | 3** **203**

arterial smooth muscle, to boost up the discharge of nitric oxide from the nitrergic nerve endings and endothelium and/or by inhibitory action on angiotensin converting enzyme (Kamrani et al., 2019; Verma et al., 2020).

The ability of *Tribulus terrestris* to relieve painful urination or dysuria, hematuria, urinary urgency, burning micturition, frequent urination, nausea, and vomiting can be explained by its bacteriostatic and bacteriocide action on a wide spectrum of urogenital bacteria, vira, and fungi (Al-Bayati et al., 2008; Shaheen et al., 2019; Parham et al., 2020). This action could be explained by the ability of Tribulus saponins to induce lysis of bacterial membranes (Al-Bayati et al., 2008).

These observations demonstrate the variability of the mechanisms mediated by Tribulus terrestris on various target organs and processes (and sometimes on the same target).

3.10.5 Effect on male reproductive processes

The best-known feature and area of Tribulus terrestris application in folk and official medicine is its ability to treat the loss of libido and infertility in animals (Haghmorad et al., 2019) and man (Sahin et al., 2016; Shahid et al., 2016; GamalEl Din, 2018; GamalEl Din et al., 2019). It is proposed that saponins like dioscine, diosgenin, and the protodioscin can have beneficial effects on libido, while phytosterols, in particular beta-sistosterols, can be beneficial for the prostate function, and, therefore, for male fecundity (Sellami et al., 2018).

Analysis of phytochemical and pharmacological studies on animals and humans revealed an important role of Tribulus in treating erectile dysfunction and sexual desire problems (Neychev and Mitev, 2016) and fertility (Kumari et al., 2018). On the other hand, the performed clinical studies provided controversial and inconclusive results concerning the applicability of this plant for treatment of male infertility. Such results could be due to clinical trials which were imperfect from the viewpoint of methodology and patient number, as well as the differences in causes of male infertility (GamalEl Din, 2018).

In contrast to Tribulus terrestris action on libido and erectile function, the ability of this plant to improve on male sperm parameters, namely number, motility and morphology, has been demonstrated in the majority of performed clinical studies (Roaiah et al., 2017; Salgado et al., 2017; Sanagoo et al., 2019). Animal and *in vitro* studies also demonstrated that Tribulus terrestris can increase sperm quantity, quality in rats (Martino-Andrade et al., 2010; Kumari and Singh, 2015; Sahin et al., 2016; de Souza et al., 2019; Salahshoor et al., 2020), rams (Kistanova et al., 2005), livestocks (Clément et al., 2012) and humans (Asadmobini. Et al., 2017; Khaleghi et al., 2017). In rams Tribulus terrestris increased fertility (Kistanova et al., 2005).

The positive influence of Tribulus on sperm has been associated with signs of activation of male reproductive organs. At least, Tribulus terrestris increased mouse testis weight and recovered the spermatogenic cycle (Kumari and Singh, 2015). Puncture vine enhanced the number of Leydig, spermatogonia and spermatid cells in rat testis (Haghmorad et al., 2019). Similarly, protodioscin, the main phytochemical agent of the Tribulus, can promote proliferation of Sertoli and germ cell proliferation, which promotes the growth and function of seminiferous tubules in men (Salgado et al., 2017). On the other hand, no influence of Tribulus terrestris extract on weight and histostructure of rat prostate, seminal vesicles and height of germinal layer of seminiferous tubule has been found (Martino-Andrade et al., 2010; Sahin et al., 2016; Salahshoor et al., 2020).

Therefore, the ability of Tribulus terrestris to improve sperm characteristics and therefore male fertility can be evident. On the other hand, Tribulus action on the state of reproductive organs has been examined only in rats, while the obtained results were sometimes contradictory. Tribulus terrestris applicability to treat psychological and physical impotence remains an open question too, which should be addressed by strong clinical studies.

3.10.6 Mechanisms of effects on male reproductive processes

Results of the basic studies indicate that Tribulus terrestris can affect male reproductive processes via the same extra- and intracellular mediators as the nonreproductive processes (see earlier).

Some (but not all) studies indicated that puncture vine can affect male reproductive processes through the up-regulation of release of **reproductive hormones**—gonadotropins and androgens. Treatment of rats with Tribulus terrestris increased level of LH (but not FSH) and testosterone in their plasma (Ghosian Moghaddam et al., 2013; Haghmorad et al., 2019; Salahshoor et al., 2020). Sahin et al. (2016) observed the stimulatory action of Tribulus extract on rat plasma testosterone, but not LH or FSH. Some studies showed the ability of Tribulus terrestris to increase dihydrotestosterone (Salgado et al., 2017) and testosterone (Sahin et al., 2016; Roaiah et al., 2017; Sellami et al., 2018; GamalEl Din, 2018; GamalEl Din et al., 2019) levels in men's plasma. Other studies however did not detect Tribulus terrestris influence on androgen release in rat (Martino-Andriade et al., 2010; Ghosian Moghaddam et al., 2013) or humans (Pokrywka et al., 2014; Neychev and Mitev, 2016; Ma et al., 2017; Santos et al., 2019; Kuchakulla et al., 2020), although the puncture vine influence on libido and fecundity occurred. These observations suggest that "empirical evidence to support the hypothesis that this desirable effects are due to androgen enhancing properties of Tribulus terrestris is, at best, inconclusive" (Neychev and Mitev, 2016), and androgens are not (or not always) mediators of male reproduction-stimulating activity of Tribulus terrestris.

More probable appears the hypothesis explaining Tribulus terrestris effect on **brain structures**. At least, treatments with Tribulus terrestris extract promoted nicotinamide adenine dinucleotide phosphate-diaphorase activity and androgen receptor immunoreactivity in rat brain (Gauthman and Adaikan, 2005). These structures could be central mediators of puncture vine action on sexual behavior. Therefore, this plant could promote androgen-dependent events not (or not only) via androgen production, but also through androgen reception and effects in the brain.

Tribulus can improve sperm production through activation of proliferation of **testicular cells and spermatogenesis**, which has been reported after Tribulus terrestris or its constituent protodioscin in mice (Kumari and Singh, 2015), rat (Haghmorad et al., 2019) and man (Salgado et al., 2017).

Tribulus terrestris-induced stimulation of rat reproductive behavior and testosterone release were associated with reduction of transcription factor **NF-κB** and increased the levels of **NF-E2-related factor 2** (Nrf2) and **heme-oxygenase-1** (HO-1) accumulation in their male reproductive organs (Sahin et al., 2016). It suggests that these signaling substances could be intracellular mediators of puncture vine on libido or androgen release. Furthermore, there is emerging compelling evidence from experimental studies in animals for possible endothelium and nitric oxide-dependent mechanisms underlying tribulus terrestris aphrodisiac and proerectile activities (Neychev and Mitev, 2016). It is still to be examined whether intracellular signaling mechanisms mediate puncture vine action on human male reproductive processes.

The stimulatory action of Tribulus terrestris on male sperm quality could be due to **antioxidative properties** of some plant molecules (see above). Addition of antioxidants usually increases sperm quality. Tribulus terrestris increased the concentration of antioxidative enzymes and the antioxidant activity in mice testis, as well as the sperm resistance to oxidative stress (Kumari and Singh, 2015; Salahshoor et al., 2020). Clément et al. (2012) explained the benefits of Tribulus terrestris extract for livestock sperm by antioxidant properties of puncture vine constituents.

Therefore, Tribulus terrestris can improve various male reproductive parameters via brain structures and through direct action on the testis and sperm. This effect can be mediated by stimulatory action of Tribulus action on ovarian gonadotropins, androgens, androgen receptors, several intracellular signaling pathways and by the ability of Tribulus to scavenger reactive oxygen species.

3.10.7 Effects on female reproductive processes

The effect of Tribulus terrestris on female reproductive processes and its application for treatment of female reproductive disorders are studied much less than this plant in relation to male reproduction. Nevertheless, the few available publications indicate the influence of puncture vine on female reproductive hormones, behavior, ovarian, oviductal and uterus state, ovarian cycle, ovulation and fecundity.

Clinical studies demonstrated the ability of Tribulus terrestris to promote libido and other signs of sexual behavior and to prevent their decline in pre- and postmenopausal women (Mazaro-Costa et al., 2010; Martimbianco et al., 2020).

Abadjieva and Kistanova (2016) reported that feeding with Tribulus terrestris altered the expression of bone morphogenetic protein (BMP) 15 and growth differentiation factor (GDF) 9 in rabbit ovarian structures. Tribulus terrestris caused a decrease in the BMP15 mRNA level in the oocytes and an increase in the cumulus cells. The GDF9 mRNA level increased significantly in both oocytes and cumulus cells.

An *in vitro* study showed the ability of Tribulus terrestris extract to promote accumulation of both proliferation and apoptosis markers in cultured porcine ovarian granulosa cells. The stimulatory action of Tribulus on both proliferation and apoptosis could indicate the ability of this plant to boost ovarian cell turnover. Furthermore, in these experiments, Tribulus prevented and even reversed the stimulatory action of metabolic hormone ghrelin on apoptosis marker (Sirotkin et al., 2020).

Experiments on rats with polycystic ovarian cysts (Dehghan et al., 2012) and with polycystic ovarian syndrome (Sandeep et al., 2015; Saiyed et al., 2016), as well as on women with oligo/anovular infertility (Arentz et al., 2014) demonstrated the ability of Tribulus terrestris to induce their ovarian follicle development, ovulations, and fertility. Tribulus terrestris given together with *Withania somnifera*, prevented the signs of polycystic ovarian syndrome in rats—increase in estradiol and testosterone (but not in LH) release, increase in ovarian (but not uterus) weight and prolongation of the estrous cycle (Saiyed et al., 2016). Experiments of Dehghan et al. (2012) on rats demonstrated the applicability of Tribulus terrestris for luteinization and treatment of ovarian cysts.

Some animal studies demonstrated the ability of Tribulus terrestris extract to promote the development of uterine and vaginal tissue in rats (Esfandiari et al., 2011), but other studies did not show this action in rats (Martino-Andrade et al., 2010; Saiyed et al., 2016) or women (Arentz et al., 2014).

Therefore, the available publications suggest the ability of Tribulus terrestris to promote female reproductive functions on various levels. It can promote female sexual desire, to alter release of pituitary gonadotropins and ovarian steroid hormones, BMP15 and GDF9, to promote ovarian cell proliferation and apoptosis, to promote ovarian cycle, development of uterus and vagina, ovulation and fecundity, as well as to mitigate the signs of polycystic ovarian syndrome and oligo/anovular infertility. Nevertheless, the available information concerning Tribulus terrestris action on animal ovary and uterus is conflicting. Furthermore, the ability of Tribulus terrestris to prevent signs of polycystic ovarian syndrome was more pronounced in animal experiments, than in clinical studies. The performed clinical trials were methodologically heterogenic, and they were performed on a relatively small number of patients, therefore they could not be considered as conclusive yet (Arentz et al., 2014).

3.10.8 Mechanisms of effects on female reproductive processes

There is evidence that Tribulus terrestris stimulates ovulation and fecundity and relieves symptoms of polycystic ovarian syndrome via promotion of release of **pituitary and ovarian hormones**—the known stimulators of reproductive processes. At least treatment of rats with Tribulus terrestris prevented the signs of induced polycystic ovarian syndrome - decrease in plasma FSH and increase in LH, testosterone and estradiol level, increase in ovarian weight and decrease in weight of uterus (Saiyed et al., 2016). Tribulus treatment in women was associated with an increase in FSH and estradiol (which is characteristic for healthy women), but not of LH and testosterone (which is characteristic for women suffering from polycystic ovarian syndrome) (Arentz et al., 2014). Other studies however did not note decrease, but increase in testosterone release in premenopausal women treated with Tribulus (Martimbianco et al., 2020).

The yeast androgen bioassay demonstrated that extract of Tribulus terrestris has an antiandrogen activity (Sandeep et al., 2015). This antiandrogen action of Tribulus could explain its medicinal effect against polycystic ovarian syndrome associated with women androgenisation.

In contrast to animals, women's sexual desire it promoted by both estrogens and androgens, which activate sexual centers in CNS (Cappelletti and Wallen, 2016). The ability of Tribulus terrestris to promote women's sexual behavior can be explained by increased production of androgen (Martimbianco et al., 2020) and/or estrogen (Arentz et al., 2014). Estrogen, whose production can be boosted by Tribulus, is also a known regulator of ovarian cell proliferation and apoptosis, as well as a promoter of ovarian follicle development, survival and oogenesis (Sirotkin, 2014). These processes can be promoted not only by estrogen, but also by **BMP15 and GDF9** (Sanfins et al., 2018), whose expression in the ovary has been affected (mainly increased) by Tribulus terrestris (Abadjieva and Kistanova, 2016).

Steroid hormones, like BMP15 and GDF9, which can regulate ovarian folliculogenesis and the resulting fecundity through an effect on ovarian cell proliferation and apoptosis, which changes under direct influence of Tribulus terrestris, have been documented (Sirotkin et al., 2020).

The similarity of stimulatory action of Tribulus terrestris and estrogen on uterine and vaginal tissue's growth and development (Esfandiari et al., 2011)

208 Contaminants and Plants Action on Female Reproduction

indicates that Tribulus contains phytoestrogens, activating estrogen receptors, which could be one of the possible mediators of Tribulus action on female reproductive system.

The involvement of other mechanisms in mediating puncture vine on female reproduction cannot be excluded either. It is very probable that **brain structures, antioxidants** a. o. can mediate Tribulus terrestris action not only on male (see earlier), but also on female reproduction.

The available data enable to hypothesize several interrelated signaling pathways mediating Tribulus terrestris action on female reproductive functions and their disorders. They can include CNS structures, pituitary gonadotropins, steroid hormones, BMP15, GDF9 and their target processes including ovarian cell proliferation, apoptosis, ovarian folliculogenesis, oogenesis, ovulation, and fecundity. The intervention of Tribulus terrestris to several regulatory pathways at once is highly probable.

3.10.9 Application in reproductive biology and medicine

The results of preclinical and clinical studies demonstrated the applicability of Tribulus terrestris for improvement of sperm quality and therefore of fertility in man and farm animals. It is not to be excluded that it could be useful also for promotion man's libido and erectile functions, although such application requires further validation yet.

The available information suggests that Tribulus terrestris could be a potent biostimulators of female reproductive processes—of sexual desire, ovarian cell turnover, and follicular development, ovulation and fecundity. Its stimulatory influence on oogenesis, uterus and vagina growth are possible too, although this influence has not been verified on women. Furthermore, a limited number of clinical trials suggested that Tribulus terrestris can be successfully used for treatment of polycystic ovarian syndrome and related infertility.

3.10.10 Conclusions and possible direction of future studies

Analysis of the available publications demonstrated the influence of Tribulus terrestris on a wide spectrum of targets and physiological processes. This effect can be mediated by multiple extra- and intracellular signaling pathways. Some performed clinical studies demonstrated the applicability of this plant and its polyphenols for treatment of a number of various disorders according to the principles of traditional oriental and modern medicine. In particular, Tribulus terrestris can be a stimulator of male and female reproductive processes at the level of central nervous system, sexual behavior, pituitary and gonadal hormones and their receptors, gonadal functions (including ovarian folliculogenesis and spermatogenesis), improvement of the quality and quantity of gametes (at least of sperm) and fecundity. This ability of puncture vine is applicable for improvement of man's sexual desire and sperm quality *in vivo* and *in vitro*, as well as of women's libido, activation of women

Food/medicinal herbs and their influence Chapter | 3 **209**

reproductive organs, fecundity and treatment of infertility, especially related to polycystic ovarian syndrome.

Nevertheless, understanding the biological role and application of puncture vine requires more profound studies. Although the main studies of this plant were aimed at its medical application, the physiological and medicinal effects of Tribulus terrestris were expressed better in animal experiments than in clinical trials, whose protocols and results are variable and therefore less conclusive. In the main cases, it remained unknown which plant constituent was responsible for a particular effect. Often the mediators of puncture vine effects have been only hypothesized on the basis of their changes after treatment with this plant, but their mediatory role remains to be demonstrated. A wide variety of possible factors mediating Tribulus effect on various targets and their possible interrelationships complicate understanding the mechanism of action of this plant and the causes of variability in its observed effects. Finally, the medicinal application of puncture vine requires more clinical tests on a larger number of patients with an adequate protocol and control groups. Nevertheless, the available data listed here demonstrate the biological and medicinal potency of this plant, the study and application of which is worth further efforts.

References

Abadjieva D, Kistanova E. Tribulus terrestris alters the expression of growth differentiation factor 9 and bone morphogenetic protein 15 in rabbit ovaries of mothers and F1 female offspring. PLoS One 2016;11(2):e0150400. https://doi.org/10.1371/journal.pone.0150400.

Abarikwu SO, Onuah CL, Singh SK. Plants in the management of male infertility. Andrologia 2020;52(3):e13509. https://doi.org/10.1111/and.13509.

Al-Bayati FA, Al-Mola HF. Antibacterial and antifungal activities of different parts of Tribulus terrestris L. growing in Iraq. J Zhejiang Univ Sci B 2008;9(2):154−9. https://doi.org/10.1631/jzus.B0720251.

Arentz S, Abbott JA, Smith CA, Bensoussan A. Herbal medicine for the management of polycystic ovary syndrome (PCOS) and associated oligo/amenorrhoea and hyperandrogenism; a review of the laboratory evidence for effects with corroborative clinical findings. BMC Compl Alternative Med 2014;14:511. https://doi.org/10.1186/1472-6882-14-511.

Asadmobini A, Bakhtiari M, Khaleghi S, Esmaeili F, Mostafaei A. The effect of Tribulus terrestris extract on motility and viability of human sperms after cryopreservation. Cryobiology 2017;75:154−9. https://doi.org/10.1016/j.cryobiol.2017.02.005.

Basaiyye SS, Naoghare PK, Kanojiya S, Bafana A, Arrigo P, Krishnamurthi K, Sivanesan S. Molecular mechanism of apoptosis induction in Jurkat E6-1 cells by *Tribulus terrestris* alkaloids extract. J Tradit Complement Med 2017;8(3):410−9. https://doi.org/10.1016/j.jtcme.2017.08.014.

Cappelletti M, Wallen K. Increasing women's sexual desire: the comparative effectiveness of estrogens and androgens. Horm Behav 2016;78:178−93. https://doi.org/10.1016/j.yhbeh.2015.11.003.

Chen Y, Quinn JC, Weston LA, Loukopoulos P. The aetiology, prevalence and morbidity of outbreaks of photosensitisation in livestock: a review. PLoS One 2019;14(2):e0211625. https://doi.org/10.1371/journal.pone.0211625.

Chhatre S, Nesari T, Somani G, Kanchan D, Sathaye S. Phytopharmacological overview of tribulus terrestris. Pharmacogn Rev 2014;8(15):45−51. https://doi.org/10.4103/0973-7847.125530.

Clément C, Witschi U, Kreuzer M. The potential influence of plant-based feed supplements on sperm quantity and quality in livestock: a review. Anim Reprod Sci 2012;132(1-2):1−10. https://doi.org/10.1016/j.anireprosci.2012.04.002.

de Souza DB, Buys-Goncalves GF. Editorial comment: improvement of fertility parameters with tribulus terrestris and anacyclus pyrethrum treatment in male rats. Int Braz J Urol 2019;45(5):1055−6. https://doi.org/10.1590/S1677-5538.IBJU.2018.0843.1.

Dehghan A, Esfandiari A, Bigdeli SM. Alternative treatment of ovarian cysts with *Tribulus terrestris* extract: a rat model. Reprod Domest Anim 2012;47(1):e12−5. https://doi.org/10.1111/j.1439-0531.2011.01877.x.

Esfandiari A, Dehghan A, Sharifi S, Najafi B, Vesali E. Effect of *Tribulus terrestris* extract on ovarian activity in immature Wistar rat: a histological evaluation. J Anim Vet Adv 2011;10(7):883−6.

GamalEl Din SF. Role of tribulus terrestris in male infertility: is it real or fiction? J Diet Suppl 2018;15(6):1010−3. https://doi.org/10.1080/19390211.2017.1402843.

GamalEl Din SF, Abdel Salam MA, Mohamed MS, Ahmed AR, Motawaa AT, Saadeldin OA, Elnabarway RR. Tribulus terrestris versus placebo in the treatment of erectile dysfunction and lower urinary tract symptoms in patients with late-onset hypogonadism: a placebo-controlled study. Urologia 2019;86(2):74−8. https://doi.org/10.1177/0391560318802160.

Gautam M, Ramanathan M. Saponins of *Tribulus terrestris* attenuated neuropathic pain induced with vincristine through central and peripheral mechanism. Inflammopharmacology 2019;27(4):761−72. https://doi.org/10.1007/s10787-018-0502-0.

Gauthaman K, Adaikan PG. Effect of *Tribulus terrestris* on nicotinamide adenine dinucleotide phosphate-diaphorase activity and androgen receptors in rat brain. J Ethnopharmacol 2005;96(1-2):127−32. https://doi.org/10.1016/j.jep.2004.08.030. PMID.

Ghosian Moghaddam MH, Khalili M, Maleki M, Ahmad Abadi ME. The effect of oral feeding of tribulus terrestris L. On sex hormone and gonadotropin levels in addicted male rats. Int J Fertil Steril 2013;7(1):57−62.

Haghmorad D, Mahmoudi MB, Haghighi P, Alidadiani P, Shahvazian E, Tavasolian P, Hosseini M, Mahmoudi M. Improvement of fertility parameters with *Tribulus terrestris* and *Anacyclus pyrethrum* treatment in male rats. Int Braz J Urol 2019;45(5):1043−54. https://doi.org/10.1590/S1677-5538.IBJU.2018.0843.

Hong CH, Hur SK, Oh OJ, Kim SS, Nam KA, Lee SK. Evaluation of natural products on inhibition of inducible cyclooxygenase (COX-2) and nitric oxide synthase (iNOS) in cultured mouse macrophage cells. J Ethnopharmacol 2002;83(1-2):153−9. https://doi.org/10.1016/s0378-8741(02)00205-2.

Kamrani Rad SZ, Javadi B, Hayes AW, KarimI G. Potential angiotensin converting enzyme (ACE) inhibitors from Iranian traditional plants described by Avicenna's Canon of Medicine. Avicenna J Phytomed 2019;9(4):291−309.

Khaleghi S, Bakhtiari M, Asadmobini A, Esmaeili F. *Tribulus terrestris* extract improves human sperm parameters in vitro. J Evid Based Complement Altern Med 2017;22(3):407−12. https://doi.org/10.1177/2156587216668110.

Kistanova E, Zlatev H, Karcheva V, Kolev A. Effect of plant *Tribulus terrestris* extract on reproductive performances of rams. Biotechnol Anim Husb 2005;21:55−63.

Kuchakulla M, Narasimman M, Soni Y, Leong JY, Patel P, Ramasamy R. A systematic review and evidence-based analysis of ingredients in popular male testosterone and erectile dysfunction supplements. Int J Impot Res 2020. https://doi.org/10.1038/s41443-020-0285-.

Kumari M, Singh P. Tribulus terrestris ameliorates metronidazole-induced spermatogenic inhibition and testicular oxidative stress in the laboratory mouse. Indian J Pharmacol 2015;47(3):304−10. https://doi.org/10.4103/0253-7613.157129.

Kumari M, Singh P. *Tribulus terrestris* improves metronidazole-induced impaired fertility in the male mice. Afr Health Sci 2018;8(3):645−52. https://doi.org/10.4314/ahs.v18i3.22.

Ma Y, Guo Z, Wang X. *Tribulus terrestris* extracts alleviate muscle damage and promote anaerobic performance of trained male boxers and its mechanisms: roles of androgen, IGF-1, and IGF binding protein-3. J Sport Health Sci 2017;6(4):474−81. https://doi.org/10.1016/j.jshs.2015.12.003.

Marques MAA, Lourenço BHLB, Reis MP, Pauli KB, Soares AL, Belettini ST, Donadel G, Palozi RAC, Froehlich DL, Lívero FADR, Gasparotto Junior A, Lourenço ELB. Osteoprotective effects of tribulus terrestris L.: relationship between dehydroepiandrosterone levels and Ca^{2+}-sparing effect. J Med Food 2019;22(3):241−7. https://doi.org/10.1089/jmf.2018.0090.

Martimbianco ALC, Pacheco RL, Vilarino FL, Latorraca COC, Torloni MR, Riera R. Tribulus terrestris for female sexual dysfunction: a systematic review. Rev Bras Ginecol Obstet 2020;42(7):427−35. https://doi.org/10.1055/s-0040-1712123.

Martino-Andrade AJ, Morais RN, Spercoski KM, Rossi SC, Vechi MF, Golin M, Lombardi NF, Greca CS, Dalsenter PR. Effects of *Tribulus terrestris* on endocrine sensitive organs in male and female Wistar rats. J Ethnopharmacol 2010;127(1):165−70. https://doi.org/10.1016/j.jep.2009.09.031.

Mazaro-Costa R, Andersen ML, Hachul H, Tufik S. Medicinal plants as alternative treatments for female sexual dysfunction: utopian vision or possible treatment in climacteric women? J Sex Med 2010;7(11):3695−714. https://doi.org/10.1111/j.1743-6109.2010.01987.x.

Neychev V, Mitev V. Pro-sexual and androgen enhancing effects of *Tribulus terrestris* L.: fact or Fiction. J Ethnopharmacol 2016;179:345−55. https://doi.org/10.1016/j.jep.2015.12.055.

Neychev VK, Nikolova E, Zhelev N, Mitev VI. Saponins from *Tribulus terrestris* L are less toxic for normal human fibroblasts than for many cancer lines: influence on apoptosis and proliferation. Exp Biol Med 2007;232(1):126−33.

Parham S, Kharazi AZ, Bakhsheshi-Rad HR, Nur H, Ismail AF, Sharif S, RamaKrishna S, Berto F. Antioxidant, antimicrobial and antiviral properties of herbal materials. Antioxidants 2020;9(12):1309. https://doi.org/10.3390/antiox9121309.

Pokrywka A, Obmiński Z, Malczewska-Lenczowska J, Fijałek Z, Turek-Lepa E, Grucza R. Insights into supplements with *Tribulus terrestris* used by athletes. J Hum Kinet 2014;41:99−105. https://doi.org/10.2478/hukin-2014-0037.

Ranjithkumar R, Alhadidi Q, Shah ZA, Ramanathan M. Tribulusterine containing *Tribulus terrestris* extract exhibited neuroprotection through attenuating stress kinases mediated inflammatory mechanism: in vitro and in vivo studies. Neurochem Res 2019;44(5):1228−42. https://doi.org/10.1007/s11064-019-02768-7.

Reshma PL, Binu P, Anupama N, Vineetha RC, Abhilash S, Nair RH, Raghu KG. Pretreatment of *Tribulus terrestris* L. causes anti-ischemic cardioprotection through MAPK mediated antiapoptotic pathway in rat. Biomed Pharmacother 2019;111:1342−52. https://doi.org/10.1016/j.biopha.2019.01.033.

Roaiah MF, Elkhayat YI, Abd El Salam MA, Din SFG. Prospective analysis on the effect of botanical medicine (*Tribulus terrestris*) on serum testosterone level and semen parameters in males with unexplained infertility. J Diet Suppl 2017;14(1):25−31. https://doi.org/10.1080/19390211.2016.1188193.

Sahin K, Orhan C, Akdemir F, Tuzcu M, Gencoglu H, Sahin N, Turk G, Yilmaz I, Ozercan IH, Juturu V. Comparative evaluation of the sexual functions and NF-κB and Nrf2 pathways of some aphrodisiac herbal extracts in male rats. BMC Compl Alternative Med 2016;16(1):318. https://doi.org/10.1186/s12906-016-1303-x.

Saiyed A, Jahan N, Makbul SAA, Ansari M, Bano H, Habib SH. Effect of combination of *Withania somnifera* Dunal and *Tribulus terrestris* Linn on letrozole induced polycystic ovarian syndrome in rats. Integr Med Res 2016;5(4):293−300. https://doi.org/10.1016/j.imr.2016.10.002.

212 Contaminants and Plants Action on Female Reproduction

Salahshoor MR, Abdolmaleki A, Faramarzi A, Jalili C, Shiva R. Does *Tribulus terrestris* improve toxic effect of Malathion on male reproductive parameters? J Pharm BioAllied Sci 2020;12(2):183–91. https://doi.org/10.4103/jpbs.JPBS_224_19.

Šalamon I, Habán M, Baranec T, Habánová M, Knoll M. The occurrence of puncture vine (*Tribulus terrestris*) and its metabolic characteristics in Slovakia. Biologia 2006;61(1):25–30. https://doi.org/10.2478/s11756-006-0004.

Salgado RM, Marques-Silva MH, Gonçalves E, Mathias AC, Aguiar JG, Wolff P. Effect of oral administration of *Tribulus terrestris* extract on semen quality and body fat index of infertile men. Andrologia 2017;49(5). https://doi.org/10.1111/and.12655.

Sanagoo S, Sadeghzadeh Oskouei B, Gassab Abdollahi N, Salehi-Pourmehr H, Hazhir N, Farshbaf-Khalili A. Effect of *Tribulus terrestris* L. on sperm parameters in men with idiopathic infertility: a systematic review. Complement Ther Med 2019;42:95–103. https://doi.org/10.1016/j.ctim.2018.09.015.

Sandeep PM, Bovee TF, Sreejith K. Anti-androgenic activity of nardostachys jatamansi DC and *Tribulus terrestris* L. And their beneficial effects on polycystic ovary syndrome-induced rat models. Metab Syndr Relat Disord 2015;13(6):248–54. https://doi.org/10.1089/met.2014.0136.

Sanfins A, Rodrigues P, Albertini DF. GDF-9 and BMP-15 direct the follicle symphony. J Assist Reprod Genet 2018;35(10):1741–50. https://doi.org/10.1007/s10815-018-1268-4.

Santos HO, Howell S, Teixeira FJ. Beyond tribulus (*Tribulus terrestris* L.): the effects of phyto-therapics on testosterone, sperm and prostate parameters. J Ethnopharmacol 2019;235:392–405. https://doi.org/10.1016/j.jep.2019.02.033.

Sellami M, Slimeni O, Pokrywka A, Kuvačić G, Hayes D, Milic M, Padulo J. Herbal medicine for sports: a review. J Int Soc Sports Nutr 2018;15:14. https://doi.org/10.1186/s12970-018-0218-y.

Shaheen G, Akram M, Jabeen F, Ali Shah SM, Munir N, Daniyal M, Riaz M, Tahir IM, Ghauri AO, Sultana S, Zainab R, Khan M. Therapeutic potential of medicinal plants for the management of urinary tract infection: a systematic review. Clin Exp Pharmacol Physiol 2019;46(7):613–24. https://doi.org/10.1111/1440-1681.13092.

Shahid M, Riaz M, Talpur MM, Pirzada T. Phytopharmacology of *Tribulus terrestris*. J Biol Regul Homeost Agents 2016;30(3):785–8.

Sirotkin AV, Alexa R, Harrath AH. Puncturevine (*Tribulus terrestris* L.) affects the proliferation, apoptosis, and ghrelin response of ovarian cells. Reprod Biol 2020;20(1):33–6. https://doi.org/10.1016/j.repbio.2019.12.009.

Ukani MD, Nanavati DD, Mehta NK. A review on the ayurvedic herb *Tribulus terrestris* L. Ancient Sci Life 1997;17(2):144–50.

Verma T, Sinha M, Bansal N, Yadav SR, Shah K, Chauhan NS. Plants used as antihypertensive. Nat Prod Bioprospect 2020. https://doi.org/10.1007/s13659-020-00281-x.

Wu Y, Yang H, Wang X. The function of androgen/androgen receptor and insulin growth factor-1/insulin growth factor-1 receptor on the effects of *Tribulus terrestris* extracts in rats undergoing high intensity exercise. Mol Med Rep 2017;16(3):2931–8. https://doi.org/10.3892/mmr.2017.6891.

Yuan Z, Du W, He X, Zhang D, He W. *Tribulus terrestris* ameliorates oxidative stress-induced ARPE-19 cell injury through the PI3K/Akt-Nrf2 signaling pathway. Oxid Med Cell Longev 2020;2020:7962393. https://doi.org/10.1155/2020/7962393.

Zhu W, Du Y, Meng H, Dong Y, Li L. A review of traditional pharmacological uses, phyto-chemistry, and pharmacological activities of *Tribulus terrestris*. Chem Cent J 2017;11(1):60. https://doi.org/10.1186/s13065-017-0289-x.

Chapter 3.11

Rooibos (*Aspalathus linearis* (Burm.f.) R.Dahlgren)

3.11.1 Introduction

Rooibos (*Aspalathus linearis* (Burm.f.) R.Dahlgren) tea is a known health-promoting drink. Its popularity is growing now. Nevertheless, the available publications concerning chemical, physiological and medicinal effects of rooibos describe only particular aspects of this plant. The review summarizing the physiological, medicinal and reproductive effects of this plant is absent now. In the present review we have tried to summarize briefly the available information concerning rooibos provenance, constituents, properties, its action and application in the control and treatment of female reproductive and nonreproductive processes and their disorders.

3.11.2 Provenance and properties

Rooibos (*Aspalathus linearis* (Burm.f.) R.Dahlgren) (family Fabaceae) is plant with green needle-shaped leaves and small yellow flowers, which can grow up to 2 m in height. The rooibos plant is indigenous to South Africa, specifically the Cederberg mountains region around Clamwilliam, North of Cape Town. Rooibos tea is available in two forms: a "fermented" or oxidized form; and an "unfermented" or unoxidized form. The unfermented product is also referred to as green rooibos. The "fermentation" process gives fermented rooibos its distinctive reddishbrown color, while unfermented rooibos tea maintains its green color. The fermentation process is important to develop the characteristic taste and aroma of rooibos tea, traditionally consumed. However, fermentation decreased the flavonoid content and antioxidation capacity of rooibos. Consumption of a "ready-to-drink" unfermented rooibos beverage as opposed to one produced from fermented rooibos effected a 28% higher total radical-trapping antioxidant potential in the plasma of human subjects (Dludla et al., 2017; Sasaki et al., 2018).

The sensoric and physiological actions of rooibos depend on the content of phenolic compounds - dihydrochalcones (aspalathin and nothofagin), flavones (kuteolin, chrysoeriol, orientin and isoorientin, vitexin, and isovitexin), flavanols (quercetin and isoquercetin, rutin, hyperoside), and phenylpropenoids (phenylpyruvic acid-2-O-glucoside). The dihydrochalcone, aspalathin is unique to rooibos, while its 3-deoxy analogue, nothofagin, is relatively rare (Ajuwon et al., 2018). Water-based green rooibos extracts contain about three times higher levels of total phenolic compounds than fermented extract. Quantitative composition analysis of green rooibos extract indicates that aspalathin is the major flavonoid (106.2 ± 16.5 mg/g, 66.1% w/w), followed by isoorientin (13.4 ± 2.4 mg/g, 8.3%), nothofagin (11.7 ± 2.5 mg/g, 7.3%), orientin (10.6 ± 1.6 mg/g, 6.6%), queretin-3-O-robinobiside (6.4 ± 0.6 mg/g, 4.0%) and phenylpyruvic acid-2-O-glucoside (2.9 ± 0.4 mg/g, 1.8%) (Sasaki et al., 2018). Other phenolic compounds that have been isolated from rooibos are phenolic acids, lignans, (+)-catechin, coumarins, esculentin and esculin (Ajuwon et al., 2018). The leaves contain much more aspalathin than the stems (Johnson et al., 2018). The majority of aspalathin and nothofagin are either C-glucosides or O-rhamnoglucosides, which are poorly absorbed. Phase II metabolism and degradation by intestinal bacteria are important factors in their absorption and bioavailability (Johnson et al., 2018; Muller et al., 2018).

3.11.3 Physiological actions

The best-known effect of rooibos and its constituents is their influence on glucose and lipid metabolism via increased resistance to insulin. Low resistance to this hormone is associated with metabolic dysfunctions in muscle, fat, kidney, liver, and pancreatic β-cells and development of oxidative stress, type 2 diabetes mellitus, cardiovascular diseases and obesity termed the metabolic syndrome. Signs of metabolic syndrome can be prevented by aspalathin, a C-glucoside dihydrochalcone, as well as the Z-2-(β-D-glucopyranosyloxy)-3-phenylpropenoic acid (Dludla et al., 2017, 2020; Ajuwon et al., 2018; Muller et al., 2018). In addition, rooibos extract can directly promote insulin secretion (Ajuwon et al., 2018), decrease plasma glucose level (Pyrzanowska et al., 2019), increase glucose uptake by various cells (Ajuwon et al., 2018), including adipocytes, suppress accumulation of lipids and production of the hormone leptin by isolated adipocytes (Sanderson et al., 2014) and increase respiration and energy expenditure of hepatic cells (Mazibuko-Mbeje et al., 2019; Uličná et al., 2019). Aspalathin, and phenylpyruvic acid-2-O-β-D-glucoside can ameliorate hyperglycemia-induced cardiomyocyte damage *in vitro* and, therefore, probably protect hearth from hyperglycemia-induced injury (Dludla et al., 2017, 2020). Rooibos polyphenols can be natural activators of immunocompetent cells (McKay and Blumberg, 2007; Lee and Bae, 2015; Ajuwon et al., 2018) and protectors from development of Alzheimer's disease (Darvesh et al., 2010) and mycotoxin fumonisin B_1 (Sheik et al.,

2020). Rooibos flavonoids, orientin and luteolin, stimulate mineralization in human osteoblasts and therefore protection against osteoporosis (Nash et al., 2015). The consumption of rooibos could harm some indexes of liver, kidney and epididymal morphology, but improve sperm concentration, viability, motility (Opuwari et al., 2014), and memory (Pyrzanowska et al., 2019).

Therefore, the available literature documents multiple physiological, protective, and therapeutic effects of rooibos and its constituents in various organs.

3.11.4 Mechanisms of action

The mechanisms of action of rooibos and its molecules have not been detected by direct experimental studies. Nevertheless, rooibos extract and its constituents have antioxidant properties (McKay and Blumberg, 2007; Dludla et al., 2017, 2020; Ajuwon et al., 2018) and the ability to increase the level of antioxidants in blood (Opuwari et al., 2014) and in cells (Dludla et al., 2020). It indicates that the key physiological and curative effect of rooibos molecules can be mediated by deletion of reactive oxygen species and mitigation the oxidative stress.

Oxidative stress can trigger inflammatory processes. Rooibos dihydrochalcones, aspalathin and nothofagin are able to down-regulate inflammatory processes and their extracellular (cytokines tumor necrosis factor-α and interleukin-6) and intracellular (nuclear factor-κB or MAP kinase) promoters (Lee and Bae, 2015).

Oxidative stress can induce apoptosis. Rooibos and its molecules can promote apoptosis in various cell types (Ajuwon et al., 2018). Orientin can induce intrinsic apoptosis of colorectal cancer cells via regulation of intracellular regulators of apoptosis bax/bcl-2 and tumor suppressing transcription factor p53 (Thangaraj et al., 2019). On the other hand, aspalathin and phenylpyruvic acid-2-O-β-D-glucoside can independently protect cardiomyocytes from hyperglycemia-related reactive oxygen species. While aspalathin shows more potency by enhancing intracellular antioxidant defences, phenylpyruvic acid-2-O-β-D-glucoside acts more as an antiapoptotic agent (Dludla et al., 2020). The antiapoptotic action of rooibos molecules was associated with an increase in the expression of nuclear factor (erythroid-derived 2)-like 2 (*Nrf2*), an intracellular antioxidant response element (Dludla et al., 2017, 2020), which can be a mediator of antiapoptotic action of rooibos on cardyomiocytes. Nash et al. (2015) reported the ability of other rooibos flavonoids, orientin and luteolin to promote viability, mitochondrial activity, and mineralization of human osteoblasts, which was associated with changes in Wnt pathway (Duan et al., 2016), the next possible intracellular mediator of rooibos action.

In addition to the influence on oxidative stress-induced apoptosis, rooibos' constituents could down-regulate cell proliferation. Orientin can suppress proliferation of these cells by cell cycle arrest at G0/G1 phase and prevention of entry to S-phase of this cycle via down-regulation of cell cycle promoters cyclin and cyclin-dependent protein kinases (Thangaraj et al., 2019).

The inhibitory action of rooibos on adipocytes was associated with changes in the expression of AMP-activated protein kinase, peroxisome proliferator-activated receptor-alpha, peroxisome proliferator-activated receptor-gamma, SREBF1 and FASN (Sanderson et al., 2014). Aspalathin and phenylpyruvic acid-2-O-β-D-glucoside are able to alter the expression of genes for AMP-activated protein kinase, and peroxisome proliferator-activated receptor-alpha in cultured cardiomyocytes as well (Dludla et al., 2020). AMP-activated protein kinase together with phosphatidylinositol-4,5-bisphosphate 3-kinase/protein kinase B (PI3K/AKT) can mediate aspalathin action on hepatocytes (Mazibuko-Mbeje et al., 2019).

Therefore, rooibos and its constituents can affect target cells via blocking the reactive oxygen species, activation of antioxidant, antiapoptotic intracellular mechanisms and several enzyme-dependent intracellular signaling pathways.

3.11.5 Effects on female reproductive processes

The few available evidence demonstrates direct action of rooibos on basic ovarian cell functions. Additions of rooibos extract at doses 1, 10 and 100 µg/mL of medium were able to inhibit proliferation (down-regulated PCNA, cyclin B1 and their mRNAs), to promote apoptosis (accumulation of bax) and to suppress release of progesterone and leptin by cultured porcine ovarian granulosa cells (Štochmaľová et al., 2018). Subsequent studies, however, did not confirm the ability of rooibos additions at dose 10 µg/mL to affect this marker of apoptosis. Furthermore, in these studies rooibos additions promoted accumulation of PCNA, which however did not alter cell viability (Sirotkin et al., 2020a,b). Furthermore, in these studies rooibos was able to either promote (Sirotkin et al., 2020a) or reduce (Sirotkin et al., 2020b) progesterone release.

In addition, rooibos was able to mitigate adverse effects of xylene on all measured cell functions (Sirotkin et al., 2020b). Moreover, rooibos prevented the toxic effect of benzene on viability, but promoted the adverse effect of benzene on apoptosis, progesterone, and estradiol output of these cells (Sirotkin et al., 2020a).

Taken together, the *in vitro* studies performed on cultured porcine ovarian granulosa cells suggest the ability of rooibos to affect directly ovarian cell proliferation, apoptosis and release of steroid and peptide hormones and to mitigate the adverse influence of oil-related environmental contaminants on these cells.

3.11.6 Mechanisms of action on female reproductive processes

The ability of rooibos to affect markers and promoters of **proliferation** (PCNA, cyclin B1) and **apoptosis** (bax) (Štochmaľová et al., 2018; Sirotkin et al., 20120a,b) suggests that is could affect female reproduction via changes in ovarian cell proliferation and apoptosis and their equilibration, which can affect ovarian follicular growth, development, atresia and fecundity (Sirotkin, 2014). Nevertheless, rooibos influence on proliferation and apoptosis were not associated with significant changes in ovarian cell viability (Sirotkin et al., 20120a,b). The influence of rooibos on ovarian cell proliferation and apoptosis are probably due to changes in the analyzed **ovarian hormones** release: rooibos promoted release of progesterone and leptin, which are considered as stimulators of ovarian cell functions, but these changes were associated with not stimulation, but inhibition of ovarian cell proliferation and apoptosis (Štochmaľová et al., 2018). Nevertheless, it is not to be excluded that rooibos can affect other reproductive processes via changes in hormone release. Since the key rooibos molecules possess phytoestrogenic features (see earlier), their influence via steroid hormones receptors is possible too.

The rooibos constituent(s) and mechanisms of their action on ovarian cells remain unknown yet. Nevertheless, the antiapototic action of rooibos could be explained by presence of asphalatin (see above), which could prevent apoptosis of cardiomyocytes due to its antioxidant (Dludla et al., 2017, 2020) and/or antiinflammatory (Lee and Baie, 2015) properties. It is not to be excluded that rooibos can down-regulate ovarian cell apoptosis through similar mechanisms. Nevertheless, both search for rooibos active molecule(s) and mechanisms of its action on the ovary requires further, more profound, studies.

3.11.7 Application in reproductive biology and medicine

The observed influence of rooibos extract on basic ovarian cell functions indicates its potential applicability in control of female reproductive processes. The stimulatory action on PCNA (Sirotkin et al., 2020a,b) and progesterone (Sirotkin et al., 2020a) observed in some experiments suggests that rooibos can be potentially useful for promotion of ovarian functions and related female reproduction. On the other hand, the suppressive effect of rooibos on some ovarian functions observed in some experiments (Štochmaľová et al., 2018; Sirotkin et al., 2020b) indicates potential hazardous influence of rooibos on female reproductive processes, which are to be considered before consumption of rooibos tea or related products.

The ability of rooibos extract to prevent the influence of benzene (Sirotkin et al., 2020a) and xylene (Sirotkin et al., 2020b) on ovarian cells suggests that functional food containing rooibos could be a natural protector against adverse action of oil-related contaminant on female reproductive processes.

3.11.8 Conclusions and possible direction of future studies

The influence of rooibos and its various constituents on various physiological processes and dysfunctions (including metabolic disorders and cancer) is well documented. Multiple mechanisms of their action are possible. Furthermore, the direct action of rooibos on ovarian cell proliferation, apoptosis and hormones release suggests its possible influence on female reproductive processes.

Nevertheless, the data concerning physiological and therapeutic action of rooibos and its molecules are obtained in animal and *in vitro* experiments, while no sufficient epidemiological or clinical data has been published yet. The influence of rooibos on female reproduction induces more queries than answers. The related studies were performed on one object—cultured porcine ovarian granulosa cells, while the observations concerning character of rooibos action on particular ovarian cell characteristics are inconsistent and sometimes contradictory. It remains unknown what rooibos substance(s) can be responsible for the observed effects. Therefore, the currently available preliminary data require confirmation and explanation by other *in vitro* and *in vivo* animal and clinical studies to convert the current speculations concerning character, mechanisms and practical application of rooibos to logical mechanistic concept and technologies for medicine and animal production.

References

Ajuwon OR, Ayeleso AO, Adefolaju GA. The potential of South African herbal tisanes, rooibos and honeybush in the management of type 2 diabetes mellitus. Molecules 2018;23(12):3207. https://doi.org/10.3390/molecules23123207.

Darvesh AS, Carroll RT, Bishayee A, Geldenhuys WJ, Van der Schyf CJ. Oxidative stress and Alzheimer's disease: dietary polyphenols as potential therapeutic agents. Expert Rev Neurother 2010;10(5):729—45. https://doi.org/10.1586/ern.10.42.

Dludla PV, Joubert E, Muller CJF, Louw J, Johnson R. Hyperglycemia-induced oxidative stress and heart disease-cardioprotective effects of rooibos flavonoids and phenylpyruvic acid-2-*O*-β-D-glucoside. Nutr Metab 2017;14:45. https://doi.org/10.1186/s12986-017-0200-8.

Dludla PV, Muller CJF, Louw J, Mazibuko-Mbeje SE, Tiano L, Silvestri S, Orlando P, Marcheggiani F, Cirilli I, Chellan N, Ghoor S, Nkambule BB, Essop MF, Huisamen B, Johnson R. The combination effect of aspalathin and phenylpyruvic acid-2-*O*-β-D-glucoside from rooibos against hyperglycemia-induced cardiac damage: an in vitro study. Nutrients 2020;12(4):1151. https://doi.org/10.3390/nu12041151.

Duan P, Bonewald LF. The role of the wnt/β-catenin signaling pathway in formation and maintenance of bone and teeth. Int J Biochem Cell Biol 2016;77(Pt A):23—9. https://doi.org/10.1016/j.biocel.2016.05.015.

Johnson R, Beer D, Dludla PV, Ferreira D, Muller CJF, Joubert E. Aspalathin from rooibos (*Aspalathus linearis*): a bioactive C-glucosyl dihydrochalcone with potential to target the metabolic syndrome. Planta Med 2018;84(9-10):568—83. https://doi.org/10.1055/s-0044-100622.

Mazibuko-Mbeje SE, Dludla PV, Johnson R, Joubert E, Louw J, Ziqubu K, Tiano L, Silvestri S, Orlando P, Opoku AR, Muller CJF. Aspalathin, a natural product with the potential to reverse hepatic insulin resistance by improving energy metabolism and mitochondrial respiration. PLoS One 2019;14(5):e0216172. https://doi.org/10.1371/journal.pone.0216172.

McKay DL, Blumberg JB. A review of the bioactivity of South African herbal teas: rooibos (aspalathus linearis) and honeybush (cyclopia intermedia). Phytother Res 2007;21(1):1—16. https://doi.org/10.1002/ptr.1992.

Muller CJF, Malherbe CJ, Chellan N, Yagasaki K, Miura Y, Joubert E. Potential of rooibos, its major C-glucosyl flavonoids, and Z-2-(β-D-glucopyranosyloxy)-3-phenylpropenoic acid in prevention of metabolic syndrome. Crit Rev Food Sci Nutr 2018;58(2):227—46. https://doi.org/10.1080/10408398.2016.1157568.

Nash LA, Sullivan PJ, Peters SJ, Ward WE. Rooibos flavonoids, orientin and luteolin, stimulate mineralization in human osteoblasts through the Wnt pathway. Mol Nutr Food Res 2015;59(3):443—53. https://doi.org/10.1002/mnfr.201400592.

Opuwari CS, Monsees TK. In vivo effects of *Aspalathus linearis* (rooibos) on male rat reproductive functions. Andrologia 2014;46(8):867—77. https://doi.org/10.1111/and.12158.

Pyrzanowska J, Fecka I, Mirowska-Guzel D, Joniec-Maciejak I, Blecharz-Klin K, Piechal A, Wojnar E, Widy-Tyszkiewicz E. Long-term administration of Aspalathus linearis infusion affects spatial memory of adult Sprague-Dawley male rats as well as increases their striatal dopamine content. J Ethnopharmacol 2019;238:111881. https://doi.org/10.1016/j.jep.2019.111881.

Sanderson M, Mazibuko SE, Joubert E, de Beer D, Johnson R, Pheiffer C, Louw J, Muller CJ. Effects of fermented rooibos (*Aspalathus linearis*) on adipocyte differentiation. Phytomedicine 2014;21(2):109—17. https://doi.org/10.1016/j.phymed.2013.08.011.

Sasaki M, Nishida N, Shimada M. A beneficial role of rooibos in diabetes mellitus: a systematic review and meta-analysis. Molecules 2018;23(4):839. https://doi.org/10.3390/molecules23040839.

Sheik Abdul N, Marnewick JL. Fumonisin B_1 -induced mitochondrial toxicity and hepatoprotective potential of rooibos: an update. J Appl Toxicol 2020;40(12):1602—13. https://doi.org/10.1002/jat.4036.

Sirotkin AV, Macejková M, Tarko A, Fabova Z, Alrezaki A, Alwasel S, Harrath AH. Effects of benzene on gilts ovarian cell functions alone and in combination with buckwheat, rooibos, and vitex. Environ Sci Pollut Res Int 2020. https://doi.org/10.1007/s11356-020-10739-7.

Sirotkin AV, Macejková M, Tarko A, Fabova Z, Alwasel S, Harrath AH. Buckwheat, rooibos, and vitex extracts can mitigate adverse effects of xylene on ovarian cells in vitro. Environ Sci Pollut Res Int 2020. https://doi.org/10.1007/s11356-020-11082-7.

Štochmaľová A, Kádasi A, Alexa R, Bauer M, Harrath AH, Sirotkin AV. Direct effect of polyphenol-rich plants, rooibos and ginkgo, on porcine ovarian cell functions. J Anim Physiol Anim Nutr 2018;102(2):e550—7. https://doi.org/10.1111/jpn.12795.

Thangaraj K, Balasubramanian B, Park S, Natesan K, Liu W, Manju V. Orientin induces G0/G1 cell cycle arrest and mitochondria mediated intrinsic apoptosis in human colorectal carcinoma HT29 cells. Biomolecules 2019;9(9):418. https://doi.org/10.3390/biom9090418.

Uličná O, Vančová O, Kucharská J, Janega P, Waczulíková I. Rooibos tea (*Aspalathus linearis*) ameliorates the CCl4-induced injury to mitochondrial respiratory function and energy production in rat liver. Gen Physiol Biophys 2019;38(1):15—25. https://doi.org/10.4149/gpb_2018037.

220 Contaminants and Plants Action on Female Reproduction

Chapter 3.12

Tea (*Camelia sinensis* L.)

3.12.1 Introduction

Tea is one of the most popular drink in the world, which benefits for health are well known (see Khan and Mukhtar, 2018; Tang et al., 2019; Abe and Inoue, 2020 for review). The data concerning its influence on female reproduction are however insufficient and contradictory. To our knowledge, they have not been reviewed yet. This chapter reviews provenance, sorts, chemical composition, and properties of tea (*Camelia sinensis* L.), its general health effects, as well as the currently available knowledge concerning both inhibitory and stimulatory action of tea and its constituents on female reproductive processes.

3.12.2 Provenance and properties

Tea tree (*Camelia sinensis* L.) is a bush or a small tree from the Theaceae family grown in tropic and subtropical regions around the world. In Southeast Asia, the most common cultivar is *Camellia sinensis* var. *sinensis*, in India and on Ceylon—*Camellia sinensis* var. *assamica*. The individual varieties of tea differ in their appearance, size and even the chemical composition of the leaf. For production of beverages, food, and medication, young leaves are used. The 2—3 top young leaves are considered the most tender and valuable. The final appearance, taste and biological effects of tea beverage are decided by the method of the leaf's processing. Based on the processing method (fermentation in the presence of bacteria or oxidation) and the taste, color, aroma and biological effect of the tea, we recognize white, green, matcha, black, oolong, and Pu-erh tea (Khan and Mukhtar, 2018; Rothenberg et al., 2018; Tang et al., 2019).

Leaves of *white* tea are harvested before they unfurl and then they are wilted. They are dried slowly in direct sunlight. They undergo no further mechanical processing and are usually not fermented at all. Due to that, the brewed tea remains light, almost colourless.

In production of *green t*ea, the harvested tea leaves are wilted. Then they are rapidly heated to destroy the enzymes which cause fermentation. The tea is

Food/medicinal herbs and their influence Chapter | 3 **221**

then dried in sun or in an oven and sorted. Therefore, in green tea, like in white, almost no fermentation takes place.

Matcha is green tea of high quality milled into fine powder, which is used in Japanese tea ceremony.

Black tea acquires its color through fermentation, thanks to which tea leaves turn reddish or even dark brown and the color of the brew ranges from honey to brown. The entire process developed in the past because tea leaves traveled a long time from the East to Europe. In Asia, nonfermented tea is preferred.

*Oolong i*s somewhere between green and black tea. It is referred to also as half-fermented or blue or blue-green tea.

Pu-erh is a tea prepared from enormous green leaves, pressed and long fermented by bacteria and fungi until it acquires reddish-brown color. This tea is very low in tannins. Its brew is dark brown with earthy or even smoky taste. It is similar to black tea. Pu-erh is suitable for long transport, during which it matures.

Despite most of Europe and the New World being accustomed to black tea, more antioxidants and tannins are present in nonoxidated and nonfermented white and green teas (Prasadh et al., 2019), and they also have a stronger benefit to health.

From the standpoint of biology and medicine, the most important are tea metabolites—alkaloid theobromine, its metabolite caffeine and polyphenols (theaflavins), catechins—epigallocatechin, epicatechin and their metabolites epigallocatechin gallate, epicatechin gallate, gallocatechins, and gallocatechin gallate. Gallates contain biologically active ethers of gallic acid. Tea contains also smaller amounts of other biologically active polyphenols—quercetin, kaempferol and myricetin. Black tea contains also polyphenolic compounds theaflavins and thearubigins (Saeed et al., 2017; Khan and Mukhtar, 2018; Rothenberg et al., 2018; Tang et al., 2019). In addition to that, tea contains microelements borate, cobalt, copper, iron, manganese, molybdenite and lead (Karak et al., 2017), pigments, polysaccharides, alkaloids, free amino acids, and saponins (Tang et al., 2019).

3.12.3 Physiological actions

Many studies have demonstrated that tea can affect a wide array of physiological processes and express a number of positive effects on health. Its antioxidant, antiinflammatory, immunoregulatory, anticancer, cardiovascular-protective, antidiabetic, antiobesity, and hepatoprotective properties are well documented (Khan and Mukhtar, 2018; Tang et al., 2019; Abe and Inoue, 2020).

Polyphenols of green tea are efficient against chronic inflammatory conditions of liver, gastrointestinal tract, and against neurodegenerative diseases (Oz, 2017; Prasanth et al., 2019).

Tea molecules at the right ratio counteract anxiety and stress (Unno et al., 2018). A link has been determined between green tea consumption and memory of older men (Xu et al., 2018). Tea prevents the aging processes in skin (Prasanth et al., 2019).

Tea polyphenols have antioxidant properties, which have a protective function against oxidative stress and occurrence of cancer (Yang et al., 2016; Rothenberg et al., 2018; Abe and Inoue, 2020). They can prevent stress induced by ultraviolet light and heavy metals (Prasanth et al., 2019). A number of meta-analyses observed an inverse association between drinking of green tea and occurrence of various kinds of cancers (Abe and Inoue, 2020). Tea catechins can have an antitumor effect: they inhibit multiplication of cells, trigger their apoptosis (death) and autophagy and reduce vitality of cancer cells (Singh et al., 2018). They can even enhance the therapeutic effects of medicine against tumor diseases and simultaneously mitigate their adverse side-effects (Cao et al., 2016; Bedrood et al., 2018). On the other hand, epigallocatechin gallate can produce *in vitro* free radicals and consequently lower the vitality of regular as well as cancer cells (Yang et al., 2016).

Most studies carried out on animals and humans report that tea (extract from tea tree leaves) reduces manifestations of metabolic syndrome, diabetes and cardiovascular diseases (including reducing blood pressure and the risk of heart attack) (Yang et al., 2018; Abe and Inoue, 2020). Other studies demonstrated positive effect of tea on decreasing insulin in blood, but not on the concentration of glucose, triacylglycerols, fatty acids and hormones of adipose tissue (Mielgo-Ayuso et al., 2014; Li et al., 2016). Yang et al. (2016), however, report that the reduced glucose levels in blood can be caused by the action of green tea polyphenols on glucose production in liver. Both green and black tea are efficient in reduction the fat stores as antiobesity functional food (Park et al., 2016; Yang et al., 2016, 2018; Rothenberg et al., 2018; Silvester et al., 2019). Some studies indicated the association between green tea and cognitive outcomes, dental health, injuries and respiratory diseases (Abe and Inoue, 2020). Tea can mitigate signs of osteoporosis (Das et al., 2004). Finally, tea polyphenols are natural antibiotics, antimycotics and antivirotics (Tang et al., 2019).

3.12.4 Mechanisms of action

These mechanisms of green tea catechins action on healthy and cancer cells include antioxidant activity, cell cycle regulation, receptor tyrosine kinase pathway inhibition, immune system modulation, and epigenetic modification control (Shirakami and Shimizu, 2018).

Epigallocatechin gallate can inhibits receptor tyrosine kinase, transcription factors AP-1 and NF-kappaB, thus inhibiting cell proliferation and enhancing apoptosis (Shirakami and Shimizu, 2018).

Furthermore, tea polyphenols have high antioxidant activity. They reduce accumulation of reactive oxygen species and prevent therefore oxidative stress and resulting DNA damage, autophagy and cell death. In addition, they affect several protein kinases—glycogen synthase kinase-3β, nonreceptor tyrosine kinase, which is involved in cell proliferation and differentiation (Prasanth et al., 2019).

Antiinflammatory effect of polyphenols can be explained by their inhibitory action on inflammation-promoting transcription factors AP-1 and NF-kappaB and stimulatory action on gastrointestinal microbiota, which is responsible for immune responses (Yang et al., 2016; Shirakami and Shimizu, 2018).

Tea catechins and theaflavins could be able to affect energetic balance—lower glucose, lipid, and uric acid levels. These activities are mediated by pharmacological mechanisms such as enzymatic inhibition of enzymes involved in cholesterol and uric acid metabolism and interaction with glucose transporters (Peluso and Serafini, 2017). Tea can affect metabolic state by using other mechanisms: by a combination of carbohydrate digestive enzyme inhibition and subsequent reactions of undigested carbohydrates with gut microbiota within the colon. The colon bacteria produce short-chain fatty acids, which enhance lipid metabolism through AMP-activated protein kinase activation. Furthermore, activation of this kinase decreases gluconeogenesis and fatty acid synthesis and increases catabolism, leading to body weight reduction and metabolic syndrome alleviation (Yang et al., 2016; Rothenberg et al., 2018). Moreover, teas reduce food consumption probably via neuroendocrine metabolic regulators, emulsion and absorption of lipids and protein in gastrointestinal system. Furthermore, they reduce total lipid production and promote conversion of white adipose tissue to brown, increase its oxidation, burning and expenditure of energy through heat production. Finally, they affect gastrointestinal microbiota (lacto- and bifidobacteria), which are responsible not only for food digestion, but also for immune reaction (Rothenberg et al., 2018). Different tea polyphenols, epigallocatechin and caffeine from tea leaves have independent mechanisms of effect but synergic effect on metabolism and weight loss (Janssens et al., 2016; Türközü and Tek, 2017; Vázquez Cisneros et al., 2017).

3.12.5 Effects on female reproductive processes

3.12.5.1 Effect on ovarian and reproductive state

Kao et al. (2000) reported an inhibitory action of epigallocatechin gallate injections on murine ovarian growth. In experiments of Luo et al. (2008) on rats, tea polyphenols inhibited ovarian folliculogenesis—the transition from primordial to developing follicles, extended the entire growth phase of a follicle, and reduced dominant follicle numbers per cycle. Feeding green tea to rabbits reduced conception and kindling rate, the number of liveborn and weaned pups (Baláži et al., 2019).

On the other hand, some observations demonstrated a stimulatory action of tea and their molecules on female reproductive processes and their ability to treat some reproductive disorders inducing infertility. For example, Luo et al. (2008) reported that tea polyphenols can increase the reserve of germ cells (probably due to inhibition of their recruitment for ovarian follicle development), to inhibit follicle atresia during ovarian development from birth to early aged, and retard climacterium in rats. These changes could be a result of impaired development and selection of ovarian follicles and the corresponding reproductive aging. Baláži et al. (2019) reported that feeding rabbits tea had the ability to increase their ovarian length and diameter of periovulatory (but not of primary and secondary) growing follicles. These phenomena could be however explained by inhibition of follicular ovulation.

Some publications demonstrated the applicability of green tea and its polyphenols, especially epigallocatechin gallate, for prevention and treatment of human and animal ovarian cell malignant transformation (Niedzwiecki et al., 2016; Saeed et al., 2017) and to decrease risk of ovarian cancer (Abe and Inoue, 2020), symptoms of polycystic ovarian syndrome (Ghafurniyan et al., 2015; Tehrani et al., 2017), and infertility (Roychoudhury et al., 2017).

3.12.5.2 Effect on ovarian cell functions

Basini et al. (2005) reported that addition of epigallocatechin gallate can suppress respiration and proliferation of cultured porcine ovarian granulosa cells. Similar experiments of Roychoudhury et al. (2018) and Sirotkin et al. (2019) demonstrated the ability of green tea or its polyphenols to promote apoptosis (the accumulation of caspase 3, p53, and bax) and to suppress proliferation (the accumulation of PCNA and cyclin B1) in these cells. These observations suggested the inhibitory actions of green tea constituents on porcine ovarian cell cycle in the S and G2 phases and promotion of cytoplasmic apoptosis. Roychoudhury et al. (2018) reported no discernible effects of green tea on proliferation (the accumulation of PCNA and cyclin B1) in cultured porcine granulosa cells. Data concerning direct stimulatory action of tea on ovarian cells are absent in the available literature.

3.12.5.3 Effect on oocytes and embryos

Some authors reported the inhibitory action of tea polyphenol, epigallocatechin gallate on oo- and embryogenesis *in vitro*. Spinaci et al. (2008) observed that addition of epigallocatechin gallate to culture medium at high dose (25 µg/mL) was able to impair porcine oocyte fertility and embryogenesis. A similar inhibitory action on bovine oocyte maturation and fertilization was reported by Wang et al. (2007a) for gallocatechin gallate added at high (25 µM) dose.

Gallocatechin gallate at lower dose however expressed a stimulatory effect on these process. In experiments of Spinaci et al. (2008) epigallocatechin gallate at low dose (10 μg/mL) was able to increase the fertility rate of porcine oocytes matured *in vitro*. Gallocatechin gallate, increased maturation (Wang et al., 2007b) and fertility and developmental capacity (Wang et al., 2007a) of cultured bovine oocytes when added at low doses (15 μM). These effects could be due to prevention of apoptotic processes in oocytes. At least, Luo et al. (2008) reported the ability of tea polyphenol injections to reduce apoptosis in rat oocytes *in vivo*.

3.12.5.4 Effect on reproductive hormones

There is evidence for inhibitory action of tea and its constituents on hormonal regulators of reproduction. Kao et al. (2000) reported the inhibitory action of epigallocatechin gallate injections on the level of reproductive hormones (LH, testosterone, estradiol, leptin, and insulin-like growth factor I - IGF-I) in blood plasma of mice. Tehrani et al. (2017) reported the ability of green tea to decrease plasma testosterone level in women suffering from polycystic ovarian syndrome. The direct inhibitory action of tea and its constituents on the release of ovarian hormones has been demonstrated by *in vitro* experiments. Additions of epigallocatechin gallate to culture medium inhibited progesterone, estradiol, VEGF (Basini et al., 2005), and testosterone (Sirotkin et al., 2019) release by porcine granulosa cells, as well as progesterone and testosterone output by cultured rabbit ovarian fragments (Baláži et al., 2019).

Conversely, some authors reported no effects of green tea on basal and FSH-induced release of steroid hormones (progesterone and estradiol) in cultured porcine ovarian granulosa cells (Roychoudhury et al., 2017, 2018). Moreover, Sirotkin et al. (2019) reported the stimulatory action of tea polyphenols, epigallocatechin gallate and gallocatechin gallate on basal and IGF-I-stimulated progesterone secretion by these cells. Addition of gallocatechin gallate promoted progesterone and testosterone release also by cultured rabbit ovarian fragments (Baláži et al., 2019).

Taken together, the available literature demonstrates both inhibitory and stimulatory action of green tea and its constituents on almost all reproductive processes both *in vivo* and *in vitro*. The character of this action can depend on the species, experimental model, tested substances (crude tea extract, epigallocatechin gallate, gallocatechin gallate), their dose and the measured parameter. On the other hand, in some cases the authors used the same experimental approach and analyzed the same parameter reported different and even conflicting results. This fact indicates the influence of some factor(s), which are unknown yet, for example the influence of other unknown tea constituents or initial reproductive, endocrine or metabolic state of the model animal. The variation in described tea effects could indicate also multiple mechanisms of its effects proposed below.

3.12.6 Mechanisms of action on female reproductive processes

The ability of tea to affect metabolism and mortality mentioned above suggests that tea's influence on reproduction can be explained by its **nonspecific metabolic or toxic effects**. The available information concerning the ability of tea and its constituents to affect release of peptide and steroid **hormones**, the known regulators of ovarian functions and fecundity (Sirotkin, 2014) indicates that these hormones, which affect the mechanisms of green tea and its constituents on the reproductive processes, can be the mediators of tea action on reproductive processes. Furthermore, the ability of tea to influence the **response of ovarian cells to hormones** allows to propose that it can influence also hormone reception and/or postreceptory signaling pathways mediating hormones action. For example, tea contains resveratrol (see above), a polyphenol with phytoestrogenic properties (see the chapter concerning resveratrol), i.e., the ability to bind and to regulate steroid hormone receptors and steroid hormone-dependent reproductive events. Although tea expresses antiosteoporosis effect similar to estrogens (Das et al., 2004), the phytoestrogenic properties of this plant have not been strongly demonstrated yet. Furthermore, green tea extract did not affect the response of porcine ovarian cell steroidogenesis to FSH (Roychoudhury et al., 2017), which suggest that tea did not influence FSH reception and effects. Therefore, tea action on ovarian functions via changes in reception and response to hormonal regulators is not to be excluded, but it has not been experimentally demonstrated yet. The influence of tea and its polyphenols on **regulators and markers of ovarian cell cycle** (PCNA, cyclin B1) and **apoptosis** (bax. caspase 3) is well documented (see above). The proliferation:apoptosis rate defines ovarian follicular growth, atresia, selection, ovulation, as well as quality of oocytes. Therefore, the influence of tea and its constituents on ovarian cycle, folliculogenesis, oogenesis and resulted fecundity could be due to their direct or hormones-mediated influence on ovarian cell proliferation and apoptosis. The ability of tea extract to promote accumulation of **transcription factor p53** (Roychoudhury et al., 2018), which is a known suppressor of proliferation and inductor of apoptosis in both ovarian and nonovarian cells (Fu et al., 2020) indicates that tea substance could affect proliferation and apoptosis and the related reproductive processes via this transcription factor. In addition, p53 maintains genomic stability and prevent DNA damage induced by adverse environmental factors (Fu et al., 2020). It is not to be excluded that the protective action of tea against some damaging factors and disorders mentioned above could be mediated by up-regulation of these transcription factors. Finally, this protective and therapeutic action of tea against DNA damage and related illnesses (for example, ovarian malignant transformation) could be explained by high **antioxidant activities** of its constituents (see above). There are indications that tea catechins with antioxidant properties can potentially mitigate the oxidative stress-associated infertility (Roychoudhury et al., 2017). It is however necessary to note that tea action has been investigated much better than the mechanisms of this action, and these mechanisms are hypothesized mainly only based on indirect indications (ability of tea to affect regulators of reproductive events), but not on direct experimental evidence. Moreover, the present list of possible mediators of tea action on female reproduction will be surely expanded.

3.12.7 Application in reproductive biology and medicine

The well-documented influence of tea and its constituents on female reproductive processes at different regulatory levels indicates their potential applicability for control of reproduction and treatment of some reproductive disorders. The variation in character of their effects suggest their potential applicability for both up- and downregulation of particular reproductive events and their dysfunctions.

The stimulatory action of tea and its polyphenols on some ovarian functions suggests their potential applicability for prevention of reproductive aging and exhaustion of ovarian reserve (Luo et al., 2008). It could be useful for improvement of the current technology of *in vitro* oocyte maturation, fertilization and embryo production for prevention oocyte apoptosis (Luo et al., 2008) and to increase oocyte fertilization and developmental capacity (Wang et al., 2007a,b). The ability of tea polyphenols to promote basal and IGF-I-induced ovarian steroidogenesis (Bálaži et al., 2019; Sirotkin et al., 2019) suggests their potential applicability for promoting the action of ovarian hormonal stimulators and preventing the deficit of steroid hormones. Das et al. (2004) have shown that drinking of tea by rats prevented osteoporosis induced by deficit of ovarian hormones. Finally, the stimulatory action of tea on ovarian steroidogenesis and folliculogenesis and on antioxidant system suggests its potential applicability for prevention and treatment of ovarian cancer (Niedzwiecki et al., 2016; Saeed et al., 2017; Abe and Inoue, 2020), polycystic ovarian syndrome (Ghafurniyan et al., 2015; Tehrani et al., 2017), and infertility induced by oxidative stress (Roychoudhury et al., 2017).

The inhibitory action of tea on female reproductive processes can be useful for prevention of reproductive aging (Luo et al., 2008) and overproduction of testosterone during polycystic ovarian syndrome (Tehrani et al., 2017). The ability of tea to suppress respiration, proliferation, and to promote apoptosis in healthy (Basini et al., 2005; Roychoudhury et al., 2018; Sirotkin et al., 2019) and cancer (Spinella et al., 2006; Wang et al., 2014) ovarian cells suggests that it can be a promising tool to prevent ovarian cancerogenesis. Furthermore, the inhibitory action of tea on female ovarian cell functions (Basini et al., 2005; Roychoudhury et al., 2018; Sirotkin et al., 2019) and its potential toxic

3.12.8 Conclusions and possible direction of future studies

The review of the available data demonstrates an influence of tea and its polyphenols on a number of nonreproductive and female reproductive processes. They can influence female reproductive processes at various regulatory levels: from recruitment of germ cells and ovarian follicles, gonadotropin release, ovarian hormones, ovarian cell proliferation, apoptosis, oocyte maturation, up to reproductive aging. The action of tea on nonreproductive and reproductive processes can be mediated by multiple mechanisms from nonspecific metabolic, toxic, and antioxidant effects up to influence on hormones release and effects and intracellular mediators of hormones action regulating cell cycle and apoptosis. The multiple targets and mechanisms of tea action can result in multiple characters of tea action on female reproductive processes and fecundity. Both stimulatory and inhibitory influence of tea and its constituents can be applicable in human and veterinary medicine, biotechnology and assisted reproduction for the prevention of reproductive aging, *in vitro* oocyte maturation and embryo production, up- and down-regulation of hormones release, as well as for prevention of some disorders associated with changes in hormones release and ovarian cell proliferation and apoptosis including cancer.

Nevertheless, the wide application of tea in reproductive biology and medicine is limited by deficit in the available knowledge concerning the character, sites and mechanisms of the action of tea and its molecules on reproductive system. The performed studies were focused mainly on only two tea polyphenols - epigallocatechin gallate and gallocatechin gallate. The factors defining whether tea will promote or suppress reproductive functions are not clearly defined. To our knowledge, the influence of tea or its substances on central regulators of reproduction (brain nervous and neuroendocrine system) and on receptors to hormones have not been studied yet. The downstream sites and mechanisms of action of tea are studied insufficiently. The main available evidence concerning the medicinal effects of tea and tea-related substances were obtained on a limited number of animal and *in vitro* experiments and a limited number of epidemiological studies, while no serious clinical investigations of possible therapeutic action of tea have been performed yet. If these queries are addressed, tea could be used not only as a pleasant and healthy functional food, but also as a tool to control human and animal female reproductive processes and to prevent and treat their disorders.

References

Abe SK, Inoue M. Green tea and cancer and cardiometabolic diseases: a review of the current epidemiological evidence. Eur J Clin Nutr 2020. https://doi.org/10.1038/s41430-020-00710-7.

Baláži A, Sirotkin AV, Földešiová M, Makovický P, Chrastinová Ľ, Makovický P, Chrenek P. Green tea can supress rabbit ovarian functions in vitro and in vivo. Theriogenology 2019;127:72—9. https://doi.org/10.1016/j.theriogenology.2019.01.010.

Basini G, Bianco F, Grasselli F. Epigallocatechin-3-gallate from green tea negatively affects swine granulosa cell function. Domest Anim Endocrinol 2005;28:243—56.

Bedrood Z, Rameshrad M, Hosseinzadeh H. Toxicological effects of *Camellia sinensis* (green tea): a review. Phytother Res 2018;32(7):1163—80.

Cao J, Han J, Xiao H, Qiao J, Han M. Effect of tea polyphenol compounds on anticancer drugs in terms of anti-tumor activity, toxicology, and pharmacokinetics. Nutrients 2016;8(12):E762.

Chen SN, Friesen JB, Webster D, Nikolic D, van Breemen RB, Wang ZJ, Fong HH, Farnsworth NR, Pauli GF. Phytoconstituents from Vitex agnus-castus fruits. Fitoterapia 2011;82(4):528—33. https://doi.org/10.1016/j.fitote.2010.12.003.

Das AS, Mukherjee M, Mitra C. Evidence for a prospective anti-osteoporosis effect of black tea (*Camellia sinensis*) extract in a bilaterally ovariectomized rat model. Asia Pac J Clin Nutr 2004;13(2):210—6.

Fu X, Wu S, Li B, Xu Y, Liu J. Functions of p53 in pluripotent stem cells. Protein Cell 2020;11(1):71—8. https://doi.org/10.1007/s13238-019-00665-x.

Ghafurniyan H, Azarnia M, Nabiuni M, Karimzadeh L. The effect of green tea extract on reproductive improvement in estradiol valerate-induced polycystic ovarian syndrome in rat. Iran J Pharm Res 2015;14:1215—33.

Janssens PL, Hursel R, Westerterp-Plantenga MS. Nutraceuticals for body-weight management: the role of green tea catechins. Physiol Behav 2016;162:83—7.

Kao YH, Hiipakka RA, Liao S. Modulation of endocrine systems and food intake by green tea epigallocatechin gallate. Endocrinology 2000;141:980—7.

Karak T, Kutu FR, Nath JR, Sonar I, Paul RK, Boruah RK, Sanyal S, Sabhapondit S, Dutta AK. Micronutrients (B, Co, Cu, Fe, Mn, Mo, and Zn) content in made tea (*Camellia sinensis* L.) and tea infusion with health prospect: a critical review. Crit Rev Food Sci Nutr 2017;57(14):2996—3034.

Khan N, Mukhtar H. Tea polyphenols in promotion of human health. Nutrients 2018;11(1):39. https://doi.org/10.3390/nu11010039.

Li Y, Wang C, Huai Q, Guo F, Liu L, Feng R, Sun C. Effects of tea or tea extract on metabolic profiles in patients with type 2 diabetes mellitus: a meta-analysis of ten randomized controlled trials. Diabetes Metab Res Rev 2016;32(1):2—10.

Luo LL, Huang J, Fu YC, Xu JJ, Qian YS. Effects of tea polyphenols on ovarian development in rats. J Endocrinol Invest 2008;31(12):1110—8. https://doi.org/10.1007/BF03345661.

Mielgo-Ayuso J, Barrenechea L, Alcorta P, Larrarte E, Margareto J, Labayen I. Effects of dietary supplementation with epigallocatechin-3-gallate on weight loss, energy homeostasis, cardiometabolic risk factors and liver function in obese women: randomised, double-blind, placebo-controlled clinical trial. Br J Nutr 2014;111(7):1263—71.

Niedzwiecki A, Roomi MW, Kalinovsky T, Rath M. Anticancer efficacy of polyphenols and their combinations. Nutrients 2016;8:E552. https://doi.org/10.3390/nu8090552.

Oz HS. Chronic inflammatory diseases and green tea polyphenols. Nutrients 2017;9(6):E561.

230 Contaminants and Plants Action on Female Reproduction

Peluso I, Serafini M. Antioxidants from black and green tea: from dietary modulation of oxidative stress to pharmacological mechanisms. Br J Pharmacol 2017;174(11):1195−208. https://doi.org/10.1111/bph.13649.

Prasanth MI, Sivamaruthi BS, Chaiyasut C, Tencomnao T. A review of the role of green tea (*Camellia sinensis*) in antiphotoaging, stress resistance, neuroprotection, and autophagy. Nutrients 2019;11(2):474. https://doi.org/10.3390/nu11020474.

Rothenberg DO, Zhou C, Zhang L. A review on the weight-loss effects of oxidized tea polyphenols. Molecules 2018;23(5):E1176.

Roychoudhury S, Agarwal A, Virk G, Cho CL. Potential role of green tea catechins in the management of oxidative stress-associated infertility. Reprod Biomed Online 2017;34:487−98. https://doi.org/10.1016/j.rbmo.2017.02.006.

Roychoudhury S, Halenar M, Michalcova K, Nath S, Kacaniova M, Kolesarova A. Green yea extract affects porcine ovarian cell apoptosis. Reprod Biol 2018;18:94−8.

Saeed M, Naveed M, Arif M, Kakar MU, Manzoor R, Abd El-Hack ME, Alagawany M, Tiwari R, Khandia R, Munjal A, Karthik K, Dhama K, Iqbal HMN, Dadar M, Sun C. Green tea (*Camellia sinensis*) and l-theanine: medicinal values and beneficial applications in humans-A comprehensive review. Biomed Pharmacother 2017;95:1260−75. https://doi.org/10.1016/j.biopha.2017.09.024.

Shirakami Y, Shimizu M. Possible mechanisms of green tea and its constituents against cancer. Molecules 2018;23(9):2284. https://doi.org/10.3390/molecules23092284.

Silvester AJ, Aseer KR, Yun JW. Dietary polyphenols and their roles in fat browning. J Nutr Biochem 2019;64:1−12.

Singh AK, Bishayee A, Pandey AK. Targeting histone deacetylases with natural and synthetic agents: an emerging anticancer strategy. Nutrients 2018;10(6):E731.

Sirotkin AV, Kadasi A, Maruniakova N, Grossmann R, Alwasel S, Harrath AH. Influence of green tea constituents on cultured porcine luteinized granulosa cell functions. J Anim Feed Sci 2019;28:41−51. https://doi.org/10.22358/jafs/104705/2019.

Sirotkin AV. Regulators of ovarian functions. New York: Nova Publishers, Inc.; 2014. p. 194.

Spinaci M, Volpe S, De Ambrogi M, Tamanini C, Galeati G. Effects of epigallocatechin-3-gallate (EGCG) on in vitro maturation and fertilization of porcine oocytes. Theriogenology 2008;69:877−85.

Spinella F, Rosanò L, Di Castro V, Decandia S, Albini A, Nicotra MR, Natali PG, Bagnato A. Green tea polyphenol epigallocatechin-3-gallate inhibits the endothelin axis and downstream signaling pathways in ovarian carcinoma. Mol Canc Therapeut 2006;5:1483−92.

Tang GY, Meng X, Gan RY, Zhao CN, Liu Q, Feng YB, Li S, Wei XL, Atanasov AG, Corke H, Li HB. Health functions and related molecular mechanisms of tea components: an update review. Int J Mol Sci 2019;20(24):6196. https://doi.org/10.3390/ijms20246196.

Tehrani HG, Allahdadian M, Zarre F, Ranjbar H, Allahdadian F. Effect of green tea on metabolic and hormonal aspect of polycystic ovarian syndrome in overweight and obese women suffering from polycystic ovarian syndrome: a clinical trial. J Educ Health Promot 2017;6:36. https://doi.org/10.4103/jehp.jehp_67_15.

Türközü D, Tek NA. A minireview of effects of green tea on energy expenditure. Crit Rev Food Sci Nutr 2017;57(2):254−8.

Unno K, Furushima D, Hamamoto S, Iguchi K, Yamada H, Morita A, Horie H, Nakamura Y. Stress-reducing function of matcha green tea in animal experiments and clinical trials. Nutrients 2018;10(10):E1468.

Vázquez Cisneros LC, López-Uriarte P, López-Espinoza A, Navarro Meza M, Espinoza-Gallardo AC, Guzmán Aburto MB. Effects of green tea and its epigallocatechin (EGCG)

content on body weight and fat mass in humans: a systematic review. Nutr Hosp 2017;34(3):731−7.

Wang ZG, Yu SD, Xu ZR. Effect of supplementation of green tea polyphenols on the developmental competence of bovine oocytes in vitro. Braz J Med Biol Res 2007a;40(8):1079−85. https://doi.org/10.1590/s0100-879x2007000800008.

Wang ZG, Yu SD, Xu ZR. Improvement in bovine embryo production in vitro by treatment with green tea polyphenols during in vitro maturation of oocytes. Anim Reprod Sci 2007b;100(1-2):22−31. https://doi.org/10.1016/j.anireprosci.2006.06.014.

Wang F, Chang Z, Fan Q, Wang L. Epigallocatechin-3-gallate inhibits the proliferation and migration of human ovarian carcinoma cells by modulating p38 kinase and matrix metalloproteinase 2. Mol Med Rep 2014;9:1085−9. https://doi.org/10.3892/mmr.2014.1909.

Xu H, Wang Y, Yuan Y, Zhang X, Zuo X, Cui L, Liu Y, Chen W, Su N, Wang H, Yan F, Li X, Wang T, Xiao S. Gender differences in the protective effects of green tea against amnestic mild cognitive impairment in the elderly Han population. Neuropsychiatric Dis Treat 2018;14:1795−801.

Yang CS, Zhang J, Zhang L, Huang J, Wang Y. Mechanisms of body weight reduction and metabolic syndrome alleviation by tea. Mol Nutr Food Res 2016;60(1):160−74.

Yang CS, Wang H, Sheridan ZP. Studies on prevention of obesity, metabolic syndrome, diabetes, cardiovascular diseases and cancer by tea. J Food Drug Anal 2018;26(1):1−13.

Chapter 3.13

Vitex (*Vitex agnus-castus* L.)

3.13.1 Introduction

One of the most known natural regulator of reproductive processes is chaste tree (*Vitex agnus-castus* L.). Although it is known a long time as a medicinal plant and biostimulator of both male and female reproduction, the published reviews concerning this plant don't reflect the recent related findings (Daniele et al., 2005; Anonimous, 2009; Chen et al., 2011; Rani and Sharma, 2013) or describe only some (mainly clinical) aspects of reproductive effects of Vitex (Ibrahim et al., 2008; van Die et al., 2013; Dietz et al., 2016). The present review includes the recent information concerning Vitex agnus-cactus action on nonreproductive and reproductive processes, mechanisms of its action, as well as outlines the possible areas of further related studies and application of this plant in control of both male and female reproductive processes and for treatment of their disorders.

3.13.2 Provenance and properties

Vitex is the largest genus in the family Verbenaceae, which comprises 250 species distributed all over the world. The best-known species used in medicine is *Vitex agnus-castus* Linn (chaste tree), which is widespread on riverbanks and shores in the Mediterranean region, Southern Europe and in Central Asia. Vitex agnus-castus, commonly known as chaste tree or sage tree, is a beautiful little deciduous tree or a large shrub with a showy summertime flower display. Vitex agnus-castus is a sprawling plant that grows 3–6 m and about as wide. It has slender, finger-like aromatic leaves, and a tender stem. Its branched flowers are fragrant and attract pollinating bees and hummingbirds. Flowers are followed by fleshy purple-black berries that contain four ripening seeds sometimes used as seasoning, similar to black pepper. The berries were used by monks during the Middle Ages to suppress sexual desire; hence its common names—monk's pepper and chaste tree (Daniele et al., 2005; Rani and Sharma, 2013). The dried fruits are the most commonly used medicinal form, but today, Vitex is available in a range of pharmaceutical forms, including tinctures, fluid extracts, tablets, and homeopathic preparations (van Die et al., 2013).

The main chemical compound of Vitex agnus-castus are vitexin, orientin and apigenin casticin, agnuside, p-hydroxybenzoic acid, 8-di-C-glycosides alkaloids, diterpenoids and iridoid compound aucubin. Its fruit contains flavonoids, terpenoids, neolignans, and phenolic compounds, as well as one glyceride (Ibrahim et al., 2008; Rafieian-Kopaei and Movahedi. 2017), as well as essential oils (e.g., α-pinene, limonene, β-caryophyllene, sabinene, αbisabolol, 1,8-cineol[eucalyptol], and β-farnesene (Khalilzadeh et al., 2015), iridoid glycosides (e.g., agnoside, aucubin), diterpines (e.g., vitexilactone, rotundifuran), and flavonoids (e.g., apigenin, castican, orientin, isovitexin) (Drugs and Lactation Database, 2018).

3.13.3 Physiological actions

The berries are used medicinally, with their use dating back to the ancient Greeks and Romans as well as to Persian and Chinese traditional medicine (Anonimous, 2009; Ibrahim et al., 2019). In the Iranian folk medicine, Vitex agnus-cactus is used as anticonvulsant, antiepileptic, carminative, energizer, sedative, anticonvulsant, constipation, and reduction of libido (Khalilzadeh et al., 2015). In Chinese traditional medicine it is used as an antiinflammatory agent (Ibrahim et al., 2019).

The main therapeutic application of Vitex agnus-castus in modern Western medicine is focused on treatment of the consequences of female menstrual cycle and reproductive aging (see below), but recently its usefulness for treatment of a number of other disorders has been demonstrated.

According to a multitude of studies, Vitex agnus-castus can affect metabolism and impact obesity, insulin resistance and lipid-metabolism dysfunctions (Moini Jazani et al., 2019). A crude extract and a butanolic fraction of Vitex reduced adiposity index and prevented liver steatosis (Moreno et al., 2015) and fibrosis (Chan et al., 2018).

Vitex secondary metabolites (p-hydroxybenzoic acid, methyl 3,4-dihydroxybenzoate and 3,4-dihydroxybenzoic acid) (Choudhary et al., 2009) or flavone casticin (Chan et al., 2018) expressed anti-inflammatory activity *in vitro*. *In vivo* studies confirmed the ability of Vitex extract to prevent inflammation and lung injury (Ibrahim et al., 2019). Casticin prevented development of oedemas and vascular inflammations (Chan et al., 2018).

In addition, Vitex can suppress inflammation by down-regulation (Certo et al., 2017) of angiogenesis and promote wound healing by its up-regulation (Eftekhari et al., 2014).

Well known is the anelgesic action of Vitex agnus-cactus (Chan et al., 2018) and its essential oils (Khalilzadeh et al., 2015).

Extract of Vitex agnus-cactus supports osteogenesis (bone formation). Due to this activity, it can be useful for treatment of bone fractures (Eftekhari et al., 2014) and for prevention of osteopenia and osteoporosis (Sehmisch et al., 2009; Dietz et al., 2016)

Numerous studies demonstrated the ability of Vitex extract (Kuruüzüm-Uz et al., 2003; Ohyama et al., 2005; Kikuchi et al., 2014; Ilhan et al., 2020), its flavonoid casticin (Chan et al., 2018) or its essential oils (Ricarte et al., 2020) to decrease viability and proliferation and to promote apoptosis of cancer cells of various types.

Vitex glucosides (Kuruüzüm-Uz et al., 2003) and essential oils (Gonçalves et al., 2017) have high antibacterial activity.

Vitex can influence the state of pituitary cell—producers of pituitary hormones (Wuttke et al., 2003; Šošić-Jurjević et al., 2016), to increase plasma thyroid stimulating hormone and thyroid hormones, and reduce plasma adrenocorticotrophic hormone (but not corticosterone) (Šošić-Jurjević et al., 2016), prolactin (Wuttke et al., 2003) LH, and testosterone level (Nasri et al., 2007) in male rats.

The vitex-induced reduction in prolactin release can be associated with reduced milk production (Anonimous, 2009; Dietz et al., 2016). Vitex extract has been used to decrease breastmilk oversupply in Persian traditional medicine, but galactogogic effect of Vitex has not been confirmed in clinical experiments (Drug and Lactation Database, 2018)

Extract of Vitex improved rats' learning and memory (Allahtavakoli et al., 2015).

Other pharmacological properties of Vitex agnus-cactus extract include anti-asthmatic, tracheospasmolytic, antihyperprolactinemia, immunomodulatory, opioidergic, estrogenic, antiglioma, rheumatoid arthritis amelioration (Anonimous, 2009).

Therefore, constituents of Vitex agnus-cactus affect various physiological processes and organs. Their effect makes this plant a promising medicine for prevention and treatment of a wide array of disorders.

3.13.4 Mechanisms of action

Vitex agnus-cactus affected the release of both pituitary (adrenocorticotrophic hormone, thyroid stimulating hormone, prolactin, LH) and the related peripheral (thyroid, androgen) hormones (Wuttke et al., 2003; Nasri et al., 2007; Šošić-Jurjević et al., 2016), which in turn are considered the key regulators of a wide array of physiological functions from metabolism and immunity to bone functions and cell viability. It indicates that these hormones could be extracellular mediators of Vitex action on its targets. For example, the ability of vitex to reduce milk production can be due to its ability to impair prolactin release (Anonimous, 2009).

Vitex (probably its diterpene clerodadienolscan, Drugs and Lactation Database, 2018) inhibit pituitary prolaction (Wuttke et al., 2003), gonadotropin (Nasri et al., 2007; Šošić-Jurjević et al., 2016) and testosterone (Nasri et al., 2007) production via binding and down-regulation of dopamine-2 receptors in pituitary cells.

The ability of vitex essential oils to relieve pain can be explained by its activation of endogenous opioid, muscarinergic or acetylcholine receptors in the nervous system (Anonimous, 2009; Khalilzadeh et al., 2015). The binding to delta and mu opioid receptors has been demonstrated for such Vitex constituents as apigenin, 3-methylkaempferol, luteolin, and casticin (Chen et al., 2011)

Vitex flavonoids (casticin, quercetagetin, isovitexin) have been shown *in vitro* to affect estrogen-beta receptors. These pohytoestrogens can mitigate the effect of natural or induced estrogen deficit and to prevent age-dependent osteoporosis (Anonimous, 2009; Dietz et al., 2016) and neurodegenerative disorders and stroke (Alimohamadi et al., 2019). The improvement of learning and memory was also associated with an increase in expression of estrogen receptor alpha in hippocampus (Allahtavakoli et al., 2015).

The anti-cancer (Dietz et al., 2016) and neuroprotective (Alimohamadi et al., 2019) effects of Vitex extract can be explained also by its anti-inflammatory properties. Vitex isoflavone casticin can prevent inflammation via down-regulation of transcription factor NFkB, Akt and mitogen-activated protein kinase (MAPK) signaling pathways, production of inflammation-promoting cytokines, and proliferation of lymphocytes (Chan et al., 2018).

They have also antioxidant properties: inhibit oxidative enzymes (lipoxygenase) (Choudhary et al., 2009; Chan et al., 2018) and reactive oxygen species-induced proinflammatory trabscription factor NFkB (Chan et al., 2018). The antioxidant properties of Vitex are responsible for its ability to treat obesity, liver steatosis (Moreno et al., 2015), lung injury (Ibrahim et al., 2019) and other inflammations (Chan et al., 2018).

On the other hand, the ability of vitext extract to promote oxidative enzymes Heme oxygenase-1 and NADPH oxidase and to induce oxidative stress can explain the ability of this plant to induce cancer cell apoptosis and death (Kikuchi et al., 2014). The ability of Vitex extract to increase the nuclear apoptosis associated with DNA fragmentation, as well as accumulation of promoters of cytoplasmic/mitochondrial apoptosis of caspases-8, -9 and -3, Bad, FADD, TRAIL R1/DR4 and TRAIL R2/DR5. protein and to decrease the level of antiapoptotic Bcl-2, Bcl-XL and Bid proteins had been demonstrated (Ohyama et al., 2003, 2005; Chan et al., 2018; Ilhan et al., 2020).

236 Contaminants and Plants Action on Female Reproduction

In addition, Vitex molecules can suppress cancer cell proliferation via arrest of transition of G2 to M phase of mitosis. This mitotic arrest could be mediated by changes in intracellular ATP and p38 MAP kinase pathway (Kikuchi et al., 2014), as well as up-regulation of mitosis-inhibiting PI3/Akt pathway and p21 and downregulation of cyclin B, the key promoter of mitosis (Chan et al., 2018).

Therefore, various Vitex constituents can affect a number of physiological processes via multiple extra- and intracellular mediators.

3.13.5 Effects on female reproductive processes

Although the physiological and therapeutical action of Vitex agnus-cactus on a wide spectrum of processes and illnesses has been demonstrated (see above), the main studies and areas of application of this plant are focused on processes related to female reproduction and reproduction-dependent disfunctions now.

3.13.5.1 Effect on ovarian and reproductive state

High consumption of Vitex agnus-cactus by female Phayre's leaf monkeys (*Trachypithecus phayrei crepusculus*) was associated with longer ovarian cycle lengths and follicular phases, while receptive periods did not change. Nevertheless, when ovulations occurred, females were more likely to conceive. Although the influence of other factors on these processes is not to be excluded, this observation suggests that Vitex can prolong follicular phase of ovarian cycle and promote monkey fecundity (Lu et al., 2011). Interesting is that consumption of other Vitex species (*V. doniana*) by baboons worked at cycling females as a progestagene-based contraceptive—simulated pregnancy, prevented sexual swelling, association, and copulation with males (Higham et al., 2007). The stimulatory action of Vitex agnus-cactus to female reproductive system was indicated by ability of its extract to increase rat uterine weight (Ibrahim et al., 2008) and to prevent age-dependent damage in mice estrous cycle, uterus and ovarian tissues (Ahangarpour et al., 2016). Finally, plant preparation containing Vitex was able to regulate women's ovarian cycle, to stimulate ovulation and to increase the likelihood of getting pregnant (Antoine et al., 2019). On the other hand, it was able to prevent the symptoms of gonadotropin-induced ovarian hyperstimulation syndrome in women (Cahill et al., 1994).

Therefore, the vailable reports demonstrated the (mainly stimulatory) influence of Vitex on ovarian functions and fecundity.

3.13.5.2 Effect on ovarian cell functions

In vitro experiments revealed a direct effect of Vitex extract on ovarian cells. It affected not only steroid hormones' release by isolated porcine ovarian granulosa cells, but in some experiments promoted their proliferation (PCNA accumulation) (Sirotkin et al., 2020a,b), apoptosis (accumulation of cytoplasmic apoptosis marker bax) and viability (Sirotkin et al., 2020b) as well. These observations suggest that Vitex could activate ovarian cell turnover and viability. Some experiments (Sirotkin et al., 2020a, 2021), however, did not confirm all these effects, which could be due to differences in initial state of ovarian cells.

Extract of vitex was able not only to affect basic ovarian cell functions, but also to prevent the effect of xylene (Sirotkin et al., 2020a) and copper nanoparticles (Sirotkin et al., 2020b) on these cells. On the contrary, vitex promoted the action of other environmental contaminant, benzene, on these cells (Sirotkin et al., 2021).

These observations demonstrate that Vitex can directly affect proliferation, apoptosis and viability of ovarian cells, as well as modify the effects of environmental contaminants on these processes.

3.13.5.3 Effect on reproductive hormones

The observed Vitex-induced changes in monkeys' reproductive in the events described above were associated with increased excretion of progestins (Higham et al., 2007; Lu et al., 2011), indicating that Vitex can promote progestin release. The ability of Vitex agnus-cactus extract to promote ovarian speroidogenesis was confirmed by increased plasma progesterone and estrogen level in plasma of rats treated with Vitex agnus cactus extract (Ibrahim et al., 2008) and by increased estradiol (but not progesterone or testosterone) release by porcine ovarian granulosa cells cultured with Vitex extract (Sirotkin et al., 2020b). In other experiments, however, this extract was able to reduce the release of both progesterone and estradiol by these cells (Sirotkin et al., 2020a, 2021). Vitex extract was able to reduce androgen release in women suffering from polycystic ovarian syndrome (Arentz et al., 2014).

Furthermore, treatment with Vitex extract reduced LH, FSH and prolactin in rat plasma (Ibrahim et al., 2008; Ahangarpour et al., 2016). In women, it reduced the output of prolactin, which is responsible for control of both lactation (Daniele et al., 2005; Anonimous, 2009; van Die et al., 2013; Arentz et al., 2014) and ovarian functions (Bouilly et al., 2012).

The available publications suggest the ability of Vitex to down-regulate pituitary hormones regulating ovarian functions, as well as to affect (mainly promote) ovarian steroidogenesis.

3.13.6 Mechanisms of action on female reproductive processes

There are only few, variable, and inconsistent information concerning mechanisms of Vitex agnus-cactus on female reproductive processes and their disorders. This can be due to the fact that Vitex has a number of biological active constituents, which can affect various targets via different extra- and intracellular mediators. In addition, a number of studies detected a number of processes and molecules affected by Vitex, but they provided no strong evidence that these molecules are mediators of Vitex action on particular processes. Nevertheless, they enable to formulate some hypothesis concerning mediators of Vitex action on various processes and illnesses.

Vitex action on release of **LH, FSH and prolactin** (Daniele et al., 2005; Ibrahim et al., 2008; Anonimous, 2009; van Die et al., 2013; Arentz et al., 2014; Ahangarpour et al., 2016) and **steroid hormones** (Higham et al., 2007; Lu et al., 2011; Arentz et al., 2014; Sirotkin et al., 2020a,b, 2021), which are the most known regulators of ovarian cell proliferation and apoptosis, ovarian foliculo- luteo- and oogenesis, fecundity and pregnancy (Sirotkin, 2014). Ovarian cell **proliferation:apoptosis** rate defines **viability** of ovarian cell population and fate of ovarian follicle (development and ovulation or atresia and degeneration) (Sirotkin, 2014). Furthermore, these hormones, like Vitex, are regulators and medicine in cases of osteoporosis induced by deficit of steroid hormones (Diezt et al., 2016), oligo/amenorrhea, hyperandrogenism and PCOS (Arentz et al., 2014). These facts suggest that Vitex can affect these processes through changes in pituitary and ovarian hormones. In addition, prolactin can mediate Vitex action on milk production (Daniele et al., 2005; Anonimous, 2009; van Die et al., 2013; Arentz et al., 2014) and cyclic mastalgia (Ooi et al., 2020). The ability of Vitex to prevent both reproductive aging and age-dependent decline in activity of **antioxidative** enzymes (Ahangarpour et al., 2016) indicates that some effect of this plant on reproduction could be due to the antioxidative properties of the plant molecules. **Dopaminergic** compounds presented in Vitex can explain its ability to improve premenstrual mastodynia and possibly also other symptoms of the premenstrual syndrome (Wuttke et l., 2003; Rafieian-Kopaei et al., 2017) and cyclic mastalgia (Ooi et al., 2020).

Therefore, the available information suggest that Vitex can influence various female reproductive processes through hormonal regulators (gonadotropins, prolactin, steroid hormones) and antioxidative enzymes. In addition, it can relieve the pain associated with reproductive cycles via dopamine receptors.

3.13.7 Application in reproductive biology and medicine

Vitex has been shown to be efficient for relief of many female reproduction-related conditions, including menstrual disorders (amenorrhoea, dysmenorrhoea), premenstrual syndrome and associated cyclic mastalgia, corpus luteum insufficiency and low progesterone production, hyperprolactinaemia, low fertility or infertility, acne, menopause and disrupted lactation, and menopause-related complaints. Although few randomized controlled trials exist, the German Commission E has approved the use of Vitex agnus-cactus for irregularities of the menstrual cycle, premenstrual disturbances and mastodynia (Daniele et al., 2005; van Die et al., 2013; Rafieian-Kopaei et al., 2017; Chan et al., 2018; Drugs and Lactation Database, 2018).

In addition, it may have beneficial effects for women with oligo/amenorrhea, hyperandrogenism and PCOS. Vitex agnus-castus was successfully used for treatment of patients suffered for PCOS manifesting in menstrual irregularity, acne, and hirsutism. After Vitex treatment, the patient regained menstrual cyclicity, and clinical and biochemical hyperandrogenism normalized (Arentz et al., 2014; Alois et al., 2019). Nevertheless, some authors find the available claims concerning Vitex agnus-cactus efficiency for treatment of premenstrual syndrome (Verkaik et al., 2017), PCOS (Arentz et al., 2014) and mastalgia (Ooi et al., 2020) overestimated due to the limited quality of pre-clinical data, high risk of bias, high heterogeneity of clinical evidence in the performed studies.

No Vitex action on menstrual bleeding has been observed (Mollazadeh et al., 2019).

Animal studies demonstrated the potential applicability of Vitex to prevent the symptoms of reproductive aging (disruption to the estrous cycle, damage to the uterus and ovarian tissues, female sex hormone deficiency, and atrophic endometrium; Ahangarpour et al., 2016).

Furthermore, the results of *in vitro* experiments (Sirotkin et al., 2020a,b) indicate that Vitex can prevent the negative action of environmental contaminants (xylene and copper nanoparticles) on ovarian cell functions and, probably, on female reproductive processes as a whole.

Vitex appears to not have been tested on ovarian cancer cells. Nevertheless, the ability of Vitex to up-regulate apoptosis and to down-regulate proliferation of nonovarian cancer cells (Chan et al., 2018) suggests its potential applicability for prevention or treatment of ovarian cancer.

In general, Vitex is well tolerated. The most frequent adverse events are nausea, headache, gastrointestinal disturbances, menstrual disorders, acne, pruritus, and erythematous rash; urticaria, fatigue, headache, dry mouth, tachycardia, nausea, and agitation. Nevertheless, these side-effects of Vitex were mild, reversible, and infrequent (they occurred in less than 2% of patients (Daniele et al., 2005; Anonimous, 2009; Mollazadeh et al., 2019). No drug interactions were reported. Nevertheless, because of concerning safety data and possible lactation suppression, Vitex should be avoided during lactation (Daniele et al., 2005; Drugs and Lactation Database, 2018).

Therefore, the available data show that Vitex agnus-cactus and/or its constituents could regulate female reproductive functions. Furthermore, it can be a relatively safe and efficient protector and remedia for treatment of various female reproductive disorders. Although few randomized controlled trials exist, the German Commission E has approved the use of *Vitex agnus-cactus* for irregularities of the menstrual cycle, premenstrual disturbances and mastodynia. In USA is considered a medical dietary supplement, which does not require approval from the U.S. Food and Drug Administration (Chan et al., 2018; Drugs and Lactation Database, 2018).

Taken together, Vitex agnus-cactus could be a promising and already applicable phytotherapeuticum which could prevent and mitigate a number of reproductive and nonreproductive disorders. The efficiency of Vitex treatment could be additionally improved by increasing its solubility and permeability by its application in nanoemulsion form (Piazzini et al., 2017).

3.13.8 Conclusions and possible direction of future studies

The data listed in this review demonstrate that Vitex agnus-cactus and its constituents (isoflavones and essential oils) affect a number of physiological action via multiple extra- and intracellular mechanisms of action. This makes it an efficient drug in both traditional and modern medicine for treatment of a number of illnesses. The main described target of Vitex agnus-cactus is, however, female reproduction. It can upregulate ovarian cycle and fecundity via a wide spectrum of mediators from pituitary and ovarian hormones up to intracellular regulators of proliferation, apoptosis, oxidation, and dopamine receptors. These effects determine its action as protector and medicine against a number of female reproductive disorders and reproduction-related health problems.

Nevertheless, a number of missing knowledge limits the understanding and application of Vitex-agnus-cactus. The action of only a few biologically active constituents of this plant has been studied. The targets, mechanisms of their action and their interrelationships in control of female reproduction and other processes are known insufficiently. For example, vitex action on oo- and embryogenesis appears to not have been investigated. The different (sometimes even opposite) action of Vitex on some reproductive processes require explanation. The performed clinical studies and their results are too variable and nonconclusive to enable wider therapeutical application of Vitex preparation for treatment of dysfunctions other than menstrual and breast disturbances. Vitex usage in assisted reproduction and farm animal production can be promising, but its application in these areas remains undiscovered as of yet. Therefore, Vitex agnus-cactus, as a regulator of numerous reproductive and nonreproductive processes, requires further basic and applied studies.

References

Ahangarpour A, Najimi SA, Farbood Y. Effects of Vitex agnus-castus fruit on sex hormones and antioxidant indices in a d-galactose-induced aging female mouse model. J Chin Med Assoc 2016;79(11):589−96. https://doi.org/10.1016/j.jcma.2016.05.006.

Alimohamadi R, Fatemi I, Naderi S, Hakimizadeh E, Rahmani MR, Allahtavakoli M. Protective effects of Vitex agnus-castus in ovariectomy mice following permanent middle cerebral artery occlusion. Iran J Basic Med Sci 2019;22(9):1097−101. https://doi.org/10.22038/ijbms.2019.31692.7625.

Food/medicinal herbs and their influence **Chapter | 3 241**

Allahtavakoli M, Honari N, Pourabolli I, Kazemi Arababadi M, Ghafarian H, Roohbakhsh A, Esmaeili Nadimi A, Shamsizadeh A. Vitex agnus castus extract improves learning and memory and increases the transcription of estrogen receptor α in Hippocampus of ovariectomized rats. Basic Clin Neurosci 2015;6(3):185−92.

Alois M, Estores IM. Hormonal regulation in pcos using acupuncture and herbal supplements: a case report and review of the literature. Integr Med 2019;18(5):36−9.

Anonimous. Vitex agnus-castus. Monogr Altern Med Rev 2009;14(1):67−71.

Antoine E, Chirila S, Teodorescu C. A patented blend consisting of a combination of *Vitex agnus-castus* extract, *Lepidium meyenii* (maca) extract and active folate, a nutritional supplement for improving fertility in women. Maedica 2019;14(3):274−9. https://doi.org/10.26574/maedica.2019.14.3.274.

Arentz S, Abbott JA, Smith CA, Bensoussan A. Herbal medicine for the management of polycystic ovary syndrome (PCOS) and associated oligo/amenorrhoea and hyperandrogenism; a review of the laboratory evidence for effects with corroborative clinical findings. BMC Compl Alternative Med 2014b;14:511. https://doi.org/10.1186/1472-6882-14-511.

Bouilly J, Sonigo C, Auffret J, Gibori G, Binart N. Prolactin signaling mechanisms in ovary. Mol Cell Endocrinol 2012;356(1-2):80−7. https://doi.org/10.1016/j.mce.2011.05.004.

Cahill DJ, Fox R, Wardle PG, Harlow CR. Multiple follicular development associated with herbal medicine. Hum Reprod 1994;9(8):1469−70. https://doi.org/10.1093/oxfordjournals.humrep.a138731.

Certo G, Costa R, D'Angelo V, Russo M, Albergamo A, Dugo G, Germanò MP. Anti-angiogenic activity and phytochemical screening of fruit fractions from Vitex agnus castus. Nat Prod Res 2017;31(24):2850−6. https://doi.org/10.1080/14786419.2017.1303696.

Chan EWC, Wong SK, Chan HT. Casticin from Vitex species: a short review on its anticancer and anti-inflammatory properties. J Integr Med 2018;16(3):147−52. https://doi.org/10.1016/j.joim.2018.03.001.

Chen SN, Friesen JB, Webster D, Nikolic D, van Breemen RB, Wang ZJ, Fong HH, Farnsworth NR, Pauli GF. Phytoconstituents from Vitex agnus-castus fruits. Fitoterapia 2011;82(4):528−33. https://doi.org/10.1016/j.fitote.2010.12.003.

Choudhary MI, Azizuddin JS, Nawaz SA, Khan KM, Tareen RB, Atta-ur-Rahman. Antiinflammatory and lipoxygenase inhibitory compounds from Vitex agnus-castus. Phytother Res 2009;23(9):1336−9. https://doi.org/10.1002/ptr.2639.

Daniele C, Thompson Coon J, Pittler MH, Ernst E. Vitex agnus castus: a systematic review of adverse events. Drug Saf 2005;28(4):319−32. https://doi.org/10.2165/00002018-200528040-00004.

Dietz BM, Hajirahimkhan A, Dunlap TL, Bolton JL. Botanicals and their bioactive phytochemicals for women's health. Pharmacol Rev 2016;68(4):1026−73. https://doi.org/10.1124/pr.115.010843.

Drugs and Lactation Database (LactMed) [Internet]. Chasteberry. Bethesda (MD): National Library of Medicine (US); 2018. PMID: 30000866.

Eftekhari MH, Rostami ZH, Emami MJ, Tabatabaee HR. Effects of "*Vitex agnus castus*" extract and magnesium supplementation, alone and in combination, on osteogenic and angiogenic factors and fracture healing in women with long bone fracture. J Res Med Sci 2014;19(1):1−7.

Gonçalves R, Ayres VFS, Carvalho CE, Souza MGM, Guimarães AC, Corrêa GM, Martins CHG, Takeara R, Silva EO, Crotti AEM. Chemical composition and antibacterial activity of the essential oil of *Vitex agnus-castus* L. (Lamiaceae). An Acad Bras Cienc 2017;89(4):2825−32. https://doi.org/10.1590/0001-3765201720170428.

Higham JP, Ross C, Warren Y, Heistermann M, MacLarnon AM. Reduced reproductive function in wild baboons (*Papio hamadryas anubis*) related to natural consumption of the African black plum (*Vitex doniana*). Horm Behav 2007;52(3):384—90. https://doi.org/10.1016/j.yhbeh.2007.06.003.

Ibrahim NA, Shalaby AS, Farag RS, Elbaroty GS, Nofal SM, Hassan EM. Gynecological efficacy and chemical investigation of *Vitex agnus-castus* L. fruits growing in Egypt. Nat Prod Res 2008;22(6):537—46. https://doi.org/10.1080/14786410701592612.

Ibrahim SRM, Ahmed N, Almalki S, Alharbi N, El-Agamy DS, Alahmadi LA, Saubr MK, Elkablawy M, Elshafie RM, Mohamed GA, El-Kholy MA. *Vitex agnus-castus* safeguards the lung against lipopolysaccharide-induced toxicity in mice. J Food Biochem 2019;43(3):e12750. https://doi.org/10.1111/jfbc.12750.

Ilhan S. Essential oils from *Vitex agnus castus* L. Leaves induces caspase-dependent apoptosis of human multidrug-resistant lung carcinoma cells through intrinsic and extrinsic pathways. Nutr Cancer 2020:1—9. https://doi.org/10.1080/01635581.2020.

Khalilzadeh E, Vafaei Saiah G, Hasannejad H, Ghaderi A, Ghaderi S, Hamidian G, Mahmoudi R, Eshgi D, Zangisheh M. Antinociceptive effects, acute toxicity and chemical composition of Vitex agnus-castus essential oil. Avicenna J Phytomed 2015;5(3):218—30.

Kikuchi H, Yuan B, Yuhara E, Imai M, Furutani R, Fukushima S, Hazama S, Hirobe C, Ohyama K, Takagi N, Toyoda H. Involvement of histone H3 phosphorylation via the activation of p38 MAPK pathway and intracellular redox status in cytotoxicity of HL-60 cells induced by Vitex agnus-castus fruit extract. Int J Oncol 2014;45(2):843—52. https://doi.org/10.3892/ijo.2014.2454.

Kuruüzüm-Uz A, Ströch K, Demirezer LO, Zeeck A. Glucosides from vitex agnus-castus. Phytochemistry 2003;63(8):959—64. https://doi.org/10.1016/s0031-9422(03)00285-1.

Lu A, Beehner JC, Czekala NM, Koenig A, Larney E, Borries C. Phytochemicals and reproductive function in wild female Phayre's leaf monkeys (*Trachypithecus phayrei crepusculus*). Horm Behav 2011;59(1):28—36. https://doi.org/10.1016/j.yhbeh.2010.09.012.

Moini Jazani A, Nasimi Doost Azgomi H, Nasimi Doost Azgomi A, Nasimi Doost Azgomi R. A comprehensive review of clinical studies with herbal medicine on polycystic ovary syndrome (PCOS). Daru 2019;27(2):863—77. https://doi.org/10.1007/s40199-019-00312-0.

Mollazadeh S, Mirghafourvand M, Abdollahi NG. The effects of *Vitex agnus-castus* on menstrual bleeding: a systematic review and meta-analysis. J Compl Integr Med 2019;17(1). https://doi.org/10.1515/jcim-2018-0053.

Moreno FN, Campos-Shimada LB, da Costa SC, Garcia RF, Cecchini AL, Natali MR, Vitoriano Ade S, Ishii-Iwamoto EL, Salgueiro-Pagadigorria CL. *Vitex agnus-castus* L. (verbenaceae) improves the liver lipid metabolism and redox state of ovariectomized rats. Evid Based Complement Alternat Med 2015;2015:212378. https://doi.org/10.1155/2015/212378.

Nasri S, Oryan S, Rohani AH, Amin GR. The effects of Vitex agnus castus extract and its interaction with dopaminergic system on LH and testosterone in male mice. Pak J Biol Sci 2007;10(14):2300—7. https://doi.org/10.3923/pjbs.2007.2300.2307.

Ohyama K, Akaike T, Hirobe C, Yamakawa T. Cytotoxicity and apoptotic inducibility of Vitex agnus-castus fruit extract in cultured human normal and cancer cells and effect on growth. Biol Pharm Bull 2003;26(1):10—8. https://doi.org/10.1248/bpb.26.10.

Ohyama K, Akaike T, Imai M, Toyoda H, Hirobe C, Bessho T. Human gastric signet ring carcinoma (KATO-III) cell apoptosis induced by Vitex agnus-castus fruit extract through intracellular oxidative stress. Int J Biochem Cell Biol 2005;37(7):1496—510. https://doi.org/10.1016/j.biocel.2005.02.016.

Ooi SL, Watts S, McClean R, Pak SC. Vitex agnus-castus for the treatment of cyclic mastalgia: a systematic review and meta-analysis. J Womens Health 2020;29(2):262–78. https://doi.org/10.1089/jwh.2019.7770.

Piazzini V, Monteforte E, Luceri C, Bigagli E, Bilia AR, Bergonzi MC. Nanoemulsion for improving solubility and permeability of Vitex agnus-castus extract: formulation and in vitro evaluation using PAMPA and Caco-2 approaches. Drug Deliv 2017;24(1):380–90. https://doi.org/10.1080/10717544.2016.1256002.

Rafieian-Kopaei M, Movahedi M. Systematic review of premenstrual, postmenstrual and infertility disorders of *Vitex agnus castus*. Electron Physician 2017;9(1):3685–9. https://doi.org/10.19082/3685.

Rani A, Sharma A. The genus Vitex: a review. Pharmacogn Rev 2013;7(14):188–98. https://doi.org/10.4103/0973-7847.120522.

Ricarte LP, Bezerra GP, Romero NR, Silva HCD, Lemos TLG, Arriaga AMC, Alves PB, Santos MBD, Militão GCG, Silva TDS, Braz-Filho R, Santiago GMP. Chemical composition and biological activities of the essential oils from *Vitex-agnus castus*, *Ocimum campechianum* and *Ocimum carnosum*. An Acad Bras Cienc 2020;92(1):e20180569. https://doi.org/10.1590/0001-3765202020180569.

Sehmisch S, Boeckhoff J, Wille J, Seidlova-Wuttke D, Rack T, Tezval M, Wuttke W, Stuermer KM, Stuermer EK. Vitex agnus castus as prophylaxis for osteopenia after orchidectomy in rats compared with estradiol and testosterone supplementation. Phytother Res 2009;23(6):851–8. https://doi.org/10.1002/ptr.2711.

Sirotkin AV, Macejková M, Tarko A, Fabova Z, Alwasel S, Harrath AH. Buckwheat, rooibos, and vitex extracts can mitigate adverse effects of xylene on ovarian cells in vitro. Environ Sci Pollut Res Int 2020. https://doi.org/10.1007/s11356-020-11082-7. Epub ahead of print. PMID: 33033927.

Sirotkin AV, Radosová M, Tarko A, Fabova Z, Martín-García I, Alonso F. Abatement of the stimulatory effect of copper nanoparticles supported on titania on ovarian cell functions by some plants and phytochemicals. Nanomaterials 2020b;10(9):1859. https://doi.org/10.3390/nano10091859.

Sirotkin AV, Macejková M, Tarko A, Fabova Z, Alrezaki A, Alwasel S, Harrath AH. Effects of benzene on gilts ovarian cell functions alone and in combination with buckwheat, rooibos, and vitex. Environ Sci Pollut Res Int 2021;28(3):3434–44. https://doi.org/10.1007/s11356-020-10739-7.

Šošić-Jurjević B, Ajdžanović V, Filipović B, Trifunović S, Jarić I, Ristić N, Milošević V. Functional morphology of pituitary -thyroid and -adrenocortical axes in middle-aged male rats treated with Vitex agnus castus essential oil. Acta Histochem 2016;118(7):736–45. https://doi.org/10.1016/j.acthis.2016.07.007.

van Die MD, Burger HG, Teede HJ, Bone KM. *Vitex agnus-castus* extracts for female reproductive disorders: a systematic review of clinical trials. Planta Med 2013;79(7):562–75. https://doi.org/10.1055/s-0032-1327831.

Verkaik S, Kamperman AM, van Westrhenen R, Schulte PFJ. The treatment of premenstrual syndrome with preparations of Vitex agnus castus: a systematic review and meta-analysis. Am J Obstet Gynecol 2017;217(2):150–66. https://doi.org/10.1016/j.ajog.2017.02.028.

Wuttke W, Jarry H, Christoffel V, Spengler B, Seidlová-Wuttke D. Chaste tree (*Vitex agnus-castus*) —pharmacology and clinical indications. Phytomedicine 2003;10(4):348–57. https://doi.org/10.1078/094471103322004866.

Chapter 4

Plant molecules and their influence on health and female reproduction

Chapter 4.1

Amygdalin

4.1.1 Introduction

Natural substances and alternative medications such as apricot seeds and their constituent amygdalin gained huge popularity in treating various diseases due to wide availability and relatively low cost (Salama et al., 2019).

Wild apricot (*Prunus armeniaca* L.) is an important fruit tree species. It has been used in folk medicine as a remedy for various diseases (Rai et al., 2016). Amygdalin, a biomolecule presents in the seeds of apricot and other fruits, gained wide popularity owing to its purported anticancer activity (Salama et al., 2019). Around 5000 years ago, the Egyptian papyri mentioned the beneficial use of the bitter almonds derivatives in treating skin tumors. Ancient Romans and Greeks connected some therapeutic properties to those derivatives, too (Enculescu, 2009). Thereafter, those derivatives came to be known as amygdalin, vitamin B17, or laetrile, and bitter almonds are considered one of the richest sources (Yamshanov et al., 2016; Del Cueto et al., 2017; Lv et al., 2017). Interestingly, some isolated populations and tribes all over the world such as the Abkhazians, the Hopi and Navajo Indians, the Hunzas, the Eskimos, and the Karakorum did not have the incidence of cancer cases. It turned out that they had in common a diet rich in amygdalin (Enculescu, 2009). Consequently, many researchers and scientists all over the globe carried out various studies and clinical trials to prove the anticancer activity of amygdalin. Despite the beneficial effect of cyanide against cancer, it may cause several harmful side effects and lead to toxicity (Milazzo and

Environmental Contaminants and Medicinal Plants Action on Female Reproduction
https://doi.org/10.1016/B978-0-12-824292-6.00004-0
Copyright © 2022 Elsevier Inc. All rights reserved.

245

246 Contaminants and Plants Action on Female Reproduction

Horneber, 2015; Salama et al., 2019), especially from amygdalin tablets ingestion (Dang et al., 2017).

The present review attempted to summarize the recent information concerning the properties, metabolism, physiological role, toxicity, and therapeutic potential of amygdalin with an emphasis on its action on female reproductive processes at various regulatory levels.

4.1.2 Provenance and properties

Amygdalin (C20H27NO11; Fig. 4.1) is an aromatic cyanogenic compound belonging to the subclass of carbohydrates and carbohydrate conjugates. The structure of amygdalin comprises one unit of benzaldehyde, one unit of hydrocyanic acid, and two units of glucose (Fukuda et al., 2003; Shi et al., 2019). Having a molecular weight of 457.432 g/mol (Chang et al., 2006), amygdalin is colorless with a melting point of 213°C, insoluble in nonpolar solvents like chloroform, and is highly soluble in ethanol and moderately soluble in water (Salama et al., 2019). It has D-mandelonitrile-beta-D-gentiobioside structure (Chang et al., 2006). However, the active form of amygdalin is R-amygdalin which is a right-handed structure which is its natural form (Salama et al., 2019). Although known as vitamin B-17 or laetrile, from a biomedical point of view they are not the same product as amygdalin. In fact, amygdalin is a cyanogenic glucoside, and its purified form is called laetrile which refers to the terms levorotatory and mandelonitrile. Laetrile is a semisynthetic cyanogenic

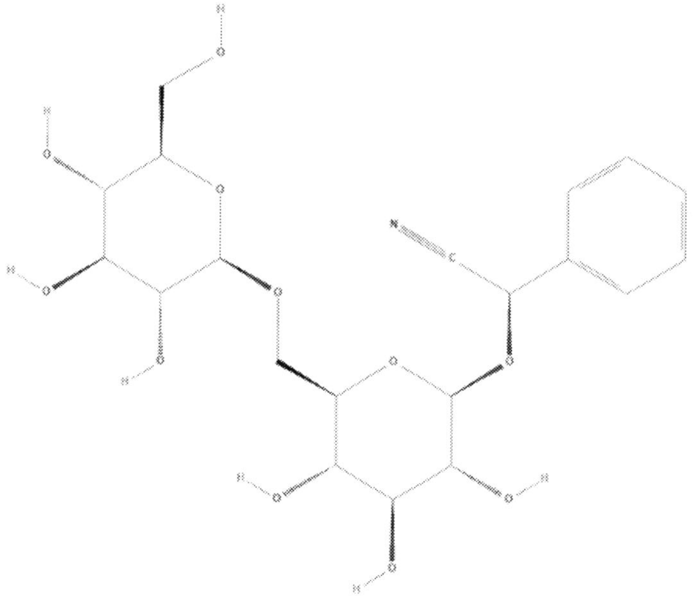

FIGURE 4.1 The chemical structures of amygdalin (PubChem, 2021).

Plant molecules and their influence on health **Chapter | 4 247**

glucuronide, and therefore it is structurally different from amygdalin (PDQ, 2002; Sauer et al., 2015; Shalayel, 2017; Salama et al., 2019). Vitamin B17 is also a misnomer as amygdalin is not considered as a vitamin in strict sense (European Food Safety Authoriry, 2016).

Amygdalin is abundantly present in the kernels of various species of Rosaceae family such as in the bitter seeds of apricots, apples, almonds, peaches, cherries, plums, grains, millets, sprouts, and nuts (Chang et al., 2006; Hwang et al., 2008; Jiagang et al., 2011; Song and Xu, 2014; Bolarinwa et al., 2014, 2015; Kolesar et al., 2015, 2018; Halenár et al., 2017; Lv et al., 2017; Kopčeková et al., 2018; Ayaz et al., 2020). Seeds contain amygdalin depending on the variety: apricot kernels and bitter almond kernels have a 3% −4% content of amygdalin by weight and it may even rise up to 8% in apricot seeds (Jiagang et al., 2011). Amygdalin content is particularly high (more than 5%) in bitter apricot cultivars (Kolesar et al., 2015; Halenár et al., 2017, 2018; Kopčekova et al., 2018; Kolesarova et al., 2020). Amygdalin contents of seeds from 15 varieties of apples ranged from 1 mg/g to 4 mg/g, however, the content is low in commercially available apple juice, ranging from 0.01 to 0.04 mg/mL for pressed apple juice and 0.001−0.007 mg/mL for long-life apple juice (Bolarinwa et al., 2015).

Amygdalin is degraded to cyanide by chewing or grinding (EFSA, 2016). Orally administered amygdalin is degraded into prunasin as the major metabolite by digestive enzymes after passing through the salivary and gastrointestinal phases. Prunasin is degraded into the mandelonitrile by β-glucosidase and then hydroxylated across the small intestinal wall, producing hydroxymandelonitrile (149 Da) (Shim and Kwon, 2010) under the glucosidase action, such as amygdalase and prunase, and ultimately decomposed into benzaldehyde and hydrogen cyanide (HCN) (Suchard et al., 1998; Chang et al., 2005; Do et al., 2006). Degradation of 1 g of amygdalin liberates 59 mg HCN which is present in its dissociated form as cyanide (EFSA, 2016; Halenar et al., 2017). Cyanide is of high acute toxicity in humans (EFSA, 2016).

Amygdalin itself is nontoxic, but its product HCN decomposed by some enzymes is a poisonous substance (Suchard et al., 1998). Recent studies have shown that HCN is released in normal cells, and therefore it may not be safe for the human body (Liczbiński and Bukowska, 2018). Serious side effects are caused by cyanide compounds liberated after amygdalin degradation (Sauer et al., 2015; Milazzo et al., 2015). Despite this fact, animal data does not provide any solid basis to support acute human health hazard assessment. Moreover, Bolarinwa et al. (2015) have reported that amygdalin contents of commercially available apple juices are unlikely to present health problems to consumers. The Panel on Contaminants in the Food Chain (CONTAM) of European Food Safety Authority (EFSA) (2016) concluded that the lethal dose is reported to be 0.5−3.5 mg/kg of body weight (b.w.). An acute reference dose of 20 μg/kg b.w. is derived from exposure of 0.105 mg/kg b.w. associated with a nontoxic blood cyanide level of 20 μM, applying an uncertainty factor

248 Contaminants and Plants Action on Female Reproduction

of 1.5 to account for toxicokinetic and of 3.16 to account for toxicodynamic interindividual differences (EFSA, 2016). The highest dose of amygdalin that does not cause any unacceptable side effects in mice, rabbits, and dogs is 3 g/kg injected intravenously and intramuscularly, and 0.075 g/kg when given orally. Also, the maximum tolerated dose of human intravenous injection of amygdalin is around 0.07 g/kg. Additionally, after treating mice by inhibiting the intestinal bacteria, the oral administration of 300 mg/kg did not lead to death, on the other hand, the mortality increased by 60% using the same dose in untreated mice. Moreover, systemic toxicity has been reported in humans following oral administration of amygdalin at a dose of 4 g per day through a period of half a month, or a month of intravenous injection. Nevertheless, after the cessation of amygdalin intake or when the daily oral dose is reduced to $0.6 \sim 1$ g, the toxicity disappeared. Furthermore, the response of the digestive system toxicity is more frequent and accompanied by changes of a trial premature beats (Song and Xu, 2014).

4.1.3 Physiological and therapeutic actions

Cyanogenic glycosides as natural plant toxicants (Bolarinwa et al., 2015) including amygdalin can cause potential negative effect on animals (Kolesar et al., 2018). Therefore, amygdalin present in apricot seeds might present a potential risk for animal health (Kolesar et al., 2018; Kolesarova et al., 2020). On the other hand, the beneficial properties of bitter apricot seeds with the content of amygdalin (Kopčeková et al., 2018) and therapeutic actions (Lv et al., 2017; Dogru et al., 2017; Salama et al., 2019; Elsaed, 2019) were confirmed. Numerous previous studies have documented that amygdalin has the antioxidative, antibacterial, antiinflammatory, immunoregulatory activities (Shi et al., 2019), antitumor effects (Ayaz et al., 2020; Zhou et al., 2020; Shi et al., 2019; Song and Xu, 2014; Makarević et al., 2014a,b) as well as improves the immune function of organisms (Baroni et al., 2005), and oxidative balance (Albogami et al., 2020). Additionally, *Semen armeniacae amarum* with amygdalin content exerts antitussive, antiasthmatic effects (Do et al., 2006) and analgesic properties (Hwang et al., 2008).

In vivo study was designed to reveal whether pure amygdalin or apricot seeds induce changes in overall health status of rabbit as a biological model. Short-term application of pure amygdalin intramuscularly injected or oral consumption of crushed bitter apricot seeds (*Prunus armeniaca* L.) did not represent a risk for animal health from the perspective of biochemical and hematological parameters, as well as antioxidant enzymes activity (Kovacikova et al., 2019), and endocrine profile (Halenar et al., 2017). Diverse studies have reported on the beneficial properties of bitter apricot seeds with the content of amygdalin on risk of cardiovascular diseases. Changes in lipid profile, such as regular intake of bitter apricot seeds, may be considered potentially useful for the prevention of cardiovascular diseases (Kopčeková et al., 2018). Amygdalin possesses therapeutic effect against atherosclerosis (Lv et al., 2017). The antioxidant activity of amygdalin is associated with a decrease in reactive oxygen species (ROS) production and protein and lipid oxidation. On the other hand, long-term administration of amygdalin at high doses can increase protein oxidation and lipid peroxidation as a result of its ability to compromise oxidative balance in the testicular tissue, as well as increase the ROS production, which can lead to cellular oxidative stress (Duracka et al., 2016). Long-term consumption of bitter apricot seeds with the amygdalin content administered orally might affect the liver microscopic structure in rabbits. However, the toxic effect could not be accurately corroborated, as in many cases changes were dose-dependent and not recorded at the highest dose (Kolesárová et al., 2020). Amygdalin possesses therapeutic effect in the management of autoimmune hepatitis (Elsaed, 2019). Amygdalin may have a protective effect in hyperoxia-induced premature lung injury (Chang et al., 2005).

Although amygdalin has a clear pharmacological activity, there is still little in-depth research on the pharmacological mechanism of this compound (Song et al., 2014). Amygdalin as a therapeutic agent is not used in more countries owing to insufficient clinical verification of its therapeutic efficacy and adverse side effects. It was suggested targeted enzyme/prodrug strategies as a means to improve the tumor selectivity of therapeutics with decreased side effects (Rooseboom et al., 2004). The character of amygdalin effect remains controversial; however, amygdalin has an important application value to systematically investigate the mechanism of its pharmacological activity and develop antitumor drugs.

4.1.4 Mechanisms of action

The toxicity of cyanide is largely attributed to the cessation of aerobic cell metabolism, which causes central nervous system and cardiovascular dysfunctions, by cellular hypoxia (Coentrão and Moura, 2011). Cyanide reversibly binds to ferric ion in cytochrome oxidase a3 within the mitochondria, effectively halting cellular respiration by blocking the reduction of oxygen to water (Chaouali et al., 2013; Hamel, 2011). Amygdalin-induced ROS production, and the subsequent benzaldehyde overproduction, may trigger protein oxidation before the lipid peroxidation process. Although low and medium doses (50 and 100 mg/kg) of orally administered amygdalin induce no toxicity in mice, amygdalin at a high dose (200 mg/kg) is capable of inducing toxicity and exerts negative effects on the oxidative balance of hepatic and testicular tissues with an obvious effect on the histopathological level in mice (Albogami et al., 2020).

The toxic effect of cyanide is associated with the cytochrome oxidase terminal in the mitochondrial respiratory pathway, which leads to creating hindrance in the ability of cells to use oxygen (Opyd et al., 2017). The following data gives the information that amygdalin can act as a prooxidant, modulating the oxidative balance instead of ROS overproduction (Yiğit et al., 2009). Unlike normal cells, in cancer cells, the high levels of ROS result in mitochondrial dysfunction and increased metabolism. This mechanism is related to the antitumor effects of amygdalin by triggering several ROS-induced cell death pathways of cancer cells (Galadari et al., 2017). The effects of amygdalin on oxidative stress parameters and oxidant and/or antioxidant properties of amygdalin and bitter or sweet apricot kernels have been described by previous authors (Yiğit et al., 2009; Abboud et al., 2018; Albogami et al., 2020; Duracka 2016). Intramuscular amygdalin administration cause changes in the oxidative profile of rabbits and may affect the testicular tissue in dose-dependent manner, acting as an antioxidant at low doses (0.6 mg/kg b.w.) while high doses (3.0 mg/kg b.w.) may compromise the delicate oxidative balance in male reproductive structures. The highest efficacy of orally administered amygdalin was noted at a moderate dose of 100 mg/kg by enhancement of antioxidant enzyme activities, including upregulation of the expression of glutathione peroxidase and superoxide dismutase, and decrease in lipid peroxidation levels in hepatic and testicular tissues thereby improving oxidative balance (Albogami et al., 2020).

The effect of amygdalin in the spinal cord is associated with suppression of proinflammatory cytokine release. Amygdalin also inhibits of a proto-oncogene c-Fos, tumor necrosis factor α (TNF-α), and interleukin-1β (IL-1β). However, the active form of amygdalin -1β (IL-1β) expression, which resulted in antiinflammatory and analgesic properties (Hwang et al., 2008).

The mechanism of action of antitumor activity is the ability of amygdalin to induce cell apoptosis through regulation apoptotic proteins, increasing the level of proapoptotic Bax proteins, inducing of activity of caspase-3, and decreasing the level of antiapoptotic protein B-cell lymphoma 2 (Bcl-2) (Hwang et al., 2008). Moreover, it can affect the cell cycle of cancer cells, reduce cell cycle activators, especially cyclin B, cyclin-dependent kinase 1 (Cdk1), E-cadherin, and N-cadherin and activate multiple cellular pathways, inhibit Akt/mammalian target of rapamycin (Akt-mTOR) signaling pathway thereby inhibit the proliferation of cancer cells (Makarević et al., 2014b; Qian et al., 2015; Juengel et al., 2016). Amygdalin exerted cytotoxic activities on breast cancer cells by induction of apoptosis of the cells. Amygdalin downregulated Bcl-2, upregulated Bcl-2-associated X protein (Bax), activated of caspase-3 and cleaved poly ADP-ribose polymerase (PARP). Amygdalin activated a proapoptotic signaling molecule p38 mitogen-activated protein kinases (p38 MAPK) in breast cancer cells. Treatment of amygdalin significantly inhibited the adhesion of breast cancer cells, in which integrin α5 may be involved (Lee and Moon, 2016). Also, amygdalin caused differential inhibition in the proliferation of the breast cancer cell lines (Abboud et al., 2018).

Amygdalin toxicity is related to cyanide mainly due to its affinity for the terminal cytochrome oxidase in the mitochondrial respiratory pathway, decreasing the tissue utilization of oxygen. Mechanism of amygdalin effect closely related with degradation of its molecule and from type of cells that are specialised to carry out a particular function.

4.1.5 Effects on female reproductive processes

There is evidence for the influence of amygdalin on reproductive functions in animals (Tanyildizi and Bozkurt, 2004; Halenár et al., 2013, 2016b; Kolesar et al., 2015, 2018; Duracka et al., 2016; Dogru et al., 2017), including female reproductive system (Dogru et al., 2017; Halenár et al. 2014, 2015, 2016a, 2017; Kolesar et al., 2015).

4.1.5.1 Effect on ovarian follicular cell functions

Previous studies by our group reported the *in vitro* and *in vivo* effects of amygdalin on endocrine regulation of ovarian functions (Halenár et al., 2017) and secretion activity of porcine ovarian granulosa cells focused particularly on the process of ovarian steroidogenesis (Halenár et al., 2014, 2015, 2016a, 2017). The dose-dependent modulatory effect of amygdalin was noted on steroid hormone secretion by ovarian cells (Halenár et al., 2014, 2015, 2016a). Amygdalin administration causes a dose-dependent stimulation of 17β-estradiol but not of progesterone release by porcine ovarian granulosa cells (Halenár et al., 2015). In addition, amygdalin application positively affects ovarian cell viability and stimulates testosterone release by porcine ovarian granulosa cells, thus, amygdalin may have the potential effect on cellular growth and the process of ovarian steroidogenesis *in vitro* (Halenár et al., 2016a). Furthermore, a possible involvement of natural substance amygdalin into the processes of steroidogenesis was suggested. Amygdalin alone or in combination with the mycotoxin deoxynivalenol exert the possible effects on the steroid hormone secretion (progesterone and 17β-estradiol) by porcine ovarian granulosa cells. Interestingly, solely the presence of pure amygdalin causes the stimulation of 17β-estradiol secretion, and amygdalin appears to be potential endocrine modulators in porcine ovaries. However, the results do not confirm the expected protective effect of pure amygdalin on mycotoxin-induced reprotoxicity (Halenár et al., 2014).

4.1.5.2 Effect on reproductive hormones

An *in vivo* study by our group was designed to reveal whether amygdalin is able to cause changes in the endocrine profile and thus alter the key reproductive and physiological functions, using rabbit as a biological model (Halenar et al., 2017). Plasma levels of steroid (progesterone, 17β-estradiol, testosterone), thyroid (triiodothyronine, thyroxine, thyroid-stimulating hormone), as well as anterior pituitary (prolactin, luteinizing hormone) hormones were assessed with no significant results. Intramuscular amygdalin application does not affect the plasma levels of selected endocrine regulators. Similarly, the oral form of amygdalin does not induce significant changes in the plasma levels of examined hormones (Halenar et al., 2017).

4.1.5.3 Effect on reproductive state

Amygdalin has an evident effect in the treatment of endometriosis, an aggressive disorder associated with infertility, pelvic pain, and intraabdominal adhesions in women of reproductive age (Dogru et al., 2017). The short-term intake of amygdalin at recommended doses 0.6 and 3.0 mg/kg b.w. does not represent a risk for animal health from the perspective of biochemical parameters including endocrine profile. No obvious beneficial or negative effects of amygdalin on the physiological functions of female rabbits were demonstrated and no clinically noticeable changes in the average body weight of experimental animals were observed (Halenár et al., 2017; Kovacikova et al., 2019).

Taken together, the earlier-mentioned findings indicate the influence of amygdalin on physiological processes including female reproduction at various regulatory levels via extra- and intracellular signaling pathways regulating secretory activity and steroidogenesis, as well as compromise the delicate oxidative balance in various cells and reproductive tissues.

4.1.6 Mechanisms of action on female reproductive processes

Amygdalin influences reproduction at various regulatory levels via extra- and intracellular signaling pathways regulating secretory activity, cell viability, steroidogenesis, proliferation, and apoptosis but evidence for the mechanism of amygdalin action on female reproductive functions are insufficient.

Data collected from already published articles revealed that **apoptosis** is a central process activated by amygdalin in cancer cells. It is suggested to stimulate apoptotic process by upregulating expression of proapoptotic protein Bax and caspase-3 and downregulating expression of antiapoptotic protein Bcl-2. It also promotes arrest of cell cycle in G0/G1 phase and decrease number of cells entering S and G2/M phases. Thus, it is proposed to enhance deceleration of cell cycle by blocking cell **proliferation** and growth (Saleem et al., 2018). Similarly, previous studies suggest that amygdalin regulates apoptosis-related proteins and signaling molecules and increased the expression level of proapoptotic protein Bax and decreased that of antiapoptotic Bcl-2. The level of procaspase-3 was decreased by amygdalin treatment (Lee and Moon, 2016). **β-glucosidase** can accelerate the hydrolysis of amygdalin into hydrogen cyanide, which can effectively kill tumor cells by inhibiting cytochrome C oxidase in mitochondria, resulting in a significant increase in the cell mortality rate (Blaheta et al., 2016).

The data suggests possible mechanism of amygdalin action in cancer cells mainly through the induction of oxidative stress, regulation of apoptosis-related proteins, signaling molecules and increased the expression level of proapoptotic protein Bax and decreased that of antiapoptotic Bcl-2 but further studies are needed.

4.1.7 Application in reproductive biology and medicine

The ability of amygdalin to promote release of ovarian hormones and to increase ovarian cell viability (Halenár et al., 2014, 2015, 2016a, 2017; Song and Xu, 2014; see earlier) indicates that it has potential to be biostimulator of female reproduction and fertility. Moreover, its applicability for treatment of endometriosis has been demonstrated (Dogru et al., 2017). The ability of amygdalin to suppress proliferation and viability and to promote apoptosis in different cancer cells suggests the applicability of amygdalin for treatment of cancer of reproductive system (Blaheta et al., 2016; Lee and Moon, 2016; Saleem et al., 2018). In addition, three analogues of amygdalin with cyanide group removed play a valuable role in improving the immune function of the organism (Baroni et al., 2005), which could affect reproductive functions.

Nevertheless, amygdalin application requires the optimal balance between its positive/therapeutic and negative/toxic effects. Little is known regarding the therapeutic doses of amygdalin. Treatment time and amygdalin application type can vary between humans and animal models. In addition, there have been reports of cyanide toxicity due to amygdalin uses (Albogami et al., 2020). Large doses may cause systemic toxicity, which limits its clinical application (Jaswal et al., 2018; Song et al., 2014). Specific activation of amygdalin by β-Glu in tumor tissue may be an effective method for increasing the killing effect (Rooseboom et al., 2004). The enzyme or its encoding gene is first delivered to the tumor site using a targeting carrier. After clearance of the enzyme from circulation, the prodrug is administered and then converted to an active anticancer drug, thus achieving enhanced anticancer efficacy, and decreased systemic toxicity (Sharma and Bagshawe, 2017; Zhang et al., 2015). A promising approach to reduce amygdalin toxicity is application of amygdalin analogues with removed cyanide group as safe and efficient drug (Baroni et al., 2005).

Understanding and changes in toxicity and efficacy of this compound may advance the development as a therapeutical agent and determination of safe and effective oral doses of amygdalin. Available data demonstrates that amygdalin can be a safe stimulator of female reproductive processes via the system of the hypothalamic-pituitary-ovarian axis, its hormones, growth factors and their receptors, and through the modulation of the oxidative balance. Furthermore, it can be a promising drug for prevention and treatment of endometriosis and malignant processes in female reproductive system and other organs.

4.1.8 Conclusions and possible direction of future studies

Amygdalin influences reproductive processes including female reproduction at various regulatory levels via extra- and intracellular signaling pathways regulating secretory activity, cell viability, steroidogenesis, proliferation, and

254 Contaminants and Plants Action on Female Reproduction

apoptosis. On the other hand, while being metabolized in the body, amygdalin releases significant amounts of cyanide, which may lead to acute health risks in those individuals who may be at risk. Despite some contradictions in the available data about benefit and toxic effects of amygdalin, its potential applicability at low doses may present as a promising tool for regulation of various reproductive processes including disease management primarily in cancer phytotherapy, animal production, medicine, and biotechnology. Furthermore, amygdalin could be promising drug for prevention and treatment of endometriosis and cancer.

However, further research involving carefully designed dose-response studies is required to overcome the possible side effects of amygdalin and assure its safety as a therapeutic agent. Furthermore, search for nontoxic amygdalin analogues or its carriers, which can bind and/or block its toxic cyanide group could be promising for safe application of this molecule in animal production and medicine.

References

Abboud MM, Awaida WA, Alkhateeb HH, Abu-Ayyad AN. Antitumor action of amygdalin on human breast cancer cells by selective sensitization to oxidative stress. Nutr Cancer 2018;71:483−90. https://doi.org/10.1080/01635581.2018.1508731.

Albogami S, Hassan A, Ahmed N, Alnefaie A, Alattas A, Alquthami L, Alharbi A. Evaluation of the effective dose of amygdalin for the improvement of antioxidant gene expression and suppression of oxidative damage in mice. PeerJ 2020;8:e9232. https://doi.org/10.7717/peerj.9232.

Ayaz Z, Zainab B, Khan S, Abbasi AM, Elshikh MS, Munir A, Al-Ghamdi AA, Alajmi AH, Alsubaie QD, Mustafa AE-ZMA. In silico authentication of amygdalin as a potent anticancer compound in the bitter kernels of family Rosaceae. Saudi J Biol Sci 2020;27:2444−51. https://doi.org/10.1016/j.sjbs.2020.06.041.

Baroni A, Paoletti I, Greco R, Satriano RA, Ruocco E, Tufano MA, Perez JJ. Immunomodulatory effects of a set of amygdalin analogues on human keratinocyte cells. Exp Dermatol 2005;14:854−9. https://doi.org/10.1111/j.1600-0625.2005.00368.x.

Blaheta RA, Nelson K, Haferkamp A, Juengel E. Amygdalin, quackery or cure? Phytomedicine 2016;23:367−76. https://doi.org/10.1016/j.phymed.2016.02.004.

Bolarinwa IF, Orfila C, Morgan MRA. Amygdalin content of seeds, kernels and food products commercially-available in the UK. Food Chem 2014;152:133−9. https://doi.org/10.1016/j.foodchem.2013.11.002.

Bolarinwa IF, Orfila C, Morgan MRA. Determination of amygdalin in apple seeds, fresh apples and processed apple juices. Food Chem 2015;170:437−42. https://doi.org/10.1016/j.foodchem.2014.08.083.

Chang L, Zhu H, Li W, Liu H, Zhang Q, Chen H. Protective effects of amygdalin on hyperoxia-exposed type II alveolar epithelial cells isolated from premature rat lungs in vitro. Zhonghua Er Ke Za Zhi 2005;43:118−23.

Chang HK, Shin MS, Yang HY, Lee JW, Kim YS, Lee MH, Kim J, Kim KH, Kim CJ. Amygdalin induces apoptosis through regulation of bax and bcl-2 expressions in human DU145 and LNCaP prostate cancer cells. Biol Pharm Bull 2006;29:1597−602. https://doi.org/10.1248/bpb.29.1597.

Chaouali N, Gana I, Dorra A, Khelifi F, Nouioui A, Masri W, Belwaer I, Ghorbel H, Hedhili A. Potential toxic levels of cyanide in almonds (*Prunus amygdalus*), apricot kernels (*Prunus armeniaca*), and almond syrup. ISRN Toxicol 2013;2013:610648. https://doi.org/10.1155/2013/610648.

Coentrão L, Moura D. Acute cyanide poisoning among jewelry and textile industry workers. Am J Emerg Med 2011;29:78—81. https://doi.org/10.1016/j.ajem.2009.09.014.

Dang T, Nguyen C, Tran PN. Physician beware: severe cyanide toxicity from amygdalin tablets ingestion. Case Rep Emerg Med 2017;2017:e4289527. https://doi.org/10.1155/2017/4289527.

Del Cueto J, Ionescu IA, Pičmanová M, Gericke O, Motawia MS, Olsen CE, Campoy JA, Dicenta F, Møller BL, Sánchez-Pérez R. Cyanogenic glucosides and derivatives in almond and sweet cherry flower buds from dormancy to flowering. Front Plant Sci 2017;8. https://doi.org/10.3389/fpls.2017.00800.

Do JS, Hwang JK, Seo HJ, Woo WH, Nam SY. Antiasthmatic activity and selective inhibition of type 2 helper T cell response by aqueous extract of semen armeniacae amarum. Immunopharmacol Immunotoxicol 2006;28:213—25. https://doi.org/10.1080/08923970600815253.

Dogru YH, Kunt Isguder C, Arici A, Zeki Ozsoy A, Bahri Delibas I, Cakmak B. Effect of amygdalin on the treatment and recurrence of endometriosis in an experimental rat study. Period Biol 2017;119:173—80. https://doi.org/10.18054/pb.v119i3.4767.

Duracka M, Tvrda E, Halenar M, Zbynovska K, Kolesar E, Lukac N, Kolesarova A. The impact of amygdalin on the oxidative profile of rabbit testicular tissue. MendelNet 2016;23:770—5.

EFSA. Acute health risks related to the presence of cyanogenic glycosides in raw apricot kernels and products derived from raw apricot kernels. EFSA J 2016;14:e04424. https://doi.org/10.2903/j.efsa.2016.4424.

Elsaed WM. Amygdalin (vitamin B17) pretreatment attenuates experimentally induced acute autoimmune hepatitis through reduction of CD4+ cell infiltration. Ann Anat 2019;224:124—32. https://doi.org/10.1016/j.aanat.2019.04.006.

Enculescu M. Vitamin B17/laetrile/amygdalin (a review). Bulletin of university of agricultural sciences and veterinary medicine Cluj-Napoca. Anim Sci Biotechnol 2009;66. https://doi.org/10.15835/buasvmcn-asb:66:1-2:3316.

Fukuda T, Ito H, Mukainaka T, Tokuda H, Nishino H, Yoshida T. Anti-tumor promoting effect of glycosides from Prunus Persica seeds. Biol Pharm Bull 2003;26:271—3. https://doi.org/10.1248/bpb.26.271.

Galadari S, Rahman A, Pallichankandy S, Thayyullathil F. Reactive oxygen species and cancer paradox: to promote or to suppress? Free Radic Biol Med 2017;104:144—64. https://doi.org/10.1016/j.freeradbiomed.2017.01.004.

Halenar M, Medvedová M, Maruniaková N, Kolesárová A. Amygdalin and its effects on animal cells. Special Issue BQRMF 2013;2:1414—23.

Halenár M, Medvedová M, Maruniaková N, Packová D, Kolesárová A. Dose-response of porcine ovarian granulosa cells to amygdalin treatment combined with deoxynivalenol. J Microbiol Biotechnol Food Sci 2014;3(2):77—9.

Halenar M, Medvedova M, Maruniakova N, Kolesarova A. Assessment of a potential preventive ability of amygdalin in mycotoxin-induced ovarian toxicity. J Environ Sci Health B 2015;50:411—6. https://doi.org/10.1080/03601234.2015.1011956.

Halenar M, Kovacikova E, Nynca A, Sadowska A. Stimulatory effect of amygdalin on the viability and steroid hormone secretion by porcine ovarian granulosa cells in vitro. J Microbiol Biotechnol Food Sci 2016a;05(1):44—6. https://doi.org/10.15414/jmbfs.2016.5.special1.44-46.

256 Contaminants and Plants Action on Female Reproduction

Halenár M, Tvrdá E, Slanina T, Ondruška Ľ, Kolesár E, Massányi P, Kolesárová A. In vitro effects of amygdalin on the functional competence of rabbit spermatozoa. Int J Anim Veterin Sci 2016b;10:712—6.

Halenar M, Chrastinova L, Ondruska L, Jurcik R, Zbynovska K, Tusimova E, Kovacik A, Kolesarova A. The evaluation of endocrine regulators after intramuscular and oral application of cyanogenic glycoside amygdalin in rabbits. Biologia 2017;72:468—74. https://doi.org/10.1515/biolog-2017-0044.

Hamel JA. Review of acute cyanide poisoning with a treatment update. Crit Care Nurse 2011;31:72—82. https://doi.org/10.4037/ccn2011799.

Hwang HJ, Kim P, Kim CJ, Lee HJ, Shim I, Yin CS, Yang Y, Hahm DH. Antinociceptive effect of amygdalin isolated from Prunus armeniaca on formalin-induced pain in rats. Biol Pharm Bull 2008;31:1559—64. https://doi.org/10.1248/bpb.31.1559.

Jaswal V, Palanivelu JCR. Effects of the gut microbiota on amygdalin and its use as an anti-cancer therapy: substantial review on the key components involved in altering dose efficacy and toxicity. Biochem Biophys Rep 2018;14:125—32. https://doi.org/10.1016/j.bbrep.2018.04.008.

Jiagang D, Li C, Wang H, Hao E, Du Z, Bao C, Lv J, Wang Y. Amygdalin mediates relieved atherosclerosis in apolipoprotein E deficient mice through the induction of regulatory T cells. Biochem Biophys Res Commun 2011;411:523—9. https://doi.org/10.1016/j.bbrc.2011.06.162.

Juengel E, Thomas A, Rutz J, Makarevic J, Tsaur I, Nelson K, Haferkamp A, Blaheta RA. Amygdalin inhibits the growth of renal cell carcinoma cells in vitro. Int J Mol Med 2016;37:526—32. https://doi.org/10.3892/ijmm.2015.2439.

Kolesar E, Halenár M, Kolesárová A, Massányi P. Natural plant toxicant — cyanogenic glycoside amygdalin: characteristic, metabolism and the effect on animal reproduction. J Microbiol Biotechnol Food Sci 2015;04(02):49—50. https://doi.org/10.15414/jmbfs.2015.4.special2.49-50.

Kolesar E, Tvrda E, Halenar M, Schneidgenova M, Chrastinova L, Ondruska L, Jurcik R, Kovacik A, Kovacikova E, Massanyi P, et al. Assessment of rabbit spermatozoa characteristics after amygdalin and apricot seeds exposure in vivo. Toxicol Rep 2018;5:679—86. https://doi.org/10.1016/j.toxrep.2018.05.015.

Kolesárová A, Džurňáková V, Michalcová K, Baldovská S, Chrastinová Ľ, Ondruška Ľ, Jurčík R, Tokárová K, Kováčiková E, Kováčik A, et al. The effect of apricot seeds on microscopic structure of rabbit liver. J Microb Biotech Food Sci 2020;10:321—4. https://doi.org/10.15414/jmbfs.2020.10.2.321-324.

Kopčeková J, Kolesárová A, Kováčik A, Kováčiková E, Gažarová M, Chlebo P, Valuch J, Kolesárová A. Influence of long-term consumption of bitter apricot seeds on risk factors for cardiovascular diseases. J Environ Sci Health B 2018;53:298—303. https://doi.org/10.1080/03601234.2017.1421841.

Kovacikova E, Kovacik A, Halenar M, Tokarova K, Chrastinova L, Ondruska L, Jurcik R, Kolesar E, Valuch J, Kolesarova A. Potential toxicity of cyanogenic glycoside amygdalin and bitter apricot seed in rabbits—health status evaluation. J Anim Physiol Anim Nutr 2019;103:695—703. https://doi.org/10.1111/jpn.13055.

Lee HM, Moon A. Amygdalin regulates apoptosis and adhesion in Hs578T triple-negative breast cancer cells. Biomol Ther 2016;24:62—6. https://doi.org/10.4062/biomolther.2015.172.

Liczbiński P, Bukowska B. Molecular mechanism of amygdalin action in vitro: review of the latest research. Immunopharmacol Immunotoxicol 2018;40:212—8. https://doi.org/10.1080/08923973.2018.1441301.

Lv J, Xiong W, Lei T, Wang H, Sun M, Hao E, Wang Z, Huang X, Deng S, Deng J, et al. Amygdalin ameliorates the progression of atherosclerosis in LDL receptor-deficient mice. Mol Med Rep 2017;16:8171—9. https://doi.org/10.3892/mmr.2017.7609.

Makarević J, Rutz J, Juengel E, Kaulfuss S, Tsaur I, Nelson K, Pfitzenmaier J, Haferkamp A, Blaheta RA. Amygdalin influences bladder cancer cell adhesion and invasion in vitro. PLoS One 2014a;9:e110244. https://doi.org/10.1371/journal.pone.0110244.

Makarević J, Rutz J, Juengel E, Kaulfuss S, Reiter M, Tsaur I, Bartsch G, Haferkamp A, Blaheta RA. Amygdalin blocks bladder cancer cell growth in vitro by diminishing cyclin A and Cdk2. PLoS One 2014b;9:e105590. https://doi.org/10.1371/journal.pone.0105590.

Milazzo S, Horneber M, Ernst E. Laetrile treatment for cancer. Cochrane Database Syst Rev 2015. https://doi.org/10.1002/14651858.CD005476.pub4.

National Center for Biotechnology Information. PubChem Compound Summary for CID 656516, Amygdalin. PubChem, https://pubchem.ncbi.nlm.nih.gov/compound/Amygdalin. Accessed June 1, 2021.

Opyd PM, Jurgoński A, Juśkiewicz J, Milala J, Zduńczyk Z, Król B. Nutritional and health-related effects of a diet containing apple seed meal in rats: the case of amygdalin. Nutrients 2017;9:1091. https://doi.org/10.3390/nu9101091.

PDQ Integrative, Alternative, and Complementary Therapies Editorial Board. Laetrile/amygdalin (PDQ®): health professional version. In: PDQ cancer information summaries. Bethesda (MD): National Cancer Institute (US); 2002.

Qian L, Xie B, Wang Y, Qian J. Amygdalin-mediated inhibition of non-small cell lung cancer cell invasion in vitro. Int J Clin Exp Pathol 2015;8:5363–70.

Rai I, Bachheti RK, Saini CK, Joshi A, Satyan RS. A review on phytochemical, biological screening and importance of Wild Apricot (*Prunus armeniaca* L.). Orient Pharm Exp Med 2016;16:1–15.

Rooseboom M, Commandeur JNM, Vermeulen NPE. Enzyme-catalyzed activation of anticancer prodrugs. Pharmacol Rev 2004;56:53–102. https://doi.org/10.1124/pr.56.1.3.

Salama R, Ramadan A, Alsanory T, Herdan M, Fathallah O, Alsanory A. Experimental and therapeutic trials of amygdalin. Int J Biochem Pharmacol 2019;1(1):21–6. https://doi.org/10.18689/ijbp-1000105.

Saleem M, Asif J, Asif M, Saleem U. Amygdalin from apricot kernels induces apoptosis and causes cell cycle arrest in cancer cells: an updated review. Anticancer Agents Med Chem 2018;18:1650–5. https://doi.org/10.2174/1871520618666180105161136.

Sauer H, Wollny C, Oster I, Tutdibi E, Gortner L, Gottschling S, Meyer S. Severe cyanide poisoning from an alternative medicine treatment with amygdalin and apricot kernels in a 4-year-old child. Wien Med Wochenschr 2015;165:185–8. https://doi.org/10.1007/s10354-014-0340-7.

Shalayel MHF. Beyond laetrile (vitamin B-17) controversy-antitumor illusion or revolution. Br Biomed Bull 2017;5(1):3.

Sharma SK, Bagshawe KD. Antibody directed enzyme prodrug therapy (ADEPT): trials and tribulations. Adv Drug Deliv Rev 2017;118:2–7. https://doi.org/10.1016/j.addr.2017.09.009.

Shi J, Chen Q, Xu M, Xia Q, Zheng T, Teng J, Li M, Fan L. Recent updates and future perspectives about amygdalin as a potential anticancer agent: a review. Cancer Med 2019;8:3004–11. https://doi.org/10.1002/cam4.2197.

Shim SM, Kwon H. Metabolites of amygdalin under simulated human digestive fluids. Int J Food Sci Nutr 2010;61:770–9. https://doi.org/10.3109/09637481003796314.

Song Z, Xu X. Advanced research on anti-tumor effects of amygdalin. J Canc Res Therapeut 2014;10(1):3–7. https://doi.org/10.4103/0973-1482.139743.

Suchard JR, Wallace KL, Gerkin RD. Acute cyanide toxicity caused by apricot kernel ingestion. Ann Emerg Med 1998;32:742–4. https://doi.org/10.1016/S0196-0644(98)70077-0.

258 Contaminants and Plants Action on Female Reproduction

Tanyildizi S, Bozkurt T. In vitro effects of linamarin, amygdalin and gossypol acetic acid on hyaluronidase activity, sperm motility and morphological abnormality in bull sperm. Turk J Vet Anim Sci 2004;28:819−24.

Yamshanov VA, Kovan'ko EG, Pustovalov YI. Effects of amygdaline from apricot kernel on transplanted tumors in mice. Bull Exp Biol Med 2016;160:712−4. https://doi.org/10.1007/s10517-016-3257-x.

Yiğit D, Yiğit N, Mavi A. Antioxidant and antimicrobial activities of bitter and sweet apricot (*Prunus armeniaca* L.) kernels. Braz J Med Biol Res 2009;42:346−52. https://doi.org/10.1590/S0100-879X2009000400006.

Zhang J, Kale V, Chen M. Gene-directed enzyme prodrug therapy. AAPS J 2015;17:102−10. https://doi.org/10.1208/s12248-014-9675-7.

Zhou J, Hou J, Rao J, Zhou C, Liu Y, Gao W. Magnetically directed enzyme/prodrug prostate cancer therapy based on β-glucosidase/amygdalin. Int J Nanomed 2020;15:4639−57. https://doi.org/10.2147/IJN.S242359.

Chapter 4.2

Apigenin

4.2.1 Introduction

In the recent decades, phytotherapy has faced its comeback and increased popularity. One of the promising plant molecules, which attract attention of scientists, clinical doctors, and the common public, is a flavone apigenin. The current knowledge concerning its physiological and therapeutic effects are summarized in several recent reviews (Martinez et al., 2019; Salehi et al., 2019; Imran et al., 2020), but the available data concerning its action on reproductive processes have not been reviewed yet.

The present chapter is aimed at creating a snapshot of the available information concerning production, metabolism, general physiological, and therapeutic effects of apigenin. The main purpose of this review, however, is to describe, summarize and analyze the current knowledge concerning apigenin effects on female reproductive processes and their dysfunctions, which have not been reviewed yet, but which could be useful for better understanding and application of apigenin in reproductive biology and medicine.

4.2.2 Provenance and properties

Apigenin isoflavone ($C15H10O5$, 5,7-dihydroxy-2-(4-hydroxyphenyl)chromen-4-1, Fig. 4.2) is a plant trihydroxyflavone. That is flavone substituted by hydroxy groups at positions $4'$, 5 and 7. It is produced in high amounts by celery, parsley, chamomile and a number of other food and medicinal plants. Plants synthesize it probably as a protector against harmful environmental factors and pathogens and as a regulator of metabolism (Tang et al., 2017;

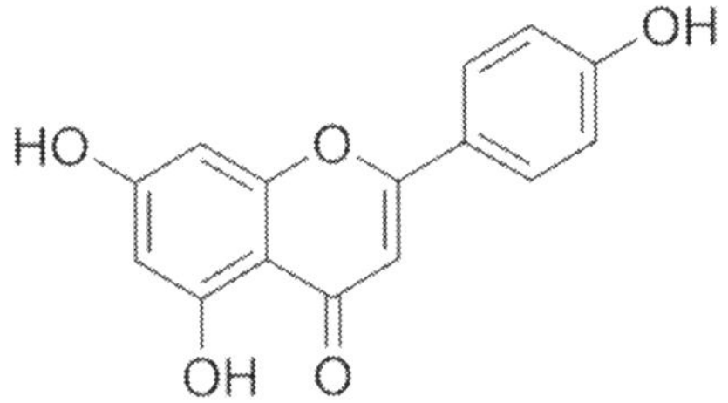

FIGURE 4.2 The chemical structure of apigenin. *From ww.reserachgate.com.*

Salehi et al., 2019). Usually, apigenin is presented in plants as O- and C-glucosides, while O-glucosides are more common, and they are usually better absorbed as C-glucosides (Hostetler et al., 2017). On the other hand, in some plants apigenin is presented also in the form of aglycone and/or its O-methyl ethers or acetylated derivatives (Salehi et al., 2019). After consumption of apigenin-containing plant, apigenin glycoside undergoes recirculation, hydrolysis, and reduction to the monoglycoside, and conjugation to form bioavailable glucuronide in liver (Tang et al., 2017), stomach (Hostetler et al., 2017), or gut microflora (Wang et al., 2019). The ingested apigenin can be either excreted unabsorbed or rapidly metabolized after absorption. Therefore, the bioavailability of natural apigenin is relatively low. The absorbed apigenin in blood circulation and tissues is mainly in the form of glucuronide, sulfate conjugates, or luteolin (Hostetler et al., 2017; Tang et al., 2017). The forms, metabolism, and therefore apigenin bioavailability and physiological effects are dependent on the plant—source of apigenin, species, and organs—the targets of apigenin and on their physiological state (Hostetler et al., 2017).

4.2.3 Physiological and therapeutical actions

The best-known and most important of the physiological and medicinal effects of apigenin are its effects on oxidative processes, viability (Yan et al., 2017; Imran et al., 2020), and inflammation (Martinez et al., 2019) in various cells and organs. These abilities determine numerous medicinal effects of apigenin. The preclinical and clinical studies demonstrated potential applicability of apigenin for treatment of numerous physiological disorders including cancer, disorders of nervous system, diabetes, cardiovascular, infectious, endocrine, and other diseases (Salehi et al., 2019).

For example, apigenin, as a dietary supplement or as adjuvant chemotherapeutic agent, was reported to suppress various human cancers *in vitro* and *in vivo* (Griffiths et al., 2016; Yan et al., 2017; Ozbey et al., 2018; Salehi et al., 2019; Imran et al., 2020; Suraweera et al., 2020; Ahmed et al., 2021). On the other hand, apigenin effects can be dose-dependent: at low concentrations it was able to promote, but in higher doses it suppressed growth and proliferation of cultured cancer cells (Suraweera et al., 2020).

Apigenin is considered as a neuroprotective molecule, which could be promising for treatment of neurodegenerative disorders like Alzheimer's and Parkinson's diseases, multiple sclerosis (Nabavi et al., 2018; Ginwala et al., 2019; Salehi et al., 2019), cerebral ischemia (Long et al., 2020), depression and insomnia (Salehi et al., 2019).

There are preclinical and clinical studies demonstrating applicability of apigenin for prevention and treatment of diabetes (Wang et al., 2014; Dewanjee et al., 2020), cardiovascular diseases (Griffiths et al., 2016) including hypertension (Ajebli et al., 2020) and endometriosis (Meresman et al., 2021).

The apigenin's antibacterial, antifungal, and antiparasitic capability has been reported too (Wang et al., 2019; Rossi et al., 2020). Furthermore, its effect on gut microflora involved in control of immune responses, nutrition, metabolism and obesity has been proposed (Wang et al., 2019). Due to its antimicrobial properties, apigenin was proposed as an alternative to antibiotics in animal production (Rossi et al., 2020).

Finally, apigenin could promote testosterone production by testicular Leydig cells. This effect indicates its potential applicability for masculine reproductive aging and age-dependent hypogonadism (Martin et al., 2020).

Similar antioxidative, antiinflammatory, anticancer, neuron-protective, hepato-protective, and cardioprotective properties, as well as the ability to reduce fat stores and glucose metabolism, were reported for vitexin, an apigenin-8-C-glucoside (Peng et al., 2021). It is however possible that these effects were induced not by the apigenin-8-C-glucoside itself, but by its metabolite apigenin or by other metabolites.

The toxicity of apigenin for healthy cells is relatively low (Matsuo et al., 2005; Yan et al., 2017; Ginwala et al., 2019). No mutagenic effect of apigenin has been found (Ginwala et al., 2019). On the other hand, apigenin can interact with other drugs, which should be taken into account before its application (Tang et al., 2017).

The limited absorption and bioavailability of natural apigenin forced the development of its novel carriers or chemical analogues with higher bioavailability, reduced degradation and increased biological efficiency (Tang et al., 2017). Such nanoformulation of apigenin has been developed for treatment of cancer (Pal et al., 2017; Tang et al., 2017; Qi et al., 2019), diabetes (Dewanjee et al., 2020) and cerebral ischemia (Long et al., 2020).

Therefore, the available reports demonstrated the abilities of apigenin to reduce oxidative stress, inflammation, and cell proliferation and to promote cell apoptosis. Due to these properties, apigenin can be applicable for treatment of numerous physiological disorders including cancer, disorders of nervous system, diabetes and other metabolic disorders, infections and testicular androgen deficiency.

4.2.4 Mechanisms of action

The most of numerous physiological and medicinal effects of apigenin are defined by its abilities to reduce reactive oxygen species/oxidative stress and cell proliferation, to promote their apoptosis (Yan et al., 2017; Imran et al., 2020), and to suppress inflammatory processes (Martinez et al., 2019) in various organs.

For example, apigenin was reported to suppress various kinds of cancers by multiple mechanisms, such as triggering cell apoptosis and autophagy, inducing cell cycle arrest, suppressing cell migration and invasion, stimulating an immune response and reducing cancer cell viability and resistance to chemo- and radiotherapeutical agents (Yan et al., 2017; Imran et al., 2020).

Apigenin effects on cancer cell proliferation and apoptosis can be mediated by the PI3K/AKT, MAPK/ERK, JAK/STAT, NF-κB, Wnt/β-catenin, and Nrf2/ARE pathways (Yan et al., 2017; Ozbey et al., 2018; Imran et al., 2020; Suraweera et al., 2020; Ahmed et al., 2021), which in turn upregulate the promoters of extrinsic (caspase-3, caspase-8, and TNF-α) and intrinsic (cytochrome c, Bax, and caspase-3) apoptosis pathways and downregulate the regulators of cancer cell movement and invasion (matrix metallopeptidases-2, -9, snail, and slug) (Imran et al., 2020). Furthermore, the molecular mechanism of inflammation driven cancer is the complex interplay between oncogenic and tumor suppressive transcription factors, which include FOXM1, NF-kB, STAT3, Wnt/β- Catenin, HIF-1α, NRF2, androgen, and estrogen receptors (Ginwala et al., 2019; Martinez et al., 2019). In particular, the antiinflammatory action of apigenin could be due to its downregulation of transcription factors NF-κB and NLRP3, which modulate differentiation, proliferation, activation of immune cells and enhance regulatory T-cell generation. Moreover, apigenin inhibits production of proinflammatory cytokines, tumor necrosis factor-α (TNF-α), interleukines (IL) IL-1β, IL-2, IL-6, IL-17A, and reactive proinflammatory oxygen and nitrogen species (Martinez et al., 2019). In addition, micro-RNAs can be mediators of apigenin on cancer cells as well. At least, apigenin can upregulate microRNAs miR-520b and miR-101, which in turn can inhibit tumor growth and promote the anticancer effects of miR-423-5p inhibitors or miR-138 mimics (Ozbey et al., 2018).

The neuroprotective effects of apigenin, including its benefits against Alzheimer's and Parkinson's diseases and multiple sclerosis, could be due to its antioxidant and antiinflammatory properties, including suppression of activity of circulating lymphocytes, monocytes/macrophages, and dendritic cells, causing edema, further inflammation, and demyelination (Nabavi et al., 2018; Ginwala et al., 2019).

Therefore, a number of physiological and therapeutic effects of apigenin could be mediated by the same or similar mechanisms. Nevertheless, some specific apigenin effects can be mediated by mediators specific for particular process. For example, the antihypertensive vasorelaxing effects of apigenin can be mediated by its direct influence on intracellular signaling molecules in vascular cells and extracellular regulators of renin-angiotensin system or diuresis (Ajebli et al., 2020). The antidiabetic action of apigenin can be due to its ability to inhibit nitric oxide and α-glucosidase involved in the digestion of carbohydrates (Wang et al., 2014). The stimulatory action of apigenin on testicular steroidogenesis can be mediated by promotion of expression of steroidogenic acute regulatory protein (StAR), which upregulates the entry of cholesterol into the mitochondria, leading to increased testosterone production in Leydig cells (Martin et al., 2020). The inhibitory action of apigenin on steroid-dependent breast cancer could be mediated by its effect on production of estrogen and estrogen receptor (Scherbakov et al., 2019).

Therefore, the majority of physiological and therapeutic effects of apigenin could be mediated by the same or similar mechanisms simultaneously regulating cell proliferation, apoptosis, inflammation, accumulation of reactive oxygen species, and angiogenesis. On the other hand, apigenin action on some particular target processes could be mediated by process-specific signaling molecules regulating water and carbohydrate metabolism and steroid hormone production and reception.

4.2.5 Effects on female reproductive processes

4.2.5.1 Effect on ovarian and reproductive state

Injections of apigenin to rats increased the number of their ovarian follicles at different stages, number of ovulations/corpora lutea and ovulated oocytes (Soyman et al., 2017; Darabi et al., 2020; Talebi et al., 2020). Histological studies demonstrated that apigenin can decrease the number of ovarian cysts and the thickness of theca layer and increased the thickness of granulosa layer (Darabi et al., 2020).

Furthermore, several studies demonstrated the ability of apigenin to mitigate and prevent the morphological pathological changes in the ovary induced by cancer (Chen et al., 2012; Tavsan and Kayali, 2019; Talebi et al., 2020), polycystic ovarian syndrome (Darabi et al., 2020), and ischemia (Soyman et al., 2017).

264 Contaminants and Plants Action on Female Reproduction

Apigenin could affect not only ovarian, but also uterine status. At least it was able to inhibit contractions of isolated murine uterine tissue (Suhas et al., 2018). On the other hand, *in vivo* experiments did not reveal the influence of apigenin on rat uterine weight, epithelial cell height, and vaginal opening (Barlas et al., 2014). Jiang et al. (2018) reported the ability of apigenin to prevent rat uterine lesions induced by endometritis.

These observations indicate the stimulatory influence of apigenin on female ovarian functions and its action on uterine state. The stimulatory action of apigenin on female ovarian functions has been confirmed by apigenin influence on ovarian cell state and functions, reproductive hormones, embryogenesis and resistance of reproductive system to adverse external factors described later.

4.2.6 Effect on ovarian and uterine cell functions

There are reports of apigenin's ability to suppress apoptosis and to promote the proliferation and viability of healthy ovarian cells. At least, injections of apigenin reduced the expression of apoptotic and increased the expression of antiapoptotic markers in rat ovarian cells (Talebi et al., 2020). In some *in vitro* experiments of Sirotkin et al. (2020) apigenin addition to cultured porcine ovarian granulosa cells promoted their viability and proliferation (accumulation of PCNA), but not the accumulation of apoptosis marker bax. This effect, however, has not been observed in other similar experiments (Sirotkin et al., 2021).

In cultured ovarian cancer cells, apigenin decreased cell renewal (Tang et al., 2015), cell viability and growth, arrested cell cycle at different phases and promoted cell apoptosis (Chen et al., 2012; Suh et al., 2015; Pal et al., 2017; Ittiudomrak et al., 2019; Tavsan and Kayali, 2019; Abd Ghani et al., 2020).

The available reports suggest the stimulatory action of apigenin on healthy ovarian cells (promotion of proliferation and viability and suppression of their apoptosis) and the opposite, suppressive apigenin action on ovarian cancer cells. These observations indicate a different character and mechanisms of apigenin action on healthy and cancer ovarian cells, as well as the potential applicability of apigenin for support of healthy ovarian functions and for suppression of ovarian cancerogenesis.

Apigenin could affect not only ovarian, but also uterine cells. At least apigenin was able to inhibit contractions of isolated murine uterine tissue (Suhas et al., 2018) and to suppress proliferation of cultured uterine Ishikawa cells (Dean et al., 2018). On the other hand, *in vivo* experiments did not reveal the influence of apigenin on rat uterine weight, epithelial cell height and vaginal opening (Barlas et al., 2014). These observations indicate that apigenin could suppress uterine cell functions, although this action requires confirmation by further, *in vivo*, studies.

4.2.7 Effect on oocytes and embryos

As it was mentioned earlier, in the experiments of Talebi et al. (2020) apigenin treatment was able to increase the number of ovulated oocytes in rats, but their quality was not assessed in this study. The ability of apigenin to modulate GABA receptors in *Xenopus laevis* oocytes (Abdullah et al., 2013) indicates its potential influence on some oocyte functions, but such action has not been examined yet.

Apigenin was toxic for cultured human lung embryonic firbroblasts (Matsuo et al., 2005). On the other hand, culture of mouse embryos with apigenin increased their quality, viability and developmental capacity. These changes were associated with decreased occurrence of apoptosis in blastomeres (Safari et al., 2018). These observations suggest the positive influence of apigenin on embryogenesis.

These observations, despite their insufficiency and variability, indicate that apigenin could be a potential stimulator of oogenesis, embryogenesis and fecundity.

4.2.8 Effect on reproductive hormones

The current literature contains a number of evidence of apigenin action on hormones produced by pituitary, ovaries and immune system, which are known regulators of reproductive processes (Sirotkin, 2014).

In vivo studies demonstrated the ability of apigenin injections to decrease LH and to increase FSH level in rat plasma. These injections increased progesterone and decreased testosterone and estradiol concentration (Darabi et al., 2020).

In *in vitro* experiments of Sirotkin et al. (2020a,b), addition of apigenin to cultured porcine granulosa cells promoted estradiol and testosterone and suppressed progesterone output. In other experiments (Sirotkin et al., 2021) on the same model addition of apigenin promoted estradiol, but not progesterone and testosteron release.

Talebi et al. (2020) reported that injections of apigenin to rats increased the plasma level of anti-Mullerian hormone — marker and promoter of ovarian folliculogenesis and follicular reserve. On the other hand, Soyman et al. (2017) did not observe such apigenin influence on rat blood anti-Mullerian hormone.

Apigenin administration decreased production of the inflammatory cytokines tumor necrosis factor-α (TNF-α) and interleukin-6 (IL-6) in rat plasma (Darabi et al., 2020), but it did not affect IL-6 production by cultured ovarian cancer cells (Suh et al., 2015).

Finally, apigenin downregulated production of vascular endothelial growth factor (VEGF) by ovarian cancer cells (Chen et al., 2012).

The available data demonstrate that apigenin could stimulate the release of gonadotropins, steroid hormones and anti-Mullerian hormone, which are known as endocrine promoters and markers of ovarian follicular reserve, folliculogenesis and fecundity. On the contrary, apigenin could be an inhibitor of cytokines—physiological promoters of inflammatory processes and inhibitors of female reproductive processes (Sirotkin, 2014). These data are in line with the data on stimulatory and antiinflammatory action of apigenin on healthy ovarian cells functions. Furthermore, they indicate that these hormones could be extracellular mediators of apigenin action on healthy ovaries.

On the other side, the inability of apigenin to suppress the proinflammatory and antireproductive cytokine IL-6 (Suh et al., 2015) and to downregulate the production of endocrine promoter of tumor angiogenesis VEGF (Chen et al., 2012) indicates that apigenin might suppress ovarian cancer development through this cytokine and growth factor.

4.2.9 Effect on response to adverse external factors

In vivo experiments on rats demonstrated that apigenin is able to prevent the ovarian destruction induced by chemotherapy (Talebi et al., 2020), ischemia (Soyman et al., 2017) and the toxic influence of H_2O_2 and actinomycin (Safari et al., 2018). Moreover, *in vitro* studies revealed that apigenin can protect cultured porcine ovarian granulosa cells from the influence of benzene (Sirotkin, 2020a), toluene (Sirotkin et al., 2021) and copper nanoparticles (Sirotkin et al., 2020b). On the other hand, apigenin analogue was able to promote toxic action of chemotherapeutic agent paclitaxel on ovarian cancer cells (Pal et al., 2017).

The reports listed earlier demonstrated that apigenin can not only stimulated healthy ovarian cell functions and inhibit cancer ovarian cell functions. The present reports indicate that it can be also a natural protector of healthy ovarian cells, but, on the contrary, in cancer cells it can additionally increase the toxic effects of anticancer drug. These observations suggest the potential applicability of apigenin as a stimulator and protector of healthy ovarian cells and an inhibitor of ovarian cell functions and their responsibility to anticancer treatment.

4.2.10 Mechanisms of action on female reproductive processes

Apigenin action on nonreproductive processes could be mediated by multiple signaling molecules and signaling pathways (see earlier). The results of many *in vitro* and *in vivo* studies suggested the involvement of some (but not all) of these signaling pathways in mediating apigenin action on reproductive processes.

Administration of apigenin increased the total **antioxidant capacity** and activity of antioxidant enzyme superoxide dismutase in rat plasma and decreased its oxidative status. It indicates that one of the mechanism of positive action of apigenin on rat hormonal status and ovarian functions could be due to its antioxidant activity (Soyman et al., 2017; Darabi et al., 2020). On the contrary, apigenin induced accumulation of reactive oxygen species in cultured embryonic fibroblasts (Matsuo et al., 2005) and cultured cancer cells, which was associated with a decrease in their viability (Pal et all., 2017; Tavsan and Kayali, 2019). These observations indicate the antioxidant action of apigenin on healthy ovarian cells *in vivo* and its prooxidant toxic effect on cultured embryonal and cancer cells.

It is not to be excluded that apigenin can act as an antioxidant also on healthy gestational tissues. At least, it reduced the release of 8-isoprostane, a marker of oxidative stress, in cultures of healthy placenta and fetal membranes and myometrial cells (Lim et al., 2013). Therefore, apigenin can prevent oxidative stress in healthy reproductive tissues and promote it in ovarian cancer cells.

Apigenin addition to ovarian cancer cell culture reduced the intracellular expression of Axl and Tyro3 receptors - the promoters of cell **viability and proliferation** (Suh et al., 2015) and stabilized the antiproliferative transcription factor p53 (Chen et al., 2012). These observations demonstrate that the suppressive action of apigenin on cancer cell viability can be mediated by their action on regulators of proliferation and apoptosis.

Reactive oxygen species induce cytoplasmic/intrinsic **apoptosis** (accumulation of apoptosis marker and promoter bax and reduction in mitochondrial membrane potential) in ovarian cancer cells. This effect was promoted by apigenin analogues (Qi et al., 2019). Apigenin expresses high ability to bind intracellular inhibitors of apoptosis, bcl-2 and bcl-xl (Abd Ghani et al., 2020). In experiments of Suh et al. (2015) and Abd Ghani et al. (2020) apigenin addition to cultured ovarian cancer cells inhibited the expression of both bcl-2 and bcl-xl, as well as the phosphorylation of Akt, the next intracellular antiapoptotic molecule. In the similar experiments, apigenin promoted ovarian cancer cell apoptosis via upregulation promoters of apoptosis TRAIL (TNF receptor apoptosis-inducing ligand), transcription factor p53, caspase-3 and -9 (Chen et al., 2012; Ittiudomrak et al., 2019; Tavsan and Kayali, 2019), although the expression of apoptosis inhibitor bcl-2 was upregulated too (Ittiudomrak et al., 2019).

In contrast to cancer cells, in healthy ovarian cells apigenin did not induce accumulation of apoptosis marker and promoter bax (pig: Sirotkin et al., 2020, 2021) and even suppressed bax and promoted expression of antiapoptotic bcl-2 (rat: Talebi et al., 2020).

The available data suggest that apigenin can promote apoptosis in cancer cells without influencing it in healthy ovarian cells via up- and downregulation of various pro- and antiapoptotic molecules.

In contrast to cancer cells, in healthy porcine ovarian cells apigenin promoted the accumulation of proliferation marker and promoter (PCNA) (Sirotkin et al., 2020a,b) and viability (Sirotkin, 2020b). These observations suggest that apigenin has the opposite influence on proliferation, apoptosis and viability of healthy and cancer ovarian cells. The ability of apigenin to support healthy ovarian cells and to suppress cancer ovarian cells indicates the applicability of apigenin as a specific anticancer drug combatting cancer and not destroying, but stimulating healthy ovarian cells.

Cultured ovarian **cells renewal** is maintained by synergistic action of proteins casein kinase 2 (CK2α) and glioma-associated oncogene 1 (Gli1). Apigenin, like genomic inhibitors of their synthesis, inhibited both the expression of these proteins and the self-renewal of cultured cancer cells. These observations demonstrate that apigenin can suppress self-renewal of ovarian cancer cells via CK2α and Gli1 (Tang et al., 2015). The ability of apigenin on renewal of healthy ovarian cells seems to not have been studied yet.

Injections of apigenin decreased the concentration markers of **inflammation** in rat blood, suggesting that apigenin can suppress inflammatory processes associated with polycystic ovarian syndrome and other reproductive dysfunctions (Darabi et al., 2020). The antiinflammatory action as a mediator of anticancer effect of apigenin is confirmed by other studies, which demonstrated that apigenin decreased level of the proinflammatory cytokines TNF-α and IL-6 (Chen et al., 2012) and COX2 (Ittiudomrak et al., 2019) in rat plasma and of the proinflammatory transcription factor NF-κB in cultured ovarian cancer cells (Chen et al., 2012). On the other hand, Suh et al. (2015) failed to find apigenin action on accumulation and phosphorylation of STAT3 and interleukin 6-promoters of inflammatory processes, in cultured ovarian cancer cells. Therefore, the majority of the performed studies demonstrated the antiinflammatory action of apigenin on both healthy and cancer ovarian cells mediated by downregulation of various promoters of inflammatory processes. This action can define the ability of apigenin to suppress ovarian cancer and other inflammation-related ovarian disorders.

Apigenin exerted the antiinflammatory action also on healthy gestational tissues. Apigenin reduced expression on proinflammatory transcription factors NF-kB and Nrf2 in rat endometrium (Jinag et al., 2018). Furthermore, it inhibited binding of proinflammatory of transcription factor NF-kB to DNA, release and gene expression, cytokines IL-6 and IL-8; COX-2 and subsequent release of prostaglandins PGE2 and PGF2α and expression and activity of matrix-degrading enzyme matrix metalloproteinase (MMP)-9 in cultured healthy human placenta, fetal membranes and myometrial cells (Lim et al., 2013).

Therefore, apigenin action on various reproductive tissues could be due to its antiinflammatory effects.

Apigenin was able to reduce the intraovarian level of endothelium-derived relaxing factor nitric oxide in rats (Soyman et al., 2017) and to inhibit VEGF (vascular endothelial factor) production (Chen et al., 2012, see earlier). This observation indicates (but not demonstrates) that apigenin could affect ovarian functions via changes in **ovarian angiogenesis and blood flow**.

Apigenin influence of ovarian **steroid hormones** (progestagen, androgen, and estrogen) is well documented. Ovarian folliculogenesis and oogenesis (Sirotkin, 2014), gestational tissues, and many reproductive disorders (Dean et al., 2018; Kumar et al., 2018; Deligdisch-Schor, 2020) are promoted by steroid hormones via steroid hormones' receptors. An action similar to that of endogenous steroid hormones comes from phytoestrogens (Sirotkin and Harrath, 2014). The ability of apigenin to affect steroid hormones indicates that its effects on female reproductive processes could be mediated by changes in steroid hormones' release.

Furthermore, a direct action of apigenin on **steroid hormones receptors** might be proposed. Apigenin can promote the expression of **progesterone receptor** and progesterone-stimulated postreceptory transcription factor Hand2 in cultured uterine endometrial cells (Dean et al., 2018). This report demonstrates that apigenin is a phytoprogestin, which can affect female reproductive system (at least uterus) via activation of progesterone receptor and its postreceptory intracellular pathway.

Although apigenin action on estrogen receptors in reproductive tissues has not been reported yet, it was able to express an antiestrogen effect on uterus (Dean et al., 2018) and to downregulate the expression of **estrogen receptor** (ERα) in breast cancer cells (Scherbakov et al., 2019). It means that apigenin has an phytoestrogenic properties. The principal ability of apigenin to affect estrogen receptors indicates that estrogen receptors could be potential mediators of apigenin action on female reproduction too. This hypothesis requires experimental validation yet.

These data indicate that steroid hormones and their receptors could be important mediators of apigenin action on steroids-dependent female reproductive processes.

Apigenin action on uterus and its ability to treat endometriosis are mediated by estrogen receptor-a, regulators of inflammation (cyclooxygenase-2, IL-1 and IL-6, TNG-a, NF-kB), cell migration and vascularization (intercellular adhesion molecule-1, vascular endothelial growth factor, matrix metalloproteinases) as well as reactive oxygen species and apoptosis-related proteins (Meresman et al., 2021).

Taken together, the available studies indicated that the apigenin action on healthy and damaged reproductive processes could be mediated by its action on oxidative processes, cell proliferation, apoptosis, renewal and viability, ovarian blood supply, as well as on release and reception of hormones. The interplay between mediators of apigenin action requires further elucidation. Nevertheless, it might be hypothesized that antioxidative action of apigenin could prevent the inflammatory processes, apoptosis, loss of cell viability, and degenerative processes observed in ovarian cells during cancer and polycystic ovarian syndrome (Chen et al., 2012; Tavsan and Kayali, 2019; Darabi et al., 2020). Furthermore, steroid hormones are the known regulators of ovarian folliculoogenesis, gravidity, and embryogenesis (Sirotkin, 2014). Therefore, apigenin influence on steroid hormones and their receptors can mediate apigenin action on these processes too.

4.2.11 Application in reproductive biology and medicine

The currently available information demonstrates the two opposite effects of apigenin on female reproductive processes. On one side, it stimulates healthy cells and organisms—promotes ovarian folliculogenesis, ovarian, and embryonal cell proliferation and viability (Darabi et al., 2020; Sirotkin et al., 2020a,b). These effects indicate that apigenin could be a biostimulator of reproductive processes. The performed studies indicate its potential applicability for promotion of reproductive efficiency in laboratory and farm animals, for induction of ovulation and superovulation and for treatment of age-related reproductive insufficiency and infertility. Furthermore, apigenin could be a promising additive to culture medium for improvement of *in vitro* oocyte maturation, embryo production, embryo transfer, and cloning in animal biotechnology and human-assisted reproduction.

On the other side, apigenin can inhibit functions of ovarian cancer cells and mitigate pathological transformation of female reproductive system induced by environmental contaminants (Sirotkin et al., 2020a,b, 2021), toxic drugs (Talebi et al., 2020), cancer (Chen et al., 2012; Ittiudomrak et al., 2019; Tavsan and Kayali, 2019; Talebi et al., 2020), polycystic ovarian syndrome (Darabi et al., 2020), ischemia (Soyman et al., 2017), and endometriosis (Jiang et al., 2018; Meresman et al., 2021). Furthermore, animal studies indicated potential applicability of apigenin for prevention of preterm labor (Lim et al., 2013; Suhas et al., 2018). If the available results of *in vitro* and animal studies would be confirmed by sufficient clinical trials, apigenin could be efficient natural drug for prevention and treatment of several reproductive disorders.

The application of natural apigenin could be limited by its low bioavailability. This limitation could be overcome by apigenin analogues with lover degradability and higher bioavailability and biological activity. For example, several triazole analogs of the bioactive apigenin-7-methyl ether were efficient for suppression of ovarian cancer functions (Qi et al., 2019).

Apigenin as a pure drug might be principally replaced by cheaper and more accessible apigenin-containing plants. For example, an extract of medicinal plant *Senecio biafrae* Oliv. and Hiern containing a high amount of apigenin is used in folk medicine for treatment of female infertility. Its effects are similar to the effects of apigenin. At least feeding of rats with its extract increased the plasma level of FSH and estradiol, decreased level of LH, increased ovarian weight and promoted ovarian folliculogenesis (Lienou et al., 2015, 2020). These observations suggest the potential applicability of this apigenin-containing plant as promoters of ovarian functions and as replacement of more expensive apigenin. On the other hand, it remains questionable whether these effects were induced by the apigenin component of this plant, and therefore whether this plant or other apigenin-containing plants could be an adequate replacement of a pure drug.

4.2.12 Conclusions and possible direction of future studies

We have tried to outline the available information concerning production, metabolism, general physiological and therapeutic effects of apigenin and their mechanisms, especially to summarize and to analyze the current knowledge concerning apigenin effects on female reproductive processes and their dysfunctions. The currently available scientific literature indicates that apigenin is able to promote ovarian folliculogenesis, ovarian, and embryonal cell proliferation and viability, to alter release of reproductive hormones. On the other hand, apigenin can inhibit functions of ovarian cancer cells and mitigate pathological transformation of female reproductive system induced by environmental contaminants, toxic drugs, cancer, polycystic ovarian syndrome, endometriosis and ishemia. The apigenin action on both healthy and damaged reproductive processes could be mediated by its action on oxidative processes, cell proliferation, apoptosis, renewal and viability, ovarian blood supply, as well as on release and reception of hormones. These effects of apigenin could indicate its potential applicability as biostimulator of female reproductive processes and as a medicine against some reproductive disorders.

Nevertheless, the available information concerning apigenin is insufficient to generate definitive conclusions concerning its effects and application. Even the studies of the same team performed on the same model sometimes generated variable results. There are fewer studies performed on healthy organisms and cells than studies performed in conditions of induced pathology to detect not physiological, but rather therapeutical effects of apigenin. Nevertheless, even the reported medicinal studies were performed mainly on laboratory animals and cell cultures, while no adequate clinical trials have been reported yet. Even the promising testing of apigenin on farm animals is only starting now. The opposite influence of apigenin on reproduction system in conditions of health and diseases is evident and promising for application of apigenin for selective treatment of reproductive disorders, but understanding the causes and mechanisms of such differences requires further studies. The number of signaling molecules and intracellular pathways mediating apigenin action is high, but their functional interrelationships and differences in healthy and ill cells remain to be elucidated yet. The actual problem of replacement of apigenin by more bioavailable, efficient, and cheap analogue is still waiting for its solution. Nevertheless, even the contemporarily available knowledge concerning apigenin indicates biological potency and applicability of this promising molecule.

References

Abd Ghani MF, Othman R, Nordin N. Molecular docking study of naturally derived flavonoids with antiapoptotic BCL-2 and BCL-XL proteins toward ovarian cancer treatment. J Pharm BioAllied Sci 2020;12(Suppl 2):S676−80. https://doi.org/10.4103/jpbs.JPBS_272_19.

Abdullah JM, Zhang J. The GABA A receptor subunits heterologously expressed in Xenopus oocytes. Mini Rev Med Chem 2013;13(5):744−8. https://doi.org/10.2174/1389557511313050011.

Ahmed SA, Parama D, Daimari E, Girisa S, Banik K, Harsha C, Dutta U, Kunnumakkara AB. Rationalizing the therapeutic potential of apigenin against cancer. Life Sci 2021;267:118814. https://doi.org/10.1016/j.lfs.2020.118814.

Ajebli M, Eddouks M. Phytotherapy of hypertension: an updated overview. Endocr Metab Immune Disord Drug Targets 2020;20(6):812−39. https://doi.org/10.2174/1871530320666191227104648.

Barlas N, Özer S, Karabulut G. The estrogenic effects of apigenin, phloretin and myricetin based on uterotrophic assay in immature Wistar albino rats. Toxicol Lett 2014;226(1):35−42. https://doi.org/10.1016/j.toxlet.2014.01.030.

Chen SS, Michael A, Butler-Manuel SA. Advances in the treatment of ovarian cancer: a potential role of antiinflammatory phytochemicals. Discov Med 2012;13(68):7−17.

Darabi P, Khazali H, Mehrabani Natanzi M. Therapeutic potentials of the natural plant flavonoid apigenin in polycystic ovary syndrome in rat model: via modulation of pro-inflammatory cytokines and antioxidant activity. Gynecol Endocrinol 2020;36(7):582−7. https://doi.org/10.1080/09513590.2019.1706084.

Dean M, Austin J, Jinhong R, Johnson ME, Lantvit DD, Burdette JE. The flavonoid apigenin is a progesterone receptor modulator with in vivo activity in the uterus. Horm Cancer 2018;9(4):265−77. https://doi.org/10.1007/s12672-018-0333-x.

Deligdisch-Schor L. Hormone therapy effects on the uterus. Adv Exp Med Biol 2020;1242:145−77. https://doi.org/10.1007/978-3-030-38474-6_8.

Dewanjee S, Chakraborty P, Mukherjee B, De Feo V. Plant-based antidiabetic nanoformulations: the emerging paradigm for effective therapy. Int J Mol Sci 2020;21(6):2217. https://doi.org/10.3390/ijms21062217.

Ginwala R, Bhavsar R, Chigbu DI, Jain P, Khan ZK. Potential role of flavonoids in treating chronic inflammatory diseases with a special focus on the anti-inflammatory activity of apigenin. Antioxidants 2019;8(2):35. https://doi.org/10.3390/antiox8020035.

Griffiths K, Aggarwal BB, Singh RB, Buttar HS, Wilson D, De Meester F. Food antioxidants and their anti-inflammatory properties: a potential role in cardiovascular diseases and cancer prevention. Diseases 2016;4(3):28. https://doi.org/10.3390/diseases4030028.

Hostetler GL, Ralston RA, Schwartz SJ. Flavones: food sources, bioavailability, metabolism, and bioactivity. Adv Nutr 2017;8(3):423−35. https://doi.org/10.3945/an.116.012948.

Imran M, Aslam Gondal T, Atif M, Shahbaz M, Batool Qaisarani T, Hanif Mughal M, Salehi B, Martorell M, Sharifi-Rad J. Apigenin as an anticancer agent. Phytother Res 2020;34(8):1812−28. https://doi.org/10.1002/ptr.6647.

Ittiudomrak T, Puthong S, Roytrakul S, Chanchao C. α-Mangostin and apigenin induced cell cycle arrest and programmed cell death in SKOV-3 ovarian cancer cells. Toxicol Res 2019;35(2):167−79. https://doi.org/10.5487/TR.2019.35.2.167.

Jiang PY, Zhu XJ, Zhang YN, Zhou FF, Yang XF. Protective effects of apigenin on LPS-induced endometritis via activating Nrf2 signaling pathway. Microb Pathog 2018;123:139−43. https://doi.org/10.1016/j.micpath.2018.06.031.

Kumar A, Banerjee A, Singh D, Thakur G, Kasarpalkar N, Gavali S, Gadkar S, Madan T, Mahale SD, Balasinor NH, Sachdeva G. Estradiol: a steroid with multiple facets. Horm Metab Res 2018;50(5):359−74. https://doi.org/10.1055/s-0044-100920.

Lai F, Schlich M, Pireddu R, Fadda AM, Sinico C. Nanocrystals as effective delivery systems of poorly water-soluble natural molecules. Curr Med Chem 2019;26(24):4657−80. https://doi.org/10.2174/0929867326666181213095809.

Lienou LL, Telefo PB, Njimou JR, Nangue C, Bayala BR, Goka SC, Biapa P, Yemele MD, Donfack NJ, Mbemya JT, Tagne SR, Rodrigues AP. Effect of the aqueous extract of Senecio biafrae (Oliv. & Hiern) J. Moore on some fertility parameters in immature female rat. J Ethnopharmacol 2015;161:156−62. https://doi.org/10.1016/j.jep.2014.12.014.

Lienou LL, Telefo PB, Rodrigues GQ, Donfack JN, Araújo RA, Bruno JB, Njimou JR, Mbemya TG, Santos RR, Souza JF, Figueiredo JR, Rodrigues APR. Effect of different extracts and fractions of Senecio biafrae (Oliv. &Hiern) J. Moore on in vivo and in vitro parameters of folliculogenesis in experimental animals. J Ethnopharmacol 2020;251:112571. https://doi.org/10.1016/j.jep.2020.112571.

Lim R, Barker G, Wall CA, Lappas M. Dietary phytophenols curcumin, naringenin and apigenin reduce infection-induced inflammatory and contractile pathways in human placenta, foetal membranes and myometrium. Mol Hum Reprod 2013;19(7):451−62. https://doi.org/10.1093/molehr/gat015.

Long Y, Yang Q, Xiang Y, Zhang Y, Wan J, Liu S, Li N, Peng W. Nose to brain drug delivery - a promising strategy for active components from herbal medicine for treating cerebral ischemia reperfusion. Pharmacol Res 2020;159:104795. https://doi.org/10.1016/j.phrs.2020.104795.

Martin LJ, Touaibia M. Improvement of testicular steroidogenesis using flavonoids and isoflavonoids for prevention of late-onset male hypogonadism. Antioxidants 2020;9(3):237. https://doi.org/10.3390/antiox9030237.

Martínez G, Mijares MR, De Sanctis JB. Effects of flavonoids and its derivatives on immune cell responses. Recent Pat Inflamm Allergy Drug Discov 2019;13(2):84−104. https://doi.org/10.2174/1872213X13666190426164124.

Matsuo M, Sasaki N, Saga K, Kaneko T. Cytotoxicity of flavonoids toward cultured normal human cells. Biol Pharm Bull 2005;28(2):253−9. https://doi.org/10.1248/bpb.28.253.

Meresman GF, Götte M, Laschke MW. Plants as source of new therapies for endometriosis: a review of preclinical and clinical studies. Hum Reprod Update 2021;27(2):367−92. https://doi.org/10.1093/humupd/dmaa039.

Nabavi SF, Khan H, D'onofrio G, Šamec D, Shirooie S, Dehpour AR, Argüelles S, Habtemariam S, Sobarzo-Sanchez E. Apigenin as neuroprotective agent: of mice and men. Pharmacol Res 2018;128:359−65. https://doi.org/10.1016/j.phrs.2017.10.008.

Ozbey U, Attar R, Romero MA, Alhewairini SS, Afshar B, Sabitaliyevich UY, Hanna-Wakim L, Ozcelik B, Farooqi AA. Apigenin as an effective anticancer natural product: spotlight on TRAIL, WNT/β-catenin, JAK-STAT pathways, and microRNAs. J Cell Biochem 2018. https://doi.org/10.1002/jcb.27575.

Pal MK, Jaiswar SP, Dwivedi A, Goyal S, Dwivedi VN, Pathak AK, Kumar V, Sankhwar PL, Ray RS. Synergistic effect of graphene oxide coated nanotised apigenin with paclitaxel (GO-NA/PTX): a ROS dependent mitochondrial mediated apoptosis in ovarian cancer. Anticancer Agents Med Chem 2017;17(12):1721−32. https://doi.org/10.2174/1871520617666170425094549.

Peng Y, Gan R, Li H, Yang M, McClements DJ, Gao R, Sun Q. Absorption, metabolism, and bioactivity of vitexin: recent advances in understanding the efficacy of an important nutraceutical. Crit Rev Food Sci Nutr 2021;61(6):1049−64. https://doi.org/10.1080/10408398.2020.1753165.

Qi Y, Ding Z, Yao Y, Ma D, Ren F, Yang H, Chen A. Novel triazole analogs of apigenin-7-methyl ether exhibit potent antitumor activity against ovarian carcinoma cells via the induction of mitochondrial-mediated apoptosis. Exp Ther Med 2019;17(3):1670−6. https://doi.org/10.3892/etm.2018.7138.

Rossi B, Toschi A, Piva A, Grilli E. Single components of botanicals and nature-identical compounds as a non-antibiotic strategy to ameliorate health status and improve performance in poultry and pigs. Nutr Res Rev 2020;33(2):218−34. https://doi.org/10.1017/S0954422420000013.

Safari M, Parsaie H, Sameni HR, Aldaghi MR, Zarbakhsh S. Anti-oxidative and anti-apoptotic effects of apigenin on number of viable and apoptotic blastomeres, zona pellucida thickness and hatching rate of mouse embryos. Int J Fertil Steril 2018;12(3):257−62. https://doi.org/10.22074/ijfs.2018.5392.

Salehi B, Venditti A, Sharifi-Rad M, Kregiel D, Sharifi-Rad J, Durazzo A, Lucarini M, Santini A, Souto EB, Novellino E, Antolak H, Azzini E, Setzer WN, Martins N. The therapeutic potential of apigenin. Int J Mol Sci 2019;20(6):1305. https://doi.org/10.3390/ijms20061305.

Scherbakov AM, Shestakova EA, Galeeva KE, Bogush TA. BRCA1 and estrogen receptor α expression regulation in breast cancer cells. Mol Biol 2019;53(3):502−12. https://doi.org/10.1134/S0026898419030169. Russian.

Sirotkin AV, Harrath AH. Phytoestrogens and their effects. Eur J Pharmacol 2014;741:230−6. https://doi.org/10.1016/j.ejphar.2014.07.057.

Sirotkin AV, Radosová M, Tarko A, Fabova Z, Martín-García I, Alonso F. Abatement of the stimulatory effect of copper nanoparticles supported on titania on ovarian cell functions by some plants and phytochemicals. Nanomaterials 2020a;10(9):1859. https://doi.org/10.3390/nano10091859.

Sirotkin A, Záhoranska Z, Tarko A, Popovska-Percinic F, Alwasel S, Harrath AH. Plant isoflavones can prevent adverse effects of benzene on porcine ovarian activity: an in vitro study. Environ Sci Pollut Res Int 2020b;27(23):29589−98. https://doi.org/10.1007/s11356-020-09260-8.

Sirotkin A, Záhoranska Z, Tarko A, Fabova Z, Alwasel S, Halim Harrath A. Plant polyphenols can directly affect ovarian cell functions and modify toluene effects. J Anim Physiol Anim Nutr 2021;105(1):80−9. https://doi.org/10.1111/jpn.13461.

Sirotkin AV. Regulators of ovarian functions. Hauppauge, NY, USA: Nova Science Publishers, Inc.; 2014. 194pp.

Soyman Z, Kelekçi S, Sal V, Sevket O, Bayındır N, Uzun H. Effects of apigenin on experimental ischemia/reperfusion injury in the rat ovary. Balkan Med J 2017;34(5):444−9. https://doi.org/10.4274/balkanmedj.2016.1386.

Suh YA, Jo SY, Lee HY, Lee C. Inhibition of IL-6/STAT3 axis and targeting Axl and Tyro3 receptor tyrosine kinases by apigenin circumvent taxol resistance in ovarian cancer cells. Int J Oncol 2015;46(3):1405−11. https://doi.org/10.3892/ijo.2014.2808.

Suhas KS, Parida S, Gokul C, Srivastava V, Prakash E, Chauhan S, Singh TU, Panigrahi M, Telang AG, Mishra SK. Casein kinase 2 inhibition impairs spontaneous and oxytocin-induced contractions in late pregnant mouse uterus. Exp Physiol 2018;103(5):621—8. https://doi.org/10.1113/EP086826.

Suraweera T L, Rupasinghe HPV, Dellaire G, Xu Z. Regulation of nrf2/ARE pathway by dietary flavonoids: a friend or foe for cancer management? Antioxidants 2020;9(10):973. https://doi.org/10.3390/antiox9100973.

Talebi A, Hayati Roodbari N, Reza Sameni H, Zarbakhsh S. Impact of coadministration of apigenin and bone marrow stromal cells on damaged ovaries due to chemotherapy in rat: an experimental study. Int J Reprod Biomed 2020;18(7):551—60. https://doi.org/10.18502/ijrm.v13i7.7372.

Tang AQ, Cao XC, Tian L, He L, Liu F. Apigenin inhibits the self-renewal capacity of human ovarian cancer SKOV3-derived sphere-forming cells. Mol Med Rep 2015;11(3):2221—6. https://doi.org/10.3892/mmr.2014.2974.

Tang D, Chen K, Huang L, Li J. Pharmacokinetic properties and drug interactions of apigenin, a natural flavone. Expert Opin Drug Metab Toxicol 2017;13(3):323—30. https://doi.org/10.1080/17425255.2017.1251903.

Tavsan Z, Kayali HA. Flavonoids showed anticancer effects on the ovarian cancer cells: involvement of reactive oxygen species, apoptosis, cell cycle and invasion. Biomed Pharmacother 2019;116:109004. https://doi.org/10.1016/j.biopha.2019.109004.

Wang QQ, Cheng N, Yi WB, Peng SM, Zou XQ. Synthesis, nitric oxide release, and α-glucosidase inhibition of nitric oxide donating apigenin and chrysin derivatives. Bioorg Med Chem 2014;22(5):1515—21. https://doi.org/10.1016/j.bmc.2014.01.038.

Wang M, Firrman J, Liu L, Yam K. A review on flavonoid apigenin: dietary intake, ADME, antimicrobial effects, and interactions with human gut microbiota. BioMed Res Int 2019;2019:7010467. https://doi.org/10.1155/2019/7010467.

Yan X, Qi M, Li P, Zhan Y, Shao H. Apigenin in cancer therapy: anti-cancer effects and mechanisms of action. Cell Biosci 2017;7:50. https://doi.org/10.1186/s13578-017-0179-x.

Chapter 4.3

Berberine

4.3.1 Introduction

In recent years, people are going back to more traditional medicine or folk medicine to look for more natural solutions to many diseases. These natural compounds can serve as a good scaffold for rational drug design (Patwardhan and Vaidya, 2010). The naturally occurring plant alkaloid berberine is one of the phytochemicals with a broad range of biological activity, including anticancer, antiinflammatory, and antiviral activity (Warowicka et al., 2020). The diverse pharmacological properties exhibited by berberine indicate that the alkaloid has a definite potential as drug in a wide spectrum of clinical applications (Tillhon et al., 2012). Berberine that is used at the present time is synthetic in the form of chloride or sulfate (Kumar et al., 2015). In this review, we summarized knowledge about berberine effect on human health especially its effect on female reproductive system. Although berberine effects on human health are well known, effects of berberine on female reproduction are not fully understand to this day. There are some researches that have shown potential for berberine to be less invasive cancer therapy. Also, many researches nowadays are focusing on berberine effect on polycystic ovarian syndrome, where this alkaloid has shown promising results.

4.3.2 Provenance and properties

Berberine (chemical formula: $C20H18NO4$, slowly soluble in water; Fig. 4.3), a quaternary ammonium salt from the protoberberine group of isoquinoline alkaloids (Cai et al., 2016; Fang et al., 2018) of some medicinal Chinese herbs (most frequently *Berberis vulgaris*) (Mirhadi et al., 2018) including other species of Berberis (*Berberis aristata* L., *Berberis croatica* L., *Berberis aquifolium* L.), *Hydrastis* (*Hydrastis canadensis* L.) and *Coptis* (*Coptis chinensis* L., *Coptis japonica* L.) (Imenshahidi and Hosseinzadeh, 2019) is present in the roots, rhizomes, stem, and bark of *Berberis vulgaris* (Imenshahidi and Hosseinzadeh, 2019) and some plants such as huanglian (*Rhizoma coptidis*), huangbo (*Phellodendri chinensis cortex*), and sankezhen (*Berberidis radix*) (Fang et al., 2018).

Berberine has a low toxicity (Mirhadi et al., 2018; Imenshahidi and Hosseinzadeh, 2019), is well tolerated by human body (Mirhadi et al., 2018), reveals clinical benefits without major side effects (Imenshahidi and Hosseinzadeh, 2019), but high doses of berberine can cause gastrointestinal

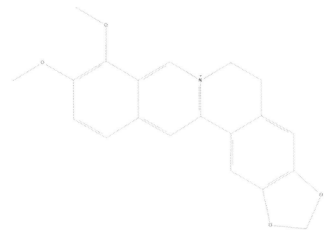

FIGURE 4.3 Chemical structure of berberine (PubChem, 2021).

side-effects (Mirhadi et al., 2018) with only mild gastrointestinal reactions (Imenshahidi and Hosseinzadeh, 2019). Some limitations including poor aqueous solubility, slight absorption, and low bioavailability have hindered berberine applications. To overcome these limitations, nanotechnology has been considered as main strategy (Mirhadi et al., 2018).

4.3.3 Physiological and therapeutic actions

Previous trials suggested a wide range of therapeutic (Imenshahidi and Hosseinzadeh, 2019) and pharmacological activities (Cai et al., 2016) including antioxidant (Cicero and Baggioni, 2016), antibacterial (Habtemariam, 2016; Fang et al., 2018; Warowicka et al., 2020), antiviral (Fang et al., 2018; Warowicka et al., 2020), antidiabetic (Warowicka et al., 2020), antilipidemic (Fang et al., 2018; Warowicka et al., 2020), glucose- and cholesterol-lowering (Ma et al., 2017; Fang et al., 2018; Warowicka et al., 2020), anticancer (Fang et al., 2018; Warowicka et al., 2020; Samadi et al., 2020), antiinflammatory (Cicero and Baggioni, 2016; Fang et al., 2018; Warowicka et al., 2020), immunomodulatory properties (Cai et al., 2016), cardiovascular (Cicero and Baggioni, 2016; Imenshahidi and Hosseinzadeh, 2019), antihypertensive (Fang et al., 2018; Warowicka et al., 2020), antiarrhythmic, antiheart failure (Fang et al., 2018), hepatoprotective (Warowicka et al., 2020; Fang et al., 2020), gastrointestinal (Chen et al., 2015; Xu et al., 2019; Imenshahidi and Hosseinzadeh, 2019; Warowicka et al., 2020), neuroprotective (Cicero and Baggioni, 2016; Imenshahidi and Hosseinzadeh, 2019) effects, analgesics (Fang et al., 2018), antidepressant (Chen et al., 2015; Warowicka et al., 2020), effects on mood disorders (Fan et al., 2019), and protective reproductive properties on spermatogenesis (Song et al., 2020).

Substantial studies suggest that berberine may be beneficial to Alzheimer's disease by limiting the pathogenesis of extracellular amyloid plaques and intracellular neurofibrillary tangles. Increasing evidence has indicated that berberine exerts a protective role in atherosclerosis related to lipid- and glucose-lowering properties, implicating that berberine has the potential to inhibit these risk factors for Alzheimer's disease (Cai et al., 2016). Moreover, preliminary clinical evidence suggests the ability of berberine to reduce endothelial inflammation improving vascular health, even in patients already affected by cardiovascular diseases (Cicero and Baggioni, 2016). In Ayurveda, berberis species have been used for the treatment of a wide range of infections such as eye, ear, and mouth infection. It is also used for quick healing of wounds, indigestion and dysentery, curing hemorrhoids. Berberine has also been used as an antidote for the treatment of scorpion sting or snakebite, it also shown signs to reduce obesity (Neag et al., 2018). Some healing effects from Ayurveda were confirmed by modern medicine. Berberine hydrochloride is well tolerated and reduces irritable bowel syndrome symptoms (Chen et al., 2015). Therefore, berberine hydrochloride is mainly used to treat gastroenteritis, bacterial diarrhea, intestinal infection, conjunctivitis, and suppurative otitis media (Fang et al., 2018). Similarly, berberine was used in traditional Chinese medicine as an antidiarrheal drug in gastrointestinal disorders (Xu et al., 2019). Berberine effect on inflammatory bowel disease (Habtemariam, 2016) and its antidiarrheal effect (Chen et al., 2015) was suggested. Potential benefit of berberine therapy could arise from its widely known antibacterial effects including activity against *Escherichia coli* (Habtemariam, 2016). Berberine targets different steps in the viral life cycle and is thus a good candidate for use in novel antiviral drugs and therapies. It has been shown that berberine reduces virus replication and targets specific interactions between the virus and its host (Warowicka et al., 2020). The most recent studies show the antiviral properties of berberine and its derivatives, which might be promising agents to be considered in future studies in the fight against the current pandemic SARS-CoV-2, the virus that causes COVID-19 (Warowicka et al., 2020).

In the available data about benefit effects of berberine, they indicate applicability in low doses as a promising tool for regulation of mainly nonreproductive and reproductive processes and treatment of their disorders primarily in phytotherapy, medicine, and biotechnology. Moreover, further research is required to find a means by which we can apply these effects on human cells and assure its safety as a dependable treatment.

4.3.4 Mechanisms of action

Many researches demonstrated the beneficial role of berberine in different neurodegenerative diseases due to its antioxidant effect. Berberine may lower oxidative stress by limiting the neuroinflammation process in Alzheimer's disease brain. The protective effects of protoberberine alkaloid against amyloid β ($A\beta$)-induced cell death in rat cortical neurons via decreasing the production of malondialdehyde (MDA) and reactive oxygen species (ROS) (Luo et al., 2012). Berberine has also notable direct antioxidant effect through metal chelation, active scavenging of reactive oxygen and nitrogenous species as well as suppression of their production. Berberine was also reported for indirect antioxidant effects through enhancement of antioxidant defenses (Habtemariam, 2016; Samadi et al., 2020). One of the cellular adaptation to oxidative stress is through regulation of antioxidant enzymes gene transcription via the nuclear factor (erythroid-derived 2)-related factor-2 (Nrf2)/heme oxygenase (HO-1) system (Habtemariam, 2016). Song et al. (2020) also demonstrated protective effect of berberine on reproductive function and spermatogenesis in diabetic rats. Diabetes mellitus can cause excessive ROS production in the body. Berberine has shown protective role in reproductive function and spermatogenesis by inhibiting reactive oxygen species pathway. Berberine has shown promising potential in further treatment. Of course, more research is needed, and the study is still ongoing.

Berberine inhibits chemotaxis, affects the expression of inflammatory mediators and regulates the phagocytic activity of neutrophils and macrophages. Berberine can induce cell cycle arrest via an effect on the expression of some signaling genes, such as nuclear factor kappa B (NF-κB), Raf/MAPK/ERK kinase (MEK)/the extracellular signal-regulated kinase (ERK), activator protein 1 (AP-1), and mitogen-activated protein kinase (MAPK) and apoptosis in murine ells in a dose-dependent manner (Yang et al., 1996; Hesari et al., 2018). Berberine can also prevent inflammation via reducing interleukin-1 (IL-1), TNF-α, interleukin 6 (IL-6), and the monocyte chemoattractant protein-1 (MCP-1), inhibiting prostaglandin E2 (PGE2) and cyclooxygenase-2 (COX-2) transcriptional activity in colon and other cancer cells, and inhibiting the activity of the AP-1 transcription factor (Fukuda et al., 1999; Hesari et al., 2018). Isoquinoline alkaloid, which is also berberine, has the ability to regulate the MEK-ERK, the adenosine monophosphate-activated protein kinase (AMPK)/mTOR, and NF-κB signaling pathways, which are necessary for viral replication (Warowicka et al., 2020). Berberine intercalates into DNA and inhibits DNA synthesis and reverse transcriptase activity. It inhibits replication of herpes simplex virus, human cytomegalovirus, human papillomavirus, and human immunodeficiency virus (Warowicka et al., 2020).

Fan et al. (2019) submitted study about pharmacological effects of berberine on mood disorders. Berberine regulates brain neurotransmitters. Norepinephrine, dopamine and serotonin are released from neurons during synaptic transmission. Berberine at low doses results in increased levels of norepinephrine, dopamine and serotonin. Berberine inhibits the release of norepinephrine through activation of adrenergic $\alpha 2$ autoreceptors and can affect dopamine in a manner that antagonizes D2 and agonizes D1 receptors (Fan et al., 2019). In addition, berberine has positive effect on reduction of

ROS, can induce cell cycle arrest via an effect on the expression of some signaling genes, regulates processes proliferation and apoptosis in pathological cells through signaling proteins and regulates synaptic transmission. More research is still needed, so we can fully understand effects of berberine and its mechanism of the action on human and animal cells.

Berberine substantially inhibited apoptosis under ischemia-reperfusion injury (Liu et al., 2019a). Berberine inhibited oxygen glucose deprivation (OGD)-inducted release of proapoptic factors from mitochondria such as apoptosis-inducing factor and cytochrome C (Zhou et al., 2008; Liu et al., 2019a). Berberine protects against hypoxia/reoxygenation (H/R)-induced apoptosis in human renal proximal tubular cells via inhibiting caspase-dependent pathway and mitochondrial dependent pathway (Yu et al., 2013). On the other hand, it was shown that berberine can inhibit proliferation, induce apoptosis and autophagy and also suppress angiogenesis and metastasis. Berberine also regulates tumor suppressor proteins such as Fas, antiproliferative and proapoptotic transcription factor p53, as well as it can suppress matrix metalloproteinase (MMP)-2 and MMP-9 expression which play critical role in tumor cell migration to form metastasis (Samadi et al., 2020).

4.3.5 Effects on female reproductive processes

Previous subchapters described positive effects of berberine on humans. Although much research is done about berberine effects on human organism, effects on female reproductive processes are still to be desired.

4.3.5.1 Effect on ovaries

Clinical, metabolic and endocrine effects of berberine in women with polycystic ovarian syndrome were studied (An et al., 2013; Shen et al., 2021). Patients with polycystic ovarian syndrome have elevated level of androgens and proandrogens androstenedione and dehydroepiandrosterone sulfate (Keefe et al., 2014). Excess androgens can be caused by insulin resistance and hyperinsulinemia (Pappalardo et al., 2017). Berberine can reduce androgen levels in woman with polycystic ovarian syndrome (Li et al., 2015). Women with polycystic ovarian syndrome that received berberine, required significantly lower follicle-stimulating hormone (FSH) dosages used for ovarian stimulation. Also, live birth rate was higher for woman that were taking berberine (An et al., 2013). Berberine reduced insulin resistance and testosterone levels in polycystic ovarian syndrome rats and also exerted protective effect on polycystic ovarian syndrome (Zhang et al., 2020; Shen et al., 2021). This action was associated with activation of apoptosis and with inhibition of regulators of cell proliferation (Zhang et al., 2020).

Among various phytochemicals, berberine, a well-known natural product, has been shown to be a promising cancer chemopreventive agent. Berberine (Li et al., 2020; Mortazavi et al., 2020; Chuang et al., 2021) and its chemical analogues (Mistry et al., 2017) have been found to efficiently inhibit viability, proliferation, and migration of ovarian cancer cells, mainly, via induction of apoptosis, suppression of epidermal growth factor receptors, and postreceptory regulators of proliferation, angiogenesis, and cell migration (Chuang et al., 2021). The structural modification of berberine showed that berberine analogues can improve its antitumor effects against gynecological cancers (Mortazavi et al., 2020).

4.3.5.2 Effect of vagina and uterus

In Ayurveda, berberis species have been used for the treatment of a wide range of infections such as treatment of vaginal and uterine disorders (Neag et al., 2018). Similarly, berberine is a promising cancer chemopreventive agent and inhibits viability, proliferation, and migration of cervical and endometrial cancer cells, mainly, via induction of apoptosis (Mortazavi et al., 2020).

Therefore, the available reports point out some therapeutic effects of berberine on disorders of female reproductive system such as polycystic ovarian syndrome, ovarian, cervical and endometrial cancer, and vaginal and uterine infections. However, further research is required to find a means by which we can apply its effects on human cells and assure its safety as a dependable treatment.

4.3.6 Mechanisms of action on female reproductive processes

4.3.6.1 Proliferation and apoptosis

Berberine suppresses proliferation and promotes apoptosis and decreases tumor growth in ovarian cancer cells (Li et al., 2020; Mortazavi et al., 2020). These changes were associated with reduced mitochondrial activity, and activation of adenosine monophosphate (AMN) accumulation and **AMN kinase signaling** (Li et al., 2020), **downregulation of promoters and markers of proliferation** (cyclin D1), and **cell migration** (MMP-2, MMP-9) (Chuang et al., 2021). Similar research was done by Liu et al. (2019b), where they determined that berberine in combination with cisplatin induces necroptosis and apoptosis in ovarian cancer cells. Study has proven that combination treatment of berberine markedly enhanced the anticancer effect of cisplatin by inducing apoptosis and necroptosis in ovarian cancer. The apoptosis involved the **caspase-dependent pathway**, while the necroptosis involved the activation of the receptor-interacting protein kinase 3 (RIPK3) and mixed lineage kinase domain-like (MLKL): **RIPK3—MLKL pathway**. Berberine analogues expressed also toxic/proapoptotic influence on cultured ovarian cancer cells (Mistry et al., 2017).

Berberine action on rats with polycystic ovarian syndrome was also associated with downregulation of intracellular promoters of **cell proliferation** (**MAPK**) and upregulation of their apoptosis (**phosphatidylinositol-3-kinase**

(PI3K)/Akt signaling pathway) (Zhang et al., 2020). This observation suggests the involvement of intracellular regulators of proliferation and apoptosis in mediating berberine action on ovarian cells during polycystic ovarian syndrome.

4.3.6.2 Oxidative stress

Berberine addition increased level of **ROS** in cultured ovarian cancer cells. This fact indicates that berberine-induced **apoptosis and death** of these cells can be due to berberine-induced oxidative stress (Li et al., 2020). On the other hand, berberine analogues, expressed toxic influence on cultured ovarian cells, showed antioxidant properties. Therefore, it is not to be excluded that berberine can inhibit ovarian cancerogenesis through reduction of reactive oxygen species and induce cancer cell death by induction of oxidative stress.

4.3.6.3 Prostaglandins

Cyclooxygenase-2 (COX-2) is a determinant enzyme in the **metabolic pathway of arachidonic acid**, which converts arachidonic acid into prostanoids, such as PGE2. Activating the cascade reaction of arachidonic acid metabolic pathway causes abnormal alterations of prostaglandin E2 (PGE2) and arachidonic acid levels in the tumor microenvironment (Yang et al., 2012). Phospholipase A2 (iPLA2) promotes activation of COX-2 enzyme which mediates conversion of arachidonic acid into PGE2. PGE2 promotes tumor growth and proliferation through enhancing the phosphorylation of focal adhesion kinase. Berberine block chemotherapy-induced activation of **caspase 3-iPLA2-AA-COX-2-PGE2 pathway** through suppressing the expression of iPLA2 and COX-2, and reverse the elevated phosphorylation of focal adhesion kinase caused by excessive level of PGE2 and therefore reverse the repopulation of ovarian cancer cells after chemotherapy (Mortazavi et al., 2020).

4.3.6.4 Hormones and growth factors

Berberine administrated to women with polycystic ovarian syndrome reduced follicle-stimulating hormone (FSH) (An et al., 2013) and androgen (Li et al., 2015) levels in plasma of woman with polycystic ovarian syndrome. It reduced testosterone secretion also in rats with induced polycystic ovarian syndrome. These rats after administration of berberine were more sensitive to insulin (Li et al., 2015). Therefore, berberine can treat the signs of polycystic ovarian syndrome via changes in release of **FSH, androgen,** and response to **insulin.** Berberine influence on cultured ovarian cancer cells was associated with downregulation of expression of epidermal growth factor (EGF) receptors and its **PI3K /Akt intracellular signaling pathway and vascular endothelial growth factor (VEGF)**, which promotes angiogenesis (Chuang et al., 2021).

The performed studies demonstrated ability of berberine to treat ovarian and endometrial cancer via prostaglandin-dependent oxidative stress,

4.3.7 Application in reproductive biology and medicine

To our knowledge, berberine effect on female reproductive processes in healthy organisms has not been studied yet. Therefore, it remains not clear, whether it could be useful as stimulator or inhibitors of reproductive processes in healthy animals or humans.

Nevertheless, the available reports demonstrate the usefulness of berberine as bioprotector for prevention, mitigation and treatment of the most common reproductive disorders—ovarian and endometrial cancer, and polycystic ovarian syndrome.

The generation and application of berberine analogues with increased efficiency could be promising for improvement of its therapeutic application.

4.3.8 Conclusions and possible direction of future studies

Little is known about the role of berberine in female reproductive processes and its mechanism action in ovarian cells and other cells of reproductive system. The character and mechanisms of action of berberine on healthy female reproductive processes remain unknown yet. Nevertheless, the performed studies demonstrated ability of berberine to treat ovarian and endometrial cancer via prostaglandin-dependent oxidative stress, apoptosis and necroptosis and EGF receptor-dependent regulators of proliferation, migration and angiogenesis in cancer cells. Furthermore, berberine can treat polycystic ovarian syndrome through reduction in FSH and androgen release and resistance to insulin, as well as through intracellular regulators of ovarian cell proliferation and apoptosis.

To our point of view, the future studies in this area could be focused

(1) on berberine action on healthy organisms,
(2) on mechanisms of berberine,
(3) on generation and application of berberine analogues or carriers with increased biological and therapeutic activity,
(4) on safety of berberine in depends on its doses and targets,
(5) on examination of applicability of berberine for control of human, farm animal or poultry reproduction, and
(6) on clinical application of berberine and its analogues for treatment of reproductive dysfunctions.

284 Contaminants and Plants Action on Female Reproduction

In this chapter we have try to summarize the available missing knowledge concerning reproductive and nonreproductive effects of berberine and its possible areas of application. We hope it can encourage further research so that we can take full advantage of berberine biological and medicinal effects.

References

An Y, Sun Z, Zhang Y, Liu B, Guan Y, Lu M. The use of berberine for women with polycystic ovary syndrome undergoing IVF treatment. Clin Endocrinol 2013;80(3):425−31. https://doi.org/10.1111/cen.12294.

Cai Z, Wang C, Yang W. Role of berberine in Alzheimer's disease. Neuropsychiat Dis Treatment 2016;12:2509. https://doi.org/10.2147/NDT.S114846.

Chen C, Tao C, Liu Z, Lu M, Pan Q, Zheng L, Fichna J. A randomized clinical trial of berberine hydrochloride in patients with diarrhea-predominant irritable bowel syndrome. Phytother Res 2015;29(11):1822−7. https://doi.org/10.1002/ptr.5475.

Chuang TC, Wu K, Lin YY, Kuo HP, Kao MC, Wang V, Hsu SC, Lee SL. Dual downregulation of EGFR and ErbB2 by berberine contributes to suppression of migration and invasion of human ovarian cancer cells. Environ Toxicol 2021;36(5):737−47. https://doi.org/10.1002/tox.23076.

Cicero AF, Baggioni A. Berberine and its role in chronic disease. Adv Exp Med Biol 2016;928:27−45. https://doi.org/10.1007/978-3-319-41334-1_2.

Fan J, Zhang K, Jin Y, Li B, Gao S, Zhu J, Cui R. Pharmacological effects of berberine on mood disorders. J Cell Mol Med 2019;23(1):21−8. https://doi.org/10.1111/jcmm.13930.

Fang LH, Wang JH, Du G, Berberine H. Natural small molecule drugs from plants. 2018. p. 371−7. https://doi.org/10.1007/978-981-10-8022-7_62.

Fang C, Schmaier AH. Novel anti-thrombotic mechanisms mediated by Mas receptor as result of balanced activities between the kallikrein/kinin and the renin-angiotensin systems. Pharmacol Res 2020;160:105096. https://doi.org/10.1016/j.phrs.2020.105096.

Fukuda K, Hibiya Y, Mutoh M, Koshiji M, Akao S, Fujiwara H. Inhibition of activator protein 1 activity by berberine in human hepatoma cells. Planta Med 1999;65(04):381−3. https://doi.org/10.1055/s-2006-960795.

Habtemariam S. Berberine and inflammatory bowel disease: a concise review. Pharmacol Res 2016;113:592−9. https://doi.org/10.1016/j.phrs.2016.09.041.

Hesari A, Ghasemi F, Cicero AF, Mohajeri M, Rezaei O, Hayat SMG, Sahebkar A. Berberine: a potential adjunct for the treatment of gastrointestinal cancers? J Cell Biochem 2018;119(12):9655−63. https://doi.org/10.1002/jcb.27392.

Imenshahidi M, Hosseinzadeh H. Berberine and barberry (*Berberis vulgaris*): a clinical review. Phytother Res 2019;33(3):504−23. https://doi.org/10.1002/ptr.6252.

Keefe CC, Goldman MM, Zhang K, Clarke N, Reitz RE, Welt CK. Simultaneous measurement of thirteen steroid hormones in women with polycystic ovary syndrome and control women using liquid chromatography-tandem mass spectrometry. PLoS One 2014;9(4):e93805. https://doi.org/10.1371/journal.pone.0093805.

Kumar A, Chopra K, Mukherjee M, Pottabathini R, Dhull DK. Current knowledge and pharmacological profile of berberine: an update. Eur J Pharmacol 2015;761:288−97. https://doi.org/10.1016/j.ejphar.2015.05.068.

Li L, Li C, Pan P, Chen X, Wu X, Ng EHY, Yang D. A single arm pilot study of effects of berberine on the menstrual pattern, ovulation rate, hormonal and metabolic profiles in anovulatory Chinese women with polycystic ovary syndrome. PLoS One 2015;10(12):e0144072. https://doi.org/10.1371/journal.pone.0144072.

Li W, Li D, Kuang H, Feng X, Ai W, Wang Y, Shi S, Chen J, Fan R. Berberine increases glucose uptake and intracellular ROS levels by promoting Sirtuin 3 ubiquitination. Biomed Pharmacother 2020;121:109563. https://doi.org/10.1016/j.biopha.2019.109563.

Liu DQ, Chen SP, Sun J, Wang XM, Chen N, Zhou YQ, Ye DW. Berberine protects against ischemia-reperfusion injury: a review of evidence from animal models and clinical studies. Pharmacol Res 2019a;148:104385. https://doi.org/10.1016/j.phrs.2019.104385.

Liu L, Fan J, Ai G, Liu J, Luo N, Li C, Cheng Z. Berberine in combination with cisplatin induces necroptosis and apoptosis in ovarian cancer cells. Biol Res 2019b;52(1):37. https://doi.org/10.1186/s40659-019-0243-6.

Luo T, Jiang W, Kong Y, Li S, He F, Xu J, Wang HQ. The protective effects of jatrorrhizine on beta-amyloid (25−35)-induced neurotoxicity in rat cortical neurons. CNS Neurol Disord Drug Targets 2012;11:1030−7. https://doi.org/10.2174/187152712804870928.

Ma YG, Liang L, Zhang YB, Wang BF, Bai YG, Dai ZJ, Wang ZW. Berberine reduced blood pressure and improved vasodilation in diabetic rats. J Mol Endocrinol 2017;59(3):191−204. https://doi.org/10.1530/JME-17-0014.

Mirhadi E, Rezaee M, Malaeekeh-Nikouei B. Nano strategies for berberine delivery, a natural alkaloid of Berberis. Biomed Pharmacother 2018;104:465−73. https://doi.org/10.1016/j.biopha.2018.05.067.

Mistry BM, Shin HS, Kcum YS, Pandurangan M, Kim DH, Moon SH, Kadam AA, Shinde SK, Patel RV. Synthesis and evaluation of antioxidant and cytotoxicity of the N-mannich base of berberine bearing benzothiazole moieties. Anticancer Agents Med Chem 2017;17(12):1652−60. https://doi.org/10.2174/1871520617666170710180549.

Mortazavi H, Nikfar B, Esmaeili SA, Rafieenia F, Saburi E, Chaichian S, Momtazi-Borojeni AA. Potential cytotoxic and anti-metastatic effects of berberine on gynaecological cancers with drug-associated resistance. Eur J Med Chem 2020;187:111951. https://doi.org/10.1016/j.ejmech.2019.111951.

National Center for Biotechnology Information. PubChem Compound Summary for CID 2353, Berberine. PubChem, https://pubchem.ncbi.nlm.nih.gov/compound/Berberine. Accessed June 1, 2021.

Neag MA, Mocan A, Echeverría J, Pop RM, Bocsan CI, Crişan G, Buzoianu AD. Berberine: botanical occurrence, traditional uses, extraction methods, and relevance in cardiovascular, metabolic, hepatic, and renal disorders. Front Pharmacol 2018;9:557. https://doi.org/10.3389/fphar.2018.00557.

Pappalardo MA, Vita R, Di Bari F, Le Donne M, Trimarchi F, Benvenga S. Gly972Arg of IRS-1 and Lys121Gln of PC-1 polymorphisms act in opposite way in polycystic ovary syndrome. J Endocrinol Invest 2017;40(4):367−76. https://doi.org/10.1007/s40618-016-0569-7.

Patwardhan B, Vaidya AD. Natural products drug discovery: accelerating the clinical candidate development using reverse pharmacology approaches. Indian J Exp Biol 2010;48:220−7.

Samadi P, Sarvarian P, Gholipour E, Asenjan KS, Aghebati-Maleki L, Motavalli R, Yousefi M. Berberine: a novel therapeutic strategy for cancer. IUBMB Life 2020;72(10):2065−79. https://doi.org/10.1002/iub.2350.

Shen HR, Xu X, Li XL. Berberine exerts a protective effect on rats with polycystic ovary syndrome by inhibiting the inflammatory response and cell apoptosis. Reprod Biol Endocrinol 2021;19(1):1−11. https://doi.org/10.1186/s12958-020-00684-y.

Song J, Gao X, Tang Z, Li H, Ruan Y, Liu Z, Jiang H. Protective effect of Berberine on reproductive function and spermatogenesis in diabetic rats via inhibition of ROS/JAK2/NFκB pathway. Andrology 2020;8(3). https://doi.org/10.1111/andr.12764.

Tillhon M, Ortiz LMG, Lombardi P, Scovassi AI. Berberine: new perspectives for old remedies. Biochem Pharmacol 2012;84(10):1260−7. https://doi.org/10.1016/j.bcp.2012.07.018.

Warowicka A, Nawrot R, Goździcka-Józefiak A. Antiviral activity of berberine. Arch Virol 2020;165(9):1935–45. https://doi.org/10.1007/s00705-020-04706-3.

Xu J, Long Y, Ni L, Yuan X, Yu N, Wu R, Zhang Y. Anticancer effect of berberine based on experimental animal models of various cancers: a systematic review and meta-analysis. BMC Cancer 2019;19(1):589. https://doi.org/10.1186/s12885-019-5791-1.

Yang IW, Chou CC, Yung BYM. Dose-dependent effects of berberine on cell cycle pause and apoptosis in Balb/c 3T3 cells. Naunyn-Schmiedeberg Arch Pharmacol 1996;354(2):102-108. https://doi.org/10.1007/BF00178709.

Yang P, Cartwright CA, Jin L, Wen S, Prokhorova IN, Shureiqi I, Kim J, et al. Arachidonic acid metabolism in human prostate cancer. Int J Oncol 2012;41(4):1495–503. https://doi.org/10.3892/ijo.2012.1588.

Yu W, Sheng M, Xu R, Yu J, Cui K, Tong J, Du H. Berberine protects human renal proximal tubular cells from hypoxia/reoxygenation injury via inhibiting endoplasmic reticulum and mitochondrial stress pathways. J Transl Med 2013;11(1):1–10. https://doi.org/10.1186/1479-5876-11-24.

Zhang N, Liu X, Zhuang L, Liu X, Zhao H, Shan Y, Liu Z, Li F, Wang Y, Fang J. Berberine decreases insulin resistance in a PCOS rats by improving GLUT4: dual regulation of the PI3K/AKT and MAPK pathways. Regul Toxicol Pharmacol 2020;110:104544. https://doi.org/10.1016/j.yrtph.2019.104544.

Zhou XQ, Zeng XN, Kong H, Sun XL. Neuroprotective effects of berberine on stroke models *in vitro* and *in vivo*. Neurosci Lett 2008;447(1):31–6. https://doi.org/10.1016/j.neulet.2008.09.064.

Chapter 4.4

Capsaicin

4.4.1 Introduction

Capsaicinoids are plant secondary metabolites. The main pungent capsaicinoid is capsaicin. Capsaicin is an active component of red chili peppers from the plant genus *Capsicum* that found uses for flavoring, coloring, and preserving food, as well as for medical purposes (Sun et al., 2016). It is used as a food additive with a unique effect on the pain sensory system (Szolcsányi, 2004; Sung et al., 2012; Chapa-Oliver and Mejía-Teniente, 2016). Moreover, capsaicin is used for therapeutic potential in the treatment of various pathological conditions (Yang et al., 2006; Zhang et al., 2008; Kim et al., 2009; Josse et al., 2010; Sharma et al., 2013; Yuan et al., 2016; Wang et al., 2017). But little is known about the effects of capsaicin on the female reproductive system. Natural bioregulators of ovarian reproduction and understanding of their mechanism of action are important for prediction and control of reproductive processes, as well as prevention or treatment of pathological reproductive conditions (Alatriste et al., 2013; Pintado et al., 2003). Dietary phytochemicals have been found to possess significant health benefits and the development of natural healthy foods and phytonutrients has received increasing attention, however to develop and design a drug with the least side effects is one of the challenging avenues in identifying potential chemotherapeutical agents (Scheau et al., 2019).

In this chapter we have tried to summarize the current knowledge concerning properties, physiological and therapeutical effects of capsaicin focused on female reproductive functions because these data are missing.

4.4.2 Provenance and properties

Capsaicin (8-methyl-N-vanillyl-6-nonenamide), C18H27NO3 (Fig. 4.4) is a crystalline alkaloid with a molecular weight of 305.42 g/mol (Johnson et al., 2007) and a homovanillic acid derivative belongs to capsaicinoids. Similar to other capsaicinoids, capsaicin contains a vanillyl group, an amide group and a fatty acid chain. Capsaicin present major pungent phytocompound extracted from natural plant sources as red chili peppers *Capsicum annuum* and *Capsicum frutescens*. Capsaicin is biosynthesized through two pathways involved in phenylpropanoid and fatty acid metabolism. The capsaicin concentration is mainly affected by several factors include genetic, environmental and crop management factors and varies in the range of 0.1%−1.0% in different peppers (Ranjan et al., 2019; Chapa-Oliver and Mejía-Teniente,

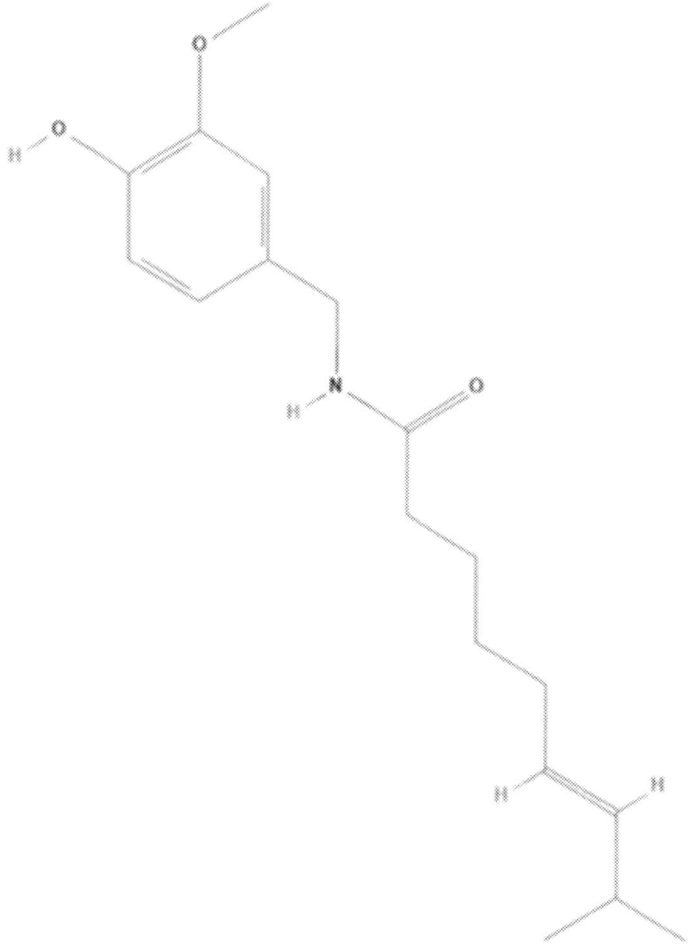

288 Contaminants and Plants Action on Female Reproduction

FIGURE 4.4 Chemical structures of capsaicin (PubChem, 2021).

2016). The content of capsaicin in *Capsicium annuum* is 1.67 mg/g dry weight, and in *Capsicum frutescens* is 0.45 mg/g dry weight (Johnson et al., 2007). Capsaicin seems to be very promising candidates for the treatment of a wide array of diseases, however, constraints such as poor aqueous solubility of capsaicin, gastric irritation, and low oral bioavailability have limited the clinical application of capsaicin (Patowary et al., 2017).

4.4.3 Physiological and therapeutic actions

Capsaicinoids and mainly capsaicin are biologically active substances exhibiting characteristics of great interest to scientific researchers, including their pharmacobiology applications (Chapa-Oliver and Mejía-Teniente, 2016). Research studies have shown multiple effects of capsaicin in a variety of physiological systems including pain, digestive, urinary and cardiovascular system (Josse et al., 2010; O'Neill et al., 2012; Sharma et al., 2013; Dow et al., 2014; Yuan et al., 2016; Patowary et al., 2017) including the prevention of cardiovascular diseases, such as atherosclerosis and coronary heart disease (Peng et al., 2010; Harada, 2009; Fattori et al., 2016). Moreover, capsaicin is used for therapeutic potential in the treatment of various conditions such as analgesics (Josse et al., 2010; Sharma et al., 2013; Srinivasan, 2015; Yuan et al., 2016; James and Drummond, 2018; Pethő et al., 2017), rheumatoid arthritis, diabetic neuropathy, as well as lipid metabolic disorders, obesity (Josse et al., 2010; Sharma et al., 2013; Srinivasan, 2015; Yuan et al., 2016), and positively affects glucose and insulin levels in humans (Yuan et al., 2016). There is persuasive experimental evidence that dietary phytosubstance capsaicin has antiantioxidant (Zimmer et al., 2012; Loizzo et al., 2015; Tanrıkulu-Küçük et al., 2019b), antiinflammatory (Zimmer et al., 2012), antibacterial (Johnson, 2007; Mokhtar et al., 2017), retinal protective, and anticancer (Friedman et al., 2018; Adaszek et al., 2018) effects against the many types of cancers (Kim et al., 2009; Sharma et al., 2013; Yang et al., 2006; Zhang et al., 2008; Wang et al., 2017) including pharyngeal (Kamaruddin et al., 2019), colon (Kim et al., 2004), pancreatic (Zhang et al., 2013), hepatocarcinoma (Jung et al., 2011), nonsmall cell lung cancer (Brown et al., 2010), bladder (Islam et al., 2019), breast (Chou et al., 2009; Wu et al., 2020), prostate (Venier et al., 2015), and ovarian (Lin et al., 2017) cancer.

It has reported that vesical instillation of capsaicin improved urinary frequency and urge incontinence in patients with hypersensitive bladders (Patowary et al., 2017). Further study has suggested the role of capsaicin in cell metabolism regulation. It was demonstrated that the capsaicin consumption is associated with a lower prevalence of obesity and can reduce insulin resistance (Kang et al., 2010). Numerous studies provide experimental evidence that dietary chili peppers and capsaicin consumption reduce body weight and capsaicin has a potential antiobesity effect (Yu et al., 2012; Ludy et al., 2012; Janssens et al., 2014; Zheng et al., 2017). High-fat diets are related to hepatic oxidant−antioxidant imbalance and the onset of liver diseases. Supplementation with capsaicin balances the hepatic oxidant-antioxidant status and may have a protective role against the development of apoptosis in fatty liver disease (Tanrıkulu-Küçük et al., 2019b). Moreover, capsaicin can inhibit acid secretion, stimulate alkali, and mucus secretion, as well as gastric mucosal blood flow which helps in the prevention and treatment of gastric ulcers. In addition, beneficial impact of capsaicin application on gastrointestinal system includes digestive stimulatory activity and modulation of intestinal ultrastructure, which leads to enhance permeability to micronutrients (Srinivasan, 2015). Capsaicin possesses antioxidant activity and have the potential in the treatment of testicular torsion (Javdan et al., 2018a). In addition, capsaicin may

290 Contaminants and Plants Action on Female Reproduction

have a role in the management of prostate cancer. The chemopreventive effect of capsaicin on prostate cancer was investigated in the transgenic adenocarcinoma of the mouse prostate model (Venier et al., 2015). Moreover, oral capsaicin administration slowed the growth of prostate cancer cells in mice (Mori et al., 2006). On the other hand, in another *in vitro* study was found that capsaicin adversely affected spermatogonial survival by inducing apoptosis to rat spermatogonial stem cells (Mizrak et al., 2008). There is an evidence suggested the health benefits of combinational use of capsaicin with other dietary phytocompounds (Clark and Lee, 2016). The combination treatment with capsaicin and curcumin offers remarkable and beneficial effects on the antioxidative defense system in testicular oxidative damage by reduction of testicular and hepatic oxidative stress and ameliorates antioxidant enzyme activities in the liver and testis (Tanrıkulu-Küçük et al., 2019a).

In an acute oral toxicity studies have been demonstrated for capsaicin oral LD50 values as low as 161.2 mg/kg (rats) and 118.8 mg/kg (mice). Topical application to human subjects resulted in side effects as transient sensations of burning and itching (Johnson, 2007). Interestingly, capsaicin plays a dual role in tumorigenesis as a carcinogen or a chemopreventive agent (Bode and Dong, 2011). The results of the long-term topical application of capsaicin demonstrated that capsaicin may increase skin carcinogenesis (Bode and Dong, 2011). The biological functions of capsaicin are affected by its concentration and the effective concentration in different cancer cells varies considerably. Therefore, well-controlled studies to assess the safety and efficacy of capsaicin and further preclinical and clinical trials to confirm the beneficial properties are required (Zhang et al., 2019). This phytosubstance has been successfully clinically applied in dermatology and capsaicin creams have been in clinical use to relieve a variety of painful conditions (Caruntu et al., 2015). However, their effectiveness in pain relief is highly debated and some adverse side effects have been reported (Srinivasan, 2015). Therefore, a real challenge is to establish effective capsaicin doses for chronic conditions, which can exert health benefits and therapeutic effects.

Although the character of the capsaicin effect remains controversial, the current review illuminates the promising use of natural phytosubstance capsaicin as a potential drug against various diseases of pain, digestive, urinary and cardiovascular system and metabolic disorders, obesity for its analgetic, antioxidant, antiinflammatory and anticancer effects.

4.4.4 Mechanisms of action

Capsaicinoids such as capsaicin are potential agonists of transient receptor potential vanilloid 1 (TRPV1) (Johnson et al., 2007) and exert the effects through the receptor-dependent pathway but also through the receptor-independent one (Srinivasan, 2015). TRPV1 plays a crucial role in the regulation of metabolic processes, including body weight, glucose and lipid metabolism, and cardiovascular systems (Sun et al., 2016; Zheng et al., 2017).

Capsaicin stimulates the release of neuropeptide calcitonin gene-related peptide (CGRP) through the activation of TRPV1 resulted in a decrease in blood pressure (Peng et al., 2010). Long-term activation of TRPV1 by capsaicin can decrease lipid storage and atherosclerotic lesions (Ma et al., 2011). Capsaicin can increase the resistance of low-density lipoprotein (LDL) to oxidation by delaying the initiation of oxidation and slowing the rate of oxidation *in vitro* (Ahuja et al., 2006). A capsaicin diet attenuates blood pressure elevation and to cause vasorelaxation via nitric oxide (NO) activation, probably through enhancing phosphorylation of Akt and endothelial nitric oxide synthase (Segawa et al., 2019). The mechanisms of action underlying the anti-obesity effects of capsaicin include an increase in lipid oxidation and inhibition of adipogenesis, activation of brown adipose tissue activity and induction of thermogenesis, reduction of appetite, and an increase of satiety regulated by neuronal circuits in the hypothalamus, and finally modulation of the function of the gastrointestinal tract and the gut microbiome (Zheng et al., 2017). Preclinical and *in vitro* research have confirmed the effectiveness of low-dose capsaicin diet in attenuating metabolic disorders mediated via activation of TRPV1, which can modulate biological processes including browning of adipocytes and activation of metabolic modulators such as AMP-activated protein kinase (AMPK), peroxisome proliferator-activated receptor α (PPAR α), and glucagon-like peptide 1 (GLP-1) (Panchal et al., 2018).

Oxidative stress plays a crucial role and is the main cause of male infertility (Villaverde et al., 2019). Testicular torsion is a serious urologic emergency that may cause a sequential chain of pathophysiological effects, including DNA damage, lipid peroxidation, acute inflammatory responses, and reactive oxygen species (ROS) generation resulted in the derangement in sperm functions and infertility (Amani et al., 2017; Çayana et al., 2014; Javdan et al., 2018a). The results of the current study showed that capsaicin reduced oxidative stress level, improved sperm quality, increased sperm concentration, motility, and reduced lipid peroxidation in rats (Hosseini et al., 2020). The improvement of sperm quality may be partly related to affinity of capsaicin to TRPV1 receptor, present on spermatozoa which can modulate Ca^{2+} level playing important role in sperm motility (Hosseini et al., 2020). Capsaicin can protect against spermatogenic cell death and testicular injury induced by scrotal hyperthermia, reduce oxidative stress and suppress apoptosis in testes (Park et al., 2017). Moreover, capsaicin reduces oxidative stress by decreasing levels of malondialdehyde (MDA), phospholipid hydroperoxide glutathione peroxidase (PHGPx), heat shock 70-kDa protein 1, and manganese superoxide dismutase (MnSOD) and suppress apoptosis induced by heat stress in testes (Park et al., 2017).

Capsaicin can alter the expression of numerous genes involved in cancer cell survival, growth arrest, angiogenesis, and metastasis (Clark and Lee, 2016). Capsaicin mediates cell cycle arrest and induce apoptosis, thus capsaicin causes cytotoxicity in many different types of cancer cells

(Kim et al., 2004; Chou et al., 2009; Brown et al., 2010; Jung et al., 2011; Lin et al., 2017; Kamaruddin et al., 2019; Wu et al., 2020). However, molecular mechanism of its action is not fully understood (Szoka and Palka, 2020).

Capsaicin targets multiple signaling pathways, acts as a tumor promotor and tumor suppressor genes in various types of cancer and noncancer models:

(1) inhibition of the viability and migration, induction of G2/M cell cycle arrest via the mitochondria-mediated intrinsic apoptotic pathway (Kamaruddin et al., 2019; Wu et al., 2020),

(2) the regulation of transcription factor expression, growth of signal transduction pathways, as well as cell cycle arrest (Bley et al., 2012; Lin et al., 2013),

(3) activation of AMP-activated protein kinase (AMPK) leads to phosphorylation of proteins involved in cell-cycle regulation and apoptosis, such as proteins p21 and p53, as well as proteins involved in autophagy (Bort et al., 2017),

(4) increasing p53 levels by interacting with proteins Bcl-2 and a cytosolic release of cytochrome c that triggers cell apoptosis (Scheau et al., 2019; Ozaki et al., 2011),

(5) induction of apoptosis in cancer cells, activation of caspase-3, -7, and -9 through an intrinsic apoptotic pathway, and subsequently, apoptotic DNA fragmentation and arrests the cell cycle at the G1 phase (Kamaruddin et al., 2019),

(6) inhibition of the phosphatidylinositol 3-kinase (PI3K) /Akt/Wnt/β-catenin signaling pathway in breast cancer, reduction of cell viability and suppression of cyclin-dependent kinase eight expression, decrease of the phosphorylation of PI3K and Akt, and downregulated Wnt and β-catenin expression (Wu et al., 2020),

(7) decreasing FBI-1, Ki-67, Bcl-2, and survivin protein expression, increasing Bax protein expression and activation of caspase-3 (Chen et al., 2021),

(8) decreasing Bax, caspase-3, increasing total antioxidant status (TAS) and Bcl-2 and downregulation overproduction of ROS in animal fatty liver model (Tanrıkulu-Küçük et al., 2019b),

(9) inhibiting the Akt/mTOR signaling pathway (Lin et al., 2017),

(10) upregulation of the activity of the signal transducer and activator of transcription 3 (p-STAT3) in liver hepatocellular carcinoma cells, and modulation of signaling pathways by the triggering of apoptosis including caspase-3 activation and production of ROS (Chen et al., 2016; Scheau et al., 2019),

(11) blockation of the activation of activator protein 1 (AP-1), nuclear factor kappa B (NF-kB), and signal transducer and activator of transcription 3 (STAT3) signaling pathways (Oyagbemi et al., 2010),

(12) downregulation of sirtuin 1 (SIRT1) deacetylase and reduction of cell migration by direct binding with SIRT1 in bladder cancer (Islam et al., 2019),

(13) stimulation of intrinsic and extrinsic pathways of apoptosis in human glioblastoma cells is not dependent on TRPV1, upregulation of the expression of peroxisome proliferator- activated receptor γ (PPAR-γ) in glioblastoma cells (Szoka and Palka, 2020),

(14) downregulation of endothelial growth factor (VEGF) in nonsmall cell lung carcinoma cells (Chakraborty et al., 2014),

(15) decreasing of the expression of proapoptotic factors, increases the expression of antiapoptotic factors, improving testicular morphology and reduction of apoptosis in *testes* (Javdan et al., 2018a),

(16) the antioxidant and anti-apoptotic effects of the capsaicin via regulation of the forkhead box protein O1 (FOXO1) pathway (Javdan et al., 2018a), and

(17) reduction of cell damage and suppression of apoptotic cell death via the mammalian target of rapamycin (mTOR) signaling pathway by over-phosphorylation of mTOR in testicular cells following torsion (Javdan et al., 2018b).

Overall, the available data demonstrate the multiple molecular targets responsible for beneficial effects and mechanism of capsacin action in cancer and noncancer cells. Moreover, the anticancer and antioxidant activity in multiple cell types by activating capsaicin receptors and regulating various intracellular signaling pathways by capsaicin was suggested.

4.4.5 Effects on female reproductive processes

4.4.5.1 Effect on ovarian and reproductive state

Both stimulatory and inhibitory action of capsaicin on ovarian and repro-ductive state has been reported. The character of its action dependent of the administrated capsaicin dose.

Administration of capsaicin (0.5—1.0 mg/kg/day) to prepubertal rats stimulate development of their ovarian follicles and prevented their atresia (Zik et al., 2010a,b; Tütüncü et al., 2016; Güler and Zik, 2018). Feeding of hens with red pepper diet (10 g/kg) increased their ovarian weight, follicle number, and induced earlier the onset of puberty (Özer et al., 2005). Moreover, capsaicin prevented the degenerative changes induced in rat ovaries by chemotherapy. This effect was associated with increase in antioxidant enzyme level in the blood and with decrease in morphological signs of inflammation and degeneration - hemorrhage, vascular congestion, and mononuclear cell infiltration (Melekoglu et al., 2018). The dietary capsaicin (150 mg/kg) pro-moted growth of duck ovarian follicles and egg production and weight. This effect was associated with increased expression of blood antioxidant enzymes (Liu et al., 2021). These reports suggest that pepper capsaicin given at low dose can promote sexual maturation and ovarian folliculogenesis and prevent ovarian follicle atresia in rats and chicken. In duck's capsaicin done it even at high dose.

On the other hand, the diet containing high (50 mg/kg) capsaicin dose did not affect terms of rat puberty (Morán et al., 2003) and courtship behavior (Pintado et al., 2003), but it lowered rat reproductive success, increased the number of ovarian atretic follicles, reduced number of ovulations and litter size, although it did not affect the courtship behavior (Pintado et al., 2003; Morales-Ledesma et al., 2015; Morán et al., 2003). Capsaicin at high doses (30 nM) induced a delay in the onset of puberty and reduced the number of both preantral and antral ovarian follicles in guinea pigs (Alatriste et al., 2013). Capsaicin at high dose (0.01−100 μM) was also toxic for cultured ovarian cancer cells (Arzuman et al., 2014; Catanzaro et al., 2014).

4.4.5.2 Effect on ovarian cell functions

Capsaicin given at low dose promoted not only rat ovarian folliculogenesis, but also reduced their apoptosis (Zik et al., 2010a,b). On the other hand, cultured rat granulosa cells after addition of high dose of curcumin, had increased occurrence of apoptosis (accumulation of cleaved caspase-3 and cleaved PARP), while their proliferation and viability were not changed (Güler and Zik, 2018). Furthermore, capsaicin given at high dose arrested the cell cycle, suppressed cell growth, and induce apoptosis in human ovarian carcinoma cells (Arzman et al., 2014; Catanzaro et al., 2014). These changes were associated with inhibition of proinflammatory transcription factors nuclear factor-κB (NF-κB) and signal transducer and activator of transcription 3 (STAT3) induced by proinflammatory interleukin 6 (IL-6) (Arzuman et al., 2014).

4.4.5.3 Effect on reproductive hormones

Administration of capsaicin to neonatal rats reduced plasma level of steroid hormones, progesterone, testosterone and estradiol, but not of gonadotropins follicle-stimulating hormone (FSH) and luteinizing hormone (LH) (Quiroz et al., 2014) indicating that capsaicin reduces not gonadotropin secretion, but response of ovarian cells to gonadotropins. These capsaicin-induced changes were associated with lower noradrenergic and serotonergic activities in the anterior hypothalamus and of the dopaminergic activity in the median hypothalamus (Quiroz et al., 2014).

Melekoglu et al. (2018) however reported the reduction level of gonadotropins FSH and LH and increase in estradiol and anti-Mullerian hormone concentration in plasma of rats treated with capsaicin.

4.4.6 Mechanisms of action on female reproductive processes

The character and mechanisms of capsaicin action on female reproductive processes are studied insufficiently, but the few available reports enable outline several possible mechanisms of their action on female reproduction.

The ability of capsaicin to activate antioxidant enzyme in rats (Melekoglu et al., 2018) and ducks (Li et al., 2021) indicate that capsaicin can protect ovarian cells from the **oxidative stress.**

Capsaicin induced accumulation of nitric oxide in the rat ovaries indicating that this signaling molecule could be one of potential mediators of capsaicin action (Zik et al., 2012). Nitric oxide is involved in upregulation of **inflammatory** processes. This fact, together with ability of capsaicin to reduce markers (NF-κB, STAT3, IL-6) (Arzuman et al., 2014) and morphological signs (Melekoglu et al., 2018) of ovarian inflammations indicate that capsaicin can affect ovarian functions as antiinflammatory agent.

The capsaicin action on ovarian follicle atresia (Morán et al., 2003; Tütüncü et al., 2016), **proliferation** and **apoptosis** of ovarian cells *in vivo* (Zik et al., 2010a,b; Güler and Zik, 2018) and *in vitro* (Catanzaro et al., 2014) suggest that changes in ovarian cell proliferation and apoptosis can mediate the influence of capsaicin on ovarian cell state and ovarian folliculogenesis.

Capsaicin was able to affect steroid **hormones** and gonadotropins and to promote release of anti-Mullerian hormone (Quiroz et al., 2014; Melekoglu et al., 2018). These hormones are important stimulators of ovarian cells and folliculogenesis (Sirotkin, 2014). Therefore, these hormones could be probably mediators of capsaicin action on ovarian functions.

Capsaicin can diminish not only plasma steroid hormones level, but also noradrenergic, serotonergic and dopaminergic activities in hypothalamus (Quiroz et al., 2014) indicating the possible involvement of **hypothalamic monoamines** in mediating some capsaicin effects.

Capsaicin can induce degeneration of ovarian sensory nerves and induce **sensory denervation** and blockage the release of neuropeptides from guinea pig and rat ovarian sensory nerves (Alatriste et al., 2013; Pintado et al., 2003; Quiroz et al., 2014; Morales-Ledesma et al., 2015). This suggests that capsaicin can affect ovarian functions *in vivo* by inducing their denervation.

These observations suggest that capsaicin influence on ovarian healthy and cancer cells could be mediated by multiple intracellular signaling pathway involving regulators of oxidative stress, inflammation, proliferation and apoptosis, hormones and neuromediators.

4.4.7 Application in reproductive biology and medicine

The available data demonstrate both stimulatory and inhibitory action of capsaicin on reproductive functions at different regulatory levels. These data enable to outline possible areas of practical application of this molecule.

The ability of capsaicin given al low doses to promote ovarian functions and fecundity (Özer et al., 2005; Zik et al., 2010a,b; Tütüncü et al., 2016; Güler and Zik, 2018) in mammals and egg production in docks (Liu et al., 2013) indicates its potential applicability for promotion of animal and human reproduction and for treatment of some cases of infertility. Moreover,

296 Contaminants and Plants Action on Female Reproduction

capsaicin can protect ovarian cells against natural and induced apoptosis and ovarian follicles against inflammation and atresia (Melekoglu et al., 2018). This indicates its applicability as natural protector against action of destructive factors (chemotherapy, aging). The ability of capsaicin to affect hypothalamic monoamines and to block innervation confirm its ability to cure pain during some reproductive and nonreproductive disorders. Finally, ability of capsaicin to suppress proliferation and viability and to promote apoptosis and death of ovarian cancer cells (Arzuman et al., 2014; Catanzaro et al., 2014) indicates its usefulness as anticancer drug. Nevertheless, the expressed dose-dependent (sometimes opposite) effect of capsaicin indicates the importance of careful dosage of capsaicin for biomedical uses.

4.4.8 Conclusions and possible direction of future studies

The data summarized here demonstrated both stimulatory and protective and inhibitory and even toxic (independent of dose) action of capsaicin on female reproductive processes—sexual maturation, ovarian folliculogenesis, ovarian cell proliferation, apoptosis, viability, and fecundity in various experimental models. These effects could be mediated by multiple intracellular signaling pathway involving regulators of oxidative stress, inflammation, proliferation and apoptosis, hormones and neuromediators. Capsaicin action on ovarian cells could be useful for stimulation and protection of ovarian cells and fecundity, as well as for prevention and treatment of some reproductive dysfunctions like cancer.

Nevertheless, this hypothesis should be verified by suitable *in vivo* animal and clinical experiments. The mediators of capsaicin action on reproductive processes are outlined here, but their mediatory role remain to be demonstrated yet. The influence of capsaicin on reproductive organs other than ovary has not been examined yet. The application of capsaicin for promotion farm animal and avian production, as protector and as drug for treatment of reproductive disorders could be promising, but it requires a number of profound studies.

References

Adaszek Ł, Slabczynska O, Łyp P, Gadomska D, Zieter J, Rozanska D, Orzelski M, Smiech A, Staniec M, Krasucka D, Winiarczyk S. Comparison of the *in vitro* anticancer effect of habanero pepper extract containing capsaicin with that of pure capsaicin in selected dog neoplastic cell lines. Turk J Vet Anim Sci 2018;42:243−50. https://doi.org/10.3906/vet-1707-6.

Ahuja KDK, Kunde DA, Ball MJ, Geraghty DP. Effects of capsaicin, dihydrocapsaicin, and curcumin on copper-induced oxidation of human serum lipids. J Agric Food Chem 2006;54(17):6436−9. https://doi.org/10.1021/jf061331j.

Alatriste V, Herrera-Camacho I, Martinez MI, Limon ID, Gonzalez-Flores O, Luna F. Sensory denervation with capsaicin reduces ovarian follicular development and delays the onset of puberty in Guinea pigs. Adv Reprod Sci 2013;1(3):29−37. https://doi.org/10.4236/arsci.2013.13005.

Amani H, Habibey R, Hajmiresmail SJ, Latifi S, Pazoki-Toroudi H, Akhavan O. Antioxidant nanomaterials in advanced diagnoses and treatments of ischemia reperfusion injuries. J Mater Chem B 2017;48(5):9452−76. https://doi.org/10.1039/C7TB01689A.

Arzuman L, Beale P, Chan C, Yu JQ, Huq F. Synergism from combinations of tris(benzimidazole) monochloroplatinum(II) chloride with capsaicin, quercetin, curcumin and cisplatin in human ovarian cancer cell lines. Anticancer Res 2014;34(10):5453−64.

Bley K, Boorman G, Mohammad B, McKenzie D, Babbar SA. Comprehensive review of the carcinogenic and anticarcinogenic potential of capsaicin. Toxicol Pathol 2012;40(6):847−73. https://doi.org/10.1177/0192623312444471.

Bode AM, Dong Z. The two faces of capsaicin. Cancer Res 2011;71(8):2809−14. https://doi.org/10.1158/0008-5472.

Bort A, Spínola E, Rodríguez-Henche N, Díaz-Laviada I. Capsaicin exerts synergistic antitumor effect with sorafenib in hepatocellular carcinoma cells through AMPK activation. Oncotarget 2017;8(50):87684−98. https://doi.org/10.18632/oncotarget.21196.

Brown KC, Witte TR, Hardman WE, Luo H, Chen YC, Carpenter AB, Lau JK, Dasgupta P. Capsaicin displays anti-proliferative activity against human small cell lung cancer in cell culture and nude mice models via the E2F pathway. PLoS One 2010;5(4):e10243. https://doi.org/10.1371/journal.pone.0010243.

Caruntu C, Negrei C, Ghita MA, Caruntu A, Badarau AI, Buraga I, Boda D, Albu A, Branisteanu D. Capsaicin, a hot topic in skin pharmacology and physiology. Farmacia 2015;63(4):487−91.

Catanzaro D, Vianello C, Ragazzi E, Caparrotta L, Montopoli M. Cell cycle control by natural phenols in cisplatin-resistant cell lines. Nat Prod Commun 2014;9(10):1465−8. https://doi.org/10.1177/1934578x1400901015.

Çayana S, Saylama B, Tiftik N, Ünal ND, Apa DD, Efesoy O, Çimen B, Bozlu M, Akbay E, Büyükafşar K. Rho-kinase levels in testicular ischemia-reperfusion injury and effects of its inhibitor, Y-27632, on oxidative stress, spermatogenesis, and apoptosis. Urology 2014;83(3):675e13−8. https://doi.org/10.1016/j.urology.2013.11.032.

Chakraborty S, Adhikary A, Mazumdar M, Mukherjee S, Bhattacharjee P, Guha D, Choudhuri T, Chattopadhyay S, Sa G, Sen A, Das T. Capsaicin-induced activation of p53-SMAR1 auto-regulatory loop downregulates VEGF in non-small cell lung cancer to restrain angiogenesis. PLoS One 2014;9(6):e99743. https://doi.org/10.1371/journal.pone.0099743.

Chapa-Oliver AM, Mejía-Teniente L. Capsaicin: from plants to a cancer-suppressing agent. Molecules 2016;21(8):931. https://doi.org/10.3390/molecules21080931.

Chen X, Tan M, Xie Z, Feng B, Zhao Z, Yang K, Hu C, Liao N, Wang T, Chen D, Xie F, Tang C. Inhibiting ROS-STAT3-dependent autophagy enhanced capsaicin-induced apoptosis in human hepatocellular carcinoma cells. Free Radic Res 2016;50(7):744−55. https://doi.org/10.3109/10715762.2016.1173689.

Chen M, Xiao C, Jiang W, Yang W, Qin Q, Tan Q, Lian B, Liang Z, Wei C. Capsaicin inhibits proliferation and induces apoptosis in breast cancer by downregulating FBI-1-Mediated NF-κB pathway. Drug Des Dev Ther 2021;15:125−40. https://doi.org/10.2147/DDDT.S269901.

Chou C-C, Wu Y-C, Wang Y-F, Chou M-J, Kuo S-J, Chen D-R. Capsaicin-induced apoptosis in human breast cancer MCF-7 cells through caspase-independent pathway. Oncol Rep 2009;21(3):665−71. https://doi.org/10.3892/or_00000269.

Clark R, Lee S-H. Anticancer properties of capsaicin against human cancer. Anticancer Res 2016;36(3):837−44.

Dow J, Simkhovich BZ, Hale SL, Kay G, Kloner RA. Capsaicin-induced cardioprotection. Is hypothermia or the salvage kinase pathway involved? Cardiovasc Drugs Ther 2014;28:295–301. https://doi.org/10.1007/s10557-014-6527-8.

Fattori V, Hohmann MSN, Rossaneis AC, Pinho-Ribeiro FA, Verri WA. Capsaicin: current understanding of its mechanisms and therapy of pain and other pre-clinical and clinical uses. Molecules 2016;21(7):844. https://doi.org/10.3390/molecules21070844.

Friedman JR, Nolan NA, Brown KC, Miles SL, Akers AT, Colclough KW, Seidler JM, Rimoldi JM, Valentovic MA, Dasgupta P. Anticancer activity of natural and synthetic capsaicin analogs. J Pharmacol Exp Therapeut 2018;364(3):462–73. https://doi.org/10.1124/jpet.117.243691.

Güler S, Zik R. Effects of capsaicin on ovarian granulosa cell proliferation and apoptosis. Cell Tissue Res 2018;372:603–9. https://doi.org/10.1007/s00441-018-2803-4.

Harada N, Okajima K. Effects of capsaicin and isoflavone on blood pressure and serum levels of insulin-like growth factor-I in normotensive and hypertensive volunteers with alopecia. Biosci Biotechnol Biochem 2009;73(6):1456–9. https://doi.org/10.1271/bbb.80883.

Hosseini M, Tavalaee M, Rahmani M, Eskandari A, Shaygannia E, Kiani-Esfahani A, Zohrabi D, Nasr-Esfahani MH. Capsaicin improves sperm quality in rats with experimental varicocele. Andrologia 2020;52(11). https://doi.org/10.1111/and.13762.

Islam A, Yang Y-T, Wu W-H, Chueh PJ, Lin M-H. Capsaicin attenuates cell migration via SIRT1 targeting and inhibition to enhance cortactin and beta-catenin acetylation in bladder cancer cells. Am J Cancer Res 2019;9(6):1172–82. PMID: 31285950.

James K, Drummond PD. Rapid induction analgesia for capsaicin-induced pain in university students: a randomized, controlled trial. Int J Clin Exp Hypn 2018;66(4):428–50. https://doi.org/10.1080/00207144.2018.1495010.

Janssens PLHR, Hursel R, Westerterp-Plantenga MS. Capsaicin increases sensation of fullness in energy balance, and decreases desire to eat after dinner in negative energy balance. Appetite 2014;77(1):44–9. https://doi.org/10.1016/j.appet.2014.02.018.

Javdan N, Ayatollahi SA, Choudhary MI, Al-Hasani S, Pazoki-Toroudi H. FOXO1 targeting by capsaicin reduces tissue damage after testicular torsion. Andrologia 2018a;50(4). https://doi.org/10.1111/and.12987.

Javdan N, Ayatollahi SA, Choudhary MI, Al-Hasani S, Kobarfard F, Athar A, Pazoki-Toroudi H. Capsaicin protects against testicular torsion injury through mTOR-dependent mechanism. Theriogenology 2018b;113:247–52. https://doi.org/10.1016/j.theriogenology.2018.03.012.

Johnson W. Final report on the safety assessment of *Capsicum annuum* extract, *Capsicum annuum* fruit extract, *Capsicum annuum* resin, *Capsicum annuum* fruit powder, *Capsicum Frutescens* fruit, *Capsicum frutescens* fruit extract, *Capsicum frutescens* resin, and Capsaicin. Int J Toxicol 2007;26(1):3–106. https://doi.org/10.1080/10915810601163939.

Josse AR, Sheerffs SS, Holwerda AM, Andrews R, Staples AW, Philips SM. Effects of capsinoid digestion on energy expenditure and lipid oxidation at rest and during exercise. Nutr Metab 2010;7:65. https://doi.org/10.1186/1743-7075-7-65.

Jung MY, Kang H-J, Moon A. Capsaicin-induced apoptosis in SK-Hep-1 hepatocarcinoma cells involves Bcl-2 downregulation and caspase-3 activation. Cancer Lett 2011;165(2):139–45. https://doi.org/10.1016/S0304-3835(01)00426-8.

Kamaruddin MF, Hossain MZ, Alabsi AM, Bakri MM. The antiproliferative and apoptotic effects of capsaicin on an oral squamous cancer cell line of asian origin, ORL-48. Medicina 2019;55(7):322. https://doi.org/10.3390/medicina55070322.

Kang J-H, Goto T, Han I-S, Kawada T, Kim YM, Yu R. Dietary capsaicin reduces obesity-induced insulin resistance and hepatic steatosis in obese mice fed a high-fat diet. Obesity 2010;18(4):780–7. https://doi.org/10.1038/oby.2009.301.

Kim C-S, Park W-H, Park J-Y, Kang J-H, Kim M-O, Kawada T, Yoo H, Han I-S, Yu R. Capsaicin, a spicy component of hot pepper, induces apoptosis by activation of the peroxisome proliferator-activated receptor in HT-29 human colon cancer cells. J Med Food 2004;7(3):267–73. https://doi.org/10.1089/jmf.2004.7.267.

Kim MY, Trudel LJ, Wogan GN. Apoptosis induced by capsaicin and resveratrol in colon carcinoma cells requires nitric oxide production and caspase activation. Anticancer Res 2009;29(10):3733–40. PMID: 19846903.

Lin C-H, Lu W-C, Wang C-W, Chan Y-C, Chen M-K. Capsaicin induces cell cycle arrest and apoptosis in human KB cancer cells. BMC Compl Altern Med 2013;13:46. https://doi.org/10.1186/1472-6882-13-46.

Lin Y-T, Wang H-C, Hsu Y-C, Cho C-L, Yang M-Y, Chien C-Y. Capsaicin induces autophagy and apoptosis in human nasopharyngeal carcinoma cells by downregulating the PI3K/AKT/mTOR pathway. Int J Mol Sci 2017;18(7):1343. https://doi.org/10.3390/ijms18071343.

Liu M, Yin Y, Ye X, Zeng M, Zhao Q, Keefe DL, Liu L. Resveratrol protects against age-associated infertility in mice. Hum Reprod 2013;28:707–17.

Loizzo MR, Pugliese A, Bonesi M, Menichini F, Tundis R. Evaluation of chemical profile and antioxidant activity of twenty cultivars from *Capsicum annuum, Capsicum baccatum, Capsicum chacoense* and *Capsicum chinense*: a comparison between fresh and processed peppers. LWT-Food Sci Technol. 2015;64(2):623–31. https://doi.org/10.1016/j.lwt.2015.06.042.

Ludy M-J, Moore GE, Mattes RD. The effects of capsaicin and capsiate on energy balance: critical review and meta-analyses of studies in humans. Chem Senses 2012;37(2):103–21. https://doi.org/10.1093/chemse/bjr100.

Ma L, Zhong J, Zhao Z, Luo Z, Ma S, Sun J, He H, Zhu T, Liu D, Zhu Z, Tepel M. Activation of TRPV1 reduces vascular lipid accumulation and attenuates atherosclerosis. Cardiovasc Res 2011;92(3):504–13. https://doi.org/10.1093/cvr/cvr245.

Melekoglu R, Ciftci O, Eraslan S, Cetin A, Basak N. Beneficial effects of curcumin and capsaicin on cyclophosphamide-induced premature ovarian failure in a rat model. J Ovarian Res 2018;11(1):33. https://doi.org/10.1186/s13048-018-0409-9.

Mizrak SC, Gadella BM, Erdost H, Ozer A, van Pelt AM, van Dissel-Emiliani FM. Spermatogonial stem cell sensitivity to capsaicin: an in vitro study. Reprod Biol Endocrinol 2008;6:52. https://doi.org/10.1186/1477-7827-6-52.

Mokhtar M, Ginestra G, Youcefi F, Filocamo A, Bisignano C, Riazi A. Antimicrobial activity of selected polyphenols and capsaicinoids identified in pepper (*Capsicum annuum* L.) and their possible mode of interaction. Curr Microbiol 2017;74(11):1253–60. https://doi.org/10.1007/s00284-017-1310-2.

Morales-Ledesma L, Trujillo A, Apolonio J. In the pubertal rat, the regulation of ovarian function involves the synergic participation of the sensory and sympathetic innervations that arrive at the gonad. Reprod Biol Endocrinol 2015;13:61. https://doi.org/10.1186/s12958-015-0062-8.

Morán C, Morales L, Razo RS, Apolonio J, Quiróz U, Chavira R, Domínguez R. Effects of sensorial denervation induced by capsaicin injection at birth or on day three of life, on puberty, induced ovulation and pregnancy. Life Sci 2003;73(16):2113–25. https://doi.org/10.1016/s0024-3205(03)00598-8.

Mori A, Lehmann S, O'Kelly J, Kumagai T, Desmond JC, Pervan M, McBride WH, Kizaki M, Koeffler HP. Capsaicin, a component of red peppers, inhibits the growth of androgen-independent, p53 mutant prostate cancer cells. Cancer Res 2006;66(6):3222–9. https://doi.org/10.1158/0008-5472.CAN-05-0087.

300 Contaminants and Plants Action on Female Reproduction

National Center for Biotechnology Information. PubChem Compound Summary for CID 1548943, Capsaicin. PubChem, https://pubchem.ncbi.nlm.nih.gov/compound/Capsaicin. Accessed June 1, 2021.

O'Neill J, Brock C, Olesen AE, Andresen T, Nilsson M, Dickenson AH, Dolphin AC. Unravelling the mystery of capsaicin: a tool to understand and treat pain. Pharmacol Rev 2012;64(4):939−71. https://doi.org/10.1124/pr.112.006163.

Oyagbemi AA, Saba AB, Azeez OI. Capsaicin: a novel chemopreventive molecule and its underlying molecular mechanisms of action. Indian J Cancer 2010;47(1):53−8. https://doi.org/10.4103/0019-509X.58860.

Ozaki T, Nakagawara A. Role of p53 in cell death and human cancers. Cancers 2011;3(1):994−1013. https://doi.org/10.3390/cancers3010994.

Özer A, Erdost H, Zõk B. Histological investigations on the effects of feeding a diet containing red hot pepper on the reproductive organs of the chicken. Phytother Res 2005;19(6):501−5. https://doi.org/10.1002/ptr.1690.

Panchal SK, Bliss E, Brown L. Capsaicin in metabolic syndrome. Nutrients 2018;10(5):630. https://doi.org/10.3390/nu10050630.

Park SG, Yon JM, Lin C, Gwon LW, Lee JG, Baek IJ, Lee BJ, Yun YW, Nam SY. Capsaicin attenuates spermatogenic cell death induced by scrotal hyperthermia through its antioxidative and anti-apoptotic activities. Andrologia 2017;48(5):e12656. https://doi.org/10.1111/and.12656.

Patowary P, Pathak MP, Zaman K, Raju PS, Chattopadhyay P. Research progress of capsaicin responses to various pharmacological challenges. Biomed Pharmacother 2017;96:1501−12. https://doi.org/10.1016/j.biopha.2017.11.124.

Peng J, Li Y-J. The vanilloid receptor TRPV1: role in cardiovascular and gastrointestinal protection. Eur J Pharmacol 2010;627(1−3):1−7. https://doi.org/10.1016/j.ejphar.2009.10.053.

Pethő G, Bölcskei K, Füredi R, Botz B, Bagoly T, Pintér E, Szolcsányi J. Evidence for a novel, neurohumoral antinociceptive mechanism mediated by peripheral capsaicin-sensitive nociceptors in conscious rats. Neuropeptides 2017;62:1−10. https://doi.org/10.1016/j.npep.2017.02.079.

Pintado CO, Pinto FM, Pennefather JN, Hidalgo A, Baamonde A, Sanchez T, Candenas ML. A role tachykinins in female mouse and rat reproductive function. Biol Reprod 2003;69(3):940−6. https://doi.org/10.1095/biolreprod.103.017111.

Quiroz U, Morales-Ladesma L, Moran C, Trujillo A, Dominguez R. Lack of sensorial innervation in the newborn female rats affects the activity of hypothalamic monoaminergic system and steroid hormone secretion during puberty. Endocrine 2014;46:309−17. https://doi.org/10.1007/s12020-013-0055-3.

Ranjan A, Ramachandran S, Gupta N, Kaushik I, Wright S, Srivastava S, Das H, Srivastava S, Prasad S, Srivastava SK. Role of phytochemicals in cancer prevention. Int J Mol Sci 2019;20(20):4981. https://doi.org/10.3390/ijms20204981.

Scheau C, Badarau IA, Caruntu C, Mihai GL, Didilescu AC, Constantin C, Neagu M. Capsaicin: effects on the pathogenesis of hepatocellular carcinoma. Molecules 2019;24(13):2350. https://doi.org/10.3390/molecules24132350.

Segawa Y, Hashimoto H, Maruyama S, Shintani M, Ohno H, Nakai Y, Osera T, Kurihara N. Dietary capsaicin-mediated attenuation of hypertension in a rat model of renovascular hypertension. Clin Exp Hypertens 2019;42(4):352−9. https://doi.org/10.1080/10641963.2019.1665676.

Sharma SK, Vij AS, Sharma M. Mechanisms and clinical uses of capsaicin. Eur J Pharmacol 2013;720(1−3):55−62. https://doi.org/10.1016/j.ejphar.2013.10.053. Epub 2013 Nov 5.

Sirotkin AV. Regulators of ovarian functions. New York: Nova Publishers Inc.; 2014. 194 pp.

Srinivasan K. Biological activities of red pepper (*Capsicum annuum*) and its pungent principle capsaicin: a review. Crit Rev Food Sci Nutr 2015;56(9):1488−500. https://doi.org/10.1080/10408398.2013.772090.

Sun F, Xiong S, Zhu Z. Dietary capsaicin protects cardiometabolic organs from dysfunction. Nutrients 2016;8(5):174. https://doi.org/10.3390/nu8050174.

Sung B, Prasad S, Yadav VR, Aggarwal BB. Cancer cell signaling pathways targeted by spice-derived nutraceuticals. Nutr Cancer 2012;64(2):173−97. https://doi.org/10.1080/01635581.2012.630551.

Szoka L, Palka J. Capsaicin upregulates pro-apoptotic activity of thiazolidinediones in glioblastoma cell line. Biomed Pharmacother 2020;132:110741. https://doi.org/10.1016/j.biopha.2020.110741.

Szolcsányi J. Forty years in capsaicin research for sensory pharmacology and physiology. Neuropeptides 2004;38(6):377−84. https://doi.org/10.1016/j.npep.2004.07.005.

Tanrıkulu-Küçük S, Başaran-Küçükgergin C, Seyithanoğlu M, Doğru-Abbasoğlu S, Koçak H, Beyhan-Özdaş S, Öner-İyidoğan Y. Effect of dietary curcumin and capsaicin on testicular and hepatic oxidant−antioxidant status in rats fed a high-fat diet. Appl Physiol Nutr Metabol 2019a;44(7):774−82. https://doi.org/10.1139/apnm-2018-0622.

Tanrıkulu-Küçük S, Başaran-Küçükgergin C, Söğüt İ, Tunçdemir M, Doğru-Abbasoğlu S, Seyithanoğlu M, Koçak H, Öner-İyidoğan Y. Dietary curcumin and capsaicin: relationship with hepatic oxidative stress and apoptosis in rats fed a high fat diet. Adv Clin Exp Med 2019b;28(8):1013−20. https://doi.org/10.17219/acem/94145.

Tütüncü S, İlhan T, Özfiliz N. Immunohistochemical expression of ghrelin in capsaicin-treated rat ovaries during the different developmental periods. Iran J Vet Res 2016;17(1):50−4.

Venier NA, Yamamoto T, Sugar LM, Adomat H, Fleshner NE, Klotz LH, Venkateswaran V. Capsaicin reduces the metastatic burden in the transgenic adenocarcinoma of the mouse prostate model. Prostate 2015;75(12):1300−11. https://doi.org/10.1002/pros.23013.

Villaverde AISB, Netherton J, Baker MA. From past to present: the link between reactive oxygen species in sperm and male infertility. Antioxidants 2019;8(12):616. https://doi.org/10.3390/antio x8120616.

Wang J, Tian W, Wang S, Wei W, Wu D, Wang H, Wang L, Yang R, Ji A, Li Y. Anti-inflammatory and retinal protective effects of capsaicin on ischaemia induced injuries through the release of endogenous somatostatin. Clin Exp Pharmacol Physiol 2017;44(7):803−14. https://doi.org/10.1111/1440-1681.12769.

Wu D, Jia H, Zhang Z, Li S. Capsaicin suppresses breast cancer cell viability by regulating the CDK8/PI3K/Akt/Wnt/β-catenin signaling pathway. Mol Med Rep 2020;22(6):4868−76. https://doi.org/10.3892/mmr.2020.11585.

Yang W, Gong X, Zhao X, An W, Wang X, Wang M. Capsaicin induces apoptosis in HeLa cells via Bax/Bcl-2 and caspase-3 pathways. Asian J Tradit Med 2006;1(3−4):159−65.

Yu Q, Wang Y, Yu Y, Li Y, Zhao S, Chen Y, Waqar AB, Fan J, Liu E. Expression of TRPV1 in rabbits and consuming hot pepper affects its body weight. Mol Biol Rep 2012;39:7583−9. https://doi.org/10.1007/s11033-012-1592-1.

Yuan L-J, Qin Y, Wang L, Zeng Y, Chang H, Wang J, Wang B, Wan J, Chen S-H, Zhang Q-Y, Zhu J-D, Zhou Y, Mi M-T. Capsaicin-containing chili improved postprandial hyperglycemia, hyperinsulinemia, and fasting lipid disorders in women with gestational diabetes mellitus and lowered the incidence of large-for-gestational-age newborns. Clin Nutr 2016;35(2):388−93. https://doi.org/10.1016/j.clnu.2015.02.011.

302 Contaminants and Plants Action on Female Reproduction

Zhang R, Humphreys I, Sahu RP, Shi Y, Srivastava SK. In vitro and in vivo induction of apoptosis by capsaicin in pancreatic cancer cells is mediated through ROS generation and mitochondrial death pathway. Apoptosis 2008;13(12):1465−78. https://doi.org/10.1007/s10495-008-0278-6.

Zhang JH, Lai FJ, Chen H, Luo J, Zhang RY, Bu HQ, Wang ZH, Lin HH, Lin SZ. Involvement of the phosphoinositide 3-kinase/Akt pathway in apoptosis induced by capsaicin in the human pancreatic cancer cell line PANC-1. Oncol Lett 2013;5(1):43−8. https://doi.org/10.3892/ol.2012.991.

Zhang S, Wang D, Huang J, Hu Y, Xu Y. Application of capsaicin as a potential new therapeutic drug in human cancers. J Clin Pharm Therapeut 2019;45(1):16−28. https://doi.org/10.1111/jcpt.13039.

Zheng J, Zheng S, Feng Q, Zhang Q, Xiao X. Dietary capsaicin and its anti-obesity potency: from mechanism to clinical implications. Biosci Rep 2017;37(3). https://doi.org/10.1042/BSR20170286. BSR20170286.

Zik B, Ozguden Akkoc CG, Tutuncu S. Sıçan ovaryumunda düşük doz capsaicinin NF-kB ve XIAP proteininin sentezlenmesi üzerine etkisi. Ankara Univ Vet Fak Derg 2010a;57:223−8. https://doi.org/10.1501/Vetfak_0000002429.

Zik B, Ozguden Akkoc CG, Tutuncu S, Tuncay I, Yilmaztepe OA, Ozencı CC. Effect of low dose capsaicin (CAP) on ovarian follicle development in prepubertal rat. Rev Med Vet 2010b;161(6):288−94.

Zik B, Altunbas K, Tutuncu S, Ozden O, Ozguden Akkoc CG, Peker S, Sevimli A. Effects of capsaicin on nitric oxide synthase isoforms in prepubertal rat ovary. Biotech Histochem 2012;87(3):218−25. https://doi.org/10.3109/10520295.2011.608716.

Zimmer AR, Leonardi B, Miron D, Schapoval E, Oliveira JRD, Gosmann G. Antioxidant and anti-inflammatory properties of *Capsicum baccatum*: from traditional use to scientific approach. J Ethnopharmacol 2012;139(1):228−33. https://doi.org/10.1016/j.jep.2011.11.005.

Chapter 4.5

Daidzein

4.5.1 Introduction

In the last decades daidzein became a subject of numerous studies due to its health benefits, mainly in prevention of cancer. The results of these epidemiological (Rienks et al., 2017; Dong et al., 2020), biomedical (Vitale et sl., 2013; Zaheer et al., 2017; Křížová et al., 2019; Mayo et al., 2019) and biochemical (Barnes et al., 2011; Zhou et al., 2019; Luca et al., 2020) studies have been described in several reviews. There are however no reviews summarizing the current available knowledge concerning daidzein action on reproductive processes. The available reviews concerning daidzein in relation to female reproduction do not describe the findings in this field obtained in the last decade (Cederroth et al., 2012), or they concern only one special aspect of daidzein—its applicability to prevent menopausal syndrome (Liu et al., 2014, 2020; Chen et al., 2019) or estrogen-related women illnesses (van Duursen, 2017). This is the review of the recent knowledge concerning the influence of dadzein and its metabolite equol on animal and human female reproductive processes (ovarian cycle, folliculo-and oogenesis), basic ovarian cell functions (viability, proliferation, apoptosis) pituitary and ovarian endocrine regulators of these functions, possible intracellular mechanisms of daidzein action on ovarian cells, the applicability of daidzein for control of animal and human female reproductive processes, as well as the queries which should be addressed to make this application more efficient.

4.5.2 Provenance and properties

Isoflavones genistein, daidzein (see Fig. 4.5), and glycitein are bioactive compounds which are present in significant quantities in legumes, mainly soybeans, green beans, mung beans. In grains (raw materials), they are present mostly as simple and complex O-glycosides, β-glucuronides and sulfate esters, highly polar, and water-soluble compounds. They are hardly absorbed by the intestinal epithelium and have weak biological activities. Thus, beans are to be processed into various food products for digestibility, taste, and bioavailability

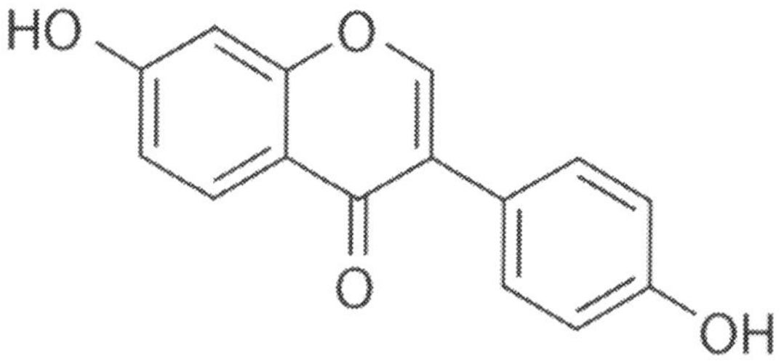

304 Contaminants and Plants Action on Female Reproduction

FIGURE 4.5 The chemical structure of daidzein. *From www.selleckchem.com.*

of nutrients and higher bioactives. Main processing steps include steaming, cooking, roasting, microbial fermentation that destroy protease inhibitors and also cleaves the glycoside bond to yield absorbable and bioactive aglycone in the processed products, such as miso, natto, soy milk, tofu; and increase shelf lives (Barnes et al., 2011; Vitale et al., 2013; Zaheer et al., 2017). Clinical studies show important differences between the biological activities of aglycone and conjugated forms of daidzein (Vitale et al., 2013).

Processed soy food products have been an integral part of regular diets in many Asia—Pacific countries for centuries, but in the last 2 decades, soy products have been successfully introduced into western diets for their health benefits (Zaheer et al., 2017).

The greatest biological and health effect has daidzein metabolite equol (Lamartiniere et al., 2002; Mayo et al., 2019; Rosenfeld, 2019; Sekikawa et al., 2019). The conversion of daidzein into equol takes place in the intestine via the action of reductase enzymes belonging to incompletely characterized members of the gut microbiota. While all animal species analyzed so far produce equol, only between one third and one half of human subjects (depending on the community) are able to do so, ostensibly those that harbor equol-producing microbes. Only 20%—30% of Westerners produce S-equol in contrast to 50%—70% in Asians. Conceivably, these subjects might be the only ones who can fully benefit from soy or isoflavone consumption (Mayo et al., 2019; Sekikawa et al., 2019). The individual and national variability in daidzein effects and effectiveness could be explained by variability in consumption of daidzein-containing food (for example, soy products) (Mayo et al., 2019; Sekikawa et al., 2019), gut microflora, and their ability to convert daidzein to equol (Rosenfeld, 2019). Besides microflora, the raw plant polyphenols can me metabolized by hepatocytes, intestine, and by some other cells including breast tumor cells. In addition, inflammatory cells produce chemical oxidants that can react with polyphenols and inactivate them (Barnes et al., 2011).

Despite low oral bioavailability, daidzein, like most polyphenols, proved significant biological effects, which brought to attention the low bioavailability/high bioactivity paradox (Luca et al., 2020).

4.5.3 Physiological and therapeutical actions

The epidemiological, animal, and *in vitro* studies demonstrated the ability of daidzein to reduce the incidence of estrogen-dependent and aging-associated disorders, such as menopause symptoms in women, osteoporosis, cardiovascular diseases, and cancer (Barnes et al., 2011; Liu et al., 2014, 2020; Vitale et al., 2013; Ronis et al., 2016; Rienks et al., 2017; Zaheer et al., 2017; Chen et al., 2019; Křížová et al., 2019; Zhou et al., 2019; Luca et al., 2020), disorders of cognitive functions, and high blood pressure (Chen et al., 2019). Higher blood and urinary daidzein and genistein concentrations were associated with lower risk of breast cancer and diabetes (Rienks et al., 2017; Zhou et al., 2019; Dong et al., 2020). Daidzein impairs glucose and lipid metabolism and vascular inflammation associated with type 2 diabetes (Das et al., 2018). It can promote detoxifying enzymes expression and activity (Zhou et al., 2019) and to prevent the toxic effects of polycyclic aromatic hydrocarbons (Ronis et al., 2016). Finally daidzein can be used for treatment of virosa, including COVID-19 (Adhikari et al., 2020).

Results of recent clinical trials and metaanalyses on the effects of equol demonstrated the preventive and curative action of equol on menopause, cardiovascular system, bone health (prevention of osteoporosis), cancer and functions of central nervous system and mental disorders (Hu et al., 2016; Mayo et al., 2019; Sekikawa et al., 2019; Liu et al., 2020). Epidemiological studies suggested the preventive action of equol against diabetes type-2 (Dong et al., 2020). Equol has also been demonstrated to modulate obesity and diabetes type-2 by controlling the glycemic index to ameliorate chronic kidney disease and to prevent skin aging (Mayo et al., 2019).

On the other hand, it should be noted that daidzein and equol effect depends on its income from food. For example, in East Asian countries, where the consumption of daidzein-containing products is 10 times higher than in the West, the incidence of cardiovascular disease, osteoporosis, mental disorders, certain types of cancer and menopausal syndrome is several times lesser than seen in the West (Mayo et al., 2019). Additional consumption of soy, daidzein and equol did not affect the expression of menopausal symptoms in Chinese women (Liu et al., 2014). It is noteworthy that risk of diabetes type-2 in Chinese population was associated with urinary level of equol, but not of daidzein (Dong et al., 2020). It suggests that antidiabetic protective effect of soy-derived food depends not (or not only) on presence of daidzein, but to its conversion by gut microflora to equol.

Nevertheless, it needs to be mentioned that the most performed clinical studies involving isoflavones have suffered from small sample sizes, short trial durations, lack of appropriate controls, the use of isoflavones from various sources, supplements with different aglycone contents, and other methodological flaws. Therefore, both specialists and the regulatory agencies usually conclude that there is still no scientifically sound evidence of isoflavones reducing the risks and symptoms of any disease (Mayo et al., 2019).

Finally, isoflavones including daidzein and its metabolites may also be considered endocrine disruptors with possible negative influences on the state of health in a certain part of the population or on the environment (Křížová et al., 2019; Mayo et al., 2019). Both *in vitro* and animal studies report isoflavones able to interfere with different checkpoints of the hypothalamic/pituitary/thyroid system. Further, the estrogenic activity of isoflavones could pose a potential hazard by promoting certain types of tumor. However, the scientific evidence supporting their having any harmful consequences is also inconclusive (Mayo et al., 2019).

4.5.4 Mechanisms of action

Daidzein, like other flavanoid isoflavones, is a phenol plant compound that, due to its molecular structure, resembles vertebrate steroid hormones. Therefore, in vertebrates it can bind steroid hormone receptors and exert estrogenic or antiestrogenic effects and, therefore, affect physiological processes and illnesses dependent of steroid hormones (Cederroth et al., 2012; Vitale et al., 2013; van Duursen, 2017; Křížová et al., 2019; Zhou et al., 2019; Liu et al., 2020). As an antioxidant, daidzein can prevent oxidative stress and the resulting disorders (Barnes et al., 2011; Ronis, 2016; Zhou et al., 2019; Luca et al., 2020). The manifestation of these illnesses could be mitigated by daidzein also due to its ability to promote antioxidant enzymes cytochromes P450 (Ronis, 2016), to inhibit tyrosine kinase, a promoter of cell proliferation (Zhou et al., 2019) and to activate Akt signaling pathway which promote DNA fragmentation and cell apoptosis (Hu et al., 2016).

Equol, a daidzein metabolite with higher biological activity, can affect these processes via direct binding of reactive oxygen species and activation of antioxidative enzymes, binding to estrogen receptors, mitogen-activated protein kinase, protein kinase B (Akt) and epidermal growth factor receptor kinase and cyclin B/CDK complex (promoters of cell proliferation), transcription factor NFkB (promoter of inflammatory processes), nitric oxide-dependent intracellular signaling pathway, transcription factors FOXO3a and p53 (promoters of apoptosis), epigenetic mechanisms, including DNA methylation, histone modification, and microRNA regulation and other intracellular signaling mechanisms (Mayo et al., 2019). In addition, both daidzein and equol can probably influence neurobehavioral processes targeting the corresponding CNS centers (Chen et al., 2019; Rosenfeld, 2019).

4.5.5 Effect on female reproductive processes

4.5.5.1 Effect on ovarian and reproductive state

Several laboratories studied the character of daidzein influence on animal reproductive system and fecundity, but the results of their studies were variable. Lamartiniere et al. (2002) failed to find any effect of daidzein on rat ovarian histomorphology and weight, fertility, numbers of male and female offspring, and anogenital distances. Kaludjerovic et al. (2012) did not find the influence of dietary daidzein also on mice ovarian weight, number of ovarian follicles, number of multiple oocyte follicles, or percent of ovarian cysts. But they observed the adverse effect of daidzein on ovariand structure and its ability to reduce number of ovarian corpus lutea. In rats daidzein suppressed follicular growth (reduced ovarian weight), but not ovarian folliculogenesis and fecundity (number of corpora lutea) or sexual behavior (lordosis quotient; Kouki et al., 2003, 2005). Talsness et al. (2015) reported the inhibitory influence of daidzein on rat ovarian folliculogenesis—it increased follicular atresia, reduced secondary and tertiary follicle numbers and probably ovarian surface epityhelium proliferation, induced cyst formation, and prolonged estrous. The ability to suppress growth and to induce atresia of murine cultured ovarian follicles was reported for quercetin metabolite equol (Mahalingam et al., 2016).

Other studies however showed the stimulatory action of daidzein on chicken ovarian germ cells (Liu et al., 2006) and chicken ovarian folliculogenesis and ovulation/egg laying rate (Liu and Zhang, 2008). Daidzein-induced promotion of rat ovarian folliculogenesis and reduction in ovarian follicular atresia were observed by Medigović et al. (2015). Dorward et al. (2007) reported that feeding of mice a mixture of daidzein and genistein promoted ovarian and uterine growth and the incidence of ovarian tumorgenesis.

The influence of daidzein on nonovarian reproductive organs remains to be established. Dietary daidzein did not affect uterine morphology and weight in rats (Lamartiniere et al., 2002) and mice (Kaludjerovic et al., 2012). No influence of daidzein on proliferation of murine endometrium and endometrioma

308 Contaminants and Plants Action on Female Reproduction

cells has been found (Takaoka et al., 2018). On the other hand, Kaludjerovic et al. (2012) reported that dietary daidzein induced hyperplasia in the murine oviduct and abnormal histomorphological changes in their uteri.

It is hypothesized that daidzein and its metabolite equol can trigger neurobehavioral programming toward reproduction, conceiving and pregnancy. The gut dysbiosis and related changes in conversion of daidzein to equol can contribute to neurobehavioral disruptions and therefore to disruption of reproductive behavior (Rosenfeld, 2019).

Therefore, the available information concerning expression and character of daidzein action on reproductive processes is inconsistent. The majority of the corresponding papers suggest an inhibitory action of this polyphenol on rodent's reproductive system. On the other hand, the stimulatory action of daidzein on rat and chicken ovarian folliculogenesis has been reported too. Although the influence of daidzein and equol on human reproductive behavior has been indicated, it remains unknown whether these polyphenols can influence human reproductive system and fecundity.

4.5.5.2 Effect on ovarian cell functions

Daidzein addition did not affect viability in cultured bovine ovarian granulosa and luteal cells (Mlynarczuk et al., 2011). In some of our experiments (Sirotkin et al., 2020b) daidzein promoted cell viability and proliferation, but not apoptosis by cultured porcine ovarian granulosa cells. In other experiments (Sirotkin et al., 2020a) daidzein reduced viability of these cells, but this effect was not associated with changes in cell proliferation or apoptosis. Daidzein was able to induce death of ovarian cancer cells both *in vivo* (Somjen et al., 2008) and *in vitro* (Somjen et al., 2008; Green et al., 2009).

Therefore, daidzein can induce death of cancer cells, but the available data concerning its action on healthy ovarian cells' viability, proliferation, and apoptosis remain inconclusive as of yet.

4.5.5.3 Effect on reproductive hormones

In contrast to other reproductive functions, the daidzein action on the release and reception of reproductive hormones is well demonstrated.

Feeding with daidzeine increased number of FSH and LH receptors in chicken ovarian follicles (Liu and Zhang, 2008). Furthermore, daidzein was able to promote the stimulatory action of LH on mice ovarian cancer development (Dorward et al., 2007). These observations indicate that daidzein can promote the reception and effect of gonadotropins on ovarian functions.

The action of daidzein on other, peripheric, hormones including hormones of ovarian origin regulating reproductive functions has been well documented, although substantial differences in the character of daidzein action even in similar experiments are evident.

In *in vivo* experiments of Medigović et al. (2015), injections of daidzein increased progesterone and estradion and decreased testosterone level in rat serum. On the contrary, in *in vivo* experiments of Lamartiniere et al. (2002), feeding rats with daidzein reduced the concentration of circulating progesterone, but not estrogen.

Some *in vitro* experiments demonstrated a direct inhibitory action of daidzein and its metabolite equol on steroid hormones' release by ovarian cells. Equol inhibited the synthesis and release of progesterone, testosterone, androstenedione, and estradiol by cultured murine ovarian follicles (Mahalingam et al., 2016). Similar inhibitory influence of daidzein addition on progesterone, but not estradiol release of cultured porcine ovarian granulosa cells was observed in *in vitro* experiments of Nynca et al. (2009, 2013). Nevertheless, similar experiments of Sirotkin et al. (2020a) failed to detect daidzein action on progesterone and estradiol release by porcine granulosa cells, but in these experiments daidzein promoted testosterone output. Other experiments on porcine granulosa cells, however, showed the upregulation of progesterone, but not of testosterone and estradiol production (Sirotkin et al., 2020b). Mlynarczuk et al. (2011) did not observe any influence of daidzein addition on progesterone and estradiol release by cultured bovine granulosa and luteal cells. In their experiments, daidzein was shown to be a potent stimulator of enzymes responsible for oxytocin synthesis and ot release of oxytocin.

Epidemiological studies of Otokozawa et al. (2015) showed association of decreased serum daidzein levels and decreased women serum adiponectin levels, and increased serum insulin levels, which could be shown to be associated with elevated risk of ovarian cancer.

Therefore, the performed studies demonstrated daidzein's influence on gonadotropin reception by ovarian cells, as well as on steroid and peptide hormones—regulators and markers of ovarian functions.

4.5.5.4 Mechanisms of action on female reproductive processes

Female reproductive functions, like other functions, are under control of a hierarchical system of exogenous and endogenous regulators—regulatory molecules produced in the hypothalamus and other areas of CNS, pituitary hormones (mainly gonadotropins), hormones produced by ovary and other peripheric organs, receptors to hormones, and intracellular mediators of their

310 Contaminants and Plants Action on Female Reproduction

action—enzymes including protein kinases, transcription factors regulating cell proliferation, apoptosis, viability, and differentiation (Sirotkin, 2014). Although the mediatory role of some of these molecules requires more strong evidence, the influence of daidzein on some (but not all) of these signaling pathways has been documented.

In the available literature we failed to find any direct evidence for daidzein action on CNS, although the influence of both daidzein and equol on CNS-dependent reproductive behavior has been documented (Rosenfeld, 2019). Furthermore, the histomorphological studies did not reveal daidzein influence on morphophysiological indexes of activity of pituitary gonadotropes and lactotropes (Medigović et al., 2015) indicating that genistein don't affect gonadotropin- and prolactin-producing pituitary cells.

The ability of daidzein to promote generation of gonadotropin receptors, which was associated with increased ovarian folliculogenesis and egg laying rate (Liu and Zhang, 2008) mentioned earlier, suggests that daidzein can promote ovarian cell function through upregulation of **gonadotropin receptors.**

The ability of dadzein to regulate the release of steroid and nonsteroid hormones (**progestagen, androgens, estrogens, oxytocin, and adipokines**) has been described earlier. These hormones are the known regulators of basic ovarian functions and fecundity (Sirotkin, 2014). Therefore, these hormones can be the extracellular mediators of daidzein action on female reproductive processes.

The similarity of daidzein and estrogen effects on rat ovarian folliculogenesis (Talsness et al., 2015) and steroidogenesis (Medigović et al., 2015) suggests the similar mechanisms of action of these molecules on adult ovarian functions via **estrogen receptors.** This hypothesis was confirmed by the ability of both daidzein and estradiol to bind to estrogen receptors in ovarian and nonovarian tissues (Vitale et al., 2013; Zaheer et al., 2017) and to upregulate the expression of estrogen receptors alpha and beta in porcine ovarian cells (Nynca et al., 2009, 2013). Moreover, blockage of estrogen receptors prevented daidzein action on proliferation of chicken ovarian germ cells (Liu et al., 2006). On the other hand, Dorward et al. (2007) reported the ability of daidzein to antagonize estrogen action on ovarian cancerogenesis. It indicates that daidzein can affect ovarian cancer development via not only up- but also by downregulation of estrogen receptors.

Several studies outlined the possible postreceptory intracellular mechanisms of daidzein action on ovarian cells. Chan et al. (2018) demonstrated that daidzein can prevent ovarian cancerogenesis via downregulation of estrogen receptors, which in turn affect **p-FAK, p-PI3K, p-AKT, p-GSK3β, p21 or** cyclin D1 signaling, and their action on **proliferation and apoptosis** (Chan et al., 2018). The stimulatory influence on apoptosis markers was reported also for daidzein metabolite equol (Mahalingam et al., 2016). Furthermore, daidzein action could be mediated also by **mitogen activating kinases**—promoters of cell proliferation and suppressors of their apoptosis (Hua et al., 2018).

The next mechanism of daidzein action on ovarian cells could be its **antioxidative effect**—ability to activate antioxidant enzymes and to block reactive oxygen species and their adverse effects on chicken ovarian germ cells (Liu et al., 2006) and rat ovaries (Medigovic et al., 2015). Moreover, these experiments indicated that daidzein can affect chicken and rat cells by using two (estrogenic and antioxidant) mechanisms at once.

4.5.5.5 Application in reproductive biology and medicine

The available information concerning daidzein effect on reproductive processes suggest possible areas of its application. Results of experiments of Liu and Zhang (2008) demonstrate the stimulatory action of daidzein on chicken ovarian follicle development and ovulation and, therefore, the applicability of daidzein as a biostimulator of poultry egg laying rate. The stimulatory action of daidzein on porcine ovarian cell viability and proliferation (Sirotkin et al., 2020b), on rat ovarian folliculogenesis and uterine functions, and its ability to prevent atresia of rat and mice ovarian follicles (Dorward et al., 2007; Medigović et al., 2015) indicates that daidzein can be a stimulator of female reproductive processes in mammals too. If the ability of daidzein to promote the activation of gonadotropin receptors not only in chicken (Liu and Zhang, 2008), but also in mammalian ovary would be detected, daidzein could be useful as an enhancer of gonadotropin action in gonadotropin-induced ovulation in animal production and assisted reproduction. Furthermore, the ability of daidzein to promote the release of ovarian estradiol (Medigović et al., 2015) and to activate estrogen receptors (Nynca et al., 2009, 2013; Vitale et al., 2013; Zaheer et al., 2017) suggests that daidzein or daidzein-containing food could be beneficial for prevention of disorders induced by a deficit of estrogens or their receptors—reproductive aging and other age-related disorders listed earlier.

On the other hand, a numerous data demonstrated the ability of daidzein (Kouki et al., 2003; Talsness et al., 2015) and equol (Mahalingam et al., 2016) to suppress functions of the ovary, as well as of healthy (Sirotkin et al., 2020a) ovarian cells. These data suggest that overconsumption of daidzein or daidzein containing food could have adverse effect on animal and human reproductive processes. On the other side, the ability of daidzein to suppress ovarian cancer cell functions demonstrate its applicability for prevention and mitigation of ovarian cancer development (Somjen et al., 2008; Green et al., 2009). The antioxidant effects of daidzein can indicate its potential applicability for prevention of not only ovarian cancer, but also of other reproductive disorders induced by oxidative stress and for increase of reproductive system's resistence to environmental factors indicating such stress. For example, daidzein was able to prevent the adverse effects of polychlorinated biphenyls on oxytocin release by cultured bovine ovarian cells, as well as on uterine cell contractions and prostaglandin F release (Kotwica et al., 2006). It is not to be excluded that daidzein could be applicable for promotion of uterine functions (including labor) and for prevention of adverse effects of other environmental contaminants and stressors on reproductive system. Furthermore, Rosenfeld (2019) suggest that daidzein and its metabolite equol can be useful for treatment of disruption of reproductive neurobehavioral programming.

Therefore, stimulatory, inhibitory and protective effects of daidzein on female reproductive processes could be principally applicable in animal production, human reproductive biology, as well as in human and veterinary medicine.

4.5.6 Conclusion and possible direction of further studies

The available publications demonstrate the influence of daidzein and its metabolite equol on various nonreproductive and reproductive processes and their disorders. Daidzein and equol can both up- and downregulate ovarian reception of gonadotropins, healthy and cancer ovarian cell proliferation, apoptosis, viability, ovarian growth, folliculo-and oogenesis, and follicular atresia, as well as to affect reproductive behavior. These effects could be mediated by daidzein and equol effect on hormones production and reception, reactive oxygen species, and intracellular regulators of proliferation and apoptosis.

Both stimulatory and inhibitory effects of daidzein and equol could be useful for stimulation of reproduction and for prevention and mitigation of cancer development and adverse effects of environmental stressors in reproductive biology and medicine.

Nevertheless, the data concerning daidzein effects are contradictory, and the data concerning equol are insufficient to generate any general conclusion concerning character of effect of these molecules on female reproduction. The causes of such differences in daidzein action observed in various experiments remain unknown. The main data was obtained in *in vitro* experiments, results of which require validation by corresponding *in vivo* studies. The *in vivo* studies were performed mainly on laboratory animals, while only few human epidemiological studies were performed. Much more is known about daidzein influence on cancer than on healthy female reproductive processes. It remains unknown whether daidzein can be an endocrine disrupter jeopardizing reproduction (Mayo et al., 2019) or reproductive behavior (Rosenfeld, 2019). Therefore, understanding and application of daidzein and its metabolites' action on female reproductive processes requires further extensive studies.

Furthermore, application of daidzein and daidzein-containing functional food is limited by low bioavailability of this isoflavone. To address this problem, numerous promising strategies, such as the use of an absorption enhancer, structural transformation (e.g., prodrugs and glycosylation), and pharmaceutical technologies (e.g., carrier complexes, nanotechnology, cocrystals), have been developed and applied to deliver poorly water-soluble flavonoids (Zhao et al., 2019). The other, more natural, approach to increase bioavailability and biological activity of isoflavones could be modification of intestine microflora metabolizing daidzein and converting it to equinol (Rosenfeld, 2019). Nevertheless, the problem of enhancement of bioavailability and efficiency of daidzein is still waiting for its solution.

Taken together, daidzein and its metabolites could be an important factor and an efficient tool to control female reproductive processes including fecundity and to prevent and treat reproductive disorders. Nevertheless, our current knowledge is too insufficient and contradictory to understand and use the influence of these flavonoids' effects on female reproductive events now. They represent a promising subject of the future basic and applied research.

References

Adhikari B, Marasini BP, Rayamajhee B, Bhattarai BR, Lamichhane G, Khadayat K, Adhikari A, Khanal S, Parajuli N. Potential roles of medicinal plants for the treatment of viral diseases focusing on COVID-19: a review. Phytother Res 2020. https://doi.org/10.1002/ptr.6893.

Barnes S, Prasain J, D'Alessandro T, Arabshahi A, Botting N, Lila MA, Jackson G, Janle EM, Weaver CM. The metabolism and analysis of isoflavones and other dietary polyphenols in foods and biological systems. Food Funct 2011;2(5):235−44. https://doi.org/10.1039/c1fo10025d.

Cederroth CR, Zimmermann C, Nef S. Soy, phytoestrogens and their impact on reproductive health. Mol Cell Endocrinol 2012;355(2):192−200. https://doi.org/10.1016/j.mce.2011.05.049.

Chan KKL, Siu MKY, Jiang YX, Wang JJ, Leung THY, Ngan HYS. Estrogen receptor modulators genistein, daidzein and ERB-041 inhibit cell migration, invasion, proliferation and sphere formation via modulation of FAK and PI3K/AKT signaling in ovarian cancer. Cancer Cell Int 2018;18:65. https://doi.org/10.1186/s12935-018-0559-2.

Chen LR, Ko NY, Chen KH. Isoflavone supplements for menopausal women: a systematic review. Nutrients 2019;11(11):2649. https://doi.org/10.3390/nu11112649.

Das D, Sarkar S, Bordoloi J, Wann SB, Kalita J, Manna P. Daidzein, its effects on impaired glucose and lipid metabolism and vascular inflammation associated with type 2 diabetes. Biofactors 2018;44(5):407−17. https://doi.org/10.1002/biof.1439.

Dong HL, Tang XY, Deng YY, Zhong QW, Wang C, Zhang ZQ, Chen YM. Urinary equol, but not daidzein and genistein, was inversely associated with the risk of type 2 diabetes in Chinese adults. Eur J Nutr 2020;59(2):719−28. https://doi.org/10.1007/s00394-019-01939-0.

314 Contaminants and Plants Action on Female Reproduction

Dorward AM, Shultz KL, Beamer WG. LH analog and dietary isoflavones support ovarian granulosa cell tumor development in a spontaneous mouse model. Endocr Relat Cancer 2007;14(2):369−79. https://doi.org/10.1677/erc.1.01232.

Green JM, Alvero AB, Kohen F, Mor G. 7-(O)-Carboxymethyl daidzein conjugated to N-t-Boc-hexylenediamine: a novel compound capable of inducing cell death in epithelial ovarian cancer stem cells. Cancer Biol Ther 2009;8(18):1747−53. https://doi.org/10.4161/cbt.8.18.9285.

Hu WS, Lin YM, Kuo WW, Pan LF, Yeh YL, Li YH, Kuo CH, Chen RJ, Padma VV, Chen TS, Huang CY. Suppression of isoproterenol-induced apoptosis in H9c2 cardiomyoblast cells by daidzein through activation of Akt. Chin J Physiol 2016;59(6):323−30. https://doi.org/10.4077/CJP.2016.BAE393.

Hua F, Li CH, Chen XG, Liu XP. Daidzein exerts anticancer activity towards SKOV3 human ovarian cancer cells by inducing apoptosis and cell cycle arrest, and inhibiting the Raf/MEK/ERK cascade. Int J Mol Med 2018;41(6):3485−92. https://doi.org/10.3892/ijmm.2018.3531.

Kaludjerovic J, Chen J, Ward WE. Early life exposure to genistein and daidzein disrupts structural development of reproductive organs in female mice. J Toxicol Environ Health 2012;75(11):649−60. https://doi.org/10.1080/15287394.2012.688482.

Kotwica J, Wróbel M, Młynarczuk J. The influence of polychlorinated biphenyls (PCBs) and phytoestrogens in vitro on functioning of reproductive tract in cow. Reprod Biol 2006;6(Suppl 1):189−94.

Kouki T, Kishitake M, Okamoto M, Oosuka I, Takebe M, Yamanouchi K. Effects of neonatal treatment with phytoestrogens, genistein and daidzein, on sex difference in female rat brain function: estrous cycle and lordosis. Horm Behav 2003;44(2):140−5. https://doi.org/10.1016/s0018-506x(03)00122-3.

Kouki T, Okamoto M, Wada S, Kishitake M, Yamanouchi K. Suppressive effect of neonatal treatment with a phytoestrogen, coumestrol, on lordosis and estrous cycle in female rats. Brain Res Bull 2005;64(5):449−54. https://doi.org/10.1016/j.brainresbull.2004.10.002.

Křížová L, Dadáková K, Kašparovská J, Kašparovský T. Isoflavones. Molecules 2019;24(6):1076. https://doi.org/10.3390/molecules24061076.

Lamartiniere CA, Wang J, Smith-Johnson M, Eltoum IE. Daidzein: bioavailability, potential for reproductive toxicity, and breast cancer chemoprevention in female rats. Toxicol Sci 2002;65(2):228−38. https://doi.org/10.1093/toxsci/65.2.228.

Liu HY, Zhang CQ. Effects of daidzein on messenger ribonucleic acid expression of gonadotropin receptors in chicken ovarian follicles. Poult Sci 2008;87(3):541−5. https://doi.org/10.3382/ps.2007-00274.

Liu H, Zhang C, Zeng W. Estrogenic and antioxidant effects of a phytoestrogen daidzein on ovarian germ cells in embryonic chickens. Domest Anim Endocrinol 2006;31(3):258−68. https://doi.org/10.1016/j.domaniend.2005.11.002.

Liu ZM, Ho SC, Woo J, Chen YM, Wong C. Randomized controlled trial of whole soy and isoflavone daidzein on menopausal symptoms in equol-producing Chinese postmenopausal women. Menopause 2014;21(6):653−60. https://doi.org/10.1097/GME.0000000000000102.

Liu ZM, Chen B, Li S, Li G, Zhang D, Ho SC, Chen YM, Ma J, Qi H, Ling WH. Effect of whole soy and isoflavones daidzein on bone turnover and inflammatory markers: a 6-month double-blind, randomized controlled trial in Chinese postmenopausal women who are equol producers. Ther Adv Endocrinol Metab 2020;11. https://doi.org/10.1177/2042018820920555.

Luca SV, Macovei I, Bujor A, Miron A, Skalicka-Woźniak K, Aprotosoaie AC, Trifan A. Bioactivity of dietary polyphenols: the role of metabolites. Crit Rev Food Sci Nutr 2020;60(4):626−59. https://doi.org/10.1080/10408398.2018.1546669.

Plant molecules and their influence on health **Chapter | 4 315**

Mahalingam S, Gao L, Gonnering M, Helferich W, Flaws JA. Equol inhibits growth, induces atresia, and inhibits steroidogenesis of mouse antral follicles in vitro. Toxicol Appl Pharmacol 2016;295:47—55. https://doi.org/10.1016/j.taap.2016.02.009.

Mayo B, Vázquez L, Flórez AB. Equol: a bacterial metabolite from the daidzein isoflavone and its presumed beneficial health effects. Nutrients 2019;11(9):2231. https://doi.org/10.3390/nu11092231.

Medigović IM, Živanović JB, Ajdžanović VZ, Nikolić-Kokić AL, Stanković SD, Trifunović SL, Milošević VL, Nestorović NM. Effects of soy phytoestrogens on pituitary-ovarian function in middle-aged female rats. Endocrine 2015;50(3):764—76. https://doi.org/10.1007/s12020-015-0691-x.

Mlynarczuk J, Wrobel MH, Kotwica J. The adverse effect of phytoestrogens on the synthesis and secretion of ovarian oxytocin in cattle. Reprod Domest Anim 2011;46(1):21—8. https://doi.org/10.1111/j.1439-0531.2009.01529.x.

Nynca A, Jablonska O, Slomczynska M, Petroff BK, Ciereszko RE. Effects of phytoestrogen daidzein and estradiol on steroidogenesis and expression of estrogen receptors in porcine luteinized granulosa cells from large follicles. J Physiol Pharmacol 2009;60(2):95—105.

Nynca A, Słonina D, Jablońska O, Kamińska B, Ciereszko RE. Daidzein affects steroidogenesis and oestrogen receptor expression in medium ovarian follicles of pigs. Acta Vet Hung 2013;61(1):85—98. https://doi.org/10.1556/AVet.2012.060.

Otokozawa S, Tanaka R, Akasaka H, Ito E, Asakura S, Ohnishi H, Saito S, Miura T, Saito T, Mori M. Associations of serum isoflavone, adiponectin and insulin levels with risk for epithelial ovarian cancer: results of a case-control study. Asian Pac J Cancer Prev 2015;16(12):4987—91. https://doi.org/10.7314/apjcp.2015.16.12.4987. PMID: 26163627.

Rienks J, Barbaresko J, Nöthlings U. Association of isoflavone biomarkers with risk of chronic disease and mortality: a systematic review and meta-analysis of observational studies. Nutr Rev 2017;75(8):616—41. https://doi.org/10.1093/nutrit/nux021.

Ronis MJ. Effects of soy containing diet and isoflavones on cytochrome P450 enzyme expression and activity. Drug Metab Rev 2016;48(3):331—41. https://doi.org/10.1080/03602532.2016.1206562.

Rosenfeld CS. Effects of phytoestrogens on the developing brain, gut microbiota, and risk for neurobehavioral disorders. Front Nutr 2019;6:142. https://doi.org/10.3389/fnut.2019.00142.

Sekikawa A, Ihara M, Lopez O, Kakuta C, Lopresti B, Higashiyama A, Aizenstein H, Chang YF, Mathis C, Miyamoto Y, Kuller L, Cui C. Effect of S-equol and soy isoflavones on heart and brain. Curr Cardiol Rev 2019;15(2):114—35. https://doi.org/10.2174/1573403X15666181205104717.

Sirotkin A, Záhoranska Z, Tarko A, Fabova Z, Alwasel S, Halim Harrath A. Plant polyphenols can directly affect ovarian cell functions and modify toluene effects. J Anim Physiol Anim Nutr 2020. https://doi.org/10.1111/jpn.13461.

Sirotkin A, Záhoranska Z, Tarko A, Popovska-Percinic F, Alwasel S, Harrath AH. Plant isoflavones can prevent adverse effects of benzene on porcine ovarian activity: an in vitro study. Environ Sci Pollut Res Int 2020b;27(23):29589—98. https://doi.org/10.1007/s11356-020-09260-8.

Sirotkin AV. Regulators of ovarian functions. 2nd ed. -. New York: Nova Science Publishers, Inc.; 2014, ISBN 978-1-62948-574-4. 194 pp.

Somjen D, Katzburg S, Nevo N, Gayer B, Hodge RP, Renevey MD, Kalchenko V, Meshorer A, Stern N, Kohen F. A daidzein-daunomycin conjugate improves the therapeutic response in an animal model of ovarian carcinoma. J Steroid Biochem Mol Biol 2008;110(1—2):144—9. https://doi.org/10.1016/j.jsbmb.2008.03.033.

Takaoka O, Mori T, Ito F, Okimura H, Kataoka H, Tanaka Y, Koshiba A, Kusuki I, Shigehiro S, Amami T, Kitawaki J. Daidzein-rich isoflavone aglycones inhibit cell growth and inflammation in endometriosis. J Steroid Biochem Mol Biol 2018;181:125–32. https://doi.org/10.1016/j.jsbmb.2018.04.004.

Talsness C, Grote K, Kuriyama S, Presibella K, Sterner-Kock A, Poça K, Chahoud I. Prenatal exposure to the phytoestrogen daidzein resulted in persistent changes in ovarian surface epithelial cell height, folliculogenesis, and estrus phase length in adult Sprague-Dawley rat offspring. J Toxicol Environ Health 2015;78(10):635–44. https://doi.org/10.1080/15287394.2015.1006711.

van Duursen MBM. Modulation of estrogen synthesis and metabolism by phytoestrogens *in vitro* and the implications for women's health. Toxicol Res 2017;6(6):772–94. https://doi.org/10.1039/c7tx00184c.

Vitale DC, Piazza C, Melilli B, Drago F, Salomone S. Isoflavones: estrogenic activity, biological effect and bioavailability. Eur J Drug Metab Pharmacokinet 2013;38(1):15–25. https://doi.org/10.1007/s13318-012-0112-y.

Zaheer K, Humayoun Akhtar M. An updated review of dietary isoflavones: nutrition, processing, bioavailability and impacts on human health. Crit Rev Food Sci Nutr 2017;57(6):1280–93. https://doi.org/10.1080/10408398.2014.989958.

Zhao J, Yang J, Xie Y. Improvement strategies for the oral bioavailability of poorly water-soluble flavonoids: an overview. Int J Pharm 2019;570:118642. https://doi.org/10.1016/j.ijpharm.2019.118642.

Zhou T, Meng C, He P. Soy isoflavones and their effects on xenobiotic metabolism. Curr Drug Metab 2019;20(1):46–53. https://doi.org/10.2174/1389200219666180427170213.

Chapter 4.6

Diosgenin

4.6.1 Introduction

Diosgenin is considered as an important biological active constituent of medicinal and functional food plants. Its medicinal (preventive and curative) effects have been described in some reviews (Chen et al., 2015; Sultana et al., 2018; Parama et al., 2020). Nevertheless, the influence and applicability of diosgenin and diosgenin-containing plants for control of female reproduction have not been summarized yet. The present chapter reviews provenance, properties, general health effects, as well as the currently available knowledge concerning the action of diosgenin, diosgenin-containing plants and some of its metabolites on female reproductive processes.

4.6.2 Provenance and properties

Diosgenin is a naturally occurring plant steroidal sapogenin (see Fig. 4.6). It is contained in relatively high amounts in the plants of *Agavaceae, Dioscoreaceae, Liliaceae, Solanaceae, Scrophulariaceae, Amaryllidaceae, Leguminosae*, and *Rhamnaceae* species. Diosgenin in commercial amounts can be isolated from tubers of wild yam (*Dioscurea villosa* Linn), rhizomes of yam (*Dioscurea*

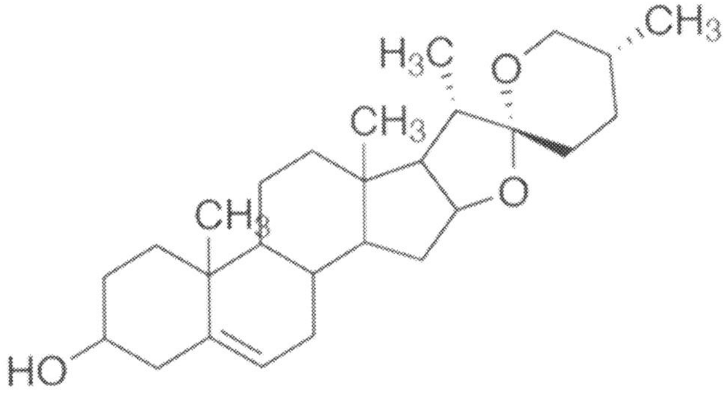

FIGURE 4.6   The chemical structure of diosgenin. *From www.researchgate.net.*

zingiberensis C. H. Wright, *Dioscorea nipponica,* Makino), seeds of fenugreek (*Trigonella foenum-graecum* Linn) or rhizomes *of* Himalayan trillium *(Trillium govanianum* Wall. ex D. Don) and ginger (*Costus speciosus* Koen ex. Retz). It is produced mainly by hydrolysis of steroidal saponins in the presence of a strong acid, base, or enzyme catalyst. In addition, the microbial transformation is a promising, specific, environmentally friendly and cheap method for the production of diosgenin, too (Cai et al., 2020). Furthermore, diosgenin can be a precursor for other biological active phytoestrogens, for example dehydroepi-androsterone and sarsasapogenin (Sultana et al., 2018; Sahu et al., 2020). and for commercial production of diosgenin-derived steroidal drugs (Sethi et al., 2018).

Diosgenin has low aqueous solubility, poor bioavailability and pharma-cokinetics, and rapid disappearance from organism in *in vivo* conditions (Cai et al., 2020).

4.6.3 Physiological action

The preclinical and clinical studies demonstrated that diosgenin and its metab-olites have anticancer, neuroprotective, antidiabetic, cardioprotective, hypo-cholesterolemic, gastro- and hepato-protective, antioxidant, antiinflammatory, antiosteoporosis, antiasthma, antiarthritis, and other positive properties (see Chen et al., 2015; Sultana et al., 2018; Parama et al., 2020 for review).

For example, diosgenin can induce cancer cell death *in vitro* and *in vivo*, reverse multi-drug resistance in cancer cells and sensitize cancer cells to standard chemotherapy (Sethi et al., 2018; Bhardwaj et al., 2021).

Diosgenin and its derivatives can be therapeutic agents for multiple disorders of central nervous system. In particular, those related to therapeutic efficacy for Parkinson's disease, Alzheimer's disease, brain injury, neuroinflammation, ischemia and stroke. They can improve learning and memory (Cai et al., 2020).

Fenugreek (Gong et al., 2016), yam (Yang et al., 2019) and their constit-uent diosgenin (Fuller et al., 2015; Gan et al., 2020) can be applicable for reduction in plasma insulin, glucose, cholesterol and lipoprotein levels and treatment of diabetes and its complications including diabetic nephropathy, diabetic liver disease, diabetic neuropathy, diabetic vascular disease, diabetic cardiomyopathy, diabetic reproductive dysfunction, and diabetic eye disease, although the results of controlled clinical trials were controversial. Diosgenin can have potential anti-obesity action, although this action requires clinical evaluation, too (Jeepipalli et al., 2020).

The preclinical *in vitro* and animal studies have shown that diosgenin has great potential in the treatment of various cardiovascular diseases, especially

Plant molecules and their influence on health **Chapter | 4** **319**

in atherosclerosis including endothelial dysfunction, lipid profile, and macrophage foam cell formation, thrombosis, and inflammation during the formation of atherosclerosis (Wu and Jiang, 2019). Diosgenin can effectively effectively improve hypertrophic cardiomyopathy, arrhythmia, myocardial I/R injury and cardiotoxicity caused by doxorubicin (Li et al., 2021).

There are reports indicating the ability of diosgenin and its steroidal metabolites to prevent osteoporosis (Pandey et al.,2018) and leishmania (Sultana et al., 2018).

In folk medicine diosgenin and diosgenin-containing plants are used as galactagogues (provide lactation aid), but the results of preclinical and few clinical studies were inconclusive (Sethi et al., 2018; Drugs and Lactation Database, 2021).

Diosgenin can be proposed as a male fertility-promoting drug, but this hypothesis has not been verified by the appropriate clinical studies (Abarikwu et al., 2020).

Clinical studies did not demonstrate substantial toxic or any other adverse effects of daidzein (Pandey et al., 2018; Cai et al., 2020; Parama et al., 2020) and daidzein-containing fenugreek (Drugs and Lactation Database, 2021).

Some physiological effects of daidzein could be not due to daidzein itself, but its metabolites. For example, daidzein metabolite dehydroepiandrosterone possesses a number of physiological and therapeutical properties similar to that of daidzein: antidiabetic, anticancer, antiallergic, antiobesity, antiaging, antidementia, antiosteoporosis, antiautoimmune disorders, and anticardiovascular disorders. Moreover, dehydroepiandrosterone in turn can be an indirect precursor to estrogen and testosterone and other steroid hormones with biological active effects similar to daidzein (Sahu et al., 2020). Another diosgenin metabolite, sarsasapogenin, had even higher antileishmanial acetyl anticholinesterase and antibutyryl cholinesterase activity than its precursor (Sultana et al., 2018).

Despite its physiological and medicinal effects, the clinical application of diosgenin is hindered by its low aqueous solubility, poor bioavailability and pharmacokinetics, and rapid biotransformation under physiological conditions (Cai et al., 2020; Parama et al., 2020). To avoid this concern, several novel diosgenin analogs and nanoformulations have been synthesized with improved pharmacokinetic profile and efficacy against cancer (Sethi et al., 2018; Bhardwaj et al., 2021), neurodegenerative (Cai et al., 2020), and cardiovascular diseases (Wu et al., 2019; Li et al., 2021).

These reports demonstrate the influence of diosgenin-containing plants, diosgenin and its metabolites on a wide array of physiological processes and illnesses, which show their high medicinal potential.

4.6.4 Mechanisms of action

Diosgenin and its metabolites can affect numerous physiological processes and illnesses via inhibition of enzymes acetyl cholinesterase, butyryl cholinesterase, and tyrosinase (Sultana et al., 2018). The *in vitro* and animal studies demonstrated that the ability of diosgenin to suppress cancer and tumor development, neurodegenerative and cardiovascular and other diseases are mediated by several common mechanisms of action.

Diosgenin and its metabolites reduce oxidative stress and peroxidation of DNA, lipids and proteins. They suppress immune response activity of immunocompetent T-cells and inflammation by inhibition of inflammatory cytokines, enzymes, adhesion molecules, PI3K/AKT/mTOR and JAK/STAT, WNT-beta-catenin intracellular pathways and transcription factor NF-κB. They activate cell death pathways (including promotion of proapoptotic bax and caspases and decrease in antiapoptotic bcl-2) and resulted cytoplasmic/mitochondrial apoptosis. Furthermore, diosgenin has a unique structural similarity to estrogen. It can promote cellular growth/differentiation through the estrogen receptor (ER) cascade, MAP kinase, and transcription factor PPARγ (see Chen et al., 2015; Sethi et al., 2018; Cai et al., 2020; Parama et al., 2020 for review). In addition, diosgenin suppress cancer cell functions via inhibition of PI3K/Akt/mTOR pathway and promotion reactive oxygen species-induced autophagy (Bhardwaj et al., 2021). Furthermore, diosgenin can prevent tumor metastasis by modulating epithelial-mesenchymal transition and actin cytoskeleton to change cellular motility, suppressing degradation of matrix barrier, and inhibiting angiogenesis (Chen et al., 2015).

Similar mediators of action have been proposed also for diosgenin-containing plants fenugreek (Fuller et al., 2015) and yam (Yang et al., 2019).

The antiobesity action of diosgenin can be due to its ability to suppress appetite at the level of CNS, to inhibit intestinal absorption of lipids, synthesis of lipids, adipogenesis and adipose tissue inflammation, and promoting fecal excretion of bile acids and triglycerides. There are indications that diosgenin can inhibit pancreatic lipase, disaccharidase enzyme, antagonistic to *in vitro* lipogenesis (Jeepipalli et al., 2020).

These data suggest the multiple mechanisms of daidzein action. Some common (antioxidant, antiinflammatory, antiproliferative, and proapoptotic) mechanisms mediate daidzein action on several processes and illnesses, but some mediators (for example, regulating adipogenesis and cell migration) are specific for particular daidzein targets.

4.6.5 Effects on female reproductive processes

4.6.5.1 Effect on ovarian and reproductive state

In experiments of Khazaei et al. (2011) the feeding of mice with fennel (*Foeniculum vulgare* Mill.) containing diosgenin increased the number of

Plant molecules and their influence on health **Chapter | 4 321**

growing, but not of small ovarian follicles. On the other hand, it remains not clear whether this effect on ovarian folliculogenesis was due to daidzein or other fennel constituents. The dietary pure diosgenin did not affect the number of follicles at each stage of folliculogenesis in murine ovaries, although it reduced the number of atretic follicles (Shen et al., 2017).

It remains not clear, whether and how daidzein can affect female reproductive organs other than the ovary. There is however one report (Lijuan et al., 2011) that an extract of wild ginger (*Costus speciosus,* Koen) rhizomes containing diosgenin promoted contraction of strips of rat uterine myometrium. On the other hand, diosgenin itself inhibited these contractions. This observation indicates a relaxing action of daidzein on uterine tension and, maybe, on the resulting reproductive processes (embryogenesis, parturition, etc.). This indication requires further experimental validation, however.

Therefore, the available information can indicate, but not demonstrate, the influence of diosgenin on ovary or uterus.

4.6.6 Effect on ovarian cell functions

Diosgenin was able to increase accumulation of both proliferation and apoptosis markers in cultured porcine ovarian granulosa cells suggesting its ability to promote ovarian cell turnover (Sirotkin et al., 2019). On the other hand, in cultured human ovarian cancer cells diosgenin-containing extract increased apoptosis and reduced viability (Xiao et al., 2012; Yang et al., 2015; Guo et al., 2018).

These observations indicate the stimulatory action of diosgenin on healthy, but the inhibitory action of this molecule on cancer ovarian cells.

4.6.7 Effect on oocytes and embryos

In traditional oriental folk medicine, many women eat diosgenin-containing fennel (Khazaei et al., 2011) or yam (Shen et al., 2017) for improvement fertility. The efficiency of this approach could be confirmed by studies of Nouri et al. (2017), which reported the ability of multinutrient containing diosgenin to improve oocyte quality, fertilizability, and pregnancy rate in women during *in vitro* fertilization program. Nevertheless, it remained unclear, whether these benefits were induced by diosgenin or another constituent of multinutrient. The dietary pure diosgenin did not affect oocyte maturation, their quality and fecundity in mice (Shen et al., 2017).

Therefore, the available data indicates the possible positive effect of diosgenin on oocytes, but this effect has not been directly demonstrated yet.

4.6.8 Effect on reproductive hormones

In *in vitro* experiments, addition of diosgenin inhibited progesterone and promoted testosterone and estradiol release in cultured porcine ovarian

322 Contaminants and Plants Action on Female Reproduction

granulosa cells and isolated follicles (Sirotkin et al., 2019, 2021). In cultured rabbit ovarian fragments, diosgenin stimulated the release of both progesterone and insulin-like growth factor I (IGF-I) (Sirotkin et al., 2019).

In *in vivo* experiments of Shen et al. (2017) feeding diosgenin to mice increased the level of anti-Mullerian hormone—marker of ovarian reserve in their plasma. No changes in expression of oocyte growth factors NOBOX, GDF9, and BMP15 involved in the control of ovarian follicle development were found.

Daidzein-containing drug reduced expression and reception of vascular endothelial growth factor (VEGF, promoter of tumor vascularization) in cultured ovarian cancer cells (Xiao et al., 2014; Guo et al., 2018).

Therefore, daidzein can influence (mostly promote) the release of hormones and growth factors—stimulators of ovarian functions (Sirotkin, 2014). On the other hand, it can suppress production and reception of VEGF, a physiological stimulator of both ovarian folliculogenesis and ovarian tumor growth.

4.6.9 Mechanisms of action on female reproductive processes

The *in vitro* experiments mentioned earlier demonstrated that diosgenin can affect ovarian and uterine cells **directly**, but not (or not only) through central (hypothalamo-pituitary) regulatory mechanisms.

The similarity of daidzein action on porcine whole ovarian follicles and isolated follicular granulosa cells (Sirotkin et al., 2019) suggests that just **granulosa cells are targets** of diosgenin action in the ovary.

The action of daidzein on ovarian hormones and growth factors listed earlier suggests that these molecules could be **extracellular mediators** of diosgenin action on female reproductive processes.

The action of daidzein on accumulation of markers and promoters of proliferation (PCNA) and apoptosis (bax) in cultured healthy porcine ovarian granulosa cells (Sirotkin et al., 2019) suggests that these molecules could be **intracellular mediators** of daidzein action on healthy ovarian cells. Studies of daidzein action on ovarian cancer cells indicated much more potential intracellular mediators of daidzein on the ovary. *In vitro* studies performed on human cancer cell lines (Xiao et al., 2012, 2014; Yang et al., 2015; Guo et al., 2018) demonstrated that the proapoptotic effect of diosgenin-containing drug on ovarian cancer cells was associated with reduction in expression of phosphoinositide 3-kinase, phosphorylated AKT and phosphorylated p38 mitogen-activated protein kinase, extracellular signal-related kinase, Src family kinase, focal adhesion kinase and IKKβ kinase, upregulation of apoptosis promoters bax and caspases 3 and 9. This observation indicates that proapoptotic and maybe antiproliferative action of diosgenin on cancer cells could be mediated by signaling pathways related to these protein kinases. Yang et al. (2015) showed also the involvement of transcription factor NF-kB, which can suppress tumor growth and angiogenesis via up regulation of proapoptotic protein bax and downregulation of antiapoptotic bcl-2 and angiogenic vascular endothelial growth factor.

Diosgenin possess antioxidative and estrogenic properties which determine its suppressive action on nonovarian cancer (Chen et al., 2015; Sethi et al., 2018; Cai et al., 2020; Parama et al., 2020). It might be proposed that these properties could determine its action on healthy female reproductive system and on its dysfunctions (cancer, polycystic ovarian syndrome, etc.) listed earlier. Nevertheless, such mechanisms of diosgenin action on female reproductive processes have not been reported yet.

The current literature suggests the existence of multiple intracellular signaling pathways regulating proliferation and apoptosis of ovarian cells and therefore the fate of ovarian follicle and whole reproductive system. On the other hand, the main information concerning intracellular mechanism of diosgenin action on the ovary has been obtained in *in vitro* experiment on ovarian cancer cell lines, while the physiological intracellular mediators of daidzein action on healthy reproductive system remain to be elucidated yet.

It is not to be excluded that the effects of diosgenin on female reproductive processes could be not due to daidzein itself, but its metabolites. At least, in experiments of Shen et al. (2017), the daidzein action on murine reproductive processes were similar to the effects of its metabolite dehydroepiandrosterone.

4.6.10 Application in reproductive biology and medicine

The *in vivo* studies listed earlier demonstrated the stimulatory action of daidzein-containing plants on ovarian folliculogenesis, oogenesis and fecundity. *In vitro* experiments revealed the ability of diosgenin to promote ovarian cell turnover and release of hormonal stimulators of ovarian functions (Sirotkin, 2014; Sirotkin et al., 2019). These observations indicate that diosgenin or diosgenin-containing plants could be a biostimulator of animal and human female reproduction. This hypothesis, however, requires validation by appropriate *in vivo* preclinical and clinical studies.

If positive influence of daidzein on oocyte maturation and quality would be confirmed, it could be used for improvement of *in vitro* maturation and

324 Contaminants and Plants Action on Female Reproduction

fertilization and embryotransfer in programs in animal production and assisted reproduction.

The ability of dietary diosgenin to prevent some age-dependent exhaustion of rat ovarian reserve and resulted fecundity (Shen et al., 2017) suggests potential applicability of diosgenin to prevent reproductive aging, to prolong reproductive cycle and to mitigate symptoms of menopause. Such effect could be examined on aged women.

The numerous reports concerning suppressive action of daidzein on cancer cells (Xiao et al., 2012, 2014; Yang et al., 2015; Guo et al., 2018) suggest its applicability for prevention and treatment of ovarian cancer. This action should be however validated by *in vivo* and clinical studies. Furthermore, the ability of daidzein to promoter ovarian folliculo- and steroidogenesis indicates that this molecule could be promising for treatment of polycystic ovarian syndrome.

Finally, the ability of diosgenin to prevent toxic effects of metal nanoparticles on cultured ovarian cells (Sirotkin et al., 2021) indicates that this molecule could be applicable as a natural protector against adverse effects of some environmental contaminants. Its ability to mitigate and prevent adverse effects of other environmental contaminants should be, however, examined in *in vitro* and *in vivo* animal and clinical studies.

Diosgenin application for improvement of reproductive processes and prevention and treatment of reproductive disorders could face the problem of its low solubility, bioavailability and rapid degradation (Cai et al., 2020). From this viewpoint, the search for more active and less degradable metabolites and analogues of daidzein could be promising for treatment not only of nonreproductive (Sethi et al., 2018; Wu et al., 2019), but also of reproductive disorders.

Therefore, the available results of *in vitro* and animal studies indicate the potential applicability of diosgenin and its metabolites or analogues for improvement animal and human reproduction and fecundity, prevention, and treatment of some reproductive disorders. These indications however require strong validation by further preclinical and clinical studies.

4.6.11 Conclusions and possible direction of future studies

The analysis of the current literature shows that diosgenin-containing plants, diosgenin and its metabolites can via numerous signaling pathways affect a wide array of nonreproductive and reproductive processes. Diosgenin can affect proliferation, apoptosis and release of hormones and growth factors by

Plant molecules and their influence on health **Chapter | 4** **325**

healthy and cancer ovarian cells and uterine contraction. Diosgenin action on ovarian folliculogenesis, oogenesis and fecundity, as well as its applicability for promotion of female reproductive processes and treatment of their disorders are possible, but they require confirmation by animal and clinical studies.

Some queries, however, are to be addressed before application of diosgenin-containing drugs. The actions of diosgenin and diosgenin-containing plants are sometimes different, suggesting either the presence of plant biologically active molecules other than diosgenin or different forms or derivates of diosgenin in plant products. A number of mediators of diosgenin action on non-reproductive processes could be involved in mediating its effects on female reproduction, but they have not been detected yet. The application of diosgenin is limited by missing or insufficient information concerning its *in vivo* action on animals and humans, as well as its low bioavailability. It is not to be excluded that replacement of diosgenin by its analogues or metabolites or by diosgenin-containing plant products could be more promising than application of diosgenin itself. Further profound studies are required to address these queries and to promote understanding and application of this promising molecule.

References

Abarikwu SO, Onuah CL, Singh SK. Plants in the management of male infertility. Andrologia 2020;52(3):e13509. https://doi.org/10.1111/and.13509.

Bhardwaj N, Tripathi N, Goel B, Jain SK. Anticancer activity of diosgenin and its semi-synthetic derivatives: role in autophagy mediated cell death and induction of apoptosis. Mini Rev Med Chem 2021. https://doi.org/10.2174/1389557521666210105111224.

Cai B, Zhang Y, Wang Z, Xu D, Jia Y, Guan Y, Liao A, Liu G, Chun C, Li J. Therapeutic potential of diosgenin and its major derivatives against neurological diseases: recent advances. Oxid Med Cell Longev 2020;2020:3153082. https://doi.org/10.1155/2020/3153082.

Chen Y, Tang YM, Yu SL, Han YW, Kou JP, Liu BL, Yu BY. Advances in the pharmacological activities and mechanisms of diosgenin. Chin J Nat Med 2015;13(8):578−87. https://doi.org/10.1016/S1875-5364(15)30053-4.

Drugs and Lactation Database (LactMed) [Internet]. Fenugreek. Bethesda (MD): National Library of Medicine (US); 2021. 2006−.

Fuller S, Stephens JM. Diosgenin, 4-hydroxyisoleucine, and fiber from fenugreek: mechanisms of actions and potential effects on metabolic syndrome. Adv Nutr 2015;6(2):189−97. https://doi.org/10.3945/an.114.007807.

Gan Q, Wang J, Hu J, Lou G, Xiong H, Peng C, Zheng S, Huang Q. The role of diosgenin in diabetes and diabetic complications. J Steroid Biochem Mol Biol 2020;198:105575. https://doi.org/10.1016/j.jsbmb.2019.105575.

Gong J, Fang K, Dong H, Wang D, Hu M, Lu F. Effect of fenugreek on hyperglycaemia and hyperlipidemia in diabetes and prediabetes: a meta-analysis. J Ethnopharmacol 2016;194:260−8. https://doi.org/10.1016/j.jep.2016.08.003.

Guo X, Ding X. Dioscin suppresses the viability of ovarian cancer cells by regulating the VEGFR2 and PI3K/AKT/MAPK signaling pathways. Oncol Lett 2018;15(6):9537—42. https://doi.org/10.3892/ol.2018.8454.

Jeepipalli SPK, Du B, Sabitaliyevich UY, Xu B. New insights into potential nutritional effects of dietary saponins in protecting against the development of obesity. Food Chem 2020;318:126474. https://doi.org/10.1016/j.foodchem.2020.126474.

Khazaei M, Montaseri A, Khazaei MR, Khanahmadi M. Study of foeniculum vulgare effect on folliculogenesis in female mice. Int J Fertil Steril 2011;5(3):122—7.

Li X, Liu S, Qu L, Chen Y, Yuan C, Qin A, Liang J, Huang Q, Jiang M, Zou W. Dioscin and diosgenin: insights into their potential protective effects in cardiac diseases. J Ethnopharmacol 2021;274:114018. https://doi.org/10.1016/j.jep.2021.114018.

Lijuan W, Kupittayanant P, Chudapongse N, Wray S, Kupittayanant S. The effects of wild ginger (*Costus speciosus* (Koen) Smith) rhizome extract and diosgenin on rat uterine contractions. Reprod Sci 2011;18(6):516—24. https://doi.org/10.1177/1933719110391278.

Liu JG, Xia WG, Chen W, Abouelezz KFM, Ruan D, Wang S, Zhang YN, Huang XB, Li KC, Zheng CT, Deng JP. Effects of capsaicin on laying performance, follicle development, and ovarian antioxidant capacity in aged laying ducks. Poult Sci 2021;100(7):101155. https://doi.org/10.1016/j.psj.2021.101155.

Nouri K, Walch K, Weghofer A, Imhof M, Egarter C, Ott J. The impact of a standardized oral multinutrient supplementation on embryo quality in in vitro fertilization/intracytoplasmic sperm injection: a prospective randomized trial. Gynecol Obstet Invest 2017;82(1):8—14. https://doi.org/10.1159/000452662.

Pandey MK, Gupta SC, Karelia D, Gilhooley PJ, Shakibaei M, Aggarwal BB. Dietary nutraceuticals as backbone for bone health. Biotechnol Adv 2018;36(6):1633—48. https://doi.org/10.1016/j.biotechadv.2018.03.014.

Parama D, Boruah M, Yachna K, Rana V, Banik K, Harsha C, Thakur KK, Dutta U, Arya A, Mao X, Ahn KS, Kunnumakkara AB. Diosgenin, a steroidal saponin, and its analogs: effective therapies against different chronic diseases. Life Sci 2020;260:118182. https://doi.org/10.1016/j.lfs.2020.118182.

Sahu P, Gidwani B, Dhongade HJ. Pharmacological activities of dehydroepiandrosterone: a review. Steroids 2020;153:108507. https://doi.org/10.1016/j.steroids.2019.108507.

Sethi G, Shanmugam MK, Warrier S, Merarchi M, Arfuso F, Kumar AP, Bishayee A. Pro-apoptotic and anti-cancer properties of diosgenin: a comprehensive and critical review. Nutrients 2018;10(5):645. https://doi.org/10.3390/nu10050645.

Shen M, Qi C, Kuang YP, Yang Y, Lyu QF, Long H, Yan ZG, Lu YY. Observation of the influences of diosgenin on aging ovarian reserve and function in a mouse model. Eur J Med Res 2017;22(1):42. https://doi.org/10.1186/s40001-017-0285-6.

Sirotkin AV, Alexa R, Alwasel S, Harrath AH. The phytoestrogen, diosgenin, directly stimulates ovarian cell functions in two farm animal species. Domest Anim Endocrinol 2019;69:35—41. https://doi.org/10.1016/j.domaniend.2019.04.002.

Sirotkin AV, Alexa R, Stochmalova A, Scsukova S. Plant isoflavones can affect accumulation and impact of silver and titania nanoparticles on ovarian cells. Endocr Regul 2021;55(1):52—60. https://doi.org/10.2478/enr-2021-0007.

Sirotkin AV. Regulators of ovarian functions. 2nd ed. New York: Nova Science Publishers, Inc.; 2014, ISBN 978-1-62948-574-4. 194 pp.

Sultana N. Microbial biotransformation of bioactive and clinically useful steroids and some salient features of steroids and biotransformation. Steroids 2018;136:76—92. https://doi.org/10.1016/j.steroids.2018.01.007.

Wu FC, Jiang JG. Effects of diosgenin and its derivatives on atherosclerosis. Food Funct 2019;10(11):7022−36. https://doi.org/10.1039/c9fo00749k.

Xiao X, Zou J, Bui-Nguyen TM, Bai P, Gao L, Liu J, Liu S, Xiao J, Chen X, Zhang X, Wang H. Paris saponin II of Rhizoma Paridis−a novel inducer of apoptosis in human ovarian cancer cells. Biosci Trends 2012;6(4):201−11. https://doi.org/10.5582/bst.2012.v6.4.201.

Xiao X, Yang M, Xiao J, Zou J, Huang Q, Yang K, Zhang B, Yang F, Liu S, Wang H, Bai P. Paris Saponin II suppresses the growth of human ovarian cancer xenografts via modulating VEGF-mediated angiogenesis and tumor cell migration. Cancer Chemother Pharmacol 2014;73(4):807−18. https://doi.org/10.1007/s00280-014-2408-x.

Yang M, Zou J, Zhu H, Liu S, Wang H, Bai P, Xiao X. Paris saponin II inhibits human ovarian cancer cell-induced angiogenesis by modulating NF-κB signaling. Oncol Rep 2015;33(5):2190−8. https://doi.org/10.3892/or.2015.3836.

Yang Q, Wang C, Jin Y, Ma X, Xie T, Wang J, Liu K, Sun H. Disocin prevents postmenopausal atherosclerosis in ovariectomized LDLR-/- mice through a PGC-1α/ERα pathway leading to promotion of autophagy and inhibition of oxidative stress, inflammation and apoptosis. Pharmacol Res 2019;148:104414. https://doi.org/10.1016/j.phrs.2019.104414.

328 Contaminants and Plants Action on Female Reproduction

Chapter 4.7

Isoquercitrin

4.7.1 Introduction

Flavonoids are a group of polyphenolic compounds and as a part of the human diet (Appleton, 2010) and as a food supplement (Michalcova et al., 2019), they have very low toxicity in addition to various beneficial health effects (Appleton, 2010; Michalcova et al., 2019). Among the numerous groups of flavonoids, the flavonols have been the focus of special attention for their potential actions on different organic reactions and biological pathways. Flavonoid glucosides such as quercetin and isoquercitrin (IQ) occur widely in the plant kingdom and are among the most common flavonoids in the human diet. Water soluble enzymatically modified isoquercitrin is generally regarded as safe for ingestion by the United States Food and Drug Administration (FDA, 2007) and is also approved in Japan as a food additive (JFA, 2007). On a daily basis, up to 4.9 mg/kg/day of enzymatically modified (α-glucosylated) IQ is acceptable (Valentová et al., 2014). IQ screening for clinical purposes has recently attracted a great deal of interest for a number of health issues including inflammation, atherosclerosis (Reuter et al., 2010), and cancers (Amado et al., 2014; Huang et al., 2014; Chen et al., 2014; Wu et al., 2017; Buonerba et al., 2018) including ovarian cancer (Michalcova et al., 2019).

4.7.2 Provenance and properties

Isoquercitrin (IQ, $C_{21}H_{20}O_{12}$, Fig. 4.7) is also sometimes called isoquercetin, which is a nearly identical quercetin-3-monoglucoside. Although they are technically different because isoquercetin has a pyranose ring whereas IQ has a furanose ring, functionally, the two molecules are indistinguishable. Published literature often considers them to be the same compound and uses the names interchangeably. In addition, quercetin and IQ differ in their structure, bioavailability, absorption, and biological actions (Murota et al., 2000; Appleton, 2010). The best-known flavonols are (1) quercetin, an aglycone molecule widely found in nature; (2) rutin, a hydrophilic molecule; and (3) isoquercitin, a naturally occurring glycoside of quercetin also known as hirsutrin, isoquercetrin, quercetin-3-glucoside (Q3G), quercetin-3-O-β-D-glucoside. The molecular formula and molar mass of IQ are C21H20O12 and 464.38 g/mol, respectively. Structurally, flavonols have one hydroxyl (OH) group on the third carbon (C3) and one carbonyl group (C=O) on the fourth carbon (C4) of the C ring (Appleton, 2010; Masuda et al., 2001). The flavonoid

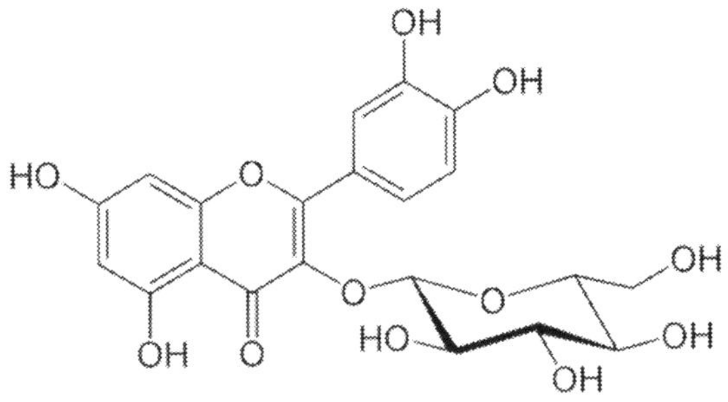

Plant molecules and their influence on health **Chapter | 4** **329**

FIGURE 4.7 The chemical structures of IQ (PubChem, 2021).

IQ (quercetin-3-O-β-D-glucopyranoside) is commonly found in medicinal herbs, fruits, vegetables and plant-derived foods and beverages. Pure IQ can now be obtained on a large scale by enzymatic rutin hydrolysis with α-L-rhamnosidase. IQ has higher bioavailability than quercetin (Valentová et al., 2014). IQ used in our previous study (Michalcova et al., 2019) was prepared selective enzymatic derhamnosylation of rutin using recombinant α-L-rhamnosidase from *Aspergillus terreus* (Weignerová et al., 2012). The detailed descriptions of quercetin and rutin are presented in the corresponding chapters.

Several studies have attempted to elucidate the pathway of IQ intestinal absorption (Orfali et al., 2016). Although small amounts of intact IQ can be found in plasma and tissues after oral application, it is extensively metabolized in the intestine and the liver (Valentová et al., 2014).

The absorption of flavonoids in the small intestine is limited owing to their molecular weight and hydrophilicity of their glycosides. Naturally occurring quercetin compounds are mainly glycosides such as IQ and are commonly found in plants and the human diet. Glycosides are not easily absorbed in the digestive tract, and most of their absorption occurs after transformation to the aglycone form (Appleton, 2010). The most widely accepted hypothesis of IQ intestinal absorption involves lactase phlorizin hydrolase (LPH) as the major step, and sodium-dependent glucose transporter 1 (SGLT1) as a second step. LPH is an extracellular enzyme localized on the outer surface of the small intestinal brush-border membrane. When quercetin-3-glucoside (Q3G) is ingested, it is first hydrolyzed in the small intestine by the lactase domain of LPH, releasing the quercetin aglycone, which then passively diffuses to the enterocity throughout the apical surface. The deglycosylation of Q3G leads to a higher concentration of the aglycone at the apical enterocyte membrane, thereby increasing the rate of absorption. A small amount of Q3G is transported by the SGLT1 present in the apical surface of the small intestine, thereby transporting the intact glycoside into the cell. In enterocytes, cytosolic β-glycosidase (CBG) hydrolyzes the intact Q3G, which is then transported via the SGLT1 route, into the quercetin aglycone. Uridine diphosphate-

330 Contaminants and Plants Action on Female Reproduction

glucuronosyltransferases (UDP-GT) glucuronidates the quercetin aglycone into quercetin glucuronides (conjugated quercetin metabolites and then it finally reaches the bloodstream (Orfali et al., 2016).

4.7.3 Physiological and therapeutic actions

Previous *in vitro* and *in vivo* studies with flavonols, particularly flavonoid IQ, have exhibited many biological activities including neuroprotective, cardioprotective, chemopreventive, antiallergic, antioxidative, antinflammatory, antiallergic, antiproliferative (Appleton, 2010; de Araújo et al., 2013; Xu et al., 2019; Qiu et al., 2019), and anticancer effects (Orfali et al., 2016; Kangawa et al., 2017; Michalcova et al., 2019) including therapy of cancers of pancreas (Chen et al., 2014), liver (Huang et al., 2014), kidney (Buonerba et al., 2018), colon (Amado et al., 2014; Kangawa et al., 2017), bladder (Wu et al., 2017), ovary (Michalcova et al., 2019), as well as reactive oxygen species (ROS)-induced diseases particularly for mesenchymal stem cell transplantation therapy (Li et al., 2016).

IQ has been able to recover H2O2-induced loss of viability in human somatic cells (Zhu et al., 2016). Similarly, IQ may promote motor functional recovery and nerve regeneration following peripheral nerve injury though inhibition of oxidative stress, which highlighted the therapeutic values of IQ as a neuroprotective drug for peripheral nerve repair applications (Qiu et al., 2019). Likewise, in hippocampal neuronal cells of rats subjected to oxygen-glucose deprivation followed by reoxygenation, pretreatment with IQ was found to promote cell viability (Chen et al., 2017a). IQ isolated from *Thuja orientalis* was able to increase cell viability in human retinal ganglion cells (Jung et al., 2010). At present, it has been found that IQ has hypoglycemic effects (Jayachandran et al., 2018; Fadul et al., 2019). Furthermore, IQ was capable of protecting the bone marrow cells by increasing their viability against H_2O_2-induced oxidative stress (Kim et al., 2019). IQ treatment was found to improve the viability of bovine sperm cells (Greifova et al., 2017).

The impact of IQ treatment on cancer cells which helped reveal the mechanisms underlying its effect on cell viability, proliferation, apoptosis, transforming growth factor beta (TGF-β) and its receptor binding including some intriguing pathways that mediate isoquercitrin-induced cancer cell death was investigated. Isolated from the root tuber of Tetrastigma hemsleyanum Diels et Gilg, IQ was found to inhibit hepatocyte growth factor/scatter factor-induced tumor cell migration colorectal cancer through inhibiting inflammation and cell proliferation (Kangawa et al., 2017).Similarly the protective effect of enzymatically modified IQ on the hepatocarcinogenic process where in tumor promotion was suppressed by cotreatment with IQ through suppression of cell proliferation activity of preneoplastic liver cells of rats was found (Hara et al., 2014). IQ isolated from *Bidens bipinnata* L. extract inhibited the progression of liver cancer *in vivo* and *in vitro*. It also promoted

Plant molecules and their influence on health **Chapter | 4** **331**

apoptosis, and inhibited cell proliferation in human liver cancer cells (Huang et al., 2014). Another therapeutic dose of IQ also promoted apoptosis in human pancreatic cancer cells and inhibited the progression of pancreatic cancer *in vivo* and *in vitro* (Chen et al., 2014). The beneficial effect of IQ on human breast cancer cells were found (Xu et al., 2019).

4.7.4 Mechanisms of action

IQ inhibits oxidative stress through reducing the production of NADPH oxidase 4 (Nox4) and dual oxidase 1 (Duox1), and promoting the expression of the nuclear factor erythroid 2-related factor 2 (Nrf2) and superoxide dismutase 2 (SOD2) in soleus muscles after sciatic nerve crush (Qiu et al., 2019).

Inhibition of gluconeogenesis in hepatocytes by IQ was related to the liver kinase B1 (LKB1) upregulation and phosphorylation of AMP-activated protein kinase α (AMPKα) (Chen et al., 2020). IQ has hypoglycemic effects, and possible mechanisms include antagonizing oxidative stress and insulin resistance (Jayachandran et al., 2018; Fadul et al., 2019), inhibiting intestinal *α-glucosidase* (Chen et al., 2017b; Renda et al., 2018), increasing endogenous glucagon-like peptide-1 (GLP-1) level (Gaballah et al., 2017).

Proliferation and Apoptosis. IQ promotes the axonal regeneration of dorsal root ganglia **neurons**, the proliferation and migration of Schwann cells, and the expression of proliferating cell nuclear antigen (PCNA) in Schwann cells *in vitro* (Qiu et al., 2019). Protective effects of IQ on neuronal cells were mediated by suppression of the H2O2-induced upregulation of pro-apoptotic cleaved caspase-3, -9, poly (ADP-ribose) polymerase (PARP), and Bax (Kim et al., 2019). IQ isolated from *Thuja orientalis* was also able to recover the death of human retinal ganglion cells by hydrogen peroxide (H_2O_2)-induced apoptosis as demonstrated by staining of dead cells for phosphatidylserine as well as the upregulation [cleaved PARP, apoptosis inducing factor (AIF), tumor protein p53 (p53), and downregulation of B-cell lymphoma 2 (Bcl-2)] of proteins associated with apoptosis and survival. It was further postulated that human retinal ganglion cells die by apoptosis following an insult of H_2O_2 and that IQ partially blunts the upregulation of cleaved PARP and p53 (Jung et al., 2010).

IQ isolated from *Bidens bipinnata* L. extract inhibited the progression of liver cancer *in vivo* and *in vitro*. It also promoted apoptosis, inhibited cell proliferation, and blocked the cell cycle in the G1 phase via the mitogen-activated protein kinase (MAPK) signaling pathway in human liver cancer cells (Huang et al., 2014).

Another therapeutic dose of IQ also promoted apoptosis in human pancreatic cancer cells and inhibited the progression of pancreatic cancer *in vivo* and *in vitro* by regulating opioid receptors and the mitogen-activated protein kinase signaling pathway (Chen et al., 2014). IQ isolated from the aerial parts of *Hyptis fasciculata*, did not affect caspase-3 dependent apoptosis in human brain cancer cells (Amado et al., 2009).

332 Contaminants and Plants Action on Female Reproduction

IQ caused downregulation of apoptotic protein expression such as cleaved caspase-9, -3, PARP, and p53 in human umbilical vein endothelial cell and lung adenocarcinoma cell. Antiapoptotic effect was further associated with the Akt /glycogen synthase kinase 3 (GSK3) signaling pathway, and IQ was recommended for clinical use owing to its capability to interfere with the progression of endothelial injury-associated cardiovascular disease (Zhu et al., 2016).

Therapeutic doses of IQ extracted from *Bidens pilosa* L. retarded proliferation and induced apoptosis by inhibiting phosphatidylinositol 3-kinase and Akt phosphorylation expression levels, and the cell cycle was arrested in the G1 phase in human bladder cancer cells (Chen et al., 2016).

Lysine-specific demethylase 1 (LSD1) has recently emerged as a therapeutic target for cancer. IQ exhibited optimal LSD1 inhibitory activity on cancer cell properties. IQ induced the expression of key proteins in the mitochondrial-mediated apoptosis pathway and caused apoptosis in LSD1-overexpressing human breast cancer cells via the inhibition of LSD1. These findings suggest that natural LSD1 inhibitors, and particularly IQ, are promising for cancer treatment (Xu et al., 2019).

Data provides evidence that the IQ promotes the axonal regeneration might be a promising candidate as a potential modulator of proliferation and apoptosis and as a potential chemoprotective agent. However, further research is essential to understand the therapeutic potential of IQ.

4.7.5 Effects on female reproductive processes

Human ovarian granulosa cells as a suitable cellular model was used for investigation of the effect of dietary bioflavonoid IQ on cell viability, survival, apoptosis, the release of 17β-estradiol, progesterone and human transforming growth factor-β2 (TGF-β2) and TGF-β2 receptor as well as the production of intracellular **reactive oxygen species (ROS)** (Kolesarova et al., not published data).

The beneficial effect of IQ on human ovarian cells were found (Michalcova et al., 2019, Kolesarova et al., not published data). The release of 17β-estradiol, progesterone, TGF-β2, and binding of TGF-β2 receptor in human ovarian granulosa cells (Kolesarova et al., not published data) (Kolesarova et al., not published data) and TGF-β1 and TGF-β1 receptors in human ovarian cancer cells (Michalcova et al., 2019) were not influenced by IQ addition. IQ treatment did not cause change in the viability, the proportion of live, dead, and apoptotic cells of human ovarian granulosa cells (Kolesarova et al., not published data) and human ovarian cancer cells (Michalcova et al., 2019). IQ exhibited antioxidative activity by hampering the production of intracellular ROS (Kolesarova et al., not published data). Similarly, lower IQ concentrations were able to exhibit beneficial effects by inhibiting the generation of intracellular ROS in human ovarian cancer cells. In contrast, elevated

Plant molecules and their influence on health **Chapter | 4** **333**

concentrations led to oxidative stress (Michalcova et al., 2019). IQ may be considered a safer alternative for managing postmenopausal osteoporosis that is a common and disabling disorder that increases the risk of bone fractures due to estrogen deprivation; this can be simulated in rats by ovariectomy (Fayed et al., 2019).

Therefore, the performed studies demonstrated the beneficial effect of IQ on human ovarian cells functions (viability, proliferation, apoptosis, secretory activity). The beneficial effect of IQ on ovarian cells may be mediated by an antioxidative pathway that involves inhibition of intracellular ROS generation, thereby limiting oxidative stress. Phytoestrogen IQ could prevent osteoporosis induced by deficit of estrogens.

4.7.6 Mechanisms of action on female reproductive processes

Little is known about the impact of isoquercitrin on the female reproductive functions. Section 4.9.5 describes the influence of quercetin on female reproductive functions. IQ exhibited antioxidative activity by hampering the **ROS production** in human ovarian granulosa and cancer cells (Michalcova et al., 2019; Kolesarova et al., not published data).

Postmenopausal osteoporosis is a common and disabling disorder that increases the risk of bone fractures due to estrogen deprivation. **Hypoxia inducible factor-1 α (HIF-1α)** expression in osteoclasts predominantly leads to its activation increasing bone resorption. Premenopausal, estrogen prevents HIF-1α expression maintaining bone density. IQ, a common edible plants phytoestrogen, is known to inhibit HIF-1α. Isoquercitrin attenuated the increased HIF-1α expression while increased that of the **vascular endothelial growth factor (VEGF)** and **β-catenin**. It also decreased the levels of the **nuclear factor κB (NF-κB)**. Therefore, IQ may affect bone via multiple extra- and intracellular signaling pathways. Due to this action, it may be considered as a safe alternative drug for managing osteoporosis (Fayed et al., 2019).

IQ treatment did not cause change in the release of TGF-β2 and binding of TGF-β2 receptor in human ovarian granulosa cells (Kolesarova et al., not published data) and TGF-β1 and TGF-β1 receptor human ovarian cancer cells (Michalcova et al., 2019). In human papilloma virus 16-infected human cervical cancer cells, IQ isolated from the leaves of *Avicennia marina* showed cytotoxicity and enhanced the mRNA expression of **tumor necrosis factor (TNF)-related apoptosis inducing ligand-receptor**, although the increase in apoptosis was meager (Arumugam et al., 2017).

Therefore, bone, ovary and uterus are sites of IQ action on female reproductive processes and their disfunctions. This action is mediated by multiple extra- and intracellular signaling molecules. Data provides evidence that the IQ as a phytoestrogen might be a promising candidate as a potential modulator of steroidogenenis proliferation and apoptosis and as a potential chemoprotective agent.

4.7.7 Application in reproductive biology and medicine

The available literature describes numerous positive and therapeutic effects of IQ on non-reproductive processes. Nevertheless, the limited available data concerning reproductive effects of IQ suggest that it, as phytoestrogen and antioxidant, could be useful for stimulation of reproductive processes in farm animals and humans, as well as for prevention and treatment of some reproductive disorders (ovarian aging, reproductive insufficiency, estrogen-dependent osteoporosis, estrogen-dependent cancers including ovarian cancer) disorders in animal production, medicine, phytotherapy, biotechnology and assisted reproduction.

IQ has a large potential as natural and cheap protector and regulator of reproductive and related processes. On the other hand, it is not to be excluded that generation and application of IQ metabolites, analogues and carriers could be more efficient, than IQ itself.

4.7.8 Conclusions and possible direction of future studies

The available data suggest IQ influence on functions of reproductive system (ovary and uterus), on ovarian-dependent processes (bone osteoporosis), which could be mediated by several extra- and intracellular signaling molecules affecting oxidative processes, proliferation, apoptosis and viability of target cells. These data indicate applicability of IQ as a promising tool for control and regulation of female reproductive and related processes and treatment of their disorders in phytotherapy, animal production, medicine, biotechnology and assisted reproduction.

Nevertheless, application of IQ is hampered by unavailability or insufficiency of information concerning its reproductive effects. Understanding the sites and mechanisms of its effects and their hierarchical interrelationships require further elucidation. The main available information concerning IQ action was obtained in *in vitro* experiments, while animal studies are rare, and clinical studies are practically absent et all. Chemical modification of IQ to affect reproductive processes have not been tested yet. This promising molecule is worth of further studies.

References

Amado NG, Cerqueira DM, Menezes FS, da Silva JF, Neto VM, Abreu JG. Isoquercitrin isolated from hyptis fasciculata reduces glioblastoma cell proliferation and changes beta-catenin cellular localization. Anti Cancer Drugs 2009;20(7):543−52. https://doi.org/10.1097/CAD.0b013e32832d1149.

Amado NG, Predes D, Fonseca BF, Cerqueira DM, Reis AH, Dudenhoeffer AC, Borges HL, Mendes FA, Abreu JG. Isoquercitrin suppresses colon cancer cell growth in vitro by targeting the Wnt/β-catenin signaling pathway. J Biol Chem 2014;289(51):35456−67. https://doi.org/10.1074/jbc.M114.621599.

Appleton J. Evaluating the bioavailability of isoquercetin. Nat Med 2010;2:1—6.

Arumugam S, Bandil K, Proksch P, Murugiyan K, Bharadwaj M. Effects of A. marina-derived isoquercitrin on TNF-related Apoptosis-Inducing Ligand Receptor (TRAIL-R) expression and apoptosis induction in cervical cancer cells. Appl Biochem Biotechnol 2017;182(2):697—707. https://doi.org/10.1007/s12010-016-2355-6.

Buonerba C, De Placido P, Bruzzese D, Pagliuca M, Ungaro P, Bosso D, Di Lorenzo G, et al. Isoquercetin as an adjunct therapy in patients with kidney cancer receiving first-line sunitinib (QUASAR): results of a phase I trial. Front Pharmacol 2018;9:189. https://doi.org/10.3389/fphar.2018.00189.

Chen Q, Li P, Xu Y, Li Y, Tang B. Isoquercitrin inhibits the progression of pancreatic cancer in vivo and in vitro by regulating opioid receptors and the mitogen-activated protein kinase signalling pathway. Oncol Rep 2014;33(2):840—8. https://doi.org/10.3892/or.2014.3626.

Chen F, Chen X, Yang D, Che X, Wang J, Li X, Zhang Z, Wang Q, Zheng W, Wang L, et al. Isoquercitrin inhibits bladder cancer progression in vivo and in vitro by regulating the PI3K/Akt and PKC signaling pathways. Oncol Rep 2016;36(1):165—72. https://doi.org/10.3892/or.2016.4794.

Chen H, Ouyang K, Jiang Y, Yang Z, Hu W, Xiong L, Wang N, Liu X, Wang W. Constituent analysis of the ethanol extracts of Chimonanthus nitens Oliv. leaves and their inhibitory effect on α-glucosidase activity. Int J Biol Macromol 2017a;98:829—36. https://doi.org/10.1016/j.ijbiomac.2017.02.044.

Chen M, Dai LH, Fei A, Pan SM, Wang HR. Isoquercetin activates the ERK1/2-NRF2 pathway and protects against cerebral ischemia-reperfusion injury in vivo and in vitro. Exp Ther Med 2017b;13(4):1353—9. https://doi.org/10.3892/etm.2017.4093.

de Araújo ME, Moreira Franco YE, Alberto TG, Sobreiro MA, Conrado MA, Priolli DG, Frankland Sawaya AC, Ruiz AL, de Carvalho JE, de Oliveira Carvalho P. Enzymatic de-glycosylation of rutin improves its antioxidant and antiproliferative activities. Food Chem 2013;141(1):266—73. https://doi.org/10.1016/j.foodchem.2013.02.127.

Fadul E, Nizamani A, Rasheed S, Adhikari A, Yousuf S, Parveen S, Gören N, Alhazmi HA, Choudhary MI, Khalid A. Anti-glycating and anti-oxidant compounds from traditionally used anti-diabetic plant Geigeria alata (DC) Oliv. & Hiern. Nat Prod Res 2019:1—9. https://doi.org/10.1080/14786419.2018.1542388.

Fayed HA, Barakat BM, Elshaer SS, Abdel-Naim AB, Menze ET. Antiosteoporotic activities of isoquercitrin in ovariectomized rats: role of inhibiting hypoxia inducible factor-1 alpha. Eur J Pharmacol 2019;865:172785. https://doi.org/10.1016/j.ejphar.2019.172785.

Gaballah HH, Zakaria SS, Mwafy SE, Tahoon NM, Ebeid AM. Mechanistic insights into the effects of quercetin and/or GLP-1 analogue liraglutide on high-fat diet/streptozotocin-induced type 2 diabetes in rats. Biomed Pharmacother 2017;9(2):331—9. https://doi.org/10.1016/j.biopha.2017.05.086.

Greifova H, Tvrda E, Lukac N. Beneficial effects of isoquercitrin on the behaviour of bovine spermatozoa in vitro. In: Presented at the 8th CASEE Conference "Sustainable development in Europe- cooperation between science and practice- What's the position of Central and South Eastern Europe?". Warsaw, Poland: Warsaw University of Life Sciences; May 2017. p. 14—6. http://www.ica-casee.eu/index.php/15-events/123-casee-conference-2017.

Hara S, Morita R, Ogawa T, Segawa R, Takimoto N, Suzuki K, Hamadate N, Hayashi SM, Odachi A, Ogiwara I, et al. Tumor suppression effects of bilberry extracts and enzymatically modified isoquercitrin in early preneoplastic liver cell lesions induced by piperonyl butoxide promotion in a two-stage rat hepato carcinogenesis model. Exp Toxicol Pathol 2014;66:225—34. https://doi.org/10.1016/j.etp.2014.02.002.

336 Contaminants and Plants Action on Female Reproduction

Huang G, Tang B, Tang K, Dong X, Deng J, Liao L, Liao Z, Yang H, He S. Isoquercitrin inhibits the progression of liver cancer in vivo and in vitro via the MAPK signalling pathway. Oncol Rep 2014;31(5):2377−84. https://doi.org/10.3892/or.2014.3099.

Japan Food Additives Association. Japanese specifications and standards for food additives. Tokyo: Japan Food Additives Association; 2007. p. 8.

Jayachandran M, Zhang T, Ganesan K, Xu B, Chung SSM. Isoquercetin ameliorates hyperglycemia and regulates key enzymes of glucose metabolism via insulin signalling pathway in streptozotocin-induced diabetic rats. Eur J Pharmacol 2018;8(29):112−20. https://doi.org/10.1016/j.ejphar.2018.04.015.

Jung SH, Kim BJ, Lee EH, Osborne NN. Isoquercitrin is the most effective antioxidant in the plant Thuja orientalis and able to counteract oxidative-induced damage to a transformed cell line (RGC-5 cells). Neurochem Int 2010;57(7):713−21. https://doi.org/10.1016/j.neuint.2010.08.005.

Kangawa Y, Yoshida T, Maruyama K, Okamoto M, Kihara T, Nakamura M, Ochiai M, Hippo Y, Hayashi S, Shibutani M. Cilostazol and enzymatically modified isoquercitrin attenuate experimental colitis and colon cancer in mice by inhibiting cell proliferation and inflammation. Food Chem Toxicol 2017;100:103−14. https://doi.org/10.1016/j.fct.2016.12.018.

Kim JH, Quilantang NG, Kim HY, Lee S, Cho EJ. Attenuation of hydrogen peroxide-induced oxidative stress in sh-sy5y cells by three flavonoids from Acer okamotoanum. Chem Paper 2019;73(5):1135−44. https://doi.org/10.1007/s11696-018-0664-7.

Li X, Jiang Q, Wang T, Liu J, Chen D. Comparison of the antioxidant effects of quercitrin and isoquercitrin: understanding the role of the 6″-oh group. Molecules 2016;21. https://doi.org/10.3390/molecules21091246.

Masuda T, Iritani K, Yonemori S, Oyama Y, Takeda Y. Isolation and antioxidant activity of galloyl flavonol glycosides from the seashore plant, *Pemphis acidula*. Biosci Biotechnol Biochem 2001;65(6):1302−9. https://doi.org/10.1271/bbb.65.1302.

Michalcova K, Roychoudhury S, Halenar M, Tvrda E, Kovacikova E, Vasicek J, Chrenek P, Baldovska S, Sanislo L, Kren V, Kolesarova A. In vitro response of human ovarian cancer cells to dietary bioflavonoid isoquercitrin. J Environ Sci Health B 2019;54:752−7. https://doi.org/10.1080/03601234.2019.1633214.

Murota K, Shimizu S, Chujo H, Moon JH, Terao J. Efficiency of absorption and metabolic conversion of quercetin and its glucosides in human intestinal cell line Caco-2. Arch Biochem Biophys 2000;384(2):391−7. https://doi.org/10.1006/abbi.2000.2123.

National Center for Biotechnology Information. PubChem Compound Summary for CID 5280804, Isoquercitrin, PubChem, https://pubchem.ncbi.nlm.nih.gov/compound/5280804 Accessed June 1, 2021.

Orfali GC, Duarte AC, Bonadio V, Martinez NP, de Araújo ME, Priviero FB, Carvalho PO, Priolli DG. Review of anticancer mechanisms of isoquercitin. World J Clin Oncol 2016;7(2):189−99. https://doi.org/10.5306/wjco.v7.i2.189.

Qiu J, Yang X, Wang L, Zhang Q, Ma W, Huang Z, Ding F, et al. Isoquercitrin promotes peripheral nerve regeneration through inhibiting oxidative stress following sciatic crush injury in mice. Ann Transl Med 2019;7(22):680. https://doi.org/10.21037/atm.2019.11.18.

Renda G, Sari S, Barut B, Šoral M, Liptaj T, Korkmaz B, Özel A, Erik İ, Söhretoğlu D. α-Glucosidase inhibitory effects of polyphenols from Geranium asphodeloides: inhibition kinetics and mechanistic insights through in vitro and in silico studies. Bioorg Chem 2018;8(1):545−52. https://doi.org/10.1016/j.bioorg.2018.09.009.

Reuter S, Gupta SC, Chaturvedi MM, Aggarwal BB. Oxidative stress, inflammation, and cancer: how are they linked? Free Radic Biol Med 2010;49(11):1603−16. https://doi.org/10.1016/j.freeradbiomed.2010.09.006.

US Food and Drug Administration. Agency response letter to GRAS notice No. GRN00220 [Alpha-glycosyl isoquercitrin]. US Food and Drug Administration Center for Food Safety and Applied Nutrition; 2007.

Valentová K, Vrba J, Bancířová M, Ulrichová J, Křen V. Isoquercitrin: pharmacology, toxicology, and metabolism. Food Chem Toxicol 2014;68:267–82. https://doi.org/10.1016/j.fct.2014.03.018.

Weignerová L, Marhol P, Gerstorferová D, Křen V. Preparatory production of quercetin-3-β-d-glucopyranoside using alkali-tolerant thermostable α-l-rhamnosidase from aspergillus terreus. Bioresour Technol 2012;115:222–7. https://doi.org/10.1016/j.biortech.2011.08.029.

Wu P, Liu S, Su J, Chen J, Li L, Zhang R, Chen T. Apoptosis triggered by isoquercitrin in bladder cancer cells by activating the AMPK-activated protein kinase pathway. Food Funct 2017;8:3707–22. https://doi.org/10.1039/C7FO00778G.

Xu X, Peng W, Liu C, Li S, Lei J, Wang Z, Kong L, Han C. Flavone-based natural product agents as new lysine-specific demethylase 1 inhibitors exhibiting cytotoxicity against breast cancer cells in vitro. Bioorg Med Chem 2019;27(2):370–4. https://doi.org/10.1016/j.bmc.2018.12.013.

Zhu M, Li J, Wang K, Hao X, Ge R, Li Q. Isoquercitrin inhibits hydrogen peroxide-induced apoptosis of ea.Hy926 cells via the pi3k/akt/gsk3β signaling pathway. Molecules 2016;21(3):356. https://doi.org/10.3390/molecules21030356.

338 Contaminants and Plants Action on Female Reproduction

Chapter 4.8

Punicalagin

4.8.1 Introduction

Pomegranate (*Punica granatum* L.) is a deciduous shrub that originally was distributed in the Mediterranean region, the southeast of Asia and the Himalaya. Pomegranate fruit contains polyphenolic compounds, including anthocyanins, flavonoids, and punicalagins, which are the most important members of ellagitannins family (Gil et al., 2000). Punicalagins impart the characteristic yellow color of pomegranate husk and are extracted with the juice during processing. Punicalagin (PUN; 2,3-hexahydroxydiphenoyl-gallagyl-D-glucose) is the most abundant polyphenol found in pomegranate (*Punica granatum* L.) husk (Yao et al., 2017). Polyphenols present in pomegranate such as gallic acid, ellagic acid, punicalagins and punicalins are the major chemical component, and pomegranate possessed the highest concentration of punicalagin α and β among the commonly consumed natural fruits (Fischer et al., 2011). Understanding mechanism action of natural regulators is important for characterization, prediction, control and regulation of physiological processes, as well as for prevention and treatment (Osler et al., 2002).

The present review discusses the evidence of the bioactivity of punicalagin present in pomegranates and summarizes the current knowledge concerning properties, benefits and mechanism of action of PUN at various regulatory levels.

4.8.2 Provenance and properties

Punicalagin (Fig. 4.8) is a part of the family of ellagitannins, which are the most abundant polyphenol found in pomegranate (*Punica granatum* L.) fruit (Yao et al., 2017). PUN is the known largest molecular weight polyphenol, which is 1084.71 g/mol (Yao et al., 2017). Chemical structure of the ellagitannin punicalagin is composed of ellagic acid moiety linked to a gallagic acid by a glucose. The glucose mutarotation at anomeric carbon induces two conformational anomers: punicalagin α and punicalagin β. PUN is water-soluble and hydrolyze into smaller phenolic compounds, such as ellagic acid (Mathon et al., 2019).

Punicalagin is reported to be broken down into different urolithins by gut microbiota, and they possess different anti-cancer effects (Gonzalez-Sarrias et al., 2014). PUN is responsible for more than 50% of anti-oxidant capacity (Seeram et al., 2005). Important role responsible for many of beneficial activities of PUN could be ascribed to its antioxidant activity (Lin et al., 2001), which has been attributed to the presence of 16 dissociable −OH groups, thus

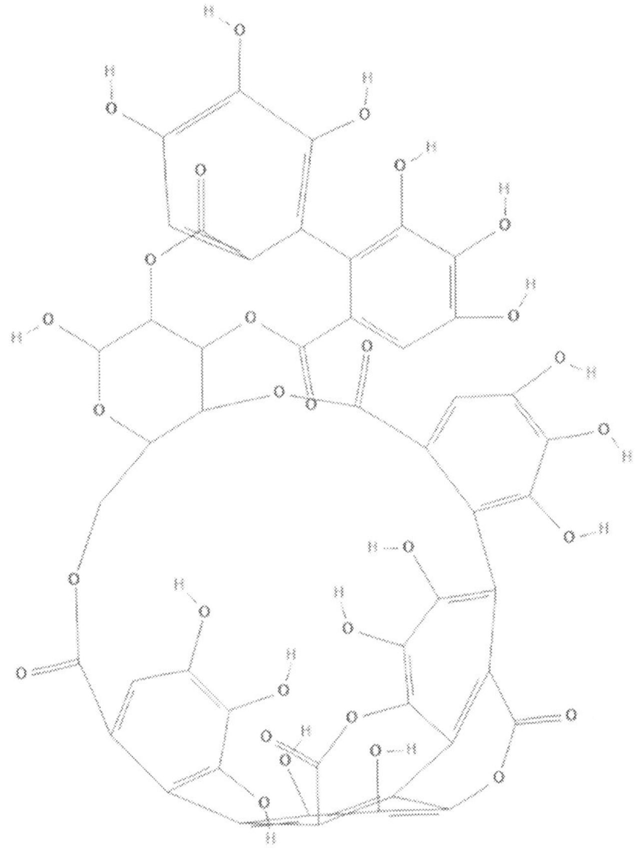

FIGURE 4.8 Chemical structure of punicalagin (PubChem, 2021).

PUN can act as a good antioxidant in living systems to stress. The ability of PUN to scavenge a variety of reactive oxygen species (ROS) and reactive nitrogen species (RNS) known to produce many diseases through cellular damage and point out its protective function (Kulkarni et al., 2007). PUN has been shown to elicit remarkable beneficial properties. Very little is known about the bioavailability of ellagic acid derivatives and ellagitannins (Clifford and Scalbert, 2000). Following bioavailability of PUN, it was shown that PUN and its metabolites were observed in feces and urine and also in plasma in rats (Cerdá et al., 2003a). The cytotoxicity studies revealed that PUN is nontoxic at low concentrations (Kulkarni et al., 2007) and oral administration of high doses of PUN to rats (for 37 days) showed no evidence of toxicity of PUN (Cerdá et al., 2003b). The health beneficial activities, coupled with bioavailability and nontoxic of PUN suggests that PUN may play an important role as a promising multifunctional molecule (Lin et al., 2001).

4.8.3 Physiological and therapeutic actions

PUN has been shown to elicit remarkable antioxidant effects (Lin et al., 2001; Seeram et al., 2005; Chen et al., 2012; Wright et al., 2014; Rao et al., 2016) including reproduction (Chen et al., 2012; Wright et al., 2014; Packova et al., 2015; Packova and Kolesarova, 2016), antiinflammatory (Xu et al., 2015; El-Missiry et al., 2015; Almowallad et al., 2020), neuroprotective (Yaidikar and Thakur, 2015), antiatherosclerotic (Almowallad et al., 2020), antidiabetic effects (El-Beih et al., 2019), cardioprotective (El-Missiry et al., 2015; Almowallad et al., 2020), hepatoprotective (Lin et al., 2001), antiviral, antifungal activities (Silva et al., 2018), antibacterial (Li et al., 2015), antigenotoxic, antimutagenic (Zahin et al., 2014), and chemopreventive and chemotherapeutic effects against many types of cancer (Gonzalez-Sarrias et al., 2014; Tang et al., 2016; Zhang et al., 2020) by inducing apoptosis and antiproliferative activity (Seeram et al., 2005; Malik et al., 2005; Syed et al., 2007; Larrosa et al., 2006; Aqil et al., 2012; Wang et al., 2013; Zahin et al., 2014; Tang et al., 2016; Zhang et al., 2020) gastrointestinal cancer (Sato et al., 2011), colon cancer (Larossa et al., 2006; Kozovska et al., 2014), thyroid carcinoma (Chen et al., 2016), osteosarcoma (Huang et al., 2020), glioblastoma (Wang et al., 2013), breast (Pan et al., 2020), ovarian (Tang et al., 2016), and cervical (Tang et al., 2017) cancer.

4.8.4 Mechanisms of action

Taken together, the available data demonstrate that PUN can modulate a wide array of cellular processes via multiple intracellular signaling pathways.

Protein disulfide-isomerase A3 (PDIA3), also known as glucose-regulated protein, a member of the protein disulfide isomerase family involved in several cellular functions, plays a critical role in human diseases and it has the potential to be a pharmacological target. PUN can bind PDIA3 and inhibit its redox activity (Giamogante et al., 2018). PUN protects against oxidative stress and excessive inflammatory diseases (Xu et al., 2015; El-Missiry et al., 2015; Almowallad et al., 2020) including tissue damage, sepsis, and endotoxemic shock (Xu et al., 2015) by affecting various mediators *in vitro* such as critical tumor/antitumor proteins, cytokines, transcription factors, and others (Xu et al., 2015).

The phosphatidylinositol 3-kinase/protein kinase B (PI3K/Akt) signaling pathway plays a critical role in modulating the nuclear factor erythroid 2-related factor 2/heme oxygenase 1 (Nrf2/HO-1) protein expression. PUN treatment can increase HO-1 expression together with its upstream mediator nuclear factor-erythroid two p45-related factor 2 (Nrf2) and inhibit lipopolysaccharide (LPS)-induced oxidative stress in macrophages by reducing ROS and NO generation and increasing superoxide dismutase (SOD) one mRNA expression, what provide new perspectives as antioxidant medicines (Xu et al., 2015).

PUN can be used as a new protein disulfide-isomerase A3 (PDIA3) inhibitor, also known as glucose-regulated protein, a member of the protein disulfide isomerase family involved in several cellular functions. It plays a critical role in human diseases, and it has the potential to be a pharmacological target. PUN can bind PDIA3 and inhibit its redox activity (Giamogante et al., 2018). PUN can affect the levels of glutamate and calcium as well as the levels of inflammatory cytokines tumor necrosis factor-α (TNF-α), IL-1β, interleukin-6 (IL-6), and supress apoptosis by downregulation of caspase-3 and Bax (Yaidikar and Thakur, 2015). On the other hand, PUN has potential role against oxidative stress induced testicular damage (Rao et al., 2016).

A neuroprotective effect of PUN and its possible prevent role in neuro-degenerative diseases related to oxidative stress were demonstrated. In addition, PUN modulates the apoptotic cascade triggered reducing Bax gene expression and caspase 3 activity (Clementi et al., 2017).

Beneficial antiatherosclerotic effects of PUN prevents cholesterol deposition and decrease of risk of atherosclerosis, which may lead to cardiovascular diseases. PUN regulates several key processes related to atherosclerosis by inhibition of interferon gamma (IFN-γ)-induced monocyte chemotactic protein-1 (MCP-1) and intercellular adhesion molecule-1 (ICAM-1) gene expression at mRNA level and reduction of monocyte chemotactic protein-1 (MCP-1) mediated monocyte migration and enhancement of cholesterol efflux from macrophages *in vitro* (Almowallad et al., 2020). Beneficial anti-atherosclerotic and antiinflammatory effects of PUN was evaluated against an atherosclerotic cell model *in vitro*. It was confirmed that PUN regulates several key processes related to atherosclerosis by inhibition of IFN-γ-induced MCP-1 and ICAM-1 gene expression at mRNA level and reduction of MCP-1 mediated monocyte migration and enhancement of cholesterol efflux from macrophages *in vitro* (Almowallad et al., 2020).

Strong therapeutic and cardioprotective potential of PUN has been shown (El-Missiry et al., 2015; Almowallad et al., 2020) by protection of the heart against apoptosis, necrosis, inflammation, and DNA damage by improving the redox state (El-Missiry et al., 2015). It modulates ocardiomyocyte apoptosis via increased Bcl-2 expression, block the increases in p53, Bax and caspases-3, 8 and 9, and ameliorated DNA damage in the heart, confirming the beneficial cardioprotective effect of PUN (El-Missiry et al., 2015). PUN prevents cholesterol deposition and decrease of risk of cardiovascular diseases (Almowallad et al., 2020).

There are scientific evidences of the cancer preventive efficacy of PUN both *in vitro* and *in vivo* animal and human models (Tang et al., 2016). The polyphenol PUN provokes the anticancer effects on various tumor cell lines by upregulation the expression levels of Bax and Bad, cleaved-poly (ADP) ribose polymerase (PARP), and cytochrome C7. PUN also promotes the activation of caspase-3 and caspase-9, and downregulates the expression of Bcl-2, Bcl-XL, and cell cycle proteins such as cyclins A, B1, D1, D2, and E6, thus regulates the cell proliferation and apoptosis (Malik et al., 2005; Syed et al., 2007; Larrosa et al., 2006; Wang et al., 2013; Tang et al., 2016).

In another study, inhibitory effect of PUN on proliferation and invasion was evaluated in one human osteoblast and human osteosarcoma cell lines. PUN

342 Contaminants and Plants Action on Female Reproduction

treatment do not only induce the degradation of nuclear factor of kappa light polypeptide gene enhancer in B-cells inhibitor alpha (IκBα) but also the nuclear translocation of p65 and downregulate interleukin (IL-6 and IL-8) levels, suggesting the inhibition of the nuclear factor-κB (NF-κB) signaling pathway (Huang et al., 2020).

PUN present in pomegranate induces human glioma cell death through both apoptotic and autophagic pathways. PUN treatment induces apoptosis through the activation of the caspase-9/caspase-3 cascade and through PARP in human primary glioblastoma cells. While PUN increases the phosphorylation of the AMP-activated protein kinase (AMPK) and p27T198, it is possible that it induces autophagic cell death through the AMPK/p27 signaling pathway (Wang et al., 2013).

Furthermore, PUN induced apoptosis via mitochondrial pathway in colon cancer cells but not in normal colon cells, PUN inhibit the cell viability via downregulation of cyclins A and B1 levels and induced apoptosis via intrinsic pathway (FAS-independent, caspase 8-independent) through Bcl-XL downregulation with mitochondrial release of cytochrome c into the cytosol, activation of initiator caspase 9 and effector caspase 3 (Larossa et al., 2006). The molecular mechanism of action of PUN on cellular process in breast cancer cells was described.

Breast cancer is one of the most common malignancies worldwide. High dose of PUN treatment inhibits viability, migration, and invasion of breast cancer cells. PUN suppresses cell viability and metastasis via regulating golgi phosphoprotein 3, and also inhibits the expressions of matrix metalloproteinase (MMP)-2, MMP-9, and N-Cadherin, and promoted E-cadherin expression (Pan et al., 2020).

Taken together, the available data demonstrate that PUN can modulate a wide array of cellular processes via multiple intracellular signaling pathways regulating oxidative stress, proliferation, apoptosis toward prevention and treatment of inflammatory, neurological, cardiovascular and cancer diseases.

4.8.5 Effects on female reproductive processes

There is an evidence of PUN effect on regulation of reproductive functions (Packova et al., 2015; Packova and Kolesarova, 2016; Tang et al., 2016, 2017; Adaramoye et al., 2017) including female reproductive processes (Packova et al., 2015; Packova and Kolesarova, 2016; Tang et al., 2016, 2017) and for improving reproductive health and fertility (Salem et al., 2020) and cancer treatment (Tang et al., 2016, 2017). PUN has a potent protective effect against gynecological alterations via effective modulation of cellular processes in reproductive system and may be clinically useful. Little is known about the impact of PUN on the productive and reproductive performance.

4.8.5.1 Effect on reproductive health and fertility

Founding of study Salem et al. (2020) has shown the positive effect of PUN alone or in combination with human chorionic gonadotropin on maiden rabbit

Plant molecules and their influence on health **Chapter | 4** **343**

does including reproductive traits, birth, and weaning weights of kits, whereas the study clearly demonstrated an improvement in rabbit doe fertility, reproductive performance, and kids survival, thus PUN can be used as promising tool for improving reproductive health and fertility in maiden rabbit does (Salem et al., 2020).

4.8.5.2 Effect on ovarian follicular cell functions

Phytonutrients PUN could play an important role as the possible modulator of process of ovarian steroidogenesis (Packova et al., 2015) and proliferation (Packova and Kolesarova, 2016). PUN has dose-dependent impact on secretion of steroid hormones progesterone and 17β-estradiol by rabbit ovarian fragments (Packova et al., 2015). On the other hand, no significant effects of PUN on secretion of steroid hormones by porcine ovarian granulosa cells were found, but the stimulatory effect of PUN on the proliferation was demonstrated (Packova and Kolesarova, 2016). Punicalagin inhibited the cell viability of ovarian carcinoma cells in a dose- and time-dependent manner (Tang et al., 2016).

4.8.5.3 Effect on uterus and pregnancy

Previous study has shown antiinflammatory activity of PUN on bovine endometrial epithelial cells. Bovine endometritis is one of the most common reproductive disorders in cattle. Results of the scientific research provide an evidence that PUN may attenuate lipopolysaccharide (LPS)-induced inflammatory injury and may have a potential option for the treatment of dairy cows with *Escherichia coli* endometritis (Lyu et al., 2017). The human placenta plays a key role in optimal pregnancy outcome. Polyphenol PUN can limit placental injury and confer protection to the exposed fetus. Excessive oxidative stress, which is presented in many complicated pregnancies, contributes to placental dysfunction. PUN reduces placental oxidative stress *in vivo* and *in vitro* and stimulus-induced apoptosis in cultured human syncytiotrophoblasts (Chen et al., 2012).

Taken together, the available data demonstrate that PUN could be involved in the regulation of processes ovarian steroidogenesis, proliferation for improving reproductive health and protection of fertility against gynecological alterations including cancer.

4.8.6 Mechanisms of action on female reproductive processes

Previous studies suggested that PUN can regulate reproductive functions through numerous signaling pathways. Molecular research showed the antiinflammatory activity of PUN in LPS-induced bovine endometrial epithelial cells (bEECs) by decreasing **LPS-induced productions of interleukin IL-1β, IL-6, IL-8**, and **TNF α** in bEECs. In addition, molecular research showed that

344 Contaminants and Plants Action on Female Reproduction

PUN inhibited the activation of the upstream mediator **NF-κB** by suppressing the production of inhibitor κBα (IκBα) and phosphorylation of p65. Results also indicated that PUN can suppress the phosphorylation of **mitogen-activated protein kinases (MAPKs)** including p38, **c-Jun N-terminal kinase (JNK)**, and **extracellular signal-regulated kinase (ERK)** (Lyu et al., 2017).

PUN also has been shown to induce apoptosis in several cancer cell lines and may exert chemopreventive and chemotherapeutic effects against ovarian carcinoma cells and cervical cancer cells (Tang et al., 2016, 2017). PUN inhibits the cell viability and the cell cycle in **G1/S phase** transition of human ovarian cancer cells. PUN suppress the **β-catenin signaling pathway**, what was demonstrated by the downregulation of β-catenin and its downstream factors including cyclin D1 and survivin (Tang et al., 2016).

PUN also suppresses the proliferation and invasion of cervical cancer cells through inhibition of the **β-catenin signaling pathway** (Tang et al., 2017). It was shown that PUN inhibited of proliferation and migration of cervical cancer cells **downregulated of MMP-2 and MMP-9, upregulated** of tissue inhibitor of **metalloproteinase (TIMP)-2** and **TIMP-3**, promoted of cell-cycle arrest in the G1 phase and induced apoptosis via alterations of **Bcl-2** and **Bax**, and downregulated of β-catenin and its downstream proteins, **cyclin D1** and **c-myc** (Tang et al., 2017).

The efficiency of PUN lying on cell viability is based on the stimulation of apoptosis via the expression of **Bcl-2 family** proteins, and **caspases** also the cell cycle regulatory proteins **p53** and **NF-κB signal pathway** in human cervical cancer cells. PUN can repress the viability of cervical cancer cells via stimulating mitochondrial mediated apoptosis, inhibit cancer cell proliferation, and induce apoptosis by suppressing NF-κB activity (Zhang et al., 2020).

PUN can affect physiological processes including reproduction at various regulatory levels by decrease of oxidative stress and regulation of cell proliferation and apoptosis. PUN exerts antiproliferative activity in cancer cells via induction of apoptosis. However, further carefully designed confirmatory studies on the potential roles of main component pomegranate—punicalagin on reproductive system are needed to reveal the exact dose-response relationship and the nature of effect. The mechanisms of the effect of PUN on steroidogenesis and embryogenesis described have not yet been elucidated.

4.8.7 Application in reproductive biology and medicine

There is a few of evidence that PUN can be used in animal production, medicine, and biotechnology as a promising tool for regulation of various reproductive processes and in prevention and treatment of many disorders primarily in cancer by designing promising candidate for anticancer drug.

PUN could play an important role as the modulator of ovarian steroidogenesis (Packova et al., 2015) and proliferation (Packova and Kolesarova, 2016) for improving reproductive functions, fertility and reproductive health.

PUN exhibits protective effect against oxidative stress, and it is a promising tool for prevention or therapy of female reproductive diseases including endometritis and cancer. In addition to the PUN effects mentioned in previous chapter, furthermore, PUN may be used as phytonutrient for the treatment of endometritis (Lyu et al., 2017) by promoting fertility and eliminating oxidative stress and inflammation. Infertility is a reproductive issue and oxidative stress, imbalance between reactive oxygen species and antioxidants in the body, is considered as an important factor that influences fertility status (Wright et al., 2014). Another study has shown that the polyphenol PUN can limit placental injury and confer protection to the exposed fetus. PUN reduces placental oxidative stress (Chen et al., 2012). Taken together, the available data demonstrate that PUN can modulate a wide array of cellular processes via multiple intracellular signaling pathways regulating oxidative stress, proliferation, apoptosis toward prevention and treatment of reproductive diseases such as tissue inflammation, cancer and others. PUN has a potent protective effect against gynecological alterations via effective modulation of cellular processes in reproductive system and may be clinically useful.

4.8.8 Conclusions and possible direction of future studies

In comparison with other physiological processes and illnesses, the available information concerning action of punicalagin substances on female reproductive processes are limited. The health beneficial activities, coupled with bioavailability and nontoxic of PUN suggests that PUN may play an important role as a promising multifunctional molecule. PUN can affect physiological processes including reproduction at various regulatory levels by decrease of oxidative stress and regulation of cell proliferation and apoptosis and prevention and treatment of many disorders, primarily in cancer. Nevertheless, the available data provides evidence that the punicalagin might be a promising candidate as a potential modulator of ovarian cell steroidogenesis, proliferation and viability and as a potential biostimulator of healthy ovarian cells and chemoprotective agent reducing viability of ovarian and cervical cancer cells. Possible direction of future studies could be targeted to design of a candidate for anticancer drug.

References

Adaramoye O, Erguen B, Nitzsche B, Höpfner M, Jung K, Rabien A. Punicalagin, a polyphenol from pomegranate fruit, induces growth inhibition and apoptosis in human PC-3 and LNCaP cells. Chem Biol Interact 2017;274:100−6. https://doi.org/10.1016/j.cbi.2017.07.009.

Almowallad S, Huwait E, Al-Massabi R, Saddeek S, Gauthaman K, Prola A. Punicalagin regulates key processes associated with atherosclerosis in THP-1 cellular model. Pharmaceuticals 2020;13(11):372. https://doi.org/10.3390/ph13110372.

Aqil F, Munagala R, Vadhanam MV, Kausar H, Jeyabalan J, Schultz DJ, Gupta RC. Antiproliferative activity and protection against oxidative DNA damage by punicalagin isolated from pomegranate husk. Food Res Int 2012;49:345−53. https://doi.org/10.1016/j.foodres.2012.07.059.

346 Contaminants and Plants Action on Female Reproduction

Cerdá B, Llorach R, Cerón JJ, Espín JC, Tomás-Baberán FA. Evaluation of the bioavailability and metabolism in the rat of punicalagin, an antioxidant polyphenol from pomegranate juice. Eur J Nutr 2003a;42:18−28. https://doi.org/10.1007/s00394-003-0396-4.

Cerdá B, Cerón JJ, Tomás-Barberán FA, Espín JC. Repeated oral administration of high doses of the pomegranate ellagitannin punicalagin to rats for 37 Days is not toxic. J Agric Food Chem 2003b;51(11):3493−501. https://doi.org/10.1021/jf020842c.

Chen B, Tuuli MG, Longtine MS, Shin JS, Lawrence R, Inder T, Nelson DM. Pomegranate juice and punicalagin attenuate oxidative stress and apoptosis in human placenta and in human placental trophoblasts. Am J Physiol Endocrinol Metab 2012;302(9):1142−52. https://doi.org/10.1152/ajpendo.00003.2012.

Clementi ME, Pani G, Sampaolese B, Tringali G. Punicalagin reduces H2O2-induced cytotoxicity and apoptosis in PC12 cells by modulating the levels of reactive oxygen species. Nutr Neurosci 2017;21(6):447−54. https://doi.org/10.1080/1028415X.2017.1306935.

Clifford MN, Scalbert A. Ellagitannins − nature, occurrence and dietary burden. J Sci Food Agric 2000;80:118−1125. https://doi.org/10.1002/(SICI)1097-0010(20000515)80:7<1118::AID-JSFA570>3.0.CO;2-9.

El-Beih NM, Ramadan G, El-Husseiny EA, Hussein AM. Effects of pomegranate aril juice and its punicalagin on some key regulators of insulin resistance and oxidative liver injury in streptozotocin-nicotinamide type 2 diabetic rats. Mol Biol Rep 2019;43:3701−11. https://doi.org/10.1007/s11033-019-04813-8.

El-Missiry MA, Amer MA, Hemicda FAE, Othman AI, Sakr DA, Abdulhadi HL. Cardio ameliorative effect of punicalagin against streptozotocin-induced apoptosis, redox imbalance, metabolic changes and inflammation. Egypt J Basic Appl Sci 2015;2:247−60. https://doi.org/10.1016/j.ejbas.2015.09.004.

Fischer UA, Carle R, Kammerer DR. Identification and quantification of phenolic compounds from pomegranate (*Punica granatum* L.) peel, mesocarp, aril and differently produced juices by HPLC−DAD−ESI/MS. Food Chem 2011;127(2):807−21. https://doi.org/10.1016/j.foodchem.2010.12.156.

Giamogante F, Marrocco I, Cervoni L, Eufemi M, Chichiarelli S, Altieri F. Punicalagin, an active pomegranate component, is a new inhibitor of PDIA3 reductase activity. Biochimie 2018;147:122−9. https://doi.org/10.1016/j.biochi.2018.01.008.

Gil MI, Tomás-Barberán FA, Hess-Pierce B, Holcroft DM, Kader AA. Antioxidant activity of pomegranate juice and its relationship with phenolic composition and processing. J Agric Food Chem 2000;48:4581−9. https://doi.org/10.1021/jf000404a.

Gonzales-Sarrias A, Gimenez-Bastida JA, Nunez-Sanchez MA, Larrosa M, Garcia-Conesa MT, Tomas-Berberan FA, Espin JC. Phase-II metabolism limits the antiproliferative activity of urolithins in human colon cancer cells. Eur J Nutr 2014;53:853−4. https://doi.org/10.1007/s00394-013-0589-4.

Huang T, Zhang X, Wang H. Punicalagin inhibited proliferation, invasion and angiogenesis of osteosarcoma through suppression of NF-κB signaling. Mol Med Rep 2020:2386−94. https://doi.org/10.3892/mmr.2020.11304.

Kozovska Z, Gabrisova V, Kucerova L. Colon cancer: cancer stem cells markers, drug resistance and treatment. Biomed Pharmacother 2014;68:911−6. https://doi.org/10.1016/j.biopha.2014.10.019.

Kulkarni AP, Mahal HS, Kapoor S, Aradhya SM. In vitro studies on the binding, antioxidant, and cytotoxic actions of punicalagin. J Agric Food Chem 2007;55:1491−500. https://doi.org/10.1021/jf0626720.

Plant molecules and their influence on health **Chapter | 4 347**

Larrosa M, Tomás-Barberán FA, Espín JC. The dietary hydrolysable tannin punicalagin releases ellagic acid that induces apoptosis in human colon adenocarcinoma Caco-2 cells by using the mitochondrial pathway. J Nutr Biochem 2006;17:611−25. https://doi.org/10.1016/j.jnutbio.2005.09.004.

Li G, Feng Y, Xu Y, Wu Q, Han Q, Liang X, Yang B, Wang X, Xia X. The anti-infective activity of punicalagin against *Salmonella enterica* subsp. enterica serovar typhimurium in mice. Food Funct 2015;6:2357−64. https://doi.org/10.1039/C5FO00053J.

Lin CC, Hsu YF, Lin TC, Hsu HY. Antioxidant and hepatoprotective effects of punicalagin and punicalin on acetaminophen-induced liver damage in rats. Phytother Res 2001;15:206−12. https://doi.org/10.1002/ptr.816.

Lyu A, Chen J-J, Wang H-C, Yu X-H, Zang Z-C, Gong P, Jiang L-S, Liu F-H. Punicalagin protects bovine endometrial epithelial cells against lipopolysaccharide-induced inflammatory injury. J Zhejiang Univ Sci B 2017;18:481−91. https://doi.org/10.1631/jzus.B1600224.

Malik A, Afaq F, Sarfaraz S, Adhami VM, Syed DN, Mukhtar H. Pomegranate fruit juice for chemoprevention and chemotherapy of prostate cancer. Proc Natl Acad Sci USA 2005;102(41):14813−8. https://doi.org/10.1073/pnas.0505870102.

Mathon C, Chater JM, Green A, Merhaut DJ, Mauk PA, Preece JE, Larive CK. Quantification of punicalagins in commercial preparations and pomegranate cultivars, by liquid chromatography−mass spectrometry. J Sci Food Agric 2019;99(8):4036−42. https://doi.org/10.1002/jsfa.9631.

National Center for Biotechnology Information. PubChem compound summary for CID 16129869, punicalagin. 2021. PubChem, https://pubchem.ncbi.nlm.nih.gov/compound/Punicalagin. Accessed June 25, 2021.

Osler M, Andreasen AH, Heitmann B, Høidrup S, Gerdes U, Jørgensen LM, Schroll M. Food intake patterns and risk of coronary heart disease: a prospective cohort study examining the use of traditional scoring techniques. Eur J Clin Nutr 2002;56:568−74. https://doi.org/10.1038/sj.ejcn.1601360.

Packova D, Kolesarova A. Do punicalagins have possible impact on secretion of steroid hormones by porcine ovarian granulosa cells? J Microbiol Biotechnol Food Sci 2016;5(1):57−9. https://doi.org/10.15414/jmbfs.2016.5.special1.57-59.

Packova D, Carbonell-Barrachina AA, Kolesarova A. Ellagitannins − compounds from pomegranate as possible effector in steroidogenesis of rabbit ovaries. Physiol Res 2015;64:583−5.

Pan L, Duan Y, Ma F, Lou L. Punicalagin inhibits the viability, migration, invasion, and EMT by regulating GOLPH3 in breast cancer cells. J Recept Signal Transduct Res 2020;40(2):173−80. https://doi.org/10.1080/10799893.2020.1719152.

Rao F, Tian H, Li W, Hung H, Sun F. Potential role of punicalagin against oxidative stress induced testicular damage. Asian J Androl 2016;18(4):627−32. https://doi.org/10.4103/1008-682X.168792.

Salem AA, El-Shahawy NA, Shabaan HM, Kobeisy M. Effect of punicalagin and human chorionic gonadotropin on body weight and reproductive traits in maiden rabbit does. Vet Anim Sci 2020;10:10140. https://doi.org/10.1016/j.vas.2020.100140.

Sato YU, Kumamoto K, Saito K, Okayama H, Hayase S, Kofunato Y, Miyamoto K, Nakamura I, Ohki S, Koyama Y. Upregulated Annexin A1 expression in gastrointestinal cancer is associated with cancer invasion and lymph node metastasis. Exp Ther Med 2011;2:239−43. https://doi.org/10.3892/etm.2011.210.

Seeram NP, Adams LS, Henning SM, Niu Y, Zhang Y, Nair MG, Heber D. In vitro anti-proliferative, apoptotic and antioxidant activities of punicalagin, ellagic acid and a total pomegranate tannin extract are enhanced in combination with other polyphenols as found in pomegranate juice. J Nutr Biochem 2005;16:360−7. https://doi.org/10.1016/j.jnutbio.2005.01.006.

Silva TC, Zara ALSA, Sá FADS, Bara MTF, Ávila RI, Costa CR, Valadares MC, Santos ASD, Freitas VAQ, Silva MDRR. Antifungal potential of punicalagin against Cryptococcus neoformans species complex. Rev Inst Med Trop Sao Paulo 2018;60:e60. https://doi.org/10.1590/S1678-9946201860060.

Syed DN, Malik A, Hadi N, Sarfaraz S, Afaq F, Mukhtar H. Photochemo preventive effect of pomegranate fruit extract on UVA-mediated activation of cellular pathways in normal human epidermal keratinocytes. Photochem Photobiol 2007;82(2):398—405. https://doi.org/10.1562/2005-06-23-RA-589.

Tang JM, Min J, Li BS, Hong SS, Liu C, Hu M, Li Y, Yang J, Hong L. Therapeutic effects of punicalagin against ovarian carcinoma cells in association with β-catenin signaling inhibition. Int J Gynecol Cancer 2016;26(9):1557—63. https://doi.org/10.1097/IGC.0000000000000805.

Tang J, Li B, Hong S, Liu C, Min J, Hu M, Li Y, Liu Y, Hong L. Punicalagin suppresses the proliferation and invasion of cervical cancer cells through inhibition of the β-catenin pathway. Mol Med Rep 2017;16(2):1439—44. https://doi.org/10.3892/mmr.2017.6687.

Wang S-G, Huang M-H, Li J-H, Lai F-I, Lee H-M, Hsu Y-N. Punicalagin induces apoptotic and autophagic cell death in human U87MG glioma cells. Acta Pharmacol Sin 2013;34:1411—9. https://doi.org/10.1038/aps.2013.98.

Wright C, Milne S, Leeson H. Sperm DNA damage caused by oxidative stress: modifiable clinical, lifestyle and nutritional factors in male infertility. Reprod Biomed Online 2014;28:684—703. https://doi.org/10.1016/j.rbmo.2014.02.004.

Xu X, Li H, Hou X, Li D, He S, Wan CH, Yin P, Liu M, Liu F, Xu J. Punicalagin induces Nrf2/HO-1 expression via upregulation of PI3K/AKT pathway and inhibits LPS-induced oxidative stress in RAW264.7 macrophages. Mediat Inflamm 2015;2015:380218. https://doi.org/10.1155/2015/380218.

Yaidikar L, Thakur S. Punicalagin attenuated cerebral ischemia-reperfusion insult via inhibition of proinflammatory cytokines, upregulation of Bcl-2, downregulation of Bax, and caspase-3. Mol Cell Biochem 2015;402:141—8. https://doi.org/10.1007/s11010-014-2321-y.

Yao X, Cheng X, Zhang L, Yu H, Bao J, Guan H, Lu R. Punicalagin from pomegranate promotes human papillary thyroid carcinoma BCPAP cell death by triggering ATM-mediated DNA damage response. Nutr Res 2017;47:63—71. https://doi.org/10.1016/j.nutres.2017.09.001.

Zahin M, Ahmad I, Gupta RC, Aqil F. Punicalagin and ellagic acid demonstrate antimutagenic activity and inhibition of benzo[a]pyrene induced DNA adducts. BioMed Res Int 2014:467465. https://doi.org/10.1155/2014/467465.

Zhang L, Chinnathambi A, Alharbi SA, Veeraraghavan VP, Mohan SK, Zhang G. Punicalagin promotes the apoptosis in human cervical cancer (ME-180) cells through mitochondrial pathway and by inhibiting the NF-kB signaling pathway. Saudi J Biol Sci 2020;27(4):1100—6. https://doi.org/10.1016/j.sjbs.2020.02.015.

Chapter 4.9

Quercetin

4.9.1 Introduction

In plants, flavonoids serve predominantly as protectors against a wide variety of environmental stresses, while in humans and animals they appear to affect a wide spectrum of biological processes. One of such plant flavonoids with potential physiological and medical action could be quercetin. Because of its putative beneficial effects in the prevention and treatment of various diseases, quercetin can be used as a medical treatment and a promising ingredient of so-called functional food. This review summarizes the available data concerning physiological, protective and therapeutic effects of quercetin, which could be useful for their better understanding and application in biology, medicine and food production.

4.9.2 Provenance and properties

Quercetin (3,3′,4′,5,7-pentahydroxyflavanone or 3,3′,4′,5,7-pentahydroxy-2-phenylchromen-4-one) (Fig. 4.9) is the major polyphenolic flavonoid found in food products, including berries, apples, cauliflower, tea, cabbage, capers, grapes, onions, shallots, tea, and tomatoes, as well as many seeds, nuts, flowers, barks, and leaves. Quercetin is also found in medicinal botanicals, including *Ginkgo biloba*, *Hypericum perforatum* (St. John's wort), and *Sambucus canadensis* (elder).

The attached glycosyl group can change the solubility, absorption, and *in vivo* effects of quercetin. As a general rule of thumb, the presence of a glycosyl group (quercetin glycoside) results in increased water solubility compared to quercetin aglycone (Kelly, 2011).

Technically, the term quercetin should be used to describe its aglycone form. Nevertheless, most of the quercetin in plants is attached to sugar moieties rather than in the free form. The most common are quercetin is linked with one or two glucose molecules (quercetin glucosides). Examples of quercetin glucosides could be hyperoside (found in St. John's wort), IQ (found in mangoes) and rutin (found in high amounts in buckwheat, citrus fruits, and *Ruta graveolens*) (Kelly, 2011; Magar et al., 2020).

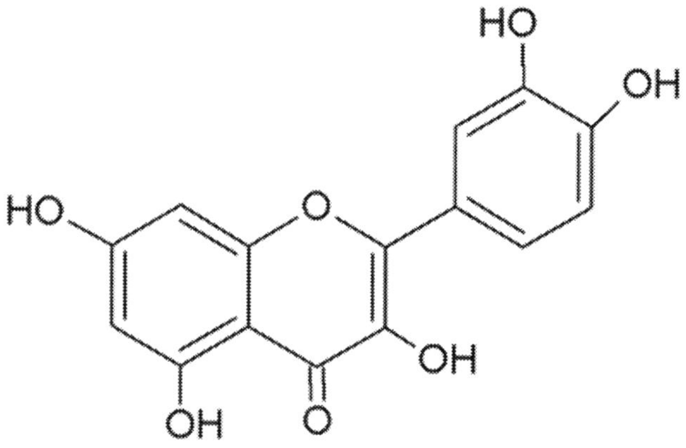

350 Contaminants and Plants Action on Female Reproduction

FIGURE 4.9 The chemical structure of quercetin. *From researchgate.net.*

The types and attachments of sugars impact bioavailability, and thus bioactivity. Generally, the attached glycosyl group can change the solubility, absorption, and *in vivo* effects (Kelly, 2011). Glucoside conjugates which are found in onions appear to have the highest bioavailability in humans. Sometimes a lower dosage from plant sources could be effective due to of its higher bioavailability compared to the aglycone form (Dabeek et al., 2019).

Initial studies suggested low bioavailability, poor aqueous solubility as well as rapid body clearance, fast metabolism and enzymatic degradation of quercetin, which hampered its application as a therapeutic agent (Massi et al., 2017). However, recent reports have shown that quercetin was detected in the plasma after food or supplements consumption and has a long half-life in human body (Maaliki et al., 2019). Quercetin has been documented to accumulate in the lungs, liver, kidneys, and small intestines, brain, heart, and spleen, and it is removed from the organism through the renal, fecal, and respiratory systems (Batiha et al., 2020).

Quercetin can be linked not only with glucose, but also with rutinose. Quercetin rutinoside or rutin is common in tea (Kelly, 2011).

Recently, pharmaceuticals have been paying attention to another quercetin analogue, taxifolin (3,5,7,3,4-pentahydroxy flavanone or dihydroquercetin). This flavonoid is commonly found in onion, milk thistle, French maritime pine bark and Douglas fir bark. It is used in various commercial preparations like Legalon, Pycnogenol, and Venoruton (Sunil and Xu, 2019) now (see later).

Absorbed quercetin is rapidly metabolized in the liver and circulates as methyl, glucuronide, and sulfate metabolites (Dabeek and Marra, 2019). Furthermore, dietary glycoside forms of quercetin are enzymatically hydrolyzed, deglycosyled and absorbed in the intestine, and are conjugated to their glucuronide/sulfate forms by phase II enzymes in epithelial cells and the liver (Murota et al., 2018).

4.9.3 Physiological actions

Animal and *in vitro* studies have demonstrated the anticancer action of quercetin (Rauf et al., 2018; Reyes-Farias et al., 2019; Shafabakhsh et al., 2019; Batiha et al., 2020; Khan et al., 2020). It has been used for the treatment of allergic (Mlcek et al., 2016; Jafarinia et al., 2020) cardiovascular diseases and inflammatory disorders including neurodegenerative disorders (Alzheimer disease) and arthritis (Costa et al., 2016; Sharma et al., 2018; Kashyap et al., 2019; Khan et al., 2019; Batiha et al., 2020; Yang et al., 2020).

Clinical studies showed a potential cardiovascular (hypotensive) effect of high intakes of quercetin-rich plants (Larson et al., 2010; Serban et al., 2016; Marunaka et al., 2017: Dabeek and Marra, 2019; Huang et al., 2020). In animal model it was able to prevent cardiac injury (Ferenczyova et al., 2020), therefore it can be considered as a natural stimulator of wound healing (Polerà et al., 2019).

Quercetin exhibits improved dyslipidemia, and, therefore, antiobesity effects in adipocyte cultures and animal models. Although quercetin supplementation does not seem to exert any beneficial effects on body weight, this polyphenol could prevent the obesity-associated mortality (Chen et al., 2016; Zhao et al., 2017; Carrasco-Pozo et al., 2019; Huang et al., 2020; Pourteymour et al., 2020; Tabrizi et al., 2020).

Quercetin may play beneficial effects on signs of type 2 diabetes (Chen et al., 2016; Eid et al., 2017; Bule et al., 2019; Shi et al., 2019; Bolouki et al., 2020; Huang et al., 2020).

Animal studies demonstrated the positive and protective effects of quercetin on bone metabolism (mainly promotion of osteogenesis and inhibition of osteolysis) (Wong et al., 2020).

Some animal and clinical studies report a protective effect of quercetin against gastric ulceration and experimental reflux esophagitis, interstitial cysts and prostatitis (Kelly, 2011).

Pharmacologically, quercetin has been examined against various microorganisms and parasites, including pathogenic bacteria, viruses, and

Plasmodium, Babesia, and *Theileria* parasites (Batiha et al., 2020) and even against COVID-19 (Colunga Biancatelli; et al., 2020). Quercetin demonstrated anti-fungal activity and ability to alleviate the toxicity of mycotoxins (Batiha et al., 2020; Yang et al., 2020).

Human experiments revealed the ability of quercetin to relieve pain (Kelly, 2011; Carullo et al., 2017) and to boost physical performance (Somerville et al., 2017) and memory (Babaei et al., 2018).

Finally, quercetin was able to prolong life span in invertebrates (Pallauf et al., 2017) and mice (Xu et al., 2018).

On the other hand, animal studies detected some adverse effect of quercetin - its nephrotoxic influence in predamaged kidney and ability to promote estrogen-related cancer. Nevertheless, quercetin is well tolerable by humans, and it is not classified as carcinogenic. More important from the medicinal viewpoint is that animal and some human studies revealed interactions between quercetin and certain drugs leading to altered drug bioavailability (Kelly, 2011; Andres et al., 2018).

The numerous physiological and medicinal effects are reported not only for quercetin, but also for its derivates. For example, dihydroquercetin (taxifolin) can affect some regulators of carcinogenesis, hepatitis and cholesterol biosynthesis (Sunil and Xu, 2019). It showed promising pharmacological activities in the management of cancer, inflammation, microbial infections, oxidative stress, cardiovascular, and liver disorders (Sunil et al., 2019). Quercetin glucoside rutin has shown multispectrum pharmacological benefits for the treatment of various chronic diseases such as cancer, diabetes, hypertension, hypercholesterolemia, neurodegenerative disorders a.o (Sharma et al., 2013; Negahdari et al., 2021)

4.9.4 Mechanisms of action

The anticancer action of quercetin can be due to its action on intracellular signaling pathways involved in control of **cell proliferation** and **apoptosis** (Farooqi et al., 2018; Sharma et al., 2018; Maaliki et al., 2019; Kashyap et al., 2019). It induced cell cycle arrest probably via suppression of its promoters **cyclin B1** and **MAPK/ERK1/2** and activation of transcription factor **p53**, prolonged **DNA repair** through and promoted apoptosis through inhibiting **survivin, activation of growth factor TGF-beta, PI3K/AKT/mTOR, Wnt/-catenin, NOTCH, SHH and Janus kinase** pathways, transcription factors **STAT** and **NFkB, STAY-3,** stimulation of **caspases** (apoptosis triggers), increase of **bax** (activator of cytoplasmic apoptosis): **bcl-1** (its inhibitor) rate. In addition, quercetin can suppress tumorgenesis modulating

vascular endothelial growth factor and its receptors—the inductors of tumor vascularisation (Farooqi et al., 2018).

On the other hand, quercetin can increase viability of healthy cells via mitigation of **oxidative stress** and other adverse factors via modulation pathways associated with biogenesis of **mitochondria**, mitochondrial membrane potential, oxidative respiration and ATP anabolism, intramitochondrial redox status, and subsequently, mitochondria-induced apoptosis (de Oliveira et al., 2016). It can eliminate reactive oxygen species directly or via activation of antioxidant enzymes (Xu et al., 2019).

This antioxidant property, as well as the inhibition of markers and promoters of **inflammation** could be responsible for cardioprotective effect of quercetin related antiinflammatory action (Carrasco-Pozo et al., 2019; Ferenczyova et al., 2020). Furthermore, the ability of quercetin to stabilize and to reduce blood pressure could be explained by improved endothelial functions (vascular compliance, peripheral vascular resistance, and total blood volume via antiinflammatory and antioxidant actions) (Larson et al., 2010; Marunaka et al., 2017). In addition it can be mediated by quercetin-induced elevation in the cytosolic Cl^- concentration by activating **Na^+-K^+-$2Cl^-$ cotransporter** 1 in renal epithelial cells (Marunaka et al., 2017).

The antioxidant and antiinflammatory properties of quercetin are responsible for its antiobesity effects. Obesity is related to inflammatory processes, which are suppressed by quercetin. This substance has been shown to aid in the attenuation **of lipid peroxidation, platelet aggregation, and capillary permeability** (Chen et al., 2016). In addition, quercetin can reduce obesity by inhibiting **adipogenesis** and **lipogenesis**, and suppressing the differentiation of preadipocytes to mature adipocytes (Zhao et al., 2017), to eliminate the senescent adipocytes and to reduce markers and promoters of adipose tissue inflammation (Xu et al., 2018). In addition, quercetin can boost the **adiponectin receptors**, which are necessary for feedback downregulation of lipogenesis (Pourteymour et al., 2020).

The antidiabetic action of quercetin could be defined by its ability to alter intestinal **glucose absorption, insulin secretory** and **insulin-sensitizing activities** as well as improved **glucose utilization** in peripheral tissues via multiple signaling pathways (Chen et al., 2016; Eid and Haddad, 2017). For example, quercetin increases the levels of **AMP-activated protein kinase.** This kinase enhances the regulation of glucose transporter 4 as a key sensor of energy, and therefore induces glucose uptake (Pourteymour et al., 2020).

The protective effect of quercetin on bone could be mediated by the ability of this molecule to inhibit **RANKL-mediated osteoclastogenesis**, osteoblast apoptosis, oxidative stress and inflammatory response while promoting osteogenesis, angiogenesis, antioxidant expression, adipocyte apoptosis and osteoclast apoptosis. The possible underlying mechanisms involved are regulation of **Wnt, NF-κB, Nrf2, SMAD**-dependent, and intrinsic and extrinsic apoptotic pathways. On the other hand, quercetin was shown to exert complex and competing actions on the MAPK signaling pathway, resulting in both stimulatory and inhibitory effects on bone cell proliferation (Wong et al., 2020).

The neuroprotective action of quercetin expressed in improvement of cognitive performance and mitigation of Alzheimer's disease mentioned earlier could be due to its ability to protect neurons from oxidative damage while reducing lipid peroxidation. In addition to its antioxidant properties, it inhibits the fibril formation of **amyloid-β proteins**, counteracting cell lyses and inflammatory cascade pathways (Khan et al., 2019). Finally, quercetin can be helpful in prevention and treatment of neurological and mental disorders due to being an efficient inhibitor of brain **monoamine oxidase** (Dhiman et al., 2019).

The antiallergic effect of quercetin can be explained by its immunimodulatory and antiinflammatory action. It can regulate **Th1/Th2** stability, and decrease the antigen-specific IgE **antibody** released by B cells (Jafarinia et al., 2020).

The mechanisms of stimulatory action of quercetin on life span require further elucidation. It is hypothesised that it could be due its energy-restriction-like effects, inhibition of insulin-like-growth-factor signaling, induction of antioxidant defense mechanisms, hormesis as well as antimicrobial properties (Pallauf et al., 2017). It has been demonstrated that a preparation containing quercetin caused selective elimination of senescent cells, decreased the number of naturally occurring senescent cells and their secretion of frailty-related proinflammatory cytokines (Xu et al., 2018).

In addition to signaling peptides involved in control of proliferation, apoptosis, inflammation and oxidative processes listed earlier, the action of quercetin on **mico RNAs** regulating these processes in healthy (Cione et al., 2019; Dostal et al., 2019) and cancer (Farooqi et al., 2018; Khan et al., 2020) cells has been reported. Moreover, a direct influence of quercetin on **DNA** has been recently detected (Wang, 2020).

Much less is known about the mechanisms of action of quercetin-related molecules and metabolites. There is evidence that dihydroquercetin can influence cancer, cardiovascular disease and liver disease. The effect of dihydroquercetin on cancerogenesis could be mediated by its action on the **antioxidant response element** and **detoxifying phase II enzymes**, inhibition of **cytochrome P(450)** and **fatty acid synthase**. Its action on TNF-alpha and NF-κB dependent transcription in hepatitis C infections has been documented.

Furthermore, dihydroquercetin action on the scavenging effect of myeloperoxidase derived reactive nitrogen species and subsequent effects on cholesterol biosynthesis as well as the effects on **apob/apoA-I, HMG-CoA reductase** and apoptosis has been reported (Weidmann, 2011).

4.9.5 Effects on female reproductive processes

4.9.5.1 Effect on ovarian and reproductive state

Dietary quercetin increased mice ovarian weight, promoted mice ovarian folliculogenesis (Beazley and Nurminskaya, 2016), increased the proportion of the ovarian primordial follicles (Shu et al., 2011; Elkady et al., 2019) and secondary follicles, decreased the number of mature follicles, atretic follicles and corpus luteums and disrupted mice estrous cycle (Shu et al., 2011). Quercetin reduced the number of atretic and cystic follicles and increased the number of healthy follicles in rats (Chen et al., 2010; Olaniyan et al., 2020) and rabbits (Naseer et al., 2017).

The positive influence of quercetin on ovarian folliculogenesis was confirmed by its influence on women with polycystic ovarian syndrome. This syndrome is characterized by growth of ovarian follicles without their ovulation and luteinization, degenerative changes in uterus and resulted lack of ovarian and menstrual cycle and infertility. Treatment of women with this syndrome with quercetin resulted improvement in ovarian morphology, folliculogenesis, and luteinization, increase number of normal follicles in ovaries, restore the anatomy of normal ovary (decreased ovarian and cystic follicle diameter, normalized the thickness of theca and granulosa layer) and improve the histological structure of the uterus (Jahan et al., 2018; Pourteymour et al., 2020).

On the other hand, in medaka (*Oryzias latipes*) increased the number of atretic ovarian follicles (Weber et al., 2002).

Dietary quercetin induced increase in litter size in young mice, but its reduction in old animals (Beazley and Nurminskaya, 2016). In hens it increased the egg laying rate and egg quality (Liu et al., 2013).

Taken together, the available *in vivo* experiments demonstrated mainly the stimulatory action of dietary quercetin on ovarian folliculogenesis, which can affect the number of ovulations and fecundity. Nevertheless, character of this effect can be species- and age-dependent.

4.9.5.2 Effect on ovarian cell functions

Dietary quercetin reduced accumulation of proliferation markers or its transcripts in mice ovaries (Shu et al., 2011), and in cultured bovine (Tarko et al.,

2018; Sirotkin et al., 2019b) and porcine (Tarko et al., 2018, 2019; Kolesarova et al., 2019; Sirotkin et al., 2019a,b) granulosa cells and human ovarian cancer cells (Scambia et al., 1990; Catanzaro et al., 2015).

Furthermore, quercetin addition reduced the expression of apoptotic markers in bovine (Tarko et al., 2018; Sirotkin et al., 2019b), porcine (Tarko et al., 2018,2019; Sirotkin et al., 2019a,b) and chicken (Jia et al., 2011) granulosa cells and in cultured mouse embryos (Sameni et al., 2018). In *in vivo* experiments, the dietary quercetin reduced the number of apoptotic ovarian cells in rabbit (Naseer et al., 2017), mice (Elkady et al., 2019) and rat (Olaniyan et al., 2020) ovaries. The reduction in markers of both proliferation and apoptosis observed in the same experiments (Tarko et al., 2018,2019; Sirotkin et al., 2019a,b) could indicate impairment of ovarian cell turnover and therefore ovarian folliculogenesis. On the other hand, in some *in vitro* studies (Kolesarova et al., 2019) no influence of quercetin addition on apoptosis in cultured healthy porcine granulosa cells have been observed. Moreover, in ovarian cancer cells, it promoted both nuclear ((Teekaraman et al., 2019)) and cytoplasmic/mitochondrial (Gong et al., 2017; Teekaraman et al., 2019) apoptosis.

Quercetin can affect not only ovarian proliferation and apoptosis, but also the response of ovarian cells to the upstream stimulator—gonadotropin. At least, FSH and IGF-I promoted proliferation and suppressed apoptosis in cultured porcine ovarian granulosa cells. Addtion of quercetin prevented these effects of FSH, but not of IGF-I (Sirotkin et al., 2019a).

Santini et al. (2009) reported that quercetin expressed antioxidant activity in cultured porcine granulosa cells. In experiments of Capcarova et al. (2015) on this model, quercetin addition activated antioxidant enzymes too, but the amount of reactive oxygen species in these cells have not been changed after quercetin addition.

These observations suggest that quercetin can suppress proliferation of both normal and cancer ovarian cells. In healthy ovarian cells it inhibits, but in cancer cells it promotes apoptosis. Furthermore, quercetin can promote antioxidative processes in ovarian cells.

4.9.5.3 Effect on oocytes and embryos

Addition of quercetin to the culture medium promoted growth and development of cultured mouse oocytes, increased their viability, reduced the occurrence of abnormalities in their development and promoted the development of their uterine receptivity markers (Pérez-Pastén et al., 2010; Wang et al., 2017; Sameni et al., 2018; Bolouki et al., 2020a). The dietary quercetin increased the number and quality of rabbit oocytes, although it did not influence their nuclear maturation rate (Naseer et al., 2017). Goat oocyte-cumulus complexes matured in the presence of quercetin have more expressed *Cumulus oophorus* expansion, retraction, and maturation rate than those counted without quercetin (Silva et al., 2018).

The supplement of culture medium with quercetin improved also development and morphological indexes of quality of murine (Wang et al., 2017), goat (Mao et al., 2018), and ovine (Karimian et al., 2018) embryos developed *in vitro*.

These observations indicate the positive influence of quercetin on quantity and quality of oocytes and embryos in various animal species. Quercetin influence on human oocytes and embryos seems to not have been examined yet.

4.9.5.4 Effect on reproductive hormones

In vivo studies on rat model showed the ability of quercetin to decrease serum LH and FSH (Cao et al., 2014), although other studies (Wang et al., 2018) did not confirm quercetin influence on FSH and LH level in rat plasma. In women, quercetin administration increased plasma LH levels (Jahan et al., 2018; Pourteymour et al., 2020).

There is a number of reports demonstrating quercetin action on steroid hormones' release, whilst the character of its action varied between the studied species (rat, pig, cow, women) and even within one species. Such variability could be due to the initial state of the ovary probably associated with stage of reproductive and ovarian cycle and related changes in production of progestagens, androgens and estrogens (Sirotkin, 2014).

The *in vivo* studies also demonstrated quercetin influence on rat plasma steroid hormones, although the character of this influence varied between experiments too. Olaniyan et al. (2020) and Pourteymour et al. (2020) reported that quercetin can decrease the activity of steroidogenic enzymes, plasma testosterone levels and to promote progesterone and estradiol release. On the other hand, Wang et al. (2018) did not find changes in steroid hormones' level in plasma of rats after quercetin treatment. Moreover, other authors (Cao et al., 2014; Jafari Khorchani et al., 2020) reported even a decrease in aromatase activity and estradiol production in rats treated with quercetin. In mice, quercetin enhanced serum estradiol level (Bolouki et al., 2020b).

In cultured porcine ovarian cells quercetin addition inhibited progesterone release (Santini et al., 2009; Sirotkin et al., 2019a, 2021) and promoted testosterone output (Sirotkin et al., 2019a). Quercetin at low doses stimulated, and given at high doses inhibited estradiol production (Santini et al., 2009). In cultured porcine ovarian follicles, quercetin reduced the release of both progesterone and testosterone (Sirotkin et al., 2019a). In other experiments on cultured porcine granulosa cells, quercetin did not affect (Tarko et al., 2019) and even promoted (Kolesarova et al., 2019) progesterone release, and it did not affect estradiol output (Kolesarova et al., 2019).

In cultured bovine granulosa cells, treatments with quercetin decreased progesterone and increased testosterone release (Tarko et al., 2018; Sirotkin et al., 2019b).

Quercetin supplementation in women was able to decrease the expression of CYP17A1 (a gene responsible for the activity of 17a-hydroxylase, a key enzyme for androgen synthesis), to reduce testosterone and to increase production of progesterone and estradiol, but not of sex hormone binding globulin (Jahan et al., 2018; Pourteymour et al., 2020). In cultured human ovarian granulosa cells quercetin inhibited progesterone production (Whitehead and Lacey, 2003).

Quercetin action on nonsteroid hormones has been documented too. Besides steroid hormones, quercetin was able to increase rat plasma level of inflammatory cytokines (Olaniyan et al., 2020), adiponectin and the expression of adipoR1 and nesfatin-1 genes (Jafari Khorchani et al., 2020). In mice, quercetin increased serum anti-Mullerian hormone level (Elkady et al., 2019). It inhibited vascular endothelial growth factor (Santini et al., 2009), leptin (Sirotkin et al., 2019a) and IGF-I (Sirotkin et al., 2019b) and stimulated oxytocin, but not prostaglandin F release (Tarko et al., 2019) by cultured porcine granulosa cells.

Quercetin was able to affect not only the basal ovarian hormones release, but also the receptors to hormones in reproductive tissues and their consequent response to hormonal regulators. For example, quercetin inhibited and even reversed the stimulatory action of FSH (but not of IGF-I) on progesterone, testosterone and leptin release by cultured porcine granulosa cells (Sirotkin et al., 2019a). Quercetin can boost alpha-receptor and, therefore, the estrogen-dependent events in women's ovary (Jahan et al., 2018; Pourteymour et al., 2020). Quercetin treatment on the contrary suppressed the expression of estrogen receptors alpha and beta and progesterone receptors in rat hypo-thalamus, pituitary and endometrium (Cao et al., 2014).

Therefore, despite some variability, the available data demonstrate quercetin influence on pituitary gonadotropin, ovarian steroidogenic enzymes and steroids, ovarian peptide hormones, growth factors and cytokines. Furthermore, it can influence receptors to hormones and response of ovarian cells to the upstream hormonal regulators. These hormones through their receptors are involved in control of ovarian folliculogenesis, metabolism, inflammatory processes, ovulation and fecundity (Sirotkin, 2014).

4.9.5.5 Effect on ovarian disfunctions

Animal and *in vitro* studies suggest that quercetin can promote **ovarian cancer** cell apoptosis and suppress their proliferation and resulting ovarian tumor

Plant molecules and their influence on health **Chapter | 4** **359**

development (Scambia et al., 1990; Gao et al., 2012; Catanzaro et al., 2015; Parvaresh et al., 2016; Gong et al., 2017; Shafabakhsh and Asemi, 2019; Khan et al., 2020). Nevertheless, the epidemiological studies commented that the potentially protective effects of quercetin are unable to significantly decrease ovarian cancer risk at levels commonly consumed in a typical diet (Parvaresh et al., 2016).

Polycystic ovarian syndrome is characterized by growth of ovarian follicles without their ovulation and luteinization, degenerative changes in uterus and resulting lack of ovarian and menstrual cycle and infertility. This syndrome is associated with metabolic disorders including diabetes. Treatment of rats (Jahan et al., 2018; Neisy et al., 2019; Jafari Khorchani et al., 2020; Olaniyan et al., 2020) and women (Pourteymour et al., 2020) suffering from this syndrome by quercetin resulted improvement in these female reproductive parameters.

Premature ovarian failure has been studied by using mice treated with cyclophosphamide, expressing ovarian toxicity attributed to oxidative stress, inflammation, and apoptosis. Quercetin restored ovarian function in the form of an increase in primordial follicles number, decrease in ovarian cell apoptosis and follicular atresia (Elkady et al., 2019).

Quercetin suppressed signs of rat **ovarian aging** manifested in decrease in ovarian follicular reserve due to suppression of follicle development and increase in follicular atresia (Chen et al., 2010), reduction in production of oocyte maturation promoting factor and oocyte maturation, promotion of oocytes apoptosis and diminished embryo yield during reproductive aging (Wang et al., 2017).

Cao et al. (2014) demonstrated the ability of quercetin to prevent both hormonal and morphological signs of **endometriosis** in rats.

These observations suggest the protective and therapeutic action of quercetin on several common female reproductive dysfunctions.

4.9.5.6 Effect on response to hazardous factors

The ability of quercetin to modify the response of ovarian cells (follicular cells and oocytes) to several hazardous factors (high temperature, chemotherapy, heavy metal, and oil-related environmental contaminant) has been demonstrated.

In rabbits, **heat stress** suppresses ovarian follicle development due to ovarian cell apoptosis, number of ovulations, as well as the quality and developmental capacity of oocytes. The dietary quercetin improved follicular development, minimized granulosa cells apoptosis, and maintained the oocyte developmental competence in rabbits subjected to this stress (Naseer et al., 2017).

The **chemotherapeutic agent** doxorubicin expresses significant reproductive toxicity. It induces a decline in the number of ovarian follicles, ovarian and its associated structures volume, the volume of the uterus and its layers.

Coadministration of quercetin together with vitamin E to doxorubicin-treated rats demonstrated an alleviative effect on most of the studied parameters (Samare-Najaf et al., 2020). There are indications of quercetin ability to alter bioavailability and, therefore, effects of other drugs used for treatment of ovarian cancer (Kelly, 2011).

Cadmium ions have toxic influence on gametes and embryo. The experiments with cultured chicken ovarian granulosa cells showed that addition of cadmium chloride decreased granulosa cell number and viability, induced chromatin condensation, DNA fragmentation (marker of nuclear apoptosis) and accumulation of several markers of cytoplasmic apoptosis. Quercetin prevented cadmium-induced cytotoxicity in granulosa cells through inhibiting apoptosis to ensure reproductive health (Jia et al., 2011). Experiments with *in vitro* maturation and insemination of goat oocytes and embryo production demonstrated protective effect of quercetin on goat oocytes, sperm and embryos cultured in presence of cadmium chloride (Mao et al., 2018).

Another environmental contaminant, **xylene**, stimulated proliferation and IGF-I release by cultured bovine granulosa cells. When administered with xylene, quercetin prevented the action of xylene on these parameters. These results demonstrated that quercetin could prevent xylene effects on the ovarian cells, which indicates the potential usefulness of quercetin for prevention of this oil-related contaminant on female reproduction (Tarko et al., 2018).

The next oil-related contaminant, **benzene**, stimulated proliferation, apoptosis, and oxytocin release and inhibited progesterone and prostaglandin F release by cultured porcine granulosa cells. Addition of quercetin promoted the inhibitory effect of benzene on progesterone release, but it did not modify benzene effect on other parameters. Therefore, no protective effect of quercetin on benzene effects on ovarian cells has been found (Tarko et al., 2019).

Quercetin promoted accumulation of toxic silver and titania nanoparticles in cultured porcine ovarian cells, but it did not affect the ability of these nanoparticles to suppress progesterone release by these cells (Sirotkin et al., 2021).

These observations demonstrate that quercetin could be a natural protector of ovarian cells against adverse effects of some hazardous drugs and environmental factors.

4.9.6 Mechanisms of action on female reproductive processes

Ovarian aging and other ovarian dysfunctions associated with increased apoptosis, inflammation and mutagenesis could be a consequence of **oxidative stress**—accumulation and destructive action of reactive oxygen species (Wang et al., 2018). Natural antioxidants including quercetin can mitigate this stress and resulting signs of ovarian cancer (Gong et al., 2017; Elkadi et al., 2019; Shafabakhsh and Asemi, 2019), polycystic ovarian syndrome (Neisy et al., 2019; Olaniyan et al., 2020; Pourteymour et al., 2020), menopause

(Wang et al., 2018), ovarian aging (Beazley et al., 2016; Wang et al., 2017; Yang et al., 2021), toxic influence of chemotherapy (Samare-Najaf et al., 2020), cadmium (Jia et al., 2011) and T-2 toxin (Capcarova et al., 2015). Antioxidant action of quercetin can mediate its stimulatory action on ovarian follicular cells (Khadrawy et al., 2019) and oocyte and embryo cultured *in vitro* (Mao et al., 2018). Quercetin can exert its antioxidant effect via both direct scavenging of the reactive oxygen species and through promotion of intracellular antioxidant enzymes (Santini et al., 2009; Khadrawy et al., 2019; Yang et al., 2021).

Prevention of oxidative stress by quercetin can reduce **apoptosis** and promote **mitochondrial activity,** which in turn improves the quality of developmental capacity of oocytes and embryos (Wang et al., 2017; Silva et al., 2018). In addition, positive action of quercetin on oocytes can be mediated by an increase in **SIRT-1** and **histone methylation** (Wang et al., 2017).

Quercetin can reduce proliferation of ovarian cancer cells via downregulation of **endoplasmic reticulum stress** (Gong et al., 2017), of the **cell cycle** promoters **cyclin B1** and **MAPK/ERK1/2** and **MAPK p44/42**. Furthermore, quercetin can induce ovarian cancer cell nuclear and cytoplasmic **apoptosis** via suppression of inhibitor of cytoplasmic apoptosis **bcl-2**, as well as through upregulation of promoters of cytoplasmic apoptosis **TGF-beta, PI3K/AKT/mTOR, transcription factors p53, STAT3 and NFkB, STAY-3, bax, bid, bad, cytochrome C and caspases** (Gao et al., 2012; Catanzaro et al., 2015; Parvaresh et al., 2016; Gong et al., 2017; Shafabakhsh and Asemi, 2019; Teekaraman et al., 2019). Changes in bcl-2, bax and **e-cadherin** can mediate the therapeutic effect of quercetin on polycystic ovarian syndrome (Olaniyan et al., 2020). The ability of quercetin to mitigate the adverse effect of ischemia-reperfusion injury on the ovary was associated with decrease in signs of nuclear (DNA fragmentation) and cytoplasmic (caspase-3) apoptosis in the ovary (Gencer et al., 2014). The protective effect if quercetin against cadmium toxicity can be also mediated by modulating intracellular regulators of apoptosis **XIAP**, bax, bcl-2, and **caspase-3** (Jia et al., 2011).

Quercetin influence on some **microRNAs** affecting proliferation of healthy ovarian cells (Khadrawy et al., 2019) and proliferation and apoptosis of ovarian cancer cell (Khan et al., 2020) suggest that this isoflavone can affect these processes via microRNAs too. In addition, the ability of quercetin to affect **estrogen receptors** in healthy (Cao et al., 2014) and cancer (Scambia et al., 1990) ovarian cells suggests that it can suppress estrogen-dependent female reproductive events and disorders via inhibition of estrogen reception and action. On the contrary to ovarian cancer cells, quercetin promoted estrogen receptor α in uterus of rats with polycystic ovarian syndrome indicating that quercetin can promote the reproductive processes and alleviate signs of this syndrome through upregulation of estrogen receptors in the uterus (Neisy et al., 2019).

Effect of quercetin on ovarian functions and polycystic ovarian syndrome can be mediated by its effect on **fat metabolism**. At least, polycystic ovarian syndrome is associated with increased fat storage, while quercetin treatment exerts a beneficial effect on both signs of this syndrome, adipokine genes and their levels and fat storage (Jafari Khorchani et al., 2020; Pourteymour et al., 2020).

Finally, the influence of quercetin on **hormonal regulators** of female reproductive processes and on **their receptors** and **response to these hormonal regulators** described earlier suggests that hormones and growth factors could be extracellular mediators of quercetin action on female reproduction and reproductive disorders.

Therefore, the available publications demonstrate a multiple sites of quercetin action on female reproductive processes and their dysfunctions, as well as a multiple mechanisms of action on these processes, which use numerous extra- and intracellular signaling molecules.

4.9.7 Application in reproductive biology and medicine

The quercetin influence on a number of nonreproductive and reproductive processes and their disorders outlined earlier suggest its applicability in various branches.

Its ability to improve reproduction in laboratory (mice: Beazley and Nurminskaya, 2016; Shu et al., 2011; Elkady et al., 2019, rat: Chen et al., 2010; Olaniyan et al., 2020) and animals (rabbits: Naseer et al., 2017) and chicken (Liu et al., 2013) suggests its applicability as a biostimulator of reproduction in laboratory, farm animal and poultry breeding.

Furthermore, quercetin was shown to be an additive useful for improvement of quality and developmental capacity of oocytes matured *in vitro* (rabbit: Naseer et al., 2017; goat: Mao et al., 2018; Silva et al., 2018; sheep: Karimian et al., 2018). It suggests its potential applicability in animal biotechnology. If this positive effects on human gametes and embryos would be observed, quercetin could be useful in assisted reproduction (*in vitro* human oocyte maturation, fertilization and embryo development).

There is more evidence for therapeutic potential of quercetin. The anticancer effects of this isoflavone (Scambia et al., 1990; Gao et al., 2012; Catanzaro et al., 2015; Parvaresh et al., 2016; Gong et al., 2017; Shafabakhsh and Asemi, 2019; Khan et al., 2020) mentioned earlier could be useful for prevention of this illness. Animal (Jahan et al., 2018; Neisy et al., 2019; Jafari Khorchani et al., 2020; Olaniyan et al., 2020) and human (Pourteymour et al., 2020) studies suggest its benefits for relief of polycystic ovarian syndrome. Studies on laboratory animals indicate the therapeutic potential of quercetin for prevention of premature ovarian failure (Elkady et al., 2019), ovarian aging (Chen et al., 2010; Wang et al., 2017) and endometriosis (Cao et al. (2014).

Some animal and *in vitro* studies demonstrated that quercetin could be a potent natural protector against the adverse influence of high temperature stress (Naseer et al., 2017), chemotherapy (Samare-Najaf et al., 2020) and such toxic environmental contaminants as cadmium (Jia et al., 2011; Mao et al., 2018), silver and titania nanoparticles (Sirotkin et al., 2021), xylene (Tarko et al., 2018) and benzene (Tarko et al., 2019).

The efficiency of quercetin application would be substantially increased by using its chemical and functional analogues with improved bioavailability or biological action. Such analogues could be natural quercetin metabolites (Magar et al., 2020), derivates like dihydroquercetin (Weidmann, 2011; Sunil et al., 2019) or rutin (Sharma et al., 2013; Negahdari et al., 2021), application of quercetin complexes with metal ions (Xu et al., 2019) or other nano-composites including nanoparticles, nanocapsules, nanoliposomes, nano-micelles, nanosuspensions and nanoemulsions (Li et al., 2018; Parhi et al., 2020). Nanodrug delivery systems have such advantages as reduced particle size, enlarged surface area, improved intracorporeal circulation and distribution of the drug, delayed drug release, increased solubility and dissolution, bioavailability, stability, permeability, and biological and therapeutic effects (Li et al., 2018). There are reports demonstrating usefulness of chemical analogue of quercetin, 3,4',7-O-trimethylquercetin, for suppression of some ovarian cancer cell functions and treatment of ovarian cancer (Yamauchi et al., 2017; Ashraf et al., 2018).

These data demonstrate the large potential of quercetin as a biostimulator and protector of female reproductive processes and medicine for treatment of several reproductive disorders.

4.9.8 Conclusions and possible direction of future studies

The available information demonstrated the ability of quercetin and its analogues to inhibit proliferation and to promote apoptosis in cancer cells, to activate regenerative processes, to treat immune, inflammatory, cardiovascular, neurodegenerative, gastric and metabolic disorders (including obesity and diabetes), to suppress microorganism, to protect bones and liver, to relieve pain and to improve physical and mental performance and to prolong life span.

The positive influence of quercetin on female reproductive processes are evident. It can promote ovarian folliculo- and oogenesis, improve quality and developmental capacity of oocytes and embryos, increase fecundity in various species. These effects can be mediated by changes in pituitary and ovarian steroid and peptide hormones, growth factors and cytokines, in their receptors and numerous postreceptory signaling pathways. Due to these effects, quercetin can be applicable as biostimulator of reproduction, for prevention, mitigation and treatment of several female reproductive disorders: ovarian cancer, polycystic ovarian syndrome, premature ovarian failure, ovarian aging, endometriosis, as well as to increase the resistance of female reproductive system to heat stress and reproductive toxicity of chemotherapeutic, cadmium ions, xylene and benzene.

On the other hand, the available knowledge is not enough to understand targets and mechanisms of quercetin action and, therefore, to arrange its adequate application. It is not always clear whether a particular effect of quercetin treatment is due to quercetin itself or to the products of its metabolization in organism. Paradoxically, the current accumulation of data concerning quercetin effects complicates their interpretation. Numerous known targets, biological effects and mechanisms of action of quercetin complicate generation of a general concept integrating the available knowledge concerning mechanisms of quercetin action and their functional interrelationships. At present, even the explanations of the mechanism of quercetin action on particular organ, process or illness are too complicated. The reports concerning the character of quercetin influence on nonreproductive and reproductive processes are sometimes contradictory. Some quercetin effects were studies only in *in vitro* or animal experiments, while quercetin action in these studies was usually more pronounced than that in clinical trials. Clinical studies were rate, their protocols were variable and the results were usually less conclusive than the results of animal and *in vitro* experiments. There are substantial contradictions in the reports concerning the character of quercetin effects even in experiments performed on the same model, which can be explained by initial variability in the state of animals or cells used in various performed studies. All these facts complicate understanding and application of quercetin effects on reproductive and nonreproductive processes and dysfunctions. Nevertheless, they demonstrate a large physiological and therapeutical potential of this isoflavone, as well as requirement of its more profound studies.

References

Andres S, Pevny S, Ziegenhagen R, Bakhiya N, Schäfer B, Hirsch-Ernst KI, Lampen A. Safety aspects of the use of quercetin as a dietary supplement. Mol Nutr Food Res 2018;62(1). https://doi.org/10.1002/mnfr.201700447.

Ashraf AHMZ, Afroze SH, Yamauchi K, Zawieja DC, Keuhl TJ, Erlandson LW, Uddin MN. Differential mechanism of action of 3,4',7-O-trimethylquercetin in three types of ovarian cancer cells. Anticancer Res 2018;38(9):5131–7. https://doi.org/10.21873/anticanres.12835.

Babaei F, Mirzababaei M, Nassiri-Asl M. Quercetin in food: possible mechanisms of its effect on memory. J Food Sci 2018;83(9):2280–7. https://doi.org/10.1111/1750-3841.14317.

Batiha GE, Beshbishy AM, Ikram M, Mulla ZS, El-Hack MEA, Taha AE, Algammal AM, Elewa YHA. The pharmacological activity, biochemical properties, and pharmacokinetics of the major natural polyphenolic flavonoid: quercetin. Foods 2020;9(3):374. https://doi.org/10.3390/foods9030374.

Beazley KE, Nurminskaya M. Effects of dietary quercetin on female fertility in mice: implication of transglutaminase 2. Reprod Fertil Dev 2016;28(7):974–81. https://doi.org/10.1071/RD14155.

Bolouki A, Zal F, Mostafavi-Pour Z, Bakhtari A. Protective effects of quercetin on uterine receptivity markers and blastocyst implantation rate in diabetic pregnant mice. Taiwan J Obstet Gynecol 2020a;59(6):927–34. https://doi.org/10.1016/j.tjog.2020.09.038.

Bolouki A, Zal F, Alaee S. Ameliorative effects of quercetin on the preimplantation embryos development in diabetic pregnant mice. J Obstet Gynaecol Res 2020b;46(5):736–44. https://doi.org/10.1111/jog.14219.

Bule M, Abdurahman A, Nikfar S, Abdollahi M, Amini M. Antidiabetic effect of quercetin: a systematic review and meta-analysis of animal studies. Food Chem Toxicol 2019;125:494—502. https://doi.org/10.1016/j.fct.2019.01.037.

Cao Y, Zhuang MF, Yang Y, Xie SW, Cui JG, Cao L, Zhang TT, Zhu Y. Preliminary study of quercetin affecting the hypothalamic-pituitary-gonadal axis on rat endometriosis model. Evid Based Complement Altern Med 2014;2014:781684. https://doi.org/10.1155/2014/781684.

Capcarova M, Petruska P, Zbynovska K, Kolesarova A, Sirotkin AV. Changes in antioxidant status of porcine ovarian granulosa cells after quercetin and T-2 toxin treatment. J Environ Sci Health B 2015;50(3):201—6. https://doi.org/10.1080/03601234.2015.982425.

Carrasco-Pozo C, Cires MJ, Gotteland M. Quercetin and epigallocatechin gallate in the prevention and treatment of obesity: from molecular to clinical studies. J Med Food 2019;22(8):753—70. https://doi.org/10.1089/jmf.2018.0193.

Carullo G, Cappello AR, Frattaruolo L, Badolato M, Armentano B, Aiello F. Quercetin and derivatives: useful tools in inflammation and pain management. Future Med Chem 2017;9(1):79—93. https://doi.org/10.4155/fmc-2016-0186.

Catanzaro D, Ragazzi E, Vianello C, Caparrotta L, Montopoli M. Effect of quercetin on cell cycle and cyclin expression in ovarian carcinoma and osteosarcoma cell lines. Nat Prod Commun 2015;10(8):1365—8.

Chen ZG, Luo LL, Xu JJ, Zhuang XL, Kong XX, Fu YC. Effects of plant polyphenols on ovarian follicular reserve in aging rats. Biochem Cell Biol 2010;88(4):737—45. https://doi.org/10.1139/O10-012.

Chen S, Jiang H, Wu X, Fang J. Therapeutic effects of quercetin on inflammation, obesity, and type 2 diabetes. Mediat Inflamm 2016;2016:9340637. https://doi.org/10.1155/2016/9340637.

Cione E, La Torre C, Cannataro R, Caroleo MC, Plastina P, Gallelli L. Quercetin, epigallocatechin gallate, curcumin, and resveratrol: from dietary sources to human MicroRNA modulation. Molecules 2019;25(1):63. https://doi.org/10.3390/molecules25010063.

Colunga Biancatelli RML, Berrill M, Catravas JD, Marik PE. Quercetin and vitamin C: an experimental, synergistic therapy for the prevention and treatment of SARS-CoV-2 related disease (COVID-19). Front Immunol 2020;11:1451. https://doi.org/10.3389/fimmu.2020.01451.

Costa LG, Garrick JM, Roquè PJ, Pellacani C. Mechanisms of neuroprotection by quercetin: counteracting oxidative stress and more. Oxid Med Cell Longev 2016;2016:2986796. https://doi.org/10.1155/2016/2986796.

Dabeek WM, Marra MV. Dietary quercetin and kaempferol: bioavailability and potential cardiovascular-related bioactivity in humans. Nutrients 2019;11(10):2288. https://doi.org/10.3390/nu11102288.

de Oliveira MR, Nabavi SM, Braidy N, Setzer WN, Ahmed T, Nabavi SF. Quercetin and the mitochondria: a mechanistic view. Biotechnol Adv 2016;34(5):532—49. https://doi.org/10.1016/j.biotechadv.2015.12.014.

Dhiman P, Malik N, Sobarzo-Sánchez E, Uriarte E, Khatkar A. Quercetin and related chromenone derivatives as monoamine oxidase inhibitors: targeting neurological and mental disorders. Molecules 2019;24(3):418. https://doi.org/10.3390/molecules24030418.

Dostal Z, Modriansky M. The effect of quercetin on microRNA expression: a critical review. Biomed Pap Med Fac Univ Palacky Olomouc Czech Repub 2019;163(2):95—106. https://doi.org/10.5507/bp.2019.030.

Eid HM, Haddad PS. The antidiabetic potential of quercetin: underlying mechanisms. Curr Med Chem 2017;24(4):355—64. https://doi.org/10.2174/0929867323666160909153707.

Elkady MA, Shalaby S, Fathi F, El-Mandouh S. Effects of quercetin and rosuvastatin each alone or in combination on cyclophosphamide-induced premature ovarian failure in female albino mice. Hum Exp Toxicol 2019;38(11):1283−95. https://doi.org/10.1177/0960327119865588.

Farooqi AA, Jabeen S, Attar R, Yaylim I, Xu B. Quercetin-mediated regulation of signal transduction cascades and microRNAs: natural weapon against cancer. J Cell Biochem 2018;119(12):9664−74. https://doi.org/10.1002/jcb.27488.

Ferenczyova K, Kalocayova B, Bartekova M. Potential implications of quercetin and its derivatives in cardioprotection. Int J Mol Sci 2020;21(5):1585. https://doi.org/10.3390/ijms21051585.

Gao X, Wang B, Wei X, Men K, Zheng F, Zhou Y, Zheng Y, Gou M, Huang M, Guo G, Huang N, Qian Z, Wei Y. Anticancer effect and mechanism of polymer micelle-encapsulated quercetin on ovarian cancer. Nanoscale 2012;4(22):7021−30. https://doi.org/10.1039/c2nr32181e.

Gencer M, Karaca T, Güngör AN, Hacıvelioğlu SÖ, Demirtaş S, Turkon H, Uysal A, Korkmaz F, Coşar E, Hancı V. The protective effect of quercetin on IMA levels and apoptosis in experimental ovarian ischemia-reperfusion injury. Eur J Obstet Gynecol Reprod Biol 2014;177:135−40. https://doi.org/10.1016/j.ejogrb.2014.03.036.

Gong C, Yang Z, Zhang L, Wang Y, Gong W, Liu Y. Quercetin suppresses DNA double-strand break repair and enhances the radiosensitivity of human ovarian cancer cells via p53-dependent endoplasmic reticulum stress pathway. Onco Targets Ther 2017;11:17−27. https://doi.org/10.2147/OTT.S147316.

Huang H, Liao D, Dong Y, Pu R. Effect of quercetin supplementation on plasma lipid profiles, blood pressure, and glucose levels: a systematic review and meta-analysis. Nutr Rev 2020;78(8):615−26. https://doi.org/10.1093/nutrit/nuz071.

Jafari Khorchani M, Zal F, Neisy A. The phytoestrogen, quercetin, in serum, uterus and ovary as a potential treatment for dehydroepiandrosterone-induced polycystic ovary syndrome in the rat. Reprod Fertil Dev 2020;32(3):313−21. https://doi.org/10.1071/RD19072.

Jafarinia M, Sadat Hosseini M, Kasiri N, Fazel N, Fathi F, Ganjalikhani Hakemi M, Eskandari N. Quercetin with the potential effect on allergic diseases. Allergy Asthma Clin Immunol 2020;16:36. https://doi.org/10.1186/s13223-020-00434-0.

Jahan S, Abid A, Khalid S, Afsar T, Qurat-Ul-Ain, Shaheen G, Almajwal A, Razak S. Therapeutic potentials of Quercetin in management of polycystic ovarian syndrome using Letrozole induced rat model: a histological and a biochemical study. J Ovarian Res 2018;11(1):26. https://doi.org/10.1186/s13048-018-0400-5.

Jia Y, Lin J, Mi Y, Zhang C. Quercetin attenuates cadmium-induced oxidative damage and apoptosis in granulosa cells from chicken ovarian follicles. Reprod Toxicol 2011;31(4):477−85. https://doi.org/10.1016/j.reprotox.2010.12.057.

Karimian M, Zandi M, Sanjabi MR, Masoumian M, Ofoghi H. Effects of grape seed extract, quercetin and vitamin C on ovine oocyte maturation and subsequent embryonic development. Cell Mol Biol 2018;64(4):98−102.

Kashyap D, Garg VK, Tuli HS, Yerer MB, Sak K, Sharma AK, Kumar M, Aggarwal V, Sandhu SS. Fisetin and quercetin: promising flavonoids with chemopreventive potential. Biomolecules 2019;9(5):174. https://doi.org/10.3390/biom9050174.

Kelly GS. Quercetin. Monograph. Altern Med Rev 2011;16(2):172−94.

Khadrawy O, Gebremedhn S, Salilew-Wondim D, Taqi MO, Neuhoff C, Tholen E, Hoelker M, Schellander K, Tesfaye D. Endogenous and exogenous modulation of Nrf2 mediated oxidative stress response in bovine granulosa cells: potential implication for ovarian function. Int J Mol Sci 2019;20(7):1635. https://doi.org/10.3390/ijms20071635.

Khan H, Ullah H, Aschner M, Cheang WS, Akkol EK. Neuroprotective effects of quercetin in Alzheimer's disease. Biomolecules 2019;10(1):59. https://doi.org/10.3390/biom10010059.

Khan K, Javed Z, Sadia H, Sharifi-Rad J, Cho WC, Luparello C. Quercetin and MicroRNA interplay in apoptosis regulation in ovarian cancer. Curr Pharmaceut Des 2020. https://doi.org/10.2174/1381612826666201019102207.

Kolesarova A, Roychoudhury S, Klinerova B, Packova D, Michalcova K, Halenar M, Kopcekova J, Mnahoncakova E, Galik B. Dietary bioflavonoid quercetin modulates porcine ovarian granulosa cell functions *in vitro*. J Environ Sci Health B 2019;54(6):533−7. https://doi.org/10.1080/03601234.2019.1586034.

Larson AJ, Symons JD, Jalili T. Quercetin: a treatment for hypertension?-A review of efficacy and mechanisms. Pharmaceuticals 2010;3(1):237−50. https://doi.org/10.3390/ph3010237.

Li Y, Yao J, Han C, Yang J, Chaudhry MT, Wang S, Liu H, Yin Y. Quercetin, inflammation and immunity. Nutrients 2016;8(3):167. https://doi.org/10.3390/nu8030167.

Li SJ, Liao YF, Du Q. [Research and application of quercetin-loaded nano drug delivery system]. Zhongguo Zhongyao Zazhi 2018;43(10):1978−84. https://doi.org/10.19540/j.cnki.cjcmm.20180312.002. Chinese.

Liu Y, Li Y, Liu HN, Suo YL, Hu LL, Feng XA, Zhang L, Jin F. Effect of quercetin on performance and egg quality during the late laying period of hens. Br Poult Sci 2013;54(4):510−4. https://doi.org/10.1080/00071668.2013.799758.

Maaliki D, Shaito AA, Pintus G, El-Yazbi A, Eid AH. Flavonoids in hypertension: a brief review of the underlying mechanisms. Curr Opin Pharmacol 2019;45:57−65. https://doi.org/10.1016/j.coph.2019.04.014.

Magar RT, Sohng JK. A review on structure, modifications and structure-activity relation of quercetin and its derivatives. J Microbiol Biotechnol 2020;30(1):11−20. https://doi.org/10.4014/jmb.1907.07003.

Mao T, Han C, Wei B, Zhao L, Zhang Q, Deng R, Liu J, Luo Y, Zhang Y. Protective effects of quercetin against cadmium chloride-induced oxidative injury in goat sperm and zygotes. Biol Trace Elem Res 2018;185(2):344−55. https://doi.org/10.1007/s12011-018-1255-8.

Marunaka Y, Marunaka R, Sun H, Yamamoto T, Kanamura N, Inui T, Taruno A. Actions of quercetin, a polyphenol, on blood pressure. Molecules 2017;22(2):209. https://doi.org/10.3390/molecules22020209.

Massi A, Bortolini O, Ragno D, Bernardi T, Sacchetti G, Tacchini M, De Risi C. Research progress in the modification of quercetin leading to anticancer agents. Molecules 2017;22(8):1270. https://doi.org/10.3390/molecules22081270.

Mlcek J, Jurikova T, Skrovankova S, Sochor J. Quercetin and its anti-allergic immune response. Molecules 2016;21(5):623. https://doi.org/10.3390/molecules21050623.

Murota K, Nakamura Y, Uehara M. Flavonoid metabolism: the interaction of metabolites and gut microbiota. Biosci Biotechnol Biochem 2018;82(4):600−10. https://doi.org/10.1080/09168451.2018.1444467.

Naseer Z, Ahmad E, Epikmen ET, Uçan U, Boyacioğlu M, İpek E, Akosy M. Quercetin supplemented diet improves follicular development, oocyte quality, and reduces ovarian apoptosis in rabbits during summer heat stress. Theriogenology 2017;96:136−41. https://doi.org/10.1016/j.theriogenology.2017.03.029.

Neisy A, Zal F, Seghatoleslam A, Alaee S. Amelioration by quercetin of insulin resistance and uterine GLUT4 and ERα gene expression in rats with polycystic ovary syndrome (PCOS). Reprod Fertil Dev 2019;31(2):315−23. https://doi.org/10.1071/RD18222.

Negahdari R, Bohlouli S, Sharifi S, Maleki Dizaj S, Rahbar Saadat Y, Khezri K, Jafari S, Ahmadian E, Gorbani Jahandizi N, Raeesi S. Therapeutic benefits of rutin and its nano-formulations. Phytother Res 2021;35(4):1719−38. https://doi.org/10.1002/ptr.6904.

Olaniyan OT, Bamidele O, Adetunji CO, Priscilla B, Femi A, Ayobami D, Okotie G, Oluwaseun I, Olugbenga E, Mali PC. Quercetin modulates granulosa cell mRNA androgen receptor gene expression in dehydroepiandrosterone-induced polycystic ovary in Wistar rats via metabolic and hormonal pathways. J Basic Clin Physiol Pharmacol 2020;31(4). https://doi.org/10.1515/jbcpp-2019-0076.

Pallauf K, Duckstein N, Rimbach G. A literature review of flavonoids and lifespan in model organisms. Proc Nutr Soc 2017;76(2):145−62. https://doi.org/10.1017/S0029665116000720.

Parhi B, Bharatiya D, Swain SK. Application of quercetin flavonoid based hybrid nanocomposites: a review. Saudi Pharmaceut J 2020a;28(12):1719−32. https://doi.org/10.1016/j.jsps.2020.10.017.

Parvaresh A, Razavi R, Rafie N, Ghiasvand R, Pourmasoumi M, Miraghajani M. Quercetin and ovarian cancer: an evaluation based on a systematic review. J Res Med Sci 2016;21:34. https://doi.org/10.4103/1735-1995.181994.

Pérez-Pastén R, Martínez-Galero E, Chamorro-Cevallos G. Quercetin and naringenin reduce abnormal development of mouse embryos produced by hydroxyurea. J Pharm Pharmacol 2010;62(8):1003−9. https://doi.org/10.1111/j.2042-7158.2010.01118.x.

Polerà N, Badolato M, Perri F, Carullo G, Aiello F. Quercetin and its natural sources in wound healing management. Curr Med Chem 2019;26(31):5825−48. https://doi.org/10.2174/0929867325666180713150626.

Pourteymour Fard Tabrizi F, Hajizadeh-Sharafabad F, Vaezi M, Jafari-Vayghan H, Alizadeh M, Maleki V. Quercetin and polycystic ovary syndrome, current evidence and future directions: a systematic review. J Ovarian Res 2020;13(1):11. https://doi.org/10.1186/s13048-020-0616-z.

Rauf A, Imran M, Khan IA, Ur-Rehman M, Gilani SA, Mehmood Z, Mubarak MS. Anticancer potential of quercetin: a comprehensive review. Phytother Res 2018;32(11):2109−30. https://doi.org/10.1002/ptr.6155.

Reyes-Farias M, Carrasco-Pozo C. The anti-cancer effect of quercetin: molecular implications in cancer metabolism. Int J Mol Sci 2019;20(13):3177. https://doi.org/10.3390/ijms20133177.

Samare-Najaf M, Zal F, Safari S, Koohpeyma F, Jamali N. Stereological and histopathological evaluation of doxorubicin-induced toxicity in female rats' ovary and uterus and palliative effects of quercetin and vitamin E. Hum Exp Toxicol 2020;39(12):1710−24. https://doi.org/10.1177/0960327120937329.

Sameni HR, Javadinia SS, Safari M, Tabrizi Amjad MH, Khanmohammadi N, Parsaie H, Zarbakhsh S. Effect of quercetin on the number of blastomeres, zona pellucida thickness, and hatching rate of mouse embryos exposed to actinomycin D: an experimental study. Int J Reprod Biomed 2018;16(2):101−8.

Santini SE, Basini G, Bussolati S, Grasselli F. The phytoestrogen quercetin impairs steroidogenesis and angiogenesis in swine granulosa cells in vitro. J Biomed Biotechnol 2009;2009:419891. https://doi.org/10.1155/2009/419891.

Scambia G, Ranelletti FO, Panici PB, Piantelli M, Bonanno G, De Vincenzo R, Ferrandina G, Rumi C, Larocca LM, Mancuso S. Inhibitory effect of quercetin on OVCA 433 cells and presence of type II oestrogen binding sites in primary ovarian tumours and cultured cells. Br J Cancer 1990;62(6):942−6. https://doi.org/10.1038/bjc.1990.414.

Serban MC, Sahebkar A, Zanchetti A, Mikhailidis DP, Howard G, Antal D, Andrica F, Ahmed A, Aronow WS, Muntner P, Lip GY, Graham I, Wong N, Rysz J, Banach M. Lipid and blood pressure meta-analysis collaboration (LBPMC) group. Effects of quercetin on blood pressure: a systematic review and meta-analysis of randomized controlled trials. J Am Heart Assoc 2016;5(7):e002713. https://doi.org/10.1161/JAHA.115.002713.

Shafabakhsh R, Asemi Z. Quercetin: a natural compound for ovarian cancer treatment. J Ovarian Res 2019;12(1):55. https://doi.org/10.1186/s13048-019-0530-4.

Sharma A, Kashyap D, Sak K, Tuli HS, Sharma AK. Therapeutic charm of quercetin and its derivatives: a review of research and patents. Pharm Pat Anal 2018;7(1):15—32. https://doi.org/10.4155/ppa-2017-0030.

Shi GJ, Li Y, Cao QH, Wu HX, Tang XY, Gao XH, Yu JQ, Chen Z, Yang Y. In vitro and in vivo evidence that quercetin protects against diabetes and its complications: a systematic review of the literature. Biomed Pharmacother 2019;109:1085—99. https://doi.org/10.1016/j.biopha.2018.10.130.

Shu X, Hu XJ, Zhou SY, Xu CL, Qiu QQ, Nie SP, Xie MY. [Effect of quercetin exposure during the prepubertal period on ovarian development and reproductive endocrinology of mice]. Yao Xue Xue Bao 2011;46(9):1051—7.

Silva AAA, Silva MNP, Figueiredo LBF, Gonçalves JD, Silva MJS, Loiola MLG, Bastos BDM, Oliveira RA, Ribeiro LGM, Barberino RS, Gouveia BB, Monte APO, Nogueira DM, Cordeiro MF, Matos MHT, Lopes Júnior ES. Quercetin influences in vitro maturation, apoptosis and metabolically active mitochondria of goat oocytes. Zygote 2018;26(6):465—70. https://doi.org/10.1017/S0967199418000485.

Sirotkin AV, Hrabovszká S, Štochmaľová A, Grossmann R, Alwasel S, Halim Harrath A. Effect of quercetin on ovarian cells of pigs and cattle. Anim Reprod Sci 2019a;205:44—51. https://doi.org/10.1016/j.anireprosci.2019.04.002.

Sirotkin AV, Štochmaľová A, Alexa R, Kádasi A, Bauer M, Grossmann R, Alrezaki A, Alwasel S, Harrath AH. Quercetin directly inhibits basal ovarian cell functions and their response to the stimulatory action of FSH. Eur J Pharmacol 2019b;860:172560. https://doi.org/10.1016/j.ejphar.2019.172560.

Sirotkin AV, Alexa R, Stochmalova A, Scsukova S. Plant isoflavones can affect accumulation and impact of silver and titania nanoparticles on ovarian cells. Endocr Regul 2021;55(1):52—60. https://doi.org/10.2478/enr-2021-0007.

Somerville V, Bringans C, Braakhuis A. Polyphenols and performance: a systematic review and meta-analysis. Sports Med 2017;47(8):1589—99. https://doi.org/10.1007/s40279-017-0675-5.

Sunil C, Xu B. An insight into the health-promoting effects of taxifolin (dihydroquercetin). Phytochemistry 2019;166:112066. https://doi.org/10.1016/j.phytochem.2019.112066.

Tabrizi R, Tamtaji OR, Mirhosseini N, Lankarani KB, Akbari M, Heydari ST, Dadgostar E, Asemi Z. The effects of quercetin supplementation on lipid profiles and inflammatory markers among patients with metabolic syndrome and related disorders: a systematic review and meta-analysis of randomized controlled trials. Crit Rev Food Sci Nutr 2020;60(11):1855—68. https://doi.org/10.1080/10408398.2019.1604491.

Tarko A, Štochmalova A, Hrabovszka S, Vachanova A, Harrath AH, Alwasel S, Grossman R, Sirotkin AV. Can xylene and quercetin directly affect basic ovarian cell functions? Res Vet Sci 2018;119:308—12. https://doi.org/10.1016/j.rvsc.2018.07.010.

Tarko A, Štochmaľová A, Jedličková K, Hrabovszká S, Vachanová A, Harrath AH, Alwasel S, Alrezaki A, Kotwica J, Baláži A, Sirotkin AV. Effects of benzene, quercetin, and their combination on porcine ovarian cell proliferation, apoptosis, and hormone release. Arch Anim Breed 2019;62(1):345—51. https://doi.org/10.5194/aab-62-345-2019.

Teekaraman D, Elayapillai SP, Viswanathan MP, Jagadeesan A. Quercetin inhibits human metastatic ovarian cancer cell growth and modulates components of the intrinsic apoptotic pathway in PA-1 cell line. Chem Biol Interact 2019;300:91—100. https://doi.org/10.1016/j.cbi.2019.01.008.

Wang H, Jo YJ, Oh JS, Kim NH. Quercetin delays postovulatory aging of mouse oocytes by regulating SIRT expression and MPF activity. Oncotarget 2017;8(24):38631—41. https://doi.org/10.18632/oncotarget.16219.

Wang J, Qian X, Gao Q, Lv C, Xu J, Jin H, Zhu H. Quercetin increases the antioxidant capacity of the ovary in menopausal rats and in ovarian granulosa cell culture in vitro. J Ovarian Res 2018;11(1):51. https://doi.org/10.1186/s13048-018-0421-0.

Wang Y. Interaction between quercetin and DNA. Sheng Wu Gong Cheng Xue Bao 2020;36(12):2877–91. https://doi.org/10.13345/j.cjb.200127. Chinese.

Weber LP, Kiparissis Y, Hwang GS, Niimi AJ, Janz DM, Metcalfe CD. Increased cellular apoptosis after chronic aqueous exposure to nonylphenol and quercetin in adult medaka (*Oryzias latipes*). Comp Biochem Physiol C Toxicol Pharmacol 2002;131(1):51–9. https://doi.org/10.1016/s1532-0456(01)00276-9.

Whitehead SA, Lacey M. Phytoestrogens inhibit aromatase but not 17beta-hydroxysteroid dehydrogenase (HSD) type 1 in human granulosa-luteal cells: evidence for FSH induction of 17beta-HSD. Hum Reprod 2003;18(3):487–94. https://doi.org/10.1093/humrep/deg125.

Wong SK, Chin KY, Ima-Nirwana S. Quercetin as an agent for protecting the bone: a review of the current evidence. Int J Mol Sci 2020;21(17):6448. https://doi.org/10.3390/ijms21176448.

Xu M, Pirtskhalava T, Farr JN, Weigand BM, Palmer AK, Weivoda MM, Inman CL, Ogrodnik MB, Hachfeld CM, Fraser DG, Onken JL, Johnson KO, Verzosa GC, Langhi LGP, Weigl M, Giorgadze N, LeBrasseur NK, Miller JD, Jurk D, Singh RJ, Allison DB, Ejima K, Hubbard GB, Ikeno Y, Cubro H, Garovic VD, Hou X, Weroha SJ, Robbins PD, Niedernhofer LJ, Khosla S, Tchkonia T, Kirkland JL. Senolytics improve physical function and increase lifespan in old age. Nat Med 2018;24(8):1246–56. https://doi.org/10.1038/s41591-018-0092-9.

Xu D, Hu MJ, Wang YQ, Cui YL. Antioxidant activities of quercetin and its complexes for medicinal application. Molecules 2019;24(6):1123. https://doi.org/10.3390/molecules24061123.

Yamauchi K, Afroze SH, Mitsunaga T, McCormick TC, Kuehl TJ, Zawieja DC, Uddin MN. 3,4',7-O-trimethylquercetin inhibits invasion and migration of ovarian cancer cells. Anticancer Res 2017;37(6):2823–9. https://doi.org/10.21873/anticanres.11633.

Yang D, Wang T, Long M, Li P. Quercetin: its main pharmacological activity and potential application in clinical medicine. Oxid Med Cell Longev 2020;2020:8825387. https://doi.org/10.1155/2020/8825387.

Yang L, Chen Y, Liu Y, Xing Y, Miao C, Zhao Y, Chang X, Zhang Q. The role of oxidative stress and natural antioxidants in ovarian aging. Front Pharmacol 2021;11:617843. https://doi.org/10.3389/fphar.2020.617843.

Zhao Y, Chen B, Shen J, Wan L, Zhu Y, Yi T, Xiao Z. The beneficial effects of quercetin, curcumin, and resveratrol in obesity. Oxid Med Cell Longev 2017;2017:1459497. https://doi.org/10.1155/2017/1459497.

Chapter 4.10

Resveratrol

4.10.1 Introduction

Understanding physiological and natural regulators of reproduction is important for characterization, prediction, and control of reproductive processes, as well as for prevention and treatment of reproductive disorders. One such promising regulator, protector and therapeutic could be resveratrol (R). The knowledge concerning involvement and application of R in control of female reproduction are summarized in several reviews, but some reviews don't reflect the recent finding in this field (Ortrega and Duleba, 2015) or they are focused on special R target cells (Jozkowiak et al., 2020) or disorders (van Duursen, 2017; Rauf et al., 2018; Banaszewska et al., 2019; Ochiai and Kuroda, 2019). In the present review we try to summarize the current (published after year 2000) knowledge concerning properties, physiological effects and application of R with emphasis to R action on female reproductive processes at various regulatory levels.

4.10.2 Provenance and properties

R (3,5,40-trihydroxy-trans-stilbene) is a stilbenoid, a derivative of stilbene. Its basic structure consists of two phenolic rings bonded together by a double styrene bond, which forms the 3,5,4′-Trihydroxystilbene. This double bond is responsible for the isometric *cis*- and *trans*-forms of resveratrol (Fig. 4.10). *Cis*- and *trans*-isomers coexist in plants and in wine. The *trans*-isomer is more predominant and stable from the steric point of view compared to the natural form. *Cis*-isomerization can occur when the *trans*-isoform is exposed to solar or artificial light or ultraviolet radiation. R is produced by more than 70 species of plants (Gambini et al., 2015) in response to stress, injury, infection or ultraviolet radiations. It plays a role as a phytoalexin, showing a capacity to inhibit the development of certain infections (Gambini et al., 2015; Pannu and Bhatnagar, 2019). The most known sources of resveratrol in food include skin of grapes, blueberries, raspberries, mulberries, and peanuts (Tian and Liu, 2020). In grape, the synthesis of R decreases regularly during the ripening

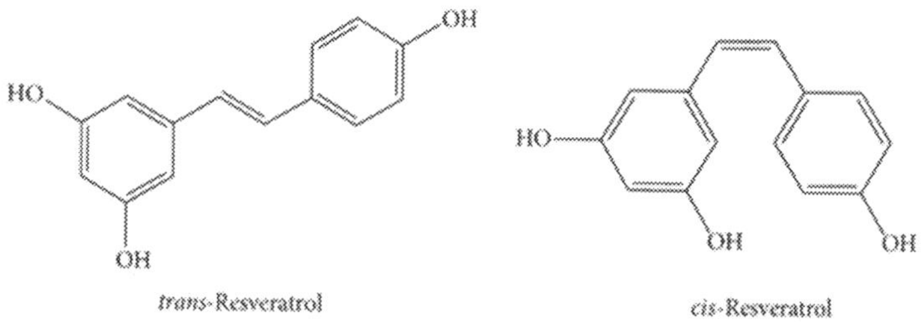

372 Contaminants and Plants Action on Female Reproduction

trans-Resveratrol *cis*-Resveratrol

FIGURE 4.10 Chemical structure of trans- and cis-resveratrol (Gambini et al., 2015).

process, this explains the increasing susceptibility of mature fruits to infection by *Botrytis cinerea* (Gambini et al., 2015). Due to poor solubility in water and rapid metabolism, R has low bioavailability in organism (Gambini et al., 2015; Intagliata et al., 2019). The solubility of R in alcohol is 1700 times higher than in water, which affects its absorption in organism and physiological effect of R (https://en.wikipedia.org/wiki/Talk:Resveratrol/Archive_1). For example, R prevented apoptosis in neuronal cells only in combination with alcohol, but not given alone (Khodaie et al., 2018). Therefore, R consummation together with alcohol, for example in red wine, is preferable compared to raw fruits. Among wines, muscadine table wines have highest resveratrol contents (https://en.wikipedia.org/wiki/Talk:Resveratrol/Archive_1). Besides alcohol, R can interact with fatty acids. 90% of free *trans*-R binds to human plasma lipoproteins; high-fat meal binds R, reduces amount of free R in blood and decreases absorption and bioavailability of dietary R (Gambin et al., 2015).

4.10.3 Physiological and therapeutic actions

Due to the properties listed earlier, R possess antiproliferative, antiaging, antiinflammatory, antineoplastic activity (Catalogna et al., 2019; Pannu and Bhatnagar, 2019; Tian and Liu, 2020). At present, R has been identified as: being cancer chemoprotective, being antiinflammatory, improving vascular function, extending the lifespan and ameliorating aging-related phenotypes, opposing the effects of a high calorie diet, mimicking the effects of calorie restriction, and improving cellular function and metabolic health in general (Springer and Moco, 2019). It is known in the prevention and reversal of numerous cardiovascular, metabolic, psychological, and reproductive disorders (Ortega and Duleba, 2015; Varoni et al., 2016; Nguen et al., 2017).

The US National library of medicine (https://medlineplus.gov/druginfo/natural/307.html) lists the clinical data concerning ability of R to treat hay fever and obesity and probably acne, age-dependent mental disorders, beta-thalassemia, cancer, kidney damages, lung diseases, osteoarthritis, polycystic ovarian syndrome, rheumatoid arthritis, inflammatory bowel disease (ulcerative colitis), aging skin, atherosclerosis, and improves cognitive functions, but not heart diseases, metabolic syndrome, and disorders in liver cholesterol metabolism disorders. Other reviews demonstrated also the therapeutic action of R on coronary heart disease (Castaldo et al., 2019), lipid profile (Asgary et al., 2019; Springer and Moco, 2019) and obesity (Mousavi et al., 2019; Tabrizi et al., 2020). They demonstrated preventive and therapeutic influence of R on cancers (Rauf et al., 2018; Catalogna et al., 2019), and various cardiovascular diseases (Catalogna et al., 2019; Dyck et al., 2019). Finally, R displays antimicrobial activity against a wide range of bacterial, viral and fungal species (Vestergaard and Ingmer, 2019).

The beneficial properties of R presented in wine can be responsible for the "French Paradox." This term refers to the fact that in north France there is a high intake of saturated fat but low mortality from coronary heart disease compared to other countries where the same high saturated fat intake exists, being the Paradox attributable to high wine consumption (Gambini et al., 2015).

Plants primarily produce R for defense against adverse environmental factors. In animals and humans too, R can not only affect various physiological processes, but also mitigate the adverse effects of some environmental factors on these processes. It is hypothesized that R is able to neutralize epigenomic aberrations induced by a wide array of environmental pollutants (Li et al., 2019). Grape and its biological active component, R, can be a protector against natural toxins (such as mycotoxins, lipopolysaccharide, and triptolide) and chemical toxins (such as antitumors, metals, and carbon tetrachloride), although not all of these protective effects against human intoxication have been confirmed by appropriate clinical trials (Tabeshpour et al., 2018). R via SIRT1 can ameliorate hyperglycemia-induced oxidative stress in renal tubuli (Wang et al., 2017). R, probably acting on oxidative processes, estrogen receptor and the postreceptory pathways, can mitigate dioxin action on carcinoma cells (Beedanagari et al., 2009), toxic effect of manganese (Latronico et al., 2013), arsenic (Cheng et al., 2014) and cadmium (Liu et al., 2015) on neuronal cells, adverse effect of mycotoxin (Sang et al., 2016) and aflatoxin (Ghadiri et al., 2019) on mammary cells and broiler chicken (Sridhar et al., 2015). Therefore, R possess a number of physiological actions which can be useful for prevention, mitigation and treatment of numerous illnesses and toxic effects of environmental contaminants.

4.10.4 Mechanisms of action

R can bind to enzymes and interfere in their catalytic mechanism. Interaction studies demonstrated the binding of R to various targets including tubulin, protein kinase C alpha (PKCα), phosphodiesterase-4D, human oral cancer cell line proteins, DNA sequences having AATT/TTAA segments, protein kinase C alpha, lysine-specific demethylase 1 (Kataria and Khatkar, 2019). Furthermore, R targets adenosine monophosphate kinase, nuclear factor-κB (NFkB), inflammatory cytokines, antioxidant enzymes—regulators of angiogenesis, apoptosis, mitochondrial biogenesis, gluconeogenesis and lipid metabolism (Pannu and Bhatnagar. 2019). These target molecules can be intracellular mediators of R action, although the mediatory role of some of them still require experimental confirmation. Furthermore, R has phytoestrogenic properties, i.e., ability to bind and to affect steroid hormones receptors and to affect steroid-dependent processes and diseases (Gambini et al., 2015; Ortega and Duleba, 2015; Varoni et al., 2016; Nguen et al., 2017; van Duursen, 2017a,b). Moreover, R can bind and affect DNA methyltransferases and associated proteins responsible for DNA methylation resulting in activation of a number of genes (Mirza et al., 2013). DNA methylation can be involved in epigenetic regulation of oxidative, metabolic, inflammatory, angiogenic, and tumorgenic processes (Li et al., 2019). Furthermore, R can affect miRNAs regulating numerous physiological and pathological processes from inflammation to cancer (Li et al., 2019). Antioxidant properties of R can define its antiaging effect and ability to affect apoptosis, growth, and cancerogenesis (Gambini et al., 2015; Ortega and Duleba, 2015; Varoni et al., 2016; Nguen et al., 2017). Finally, R is a known activator of Sirtuin 1(SIRT1), a NAD+-dependent histone deacetylase which deacetylates proteins that contribute to oxidative stress, aging, metabolism, obesity, and tumors, as well as to DNA methylation (Li et al., 2019). Therefore, R can affect a wide array of cellular processes via multiple intracellular signaling pathways.

4.10.5 Effects on female reproductive processes

4.10.5.1 Effect on ovarian and reproductive state and on ovarian cell functions

There is a number of evidences for R involvement in control of female reproduction at different regulatory levels, although the available reports concerning pattern of R action are sometimes contradictory. Some publications demonstrate inhibitory influence of R on female reproductive processes. R was able to alter sociosexual behavior (Henry and Witt, 2006), disrupt ovarian cycle (Henry and Witt, 2006; Ortega and Duleba, 2015), reduce ovarian weight (Ozgur et al., 2018), ovarian follicular growth (Cabello et al., 2015), number of ovarian follicles (Ergenoglu et al., 2015), reduce DNA synthesis (Wong et al., 2010), cell proliferation (Basini et al., 2010; Ortega and Duleba, 2015; Sirotkin, 2019a), mitochondrial activity in ovarian cells (Macedo et al., 2017), ovarian cell viability (Wong et al., 2010; Morita et al., 2012), as well as to promote apoptosis (Wong et al., 2010; Morita et al., 2012; Ortega and Duleba, 2015; Macedo et al., 2017; Hao et al., 2018; Sirotkin et al., 2019a,b) and autophagy (Sugiyama et al., 2015) in ovarian cells. Some of suppressive effects of R on ovarian cells can be associated with downregulation of their antioxidant enzymes (Macedo et al., 2017). In the endometrium, R can be suppressing decidual and senescent changes of endometrial cells, which is essential for embryo implantation and placentation. Moreover, R may also induce deacetylation of important decidual-related genes (Ochiai and Kuroda, 2019).

Other researchers however have found no evidence of R influence on CNS structures that can be responsible for reproduction (Henry and Witt, 2006), on fecundity (Bódis et al., 2019) or on ovarian cell apoptosis (Morita et al., 2012). Moreover, several authors reported that R administration had stimulatory effects on ovarian reserve and folliculogenesis (Chen et al., 2010; Kong et al., 2011; Liu et al., 2013; Ortega and Duleba, 2015; Bezerra et al., 2018; Li and Liu, 2018; Ozatik et al., 2020), on growth of cultured ovarian follicles (Hao et al., 2018), ovarian weight (Henry and Witt, 2006), on number of ovarian cells (Ortega et al., 2012) and of ovarian oocytes (Kong et al., 2011; Liu et al., 2013), oocyte maturation and embryo production (Liu et al., 2013; Wang et al., 2014), on ovarian cell proliferation (Bezerra et al., 2018; Sirotkin et al., 2019b) and on expression of anti-apoptotic peptide bcl-2 (Li and Liu, 2018). Furthermore, R administration reduced follicular atresia (Chen et al., 2010; Kong et al., 2011), expression of promoters and markers of apoptosis (Ortega et al., 2012; Bezerra et al., 2018; Li and Liu, 2018; Baláži et al., 2019; Said et al., 2019; Ozatik et al., 2020).

Therefore, the available data demonstrate R influence on ovarian cell proliferation and apoptosis and the resulted processes — ovarian cell viability, folliculogenesis, oogenesis and fecundity, but reports concerning character of R influence on these processes remain inconsistent.

4.10.5.2 Effect on reproductive hormones

Ovarian cell proliferation, apoptosis, viability, as well as ovarian folliculogenesis, ovulation and fecundity are regulated by hormones of extra- and intraovarian origin (Sirotkin, 2014; Holesh et al., 2020). No R influence on the CNS structures regulating reproduction (Henry and Witt, 2006) and key extraovarian hormonal stimulators, gonadotropins has been found (Banaszewska et al., 2016; Ozatik et al., 2020). Therefore, involvement of R on reproduction via hepatothalamo-hypophysia hormones has not been demonstrated yet.

On the other hand, R influence on peripheral and especially on ovarian hormones have been well documented, although the corresponding available information is inconsistent too. Some authors reported that administration of R led to reductions in the levels of plasma anti-Mullerian hormone, insulin-like growth factor 1 (Ergenoglu et al., 2015), insulin (Cabello et al., 2015; Banaszewska et al., 2016) and VEGF (Basini et al., 2010; Ozgur et al., 2018). In the experiments of other authors however R did not alter (Ortrega et al., 2012) or increased (Said et al., 2019) release of anti-Mullerian hormone and VEGF. R boosted release of prostaglandin F by cultured ovarian cells (Sirotkin et al., 2020).

The available data concerning the effects of R on ovarian steroidogenesis are also contradictory. Several studies showed that R promotes the expression of ovarian genes encoding steroidogenic enzymes (Morita et al., 2012), progesterone (P) (Kolesarova et al., 2012; Morita et al., 2012; Wang et al., 2014; Baláži et al., 2019), testosterone (T) (Baláži et al., 2019; Sirotkin et al., 2019a, Fig. 4.3) and estradiol (E) (Basini et al., 2010; Sirotkin, 2019a; Ozatik et al., 2020) production. Other studies however concluded that R has no effect on ovarian P release (Ortega et al., 2012). Moreover, a number of studies demonstrate the ability of R to suppress the expression of steroidogenic enzymes (Ortega et al., 2012; Cabello et al., 2015; Ortega and Duleba, 2015) and ovarian progesterone (Basini et al., 2010; Sirotkin 2019a, Sirotkin et al., 2021), testosterone (Banaszewska et al., 2016) and estradiol secretion (Kasap et al., 2016).

4.10.5.3 Possible causes of variability in resveratrol effects

Taken together, 14 available publications demonstrated an inhibitory, 3—no effect, and 14—a stimulatory action of R on ovarian growth, cycle, ovarian cell proliferation, apoptosis and oviduct functions. The similar distribution occurs in the reports concerning character of R action on ovarian hormones: 12 publications demonstrated the inhibitory action, 2—no effect, and 9 papers reported the stimulatory influence of R on ovarian hormones release. The 4 of 5 related publications suggest that R can prevent the action of hormonal stimulators on the ovarian functions, but one suggest the opposite, stimulatory action of R on gonadotropin reception.

The contradiction in reports concerning character of R action on various parameters could be due to the differences in the animal species used and

Plant molecules and their influence on health **Chapter | 4** **377**

experimental models utilized (living animals, granulosa cells, ovarian follicles or ovarian fragments, oocytes) as well as in animal age, R doses used or due to the differences in the initial state of ovarian cells, including the stage of their luteinisation. For example, in the experiments of Ortega et al. (2012) R, when added at low doses boosted, but given at high dose suppressed proliferation of ovarian cells. R administration affected reproductive behavior and ovarian cycles in adult rats, but not in their pups (Henry and Witt, 2006). In our similar experiments on porcine ovarian granulosa cells, R addition was able either reduce (Sirotkin et al., 2019a) or promote (Sirotkin et al., 2020) progesterone secretion.

Some publications described opposite R-induced changes in some parameters, which are expected to be synergists. For example, in experiments of Morita et al. (2012), R reduced ovarian cell apoptosis, but it resulted in a decrease in their viability. In experiments of Han et al. (2017) R induced apoptosis in follicular cells, but promoted growth of whole ovarian follicles. Sugiyama et al. (2015) reported that R promoted autophagy of oocytes, but increased generation of ATP, mitochondria, as well as the developmental capacity of oocytes. In some experiments of Sirotkin et al. (2019b) R boosted accumulation of both proliferation and apoptosis markers. In these cases, changes in cell apoptosis could be neutralized by the corresponding changes in cell proliferation.

Therefore, character of R action on female reproductive processes could depend of species, age, target cells and R dose. Furthermore, the existence of other unknown factors, which determine response to R, might be proposed. It might be hypothesized that some differences in character of R effect can be due to different signaling pathways, mediating R action on reproductive processes.

4.10.6 Mechanisms of action on female reproductive processes

R can affect physiological functions through numerous signaling pathways (see earlier) but the current publications indicate that only some of these pathways could mediate R action on reproduction. Moreover, the mediators of R action on female reproductive processes require further experimental validation: sometimes they have been hypothesized mainly on the basis of indirect data, because they affect reproduction and are affected by R (see review of Jozkowiak et al., 2020). R is a known promoter of SIRT1 in ovarian cells (Morita et al., 2012; Wang et al., 2014; Sugiyama et al., 2015; Li and Liu, 2018; Sirotkin et al., 2019a,b, see also Figs. 4.2 and 4.3). SIRT1 in turn can control sexual maturation and female reproduction at the hypothalamo-hypophysial system level (Kyselova et al., 2003) and by direct action on ovarian cell functions (Pavlova et al., 2013; Zhang et al., 2013; Sirotkin et al., 2014, 2015; Sirotkin, 2016). There is a close functional mutual interrelationship between SIRT-1 and inflammation-related transcription factor NF-κB in the ovary (Pavlová et al., 2013; Han et al., 2017). NF-κB is an important regulator of ovarian cell functions like proliferation, apoptosis, and release of peptide, steroid hormones and prostaglandin and response to FSH (Pavlová et al., 2011, 2013), ovarian tumorgenesis (Harrington and Annunziata, 2019), endometriosis and infertility (Donnez et al., 2016). R can either upregulate (Said et al., 2019) or downregulate (Han et al., 2017) SIRT-1/NF-κB system and NF-κB-dependent inflammatory processes. Furthermore, SIRT-1, which is under control of R, is a potent regulator of secretion of steroid hormones (Pavlova et al., 2013; Sirotkin et al., 2014), the most known regulators of ovarian functions, as well as of gonadotropins production and effects (Sirotkin, 2014; Holesh et al., 2020). In addition, R as phytoestrogen (van Duursen, 2017), can directly influence steroid hormones production, reception and metabolism and, therefore, steroids-dependent ovarian functions—ovarian cell proliferation, apoptosis, folliculogenesis and fecundity (Sirotkin, 2014; Banu et al., 2016; Holesh et al., 2020; Jozkowiak et al., 2020). R can affect peptide hormones reception (Morita et al., 2012) and ovarian cell response to these hormones (Sirotkin, 2019a). Moreover, R can affect telomerase, which is responsible for ovarian cell aging (Liu et al., 2013). Finally, R can impair oxidative stress—inductor of apoptosis, aging, and tumorgenesis of ovarian cells (Ortega and Duleba, 2015; Banaszewska et al., 2016; Banu et al., 2016; Ochiai and Kuroda, 2019). Therefore, the influence of R on female reproductive processes can be mediated by multiple interchanging signaling mechanisms. It is however necessary to keep in mind that the involvement, role and hierarchy of particular signaling pathways requires further studies.

4.10.7 Application in reproductive biology and medicine

Both inhibitory and stimulatory effects of R can be useful for its application in control of animal and human female reproductive processes, mitigation and treatment of their disorders.

The ability of R to increase the expression of ovarian LH receptors (Morita et al., 2012) indicates that R could improve induction of ovulation by gonadotropin in animal production and assisted reproduction. The stimulatory action of R on reproductive organs and their hormonal promoters listed earlier suggests that R can be applicable in animal production, biotechnology and assisted reproduction for improvement culture of ovarian cell and tissues (Sugiyama et al., 2015; Macedo et al., 2017; Bezerra et al., 2018; Hao et al., 2018), prediction of success of oocyte *in vitro* fertilization (Liu et al., 2013; Wang et al., 2014; Sugiyama et al., 2015; Bódis et al., 2019), ovarian stimulation (Cabello et al., 2015), ovarian follicular growth and oocyte maturation (Kong et al., 2011; Wang et al., 2014; Sugiyama et al., 2015), embryo production (Wang et al., 2014), improvement of laboratory (Kyselova et al., 2003), and farm (Balazi et al., 2019) animal reproduction and prevention of reproductive aging and exhaustion of ovarian follicular reserve (Liu et al., 2013; Sugiyama et al., 2015; Li and Liu, 2018; Ozatik et al., 2020).

On the other hand, the inhibitory effects of R on reproductive processes suggest that R can be potentially applicable for prevention of gonadotropin-dependent ovarian hyperstimulation (Kolesarova et al., 2012; Cabello et al., 2015; Ozgur et al., 2018), probably via mitigation the stimulatory action of FSH and its mediator, IGF-I, on basic ovarian cell functions (Sirotkin et al., 2019). Thus, R can suppress female reproductive processes and the ovarian response to endogenous and exogenous hormonal stimulators. This potentially adverse effect of R and R-containing food should be taken into account by reproduction of farm animals and humans in animal biotechnology and assisted reproduction.

Moreover, R can be useful to prevent the adverse effects of some environmental factors and pollutants. There is evidence that R acting via antioxidant enzymes and/or steroid production, reception or metabolism can protect the ovary from the toxic effect of hypoxia/ischemia (Hascalik et al., 2004), dioxin (Beedanagari et al., 2009), irradiation (Simsek et al., 2012), mangan (Latronico et al., 2013), nanoparticles (Guo et al., 2015; Sirotkin et al., 2021), hexavalent chromium (Banu et al., 2016), mycotoxins deoxynivalenol (Kolesarova et al., 2012), and zearalenone (Sang et al., 2016), chemotherapeutic cisplatin (Said et al., 2019) and benzene (Sirotkin et al., 2020). The ability of R to mitigate action of a number of adverse environmental factors on ovarian functions makes it a promising natural protector and remedy against development of environment-induced female reproductive disorders.

R is considered as a promising tool for prevention and treatment of ubiquitous reproductive disorders whose environmental trigger is unknown: infertility (Ortega and Duleba, 2015; van Duursen, 2017; Ochiai and Kuroda, 2019), age-dependent ovarian failure (Chen et al., 2010; Liu et al., 2013), ovarian cancer (Varoni et al., 2016; Rauf et al., 2018), endometriosis (Ochiai and Kuroda, 2019), hormonal dysfunctions and polycystic ovarian syndrome (Ergenoglu et al., 2015; Banaszewska et al., 2019; Ochiai and Kuroda, 2019).

These protective and curative actions of R could be due to both suppressive (for example, in cases of cancer or ovarian hyperstimulation) and stimulatory (in cases of infertility, ovarian failure, promotion of reproductive processes *in vivo* and *in vitro*) effects of R on ovarian cells and reproductive hormones.

Wide application of R is however limited by its poor bioavailability and fast elimination from the circulation (Gambini et al., 2015; Intagliata et al., 2019; Jozkowiak et al., 2020). Effectiveness of R could be improved by its incorporation into different delivery systems - liposomes, nanoemulsions, micelles, insertion into polymeric particles, solid dispersions, and nanocrystals. For example, resveratrol-bovine serum albumin nanoparticles could be applicable for induction of ovarian cancer cell death (Guo et al., 2015; Jozkowiak et al., 2020).

The other approach is chemical modifications and generation of R synthetic derivatives containing different substituents, such as methoxylic, hydroxylic groups, or halogens on the R aromatic rings with improved pharmacokinetics and stability (Chimento et al., 2019; Intagliata et al., 2019). For example, Basini et al. (2010) synthesized hydroxylated and methylated R analogues which could be applicable for suppression of malignant ovarian cell proliferation and angiogenesis/VEGF production. These analogues revealed differences in ability to inhibit cell proliferation, but not in ability to suppress VEGF and to promote steroidogenesis.

Therefore, influence of R via numerous mediators on various reproductive processes makes it promising molecule for control of reproduction and treatment of reproductive disorders in phytotherapy, animal production, medicine, biotechnology and assisted reproduction. Nevertheless, a wide application of R is limited by deficit of information based on direct experimental and clinical data concerning character and mechanisms of its action, as well as of adequate R forms and methodology of its delivery and application.

4.10.8 Conclusion and possible directions of the further studies

Taken together, the available data demonstrate numerous physiological effect of plant defense molecule R on animal and human physiological processes.

It can be useful for stimulation of reproductive processes at different regulatory levels, for prevention and mitigation of numerous illnesses, as well as of disorders induced by inadequate gonadotropin treatment and toxic environmental contaminants. These effects could be mediated by multiple intracellular mechanisms including extra- and intracellular receptors, protein kinases and other enzymes, proteins related to cell cycle, oxidative stress, DNA methylation and others.

At present a number of evidence concerning R influence on female reproductive processes including ovarian folliculo- and oogenesis, ovarian cycle, state of decidual cells, embryogenesis and fecundity. The main R action is probably not mediated by central mechanisms, but rather by ovarian peptide and steroid hormones and growth factors or their receptors and postreceptor signaling pathways. These molecules are able to affect ovarian cell differentiation, equilibration between cell proliferation, apoptosis and senescence, which in turn define ovarian cell viability, growth, development and ovulation of ovarian follicles, oogenesis and embryogenesis. It is not to be excluded that the numerous intracellular mechanisms of R action detected in nonovarian cells can mediate R action on the ovary and embryo too. Unfortunately, reproductive effects of R are studied poorly in comparison with protective and curative R action on metabolism, cancer, cardiovascular, and other nonreproductive disorders. Furthermore, the available information concerning character and, therefore, the applicability of R in reproductive biology and medicine is contradictory. From one side, a number of evidence demonstrates suppressive R action on reproduction at various regulatory levels. The similar number of publications however show the stimulatory and protective action of R, even in the same species. Understanding the causes of these differences, and search for factors, which determine character of response to R, can help to achieve the desirable R effect.

The information concerning targets and mechanisms of R action on nonreproductive processes could provide the new ideas, concepts and approaches concerning study and application of R in reproductive biology and medicine. For example, the R was able to affect not only nonreproductive (Wang et al., 2017), but also ovarian cells (Sirotkin et al., 2019a,b) via SIRT-1 and its target—mTOR system. Therefore, reproductive processes can be affected not only via R, but also by synthetic mTOR regulators with higher bioavailability, stability and efficiency, than R. On the other hand, the search for the adequate approach to increase intake and bioavailability of dietary R could be more reasonable way of its application for control of both reproductive and nonreproductive processes.

Taken together, the analysis of the available information suggests that R has a number of properties which enable its influence on various physiological processes including female reproduction at various regulatory levels via various extra- and intracellular signaling pathways. Despite insufficient information concerning factors determining character of R action, multiple R

382 Contaminants and Plants Action on Female Reproduction

effects and mechanisms of its action, and low bioavailability of natural R, the available data indicates applicability of R or its analogues with either stimulatory or inhibitory actions as a promising tool for control of various reproductive and nonreproductive processes and treatment of their disorders.

References

Asgary S, Karimi R, Momtaz S, Naseri R, Farzaei MH. Effect of resveratrol on metabolic syndrome components: a systematic review and meta-analysis. Rev Endocr Metab Disord 2019;20(2):173−86. https://doi.org/10.1007/s11154-019-09494-z.

Baláži A, Sirotkin AV, Földešiová M, et al. Green tea can supress rabbit ovarian functions in vitro and in vivo. Theriogenology 2019;127:72−9. https://doi.org/10.1016/j.theriogenology.2019.01.010.

Banaszewska B, Wrotyńska-Barczyńska J, Spaczynski RZ, Pawelczyk L, Duleba AJ. Effects of resveratrol on polycystic ovary syndrome: a double-blind, randomized, placebo-controlled trial. J Clin Endocrinol Metab 2016;101(11):4322−8. https://doi.org/10.1210/jc.2016-1858.

Banaszewska B, Pawelczyk L, Spaczynski R. Current and future aspects of several adjunctive treatment strategies in polycystic ovary syndrome. Reprod Biol 2019;19(4):309−15. https://doi.org/10.1016/j.repbio.2019.09.006.

Banu SK, Stanley JA, Sivakumar KK, Arosh JA, Burghardt RC. Resveratrol protects the ovary against chromium-toxicity by enhancing endogenous antioxidant enzymes and inhibiting metabolic clearance of estradiol. Toxicol Appl Pharmacol 2016;303:65−78. https://doi.org/10.1016/j.taap.2016.04.016.

Basini G, Tringali C, Baioni L, Bussolati S, Spatafora C, Grasselli F. Biological effects on granulosa cells of hydroxylated and methylated resveratrol analogues. Mol Nutr Food Res 2010;54(Suppl 2):S236−43. https://doi.org/10.1002/mnfr.200900320.

Beedanagari SR, Bebenek I, Bui P, Hankinson O. Resveratrol inhibits dioxin-induced expression of human CYP1A1 and CYP1B1 by inhibiting recruitment of the aryl hydrocarbon receptor complex and RNA polymerase II to the regulatory regions of the corresponding genes [published correction appears in Toxicol Sci. 2010 Aug;116(2):693] Toxicol Sci 2009;110(1):61−7. https://doi.org/10.1093/toxsci/kfp079.

Bezerra MÉS, Gouveia BB, Barberino RS, et al. Resveratrol promotes in vitro activation of ovine primordial follicles by reducing DNA damage and enhancing granulosa cell proliferation via phosphatidylinositol 3-kinase pathway. Reprod Domest Anim 2018;53(6):1298−305. https://doi.org/10.1111/rda.13274.

Bódis J, Sulyok E, Kőszegi T, Gödöny K, Prémusz V, Várnagy Á. Serum and follicular fluid levels of sirtuin 1, sirtuin 6, and resveratrol in women undergoing in vitro fertilization: an observational, clinical study. J Int Med Res 2019;47(2):772−82. https://doi.org/10.1177/0300060518811228.

Cabello E, Garrido P, Morán J, González del Rey C, Llaneza P, Llaneza-Suárez D, Alonso A, González C. Effects of resveratrol on ovarian response to controlled ovarian hyperstimulation in ob/ob mice. Fertil Steril 2015;103(2):570e1−9e1. https://doi.org/10.1016/j.fertnstert.2014.10.034.

Castaldo L, Narváez A, Izzo L, et al. Red wine consumption and cardiovascular health. Molecules 2019;24(19):3626. https://doi.org/10.3390/molecules24193626.

Catalogna G, Moraca F, D'Antona L, et al. Review about the multi-target profile of resveratrol and its implication in the SGK1 inhibition. Eur J Med Chem 2019;183:111675. https://doi.org/10.1016/j.ejmech.2019.111675.

Chen ZG, Luo LL, Xu JJ, Zhuang XL, Kong XX, Fu YC. Effects of plant polyphenols on ovarian follicular reserve in aging rats. Biochem Cell Biol 2010;88(4):737–45. https://doi.org/10.1139/O10-012.

Cheng Y, Xue J, Jiang H, et al. Neuroprotective effect of resveratrol on arsenic trioxide-induced oxidative stress in feline brain. Hum Exp Toxicol 2014;33(7):737–47. https://doi.org/10.1177/0960327113506235.

Chimento A, De Amicis F, Sirianni R, et al. Progress to improve oral bioavailability and beneficial effects of resveratrol. Int J Mol Sci 2019;20(6):1381. https://doi.org/10.3390/ijms20061381.

Donnez J, Donnez O, Orellana R, Binda MM, Dolmans MM. Endometriosis and infertility. Panminerva Med 2016;58(2):143–50.

Dyck GJB, Raj P, Zieroth S, Dyck JRB, Ezekowitz JA. The effects of resveratrol in patients with cardiovascular disease and heart failure: a narrative review. Int J Mol Sci 2019;20(4):904. https://doi.org/10.3390/ijms20040904.

Ergenoglu M, Yildirim N, Yildirim AG, et al. Effects of resveratrol on ovarian morphology, plasma anti-mullerian hormone, IGF-1 levels, and oxidative stress parameters in a rat model of polycystic ovary syndrome. Reprod Sci 2015;22(8):942–7. https://doi.org/10.1177/1933719115570900.

Gambini J, Inglés M, Olaso G, et al. Properties of resveratrol: in vitro and in vivo studies about metabolism, bioavailability, and biological effects in animal models and humans. Oxid Med Cell Longev 2015;2015:837042. https://doi.org/10.1155/2015/837042.

Ghadiri S, Spalenza V, Dellafiora L, et al. Modulation of aflatoxin B1 cytotoxicity and aflatoxin M1 synthesis by natural antioxidants in a bovine mammary epithelial cell line. Toxicol Vitro 2019;57:174–83. https://doi.org/10.1016/j.tiv.2019.03.002.

Guo L, Peng Y, Li Y, et al. Cell death pathway induced by resveratrol-bovine serum albumin nanoparticles in a human ovarian cell line. Oncol Lett 2015;9(3):1359–63. https://doi.org/10.3892/ol.2015.2851.

Han Y, Luo H, Wang H, Cai J, Zhang Y. SIRT1 induces resistance to apoptosis in human granulosa cells by activating the ERK pathway and inhibiting NF-κB signaling with anti-inflammatory functions. Apoptosis 2017;22(10):1260–72. https://doi.org/10.1007/s10495-017-1386-y.

Hao J, Tuck AR, Sjödin MOD, et al. Resveratrol supports and alpha-naphthoflavone disrupts growth of human ovarian follicles in an in vitro tissue culture model. Toxicol Appl Pharmacol 2018;338:73–82. https://doi.org/10.1016/j.taap.2017.11.009.

Harrington BS, Annunziata CM. NF-κB signaling in ovarian cancer. Cancers 2019;11(8):1182. https://doi.org/10.3390/cancers11081182.

Hascalik S, Celik O, Turkoz Y, et al. Resveratrol, a red wine constituent polyphenol, protects from ischemia-reperfusion damage of the ovaries. Gynecol Obstet Invest 2004;57(4):218–23. https://doi.org/10.1159/000076760.

Henry LA, Witt DM. Effects of neonatal resveratrol exposure on adult male and female reproductive physiology and behavior. Dev Neurosci 2006;28(3):186–95. https://doi.org/10.1159/000091916.

Holesh JE, Bass AN, Lord M. Physiology, ovulation. In: StatPearls. Treasure Island (FL): StatPearls Publishing; 2020.

Intagliata S, Modica MN, Santagati LM, Montenegro L. Strategies to improve resveratrol systemic and topical bioavailability: an update. Antioxidants 2019;8(8):244. https://doi.org/10.3390/antiox8080244.

Jozkowiak M, Hutchings G, Jankowski M, et al. The stemness of human ovarian granulosa cells and the role of resveratrol in the differentiation of MSCs-A review based on cellular and molecular knowledge. Cells 2020;9(6):1418. https://doi.org/10.3390/cells9061418.

384 Contaminants and Plants Action on Female Reproduction

Kasap E, Turan GA, Eskicioğlu F, et al. Comparison between resveratrol and cabergoline in preventing ovarian hyperstimulation syndrome in a rat model. Gynecol Endocrinol 2016;32(8):634−40. https://doi.org/10.3109/09513590.2016.1152575.

Kataria R, Khatkar A. Molecular docking, synthesis, kinetics study, structure-activity relationship and ADMET analysis of morin analogous as *Helicobacter pylori* urease inhibitors. BMC Chem 2019;13(1):45. https://doi.org/10.1186/s13065.

Khodaie N, Tajuddin N, Mitchell RM, Neafsey EJ, Collins MA. Combinatorial preconditioning of rat brain cultures with subprotective ethanol and resveratrol concentrations promotes synergistic neuroprotection. Neurotox Res 2018;34(3):749−56. https://doi.org/10.1007/s12640-018-9886-2.

Kolesarova A, Capcarova M, Maruniakova N, Lukac N, Ciereszko RE, Sirotkin AV. Resveratrol inhibits reproductive toxicity induced by deoxynivalenol. J Environ Sci Health A Tox Hazard Subst Environ Eng 2012;47(9):1329−34. https://doi.org/10.1080/10934529.2012.672144.

Kong XX, Fu YC, Xu JJ, Zhuang XL, Chen ZG, Luo LL. Resveratrol, an effective regulator of ovarian development and oocyte apoptosis. J Endocrinol Invest 2011;34(11):e374−81. https://doi.org/10.3275/7853.

Kyselova V, Peknicova J, Buckiova D, Boubelik M. Effects of p-nonylphenol and resveratrol on body and organ weight and in vivo fertility of outbred CD-1 mice. Reprod Biol Endocrinol 2003;1:30. https://doi.org/10.1186/1477-7827-1-30.

Latronico T, Branà MT, Merra E, et al. Impact of manganese neurotoxicity on MMP-9 production and superoxide dismutase activity in rat primary astrocytes. Effect of resveratrol and therapeutical implications for the treatment of CNS diseases. Toxicol Sci 2013;135(1):218−28. https://doi.org/10.1093/toxsci/kft146.

Li N, Liu L. Mechanism of resveratrol in improving ovarian function in a rat model of premature ovarian insufficiency. J Obstet Gynaecol Res 2018;44(8):1431−8. https://doi.org/10.1111/jog.13680.

Li S, Chen M, Li Y, Tollefsbol TO. Prenatal epigenetics diets play protective roles against environmental pollution. Clin Epigenet 2019;11(1):82. https://doi.org/10.1186/s13148-019-0659-4.

Liu M, Yin Y, Ye X, Zeng M, Zhao Q, Keefe DL, Liu L. Resveratrol protects against age-associated infertility in mice. Hum Reprod 2013;28(3):707−17. https://doi.org/10.1093/humrep/des437.

Liu C, Zhang R, Sun C, et al. Resveratrol prevents cadmium activation of Erk1/2 and JNK pathways from neuronal cell death via protein phosphatases 2A and 5. J Neurochem 2015;135(3):466−78. https://doi.org/10.1111/jnc.13233.

Macedo TJS, Barros VRP, Monte APO, et al. Resveratrol has dose-dependent effects on DNA fragmentation and mitochondrial activity of ovine secondary follicles cultured in vitro. Zygote 2017;25(4):434−42. https://doi.org/10.1017/S0967199417000193.

Mirza S, Sharma G, Parshad R, Gupta SD, Pandya P, Ralhan R. Expression of DNA methyltransferases in breast cancer patients and to analyze the effect of natural compounds on DNA methyltransferases and associated proteins. J Breast Cancer 2013;16(1):23−31.

Morita Y, Wada-Hiraike O, Yano T, et al. Resveratrol promotes expression of SIRT1 and StAR in rat ovarian granulosa cells: an implicative role of SIRT1 in the ovary. Reprod Biol Endocrinol 2012;10:14. https://doi.org/10.1186/1477-7827-10-14.

Mousavi SM, Milajerdi A, Sheikhi A, et al. Resveratrol supplementation significantly influences obesity measures: a systematic review and dose-response meta-analysis of randomized controlled trials. Obes Rev 2019;20(3):487−98. https://doi.org/10.1111/obr.12775.

Nguyen C, Savouret JF, Widerak M, Corvol MT, Rannou F. Resveratrol, potential therapeutic interest in joint disorders: a critical narrative review. Nutrients 2017;9(1):45. https://doi.org/10.3390/nu9010045.

Plant molecules and their influence on health **Chapter | 4 385**

Ochiai A, Kuroda K. Preconception resveratrol intake against infertility: friend or foe? Reprod Med Biol 2019;19(2):107−13. https://doi.org/10.1002/rmb2.12303.

Ortega I, Duleba AJ. Ovarian actions of resveratrol. Ann N Y Acad Sci 2015;1348(1):86−96. https://doi.org/10.1111/nyas.12875.

Ortega I, Villanueva JA, Wong DH, et al. Resveratrol reduces steroidogenesis in rat ovarian theca-interstitial cells: the role of inhibition of Akt/PKB signaling pathway. Endocrinology 2012;153(8):4019−29. https://doi.org/10.1210/en.2012-1385.

Ozatik FY, Ozatik O, Yigitaslan S, Kaygısız B, Erol K. Do resveratrol and dehydroepiandrosterone increase diminished ovarian reserve? Eurasian J Med 2020;52(1):6−11. https://doi.org/10.5152/eurasianjmed.2019.19044.

Ozgur S, Oktem M, Altinkaya SO, et al. The effects of resveratrol on ovarian hyperstimulation syndrome in a rat model. Taiwan J Obstet Gynecol 2018;57(3):383−8. https://doi.org/10.1016/j.tjog.2018.04.010.

Pannu N, Bhatnagar A. Resveratrol: from enhanced biosynthesis and bioavailability to multi-targeting chronic diseases. Biomed Pharmacother 2019;109:2237−51. https://doi.org/10.1016/j.biopha.2018.11.075.

Pavlová S, Klucska K, Vašíček D, Kotwica J, Sirotkin AV. Transcription factor NF-κB (p50/p50, p65/p65) controls porcine ovarian cells functions. Anim Reprod Sci 2011;128(1−4):73−84. https://doi.org/10.1016/j.anireprosci.2011.09.005.

Pavlová S, Klucska K, Vašíček D, et al. The involvement of SIRT1 and transcription factor NF-κB (p50/p65) in regulation of porcine ovarian cell function. Anim Reprod Sci 2013;140(3−4):180−8. https://doi.org/10.1016/j.anireprosci.2013.06.013.

Rauf A, Imran M, Butt MS, Nadeem M, Peters DG, Mubarak MS. Resveratrol as an anti-cancer agent: a review. Crit Rev Food Sci Nutr 2018;58(9):1428−47. https://doi.org/10.1080/10408398.2016.1263597.

Said RS, Mantawy EM, El-Demerdash E. Mechanistic perspective of protective effects of resveratrol against cisplatin-induced ovarian injury in rats: emphasis on anti-inflammatory and anti-apoptotic effects. Naunyn-Schmiedeberg's Arch Pharmacol 2019;392(10):1225−38. https://doi.org/10.1007/s00210-019-01662-x.

Sang Y, Li W, Zhang G. The protective effect of resveratrol against cytotoxicity induced by mycotoxin, zearalenone. Food Funct 2016;7(9):3703−15. https://doi.org/10.1039/c6fo00191b.

Simsek Y, Gurocak S, Turkoz Y, et al. Ameliorative effects of resveratrol on acute ovarian toxicity induced by total body irradiation in young adult rats. J Pediatr Adolesc Gynecol 2012;25(4):262−6. https://doi.org/10.1016/j.jpag.2012.04.001.

Sirotkin AV, Dekanová P, Harrath AH, Alwasel SH, Vašíček D. Interrelationships between sirtuin 1 and transcription factors p53 and NF-κB (p50/p65) in the control of ovarian cell apoptosis and proliferation. Cell Tissue Res 2014;358(2):627−32. https://doi.org/10.1007/s00441-014-1940-7.

Sirotkin AV, Alexa R, Dekanová P, Kádasi A, Štochmaľová A, Grossmann R, et al. The mTOR system can affect basic porcine ovarian cell functions and mediate the effect of ovarian hormonal regulators. Int J Pharmacol 2015. https://doi.org/10.3923/ijp.2015.

Sirotkin A, Alexa R, Kádasi A, Adamcová E, Alwasel S, Harrath AH. Resveratrol directly affects ovarian cell sirtuin, proliferation, apoptosis, hormone release and response to follicle-stimulating hormone (FSH) and insulin-like growth factor I (IGF-I). Reprod Fertil Dev 2019a;10. https://doi.org/10.1071/RD18425.

Sirotkin AV, Adamcova E, Rotili D, et al. Comparison of the effects of synthetic and plant-derived mTOR regulators on healthy human ovarian cells. Eur J Pharmacol 2019b;854:70−8. https://doi.org/10.1016/j.ejphar.2019.03.048.

Sirotkin A, Kádasi A, Balaží A, Kotwica J, Alwasel S, Harrath AH. The action of benzene, resveratrol and their combination on ovarian cell hormone release. Folia Biol 2020;66(2):67−71.

Sirotkin AV, Alexa R, Stochmalova A, Scsukova S. Plant isoflavones can affect accumulation and impact of silver and titania nanoparticles on ovarian cells. Endocr Regul 2021;55(1):52−60. https://doi.org/10.2478/enr-2021-0007.

Sirotkin AV. Regulators of ovarian functions. New York, USA: Nova Publishers, Inc; 2014. p. 194.

Sirotkin AV. The Role and Application of Sirtuins and mTOR Signaling in the Control of Ovarian Functions. Cells 2016;5(4):42. https://doi.org/10.3390/cells5040042.

Springer M, Moco S. Resveratrol and its human metabolites-effects on metabolic health and obesity. Nutrients 2019;11(1):143. https://doi.org/10.3390/nu11010143.

Sridhar M, Suganthi RU, Thammiaha V. Effect of dietary resveratrol in ameliorating aflatoxin B1-induced changes in broiler birds. J Anim Physiol Anim Nutr 2015;99(6):1094−104. https://doi.org/10.1111/jpn.12260.

Sugiyama M, Kawahara-Miki R, Kawana H, Shirasuna K, Kuwayama T, Iwata H. Resveratrol-induced mitochondrial synthesis and autophagy in oocytes derived from early antral follicles of aged cows. J Reprod Dev 2015;61(4):251−9. https://doi.org/10.1262/jrd.2015-001.

Tabeshpour J, Mehri S, Shaebani Behbahani F, Hosseinzadeh H. Protective effects of Vitis vinifera (grapes) and one of its biologically active constituents, resveratrol, against natural and chemical toxicities: a comprehensive review. Phytother Res 2018;32(11):2164−90. https://doi.org/10.1002/ptr.6168.

Tabrizi R, Tamtaji OR, Lankarani KB, et al. The effects of resveratrol intake on weight loss: a systematic review and meta-analysis of randomized controlled trials. Crit Rev Food Sci Nutr 2020;60(3):375−90. https://doi.org/10.1080/10408398.2018.1529654.

Tian B, Liu J. Resveratrol: a review of plant sources, synthesis, stability, modification and food application. J Sci Food Agric 2020;100:1392−404. https://doi.org/10.1002/jsfa.10152.

van Duursen MBM. Modulation of estrogen synthesis and metabolism by phytoestrogens in vitro and the implications for women's health. Toxicol Res 2017;6(6):772−94. https://doi.org/10.1039/c7tx00184c.

Varoni EM, Lo Faro AF, Sharifi-Rad J, Iriti M. Anticancer molecular mechanisms of resveratrol. Front Nutr 2016;3(8). https://doi.org/10.3389/fnut.2016.00008.

Vestergaard M, Ingmer H. Antibacterial and antifungal properties of resveratrol. Int J Antimicrob Agents 2019;53(6):716−23. https://doi.org/10.1016/j.ijantimicag.2019.02.015.

Wang F, Tian X, Zhang L, et al. Beneficial effect of resveratrol on bovine oocyte maturation and subsequent embryonic development after in vitro fertilization. Fertil Steril 2014;101(2):577−86. https://doi.org/10.1016/j.fertnstert.2013.10.041.

Wang X, Meng L, Zhao L, et al. Resveratrol ameliorates hyperglycemia-induced renal tubular oxidative stress damage via modulating the SIRT1/FOXO3a pathway. Diabetes Res Clin Pract 2017;126:172−81. https://doi.org/10.1016/j.diabres.2016.12.005.

Wong DH, Villanueva JA, Cress AB, Duleba AJ. Effects of resveratrol on proliferation and apoptosis in rat ovarian theca-interstitial cells. Mol Hum Reprod 2010;16(4):251−9. https://doi.org/10.1093/molehr/gaq002.

Chapter 4.11

Rutin

4.11.1 Introduction

Rutin (vitamin P) is a biologically active plant molecule, which can affect numerous nonreproductive and reproductive processes, and which can be promising for treatment of some disorders. Effects of rutin have been summarized in several reviews (Sharma et al., 2013; Ganeshpurkar and Saluja, 2017; Yong et al., 2020; Negahdari et al., 2021). Nevertheless, the main previous studies and the resulting reviews were focused on application of rutin as a drug for treatment of particular illnesses – diabetes (Ghorbani, 2017), cancer (Farha et al., 2020; Imani et al., 2020; Nouri et al., 2020) and neuro-degenerative disorders (Habtemariam et al., 2016; Enogieru et al., 2018; Budzynska et al., 2019). To our knowledge, the available knowledge concerning rutin action on female reproductive processes has not been summarized yet. The aim of the present review was to outline the nonreproductive effects of rutin and the mechanisms of its action and to integrate the existing knowledge concerning the action of rutin on female reproductive processes (ovarian folliculogenesis and ovarian cycle, oogenesis, embryogenesis, fecundity, ovarian cell state, reproductive hormones and their receptors) and possible signaling pathways mediating rutin action on these processes.

4.11.2 Provenance and properties

Rutin (3,30, 40,5,7-pentahydroxyflavone-3-rhamnoglucoside called also vitamin P, rutoside, quercetin-3-rutinoside, sophorin) is a flavonol-type polyphenol, which consists of the flavonol9c aglycone quercetin and the disacharide rutinose (Ganeshpurkar and Saluja, 2017; Budzynska et al., 2019, see Fig. 4.11). Rutin is abundant in numerous plants, including buckwheat, oat, tea, asparagus, apple, apricots, cherries, grapes, grapefruit, plums, oranges, pomegranate, passion flower, figs and *Ruta graveolens*, which name "rutin" comes from (Ganeshpurkar and Saluja, 2017; Enogieru et al., 2018; Latos-Brozio et al., 2019; Nouri et al., 2020).

Rutin solubility in water, accumulation in target tissues and bioavailability are relatively low (lower, than those of its analogue quercetin), and its

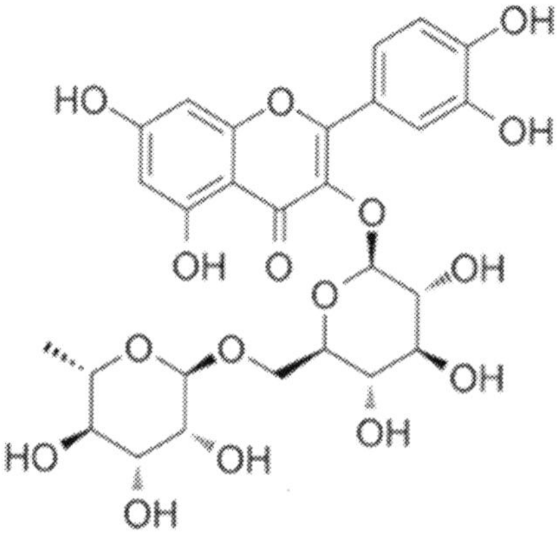

388 Contaminants and Plants Action on Female Reproduction

FIGURE 4.11 The chemical structure of rutin (Quercetin 3-O-rutinoside). *From: https://www.medchemexpress.com/Rutin.html.*

metabolism and the excretion of it and its metabolites by urine and feces are relatively high (Sharma et al., 2013; Ganeshpurkar and Saluja, 2017; Budzynska et al., 2019; Long et al., 2020; Luca et al., 2020). Already several hours after rutin ingestion, human plasma contained no rutin, but only its metabolites (glucuronides and/or sulfates) (Erlund et al., 2000). In the organism, rutin is hydrolized by rutinosidase (Suzuki et al., 2021) and metabolized by both intestine and gut microflora (Luca et al., 2020). Some rutin metabolites could have higher bioavailability and biological actions than rutin itself (Luca et al., 2020).

The aqueous solubility, bioavailability and biological and therapeutical efficiency of rutin could be increased by several approaches:

- cultivating plants containing high amount of both rutin and its hydrolizing enzyme rutinosidase, which catalyzes production of biological active rutin metabolites (Suzuki et al., 2021)
- synthesis of chemical analogues of rutin (for example, by using its glycosylation, Slámová et al., 2018; polymerization, Latos-Brozio et al., 2019, generation of its hybrid composites, Simunkova et al., 2019; Parhi et al., 2020)
- synthesis of rutin carriers—capsules (Aditya et al., 2017), liposomes (Negahdari et al., 2021) or nanocrystals (Liu et al., 2018; Lai et al., 2019)
- rutin transdermal administration, which can enable direct entry of rutin to blood and improve the uptake of rutin by target organs (Liu et al., 2018)

- rutin intranasal administration, which enables easy crossing of the blood—brain barrier and application of rutin for treatment of brain disorders (Long et al., 2020).

4.11.3 Physiological actions

The basic, preclinical and clinical studies (see reviews of Sharma et al., 2013; Ganeshpurkar and Saluja, 2017; Enogieru et al., 2018; Budzynska et al., 2019; Yong et al., 2020) indicated numerous physiological and medicinal effects of rutin. It is a good neuroprotector with sedative and anticonvulsant effects, ability to promote neural crest survival, to prevent neuroinflammation and the related dysfunctions like Alzheimer's disease, Parkinson's disease, Huntington's disease, hyperkinetic movement disorder, depressions and stroke. It has analgesic and antinociceptive actions. Rutin has antihyperglycemic and antidiabetic action. It can affect and protect the nervous system, retina, lung, cardiovascular system, spleen, liver, blood vessels and promote general reparative processes involved in wound healing. These abilities define the rutin ability to prevent and treat numerous disorders like diabetes, thyroid hypofunctions, hypercholesterinemia, hypertension, blood coagulation and thrombosis, ulcer, asthma. It protects bones against osteoporosis and osteopenia. Well-known are its anticancer, antibacterial, antifungal, antiviral, larvicidal, antimalarial, and antifatiqueal properties. It can promote diuresis, improve hair and skin, promote functions of male reproductive system and sperm quality. Negahdari et al. (2021) reviews antimicrobial, antifungal, and antiallergic actions of rutin, and Yong et al. (2020)—its applicability for treatment of varicosities, haemorrhoids and internal haemorrhage. Rutin was able to ameliorate toxic action of chemotherapy on rat reproductive system and sperm quality (Aksu et al., 2017). Finally, rutin promoted longevity, as least in *Drosophila melanogaster* (Chattopadhyay et al., 2017).

Performed studies did not show carcinogenicity of rutin. Moreover, it can increase the efficiency of anticancer chemotherapy. On the other hand, it can interfere with some drugs by induction of the drug metabolizing enzymes in the liver (Ganeshpurkar and Saluja, 2017).

As it was mentioned earlier, the biological and medicinal action of rutin can be limited by its low solubility, bioavailability and intensive metabolism (Sharma et al., 2013; Ganeshpurkar and Saluja, 2017; Budzynska et al., 2019; Long et al., 2020; Luca et al., 2020).

The available data demonstrate the multiple physiological and medicinal effects of rutin. These effects can depend on the target and state of rutin.

4.11.4 Mechanisms of action

The ability of rutin to suppress development of neurodegenerative disorders, cancer and bacterial cells could be determined by its antioxidant and antiinflammatory, and metal-chelating activities, as well as by its ability to affect regulators of cell proliferation and apoptosis (Habtemariam et al., 2016; Enogieru et al., 2018; Budzynska et al., 2019; Simunkova et al., 2019; Farha et al., 2020; Yong et al., 2020). For example, it can downregulate the proinflammatory transcription factor NF-kB, proinflammatory cytokines interleukines-1b and -18, activate antioxidant enzymes and proliferation-promoting mitogen-activated protein kinase-lipid kinase and tyrosine kinase-dependent signaling pathways, downregulation of mRNA expression of proapoptotic genes, upregulation of the ion transport and antiapoptotic genes, activation of antiproliferative and proapoptotic transcription factor p53, induction of autophagy, and restoration of the activities of mitochondrial complex enzymes (Enogieru et al., 2018; Yi et al., 2018; Simunkova et al., 2019; Farha et al., 2020; Imani et al., 2020; Nouri et al., 2020).

In addition, the anticancer action of rutin could be explained by its influence on the interplay between oncogenic and tumor suppressive transcription factors like FOXM1, NF-kB, STAT3, Wnt/β-Catenin, HIF-1α, NRF2, TNF alpha, chain 3/Beclin, androgen, and estrogen receptors (Rajagopal et al., 2018; Imani et al., 2020; Nouri et al., 2020). Furthermore, rutin can suppress tumor angiogenesis via downregulation of VEGF, growth factor stimulating vascular development (Nouri et al., 2020).

Rutin action on blood vessels and cardiovascular functions and disorders could be mediated by upregulation of prostaglandin synthesis and release (Chen et al., 2020) and via attenuating platelet aggregation and capillary permeability (Parhi et al., 2020).

The protective action of rutin on male testicular state and sperma was associated with increased activity of antioxidant enzymes and reduced apoptosis of testicular cells (Aksu et al., 2017).

The antihyperglicemic and antidiabetic effects of rutin can be explained by its ability to decrease carbohydrates' absorption from the small intestine, inhibits of tissue gluconeogenesis, and increase tissue glucose uptake, stimulate of insulin secretion from beta cells, and protect Langerhans islet against degeneration. Rutin also decreases the formation of sorbitol, reactive oxygen species, advanced glycation end-product precursors, and inflammatory cytokines. These effects are considered to be responsible also for the protective effect of rutin against hyperglycemia- and dyslipidemia-induced nephropathy, neuropathy, liver damage, and cardiovascular disorders (Ghorbani, 2017).

Rutin can exert its antiviral effects through inhibition of viral neuraminidase, proteases and DNA/RNA polymerases, as well as due to the modification of various viral proteins (Ninfali et al., 2020).

The rutin-induced improvement of Drosophila lifespan was associated with changes in expression of genes regulating cell apoptosis and viability—elevation in dFoxO, MnSod, Cat, dTsc1, dTsc2, Thor, dAtg1, dAtg5 and dAtg7, and reduced transcript levels of dTor genes (Chattopadhya et al., 2017). Nevertheless, it remains to be established whether these rutin-dependent molecules are real mediators of rutin action.

These observations demonstrate multiple mechanisms of action of rutin on target cells. Some intracellular pathways (controlling oxidative, inflammatory processes, cell proliferation, apoptosis, and viability) could be involved in mediating rutin action on several processes and dysfunctions (cancer, neurodegenerative diseases). Other mediators are specific for rutin action on particular targets (cardiovascular system and blood, carbohydrate metabolism, and viral functions).

4.11.5 Effects on female reproductive processes

4.11.5.1 Effect on ovarian and reproductive state

Previously, the antifertility action of rutin and rutin-containing plants have been hypothesized (Farnsworth et al., 1975). Nevertheless, the subsequent studies did not reveal its adverse influence on healthy female reproductive system. Moreover, several studies demonstrated its stimulatory action on female reproductive processes. In *in vivo* experiments of Jahan et al. (2016) and Hu et al. (2017), oral administration of rutin to rats promoted their estrous cycles, reduced the number of cystic and atretic ovarian follicles and increased the number of mature ovulated follicles, Corpora lutea, the number of ovulations and fertility. Culture of sheep ovarian follicles with rutin increased their morphological quality, although it did not affect their number, viability, growth and development (Lins et al., 2017).

On the other hand, the protective action of rutin against several reproductive disorders has been documented. Rutin-suppressed growth of cultured ovarian cancer cells (Luo et al., 2008). Administration of rutin prevented degeneration of ovarian follicles induced by ischemia (rat: Nayki et al., 2018), polycystic ovarian syndrome (Jahan et al., 2016; Hu et al., 2017) and chemotherapeutic agent cisplatin (mice: Lins et al., 2020). Addition of rutin was able to prevent the action of copper nanoparticles supported on titania (Sirotkin et al., 2020a) and some actions of toluene on cultured porcine granulosa cells (Sirotkin et al., 2021).

Therefore, the available data of *in vivo* and *in vitro* studies suggest the stimulatory action of rutin on healthy ovarian functions and its inhibitory and protective action on ovaries in pathological state.

4.11.5.2 Effect on ovarian cell functions

There is a limited number of evidence for direct action of rutin on viability in ovarian cells, while rutin exerted the opposite effects on healthy and cancer ovarian cells. In experiments of Sirotkin et al. (2020a, 2021) addition of rutin increased the viability of cultured porcine ovarian granulosa cells, although this effect was not associated with changes in accumulation of markers of proliferation (accumulation of PCNA) and cytoplasmic apoptosis (bax). In other similar experiments, rutin addition reduced viability and apoptosis, and promoted proliferation of cultured porcine granulosa cells (Sirotkin et al., 2020b). Native and hydrolized rutin reduced viability and proliferation of cultured human ovarian cancer cells (Luo et al., 2008; de Araújo et al., 2013).

Rutin also affects the oxidative processes in reproductive (ovarian, but not in embryonal) cells. Rutin increased the level of antioxidative enzymes and decreased the level of reactive oxygen species in rat ovaries (Jahan et al., 2016), but not in cultured sheep ovarian follicles (Lins et al., 2017). On the other hand, rutin did not affect the accumulation of reactive oxygen species in cultured human lung embryonic fibroblasts, human umbilical vein endothelial cells (Matsuo et al., 2005), or in embryonal monocytes (Zhao et al., 2004). It is not to be excluded that oxidative processes could mediate rutin action on female reproductive processes and their dysfunctions (see later).

Therefore, rutin can promote the viability of healthy cells and suppress it in ovarian cancer cells. Furthermore, it can suppress oxidative stress in some healthy ovarian cells. The current literature does not contain evidence of rutin action on oxidative processes in ovarian cancer cells.

4.11.5.3 Effect on oocytes and embryos

No rutin action on number and growth of oocytes in cultured sheep ovarian follicles were observed (Lins et al., 2017).

The available literature did not contain reports concerning rutin action on embryogenesis. No influence of rutin on Xenopus embryo development (Amado et al., 2012) or on viability of cultured human lung embryonic fibroblasts and human umbilical vein endothelial cells have been reported (Matsuo et al., 2005). On the other hand, rutin was able to affect prostaglandin F2α production by human amnion fibroblasts (Guo et al., 2014) (see later). This prostaglandin can control embryo implantation and development, therefore rutin action on embryogenesis cannot be excluded.

4.11.6 Effect on reproductive hormones

Both *in vivo* and *in vitro* experiments demonstrated rutin influence on synthesis and release of steroid hormones release and its regulator LH. Administration of rutin decreased plasma LH level, promoted the expression of ovarian steroidogenic enzymes such as P450C17, aromatase, 3β-HSD, 17β-HSD, and StAR (Hu et al., 2016), and reduced progesterone and increased testosterone and estradiol level in rat plasma (Jahan et al., 2016). *In vitro* studies with cultured rat ovarian granulosa cells demonstrated the stimulatory action of rutin on synthesis and production of FSH receptors, StAR, aromatase, estradiol and estradiol receptor beta (Wang et al., 2018). In *in vitro* experiments with cultured porcine granulosa cells, addition of rutin was able to promote estradiol and testosterone, but not progesterone release (Sirotkin et al., 2020a, 2021).

Rutin can also affects the production of cytokines and prostaglandins. Nayki et al. (2018) reported the ability of rutin administration to reduce the expression of the proinflammatory cytokines IL-1β and TNF-α in rat ovaries. Furthermore, as mentioned earlier, addition of rutin to amnion fibroblasts cultured by human activated production of prostaglandin F2α and carbonyl reductase 1, enzyme, converting prostaglandin E2 to prostaglandin F2α (Guo et al., 2014). It is highly probable that these hormones could be among the extracellular mediators of rutin action on female reproductive processes (see later).

4.11.7 Mechanisms of action on female reproductive processes

The stimulatory and protective effect of rutin on rat (Jahan et al., 2016; Nayki et al., 2018) and mice (Lins et al., 2017, 2020) folliculogenesis was associated with an increase of **antioxidant** and mitochondrial activity and a decrease in reactive oxygen species in ovarian follicles. The toxic influence of rutin on cultured cancer cells was associated with its antioxidant effect (de Araújo et al., 2013). It indicates that rutin effect on rodent and ovarian follicle development and human ovarian cancer cells could be mediated by its antioxidative action. On the other hand, rutin addition did not affect oxidative processes in cultured sheep ovarian follicles (Lins et al., 2017) indicating other mediators of rutin action on ovarian follicles in this species.

The stimulatory and protective action of rutin on rat ovaries was associated also with a decrease in markers and promoters of **inflammation**—C-protein (Jahan et al., 2016) and COX-1 (Nayki et al., 2018) (see earlier). The ability of rutin to promote these intracellular regulators of inflammation, together with its ability to increase release of proinflammatory cytokines IL-1β and TNF-α mentioned earlier (Nayki et al., 2018) suggest that rutin action on the ovary could be also due to its ability to suppress promoters of inflammation.

Furthermore, rutin administration decreased the number of rat atretic ovarian follicles (Jahan et al., 2016) and increased rat ovarian follicular cell

394 Contaminants and Plants Action on Female Reproduction

proliferation (Lins et al., 2020), decreased their **apoptosis** and PTEN and increased p-FOXO3a expression (Lins et al., 2017, 2020). These observations suggest that the stimulatory action of rutin on ovaries could be mediated by promotion of proliferation and suppression of apoptosis through changes in PTEN/FOXO3a pathway. The stimulatory action of rutin on ovarian cells via upregulation of their proliferation and downregulation of their apoptosis was confirmed by ability of rutin to inhibit proliferation of cultured human ovarian cancer cells (Luo et al., 2008) as well as to promote accumulation of PCNA and to suppress accumulation of bax in cultured porcine ovarian granulosa cells (Sirotkin et al., 2020). On the other hand, in other similar experiments, rutin was able to increase viability of cultured porcine ovarian granulosa cells without changes in these markers of proliferation and mitochondrial apoptosis (Sirotkin et al., 2021) indicating that in some conditions rutin could support ovarian cell functions via intracellular mechanisms not related to proliferation and mitochondrial apoptosis.

Furthermore, the inhibitory action of ovarian cell growth was associated with decreased VEGF expression. It suggests that the ability of rutin to suppress ovarian tumor development could be due to inhibition of tumor VEGF-induced **angiogenesis** (Luo et al., 2008).

The rutin action on gonadotropin, its receptor, steroid hormones' synthesis and reception described earlier suggests that these endocrine structures could be mediators of hormone-dependent reproductive events and disfunctions, but this hypothesis has not been verified.

Mechanisms of rutin action on oogenesis and embryogenesis remained undetected as of yet. It is known that embryogenesis can be controlled by intracellular Wnt/β-catenin signaling pathway, but no rutin action on this pathway in Xenopus embryos has been observed (Amado et al., 2012).

Therefore, the available evidence indicates that rutin action on female reproductive processes could be its ability to inhibit oxidative stress, regulators of inflammation, proliferation, apoptosis and angiogenesis. Furthermore, rutin action via several mechanisms, which in some conditions could replace each other, might be hypothesized.

4.11.8 Application in reproductive biology and medicine

The ability of rutin to stimulate and protect ovarian functions in rodents (Jahan et al., 2016; Hu et al., 2017) and sheep (Lins et al., 2017), as well as to increase viability of isolated porcine ovarian cells (Sirotkin et al., 2020a, 2021), suggests that rutin and rutin-containing plants could be a potential natural biostimulator of female reproductive processes. This action indicates potential applicability of rutin as a stimulator of fertility in farm animals and maybe humans.

On the other hand, it suggests also the applicability of rutin for prevention and treatment of such reproductive disorders as cancer (Luo et al., 2008; de Araújo et al., 2013), ischemia (Nayki et al., 2018), polycystic ovarian syndrome (Jahan et al., 2016; Hu et al., 2017), and adverse effects of chemotherapy (Lins et al., 2020), nanoparticles (Sirotkin et al., 2020a), benzene (Sirotkin et al., 2020b) and toluene (Sirotkin et al., 2021). Rutin influence on reproductive hormones (Hu et al., 2016; Jahan et al., 2016; Wang et al., 2018; Sirotkin et al., 2020a, 2021), cytokines, and prostaglandin (Guo et al., 2014; Nayki et al., 2018) indicates its potential applicability for prevention and treatment of some endocrine disorders too.

4.11.9 Conclusions and possible direction of future studies

Taken together, the available data demonstrate the stimulatory action of rutin on female reproductive processes: it can promote ovarian follicles development and ovulation, ovarian cyclicity, and the viability of ovarian cells. On the other hand, it can suppress ovarian cancer cells and tumor development by inhibition of cell proliferation and growth and activation of their apoptosis and death. Furthermore, it could be able to prevent other reproductive disorders (ischemia, polycystic ovarian syndrome, toxic effects of chemotherapy, nanoparticles, and toluene). Rutin could exert its effects via changes in the release and reception of gonadotropin, ovarian steroid hormones, prostaglandins, cytokines, VEGF, as well as in intracellular regulators and markers of oxidative and inflammatory processes, proliferation, apoptosis, and angiogenesis.

On the other hand, numerous aspects of rutin action require further clarification. For example, the available reports concerning rutin action on various models are sometimes contradictory, while the causes of such contradictions are difficult to explain. It remains unknown whether rutin influences oo- and embryogenesis, oviducts, uterus and pregnancy.

The mechanisms of rutin action require further explanations. Numerous studies demonstrated an association between rutin action on healthy and cancer ovarian cells and expression of members of several intracellular signaling pathways. Nevertheless, in some cases the mediatory role of these pathways has not been demonstrated directly. Understanding the functional interrelationships between these signaling molecules requires further elucidation. If rutin can use several mediators and change these mediators in various situations, the causes of such changes remain unknown.

Furthermore, studies indicating applicability of rutin for promotion of reproduction and for treatments of reproductive disorders have been performed only on laboratory animals and *in vitro* cultures, while *in vivo* studies on farm

396 Contaminants and Plants Action on Female Reproduction

animals or clinical trials have not been reported yet. Therefore, the applicability of rutin as a biostimulator of reproductive processes and medicinal drug requires validation by appropriate *in vivo* studies.

Nevertheless, despite such limitations, the available data demonstrate that rutin could be a promising natural regulator of ovarian functions, which can be applicable for stimulation of female reproductive processes, as well as for prevention and treatment of some reproductive disorders.

References

Aditya NP, Espinosa YG, Norton IT. Encapsulation systems for the delivery of hydrophilic nutraceuticals: food application. Biotechnol Adv 2017;35(4):450−7. https://doi.org/10.1016/j.biotechadv.2017.03.012.

Aksu EH, Kandemir FM, Özkaraca M, Ömür AD, Küçükler S, Çomaklı S. Rutin ameliorates cisplatin-induced reproductive damage via suppression of oxidative stress and apoptosis in adult male rats. Andrologia 2017;49(1). https://doi.org/10.1111/and.12593.

Amado NG, Fonseca BF, Cerqueira DM, Reis AH, Simas AB, Kuster RM, Mendes FA, Abreu JG. Effects of natural compounds on Xenopus embryogenesis: a potential read out for functional drug discovery targeting Wnt/β-catenin signaling. Curr Top Med Chem 2012;12(19):2103−13. https://doi.org/10.2174/156802612804910241.

Budzynska B, Faggio C, Kruk-Slomka M, Samec D, Nabavi SF, Sureda A, Devi KP, Nabavi SM. Rutin as neuroprotective agent: from bench to bedside. Curr Med Chem 2019;26(27):5152−64. https://doi.org/10.2174/0929867324666171003114154.

Chattopadhyay D, Chitnis A, Talekar A, Mulay P, Makkar M, James J, Thirumurugan K. Hormetic efficacy of rutin to promote longevity in *Drosophila melanogaster*. Biogerontology 2017;18(3):397−411. https://doi.org/10.1007/s10522-017-9700-1.

Chen W, Wang S, Wu Y, Shen X, Xu S, Guo Z, Zhang R, Xing D. The physiologic activity and mechanism of quercetin-like natural plant flavonoids. Curr Pharmaceut Biotechnol 2020;21(8):654−8. https://doi.org/10.2174/1389201021666200212093130.

de Araújo ME, Moreira Franco YE, Alberto TG, Sobreiro MA, Conrado MA, Priolli DG, Frankland Sawaya AC, Ruiz AL, de Carvalho JE, de Oliveira Carvalho P. Enzymatic de-glycosylation of rutin improves its antioxidant and antiproliferative activities. Food Chem 2013b;141(1):266−73. https://doi.org/10.1016/j.foodchem.2013.02.127.

Enogieru AB, Haylett W, Hiss DC, Bardien S, Ekpo OE. Rutin as a potent antioxidant: implications for neurodegenerative disorders. Oxid Med Cell Longev 2018;2018:6241017. https://doi.org/10.1155/2018/6241017.

Erlund I, Kosonen T, Alfthan G, Mäenpää J, Perttunen K, Kenraali J, Parantainen J, Aro A. Pharmacokinetics of quercetin from quercetin aglycone and rutin in healthy volunteers. Eur J Clin Pharmacol 2000;56(8):545−53. https://doi.org/10.1007/s002280000197.

Farha AK, Gan RY, Li HB, Wu DT, Atanasov AG, Gul K, Zhang JR, Yang QQ, Corke H. The anticancer potential of the dietary polyphenol rutin: current status, challenges, and perspectives. Crit Rev Food Sci Nutr 2020:1−28. https://doi.org/10.1080/10408398.2020.1829541.

Farnsworth NR, Bingel AS, Cordell GA, Crane FA, Fong HH. Potential value of plants as sources of new antifertility agents I. J Pharm Sci 1975;64(4):535−98.

Ferenczyova K, Kalocayova B, Bartekova M. Potential implications of quercetin and its derivatives in cardioprotection. Int J Mol Sci 2020;21(5):1585. https://doi.org/10.3390/ijms21051585.

Ganeshpurkar A, Saluja AK. The pharmacological potential of rutin. Saudi Pharmaceut J 2017;25(2):149−64. https://doi.org/10.1016/j.jsps.2016.04.025.

Ghorbani A. Mechanisms of antidiabetic effects of flavonoid rutin. Biomed Pharmacother 2017;96:305−12. https://doi.org/10.1016/j.biopha.2017.10.001.

Guo C, Wang W, Liu C, Myatt L, Sun K. Induction of PGF2α synthesis by cortisol through GR dependent induction of CBR1 in human amnion fibroblasts. Endocrinology 2014;155(8):3017−24. https://doi.org/10.1210/en.2013-1848.

Habtemariam S. Rutin as a natural therapy for Alzheimer's disease: insights into its mechanisms of action. Curr Med Chem 2016;23(9):860−73. https://doi.org/10.2174/092986732366616021 7124333.

Hu T, Yuan X, Ye R, Zhou H, Lin J, Zhang C, Zhang H, Wei G, Dong M, Huang Y, Lim W, Liu Q, Lee HJ, Jin W. Brown adipose tissue activation by rutin ameliorates polycystic ovary syndrome in rat. J Nutr Biochem 2017;47:21−8. https://doi.org/10.1016/j.jnutbio.2017.04.012.

Imani A, Maleki N, Bohlouli S, Kouhsoltani M, Sharifi S, Maleki Dizaj S. Molecular mechanisms of anticancer effect of rutin. Phytother Res 2020. https://doi.org/10.1002/ptr.6977.

Jahan S, Munir F, Razak S, Mehboob A, Ain QU, Ullah H, Afsar T, Shaheen G, Almajwal A. Ameliorative effects of rutin against metabolic, biochemical and hormonal disturbances in polycystic ovary syndrome in rats. J Ovarian Res 2016;9(1):86. https://doi.org/10.1186/s13048-016-0295-y.

Lai F, Schlich M, Pireddu R, Fadda AM, Sinico C. Nanocrystals as effective delivery systems of poorly water-soluble natural molecules. Curr Med Chem 2019;26(24):4657−80. https://doi.org/10.2174/0929867326666181213095809.

Latos-Brozio M, Masek A. Structure-activity relationships analysis of monomeric and polymeric polyphenols (quercetin, rutin and catechin) obtained by various polymerization methods. Chem Biodivers 2019;16(12):e1900426. https://doi.org/10.1002/cbdv.201900426.

Lins TLBG, Cavalcante AYP, Santos JMS, Menezes VG, Barros VRP, Barberino RS, Bezerra MÉS, Macedo TJS, Matos MHT. Rutin can replace the use of three other antioxidants in the culture medium, maintaining the viability of sheep isolated secondary follicles. Theriogenology 2017;89:263−70. https://doi.org/10.1016/j.theriogenology.2016.11.019.

Lins TLBG, Gouveia BB, Barberino RS, Silva RLS, Monte APO, Pinto JGC, Campinho DSP, Palheta RC, Matos MHT. Rutin prevents cisplatin-induced ovarian damage via antioxidant activity and regulation of PTEN and FOXO3a phosphorylation in mouse model. Reprod Toxicol 2020;98:209−17. https://doi.org/10.1016/j.reprotox.2020.10.001.

Liu Y, Zhao J, Wang L, Yan B, Gu Y, Chang P, Wang Y. Nanocrystals technology for transdermal delivery of water-insoluble drugs. Curr Drug Deliv 2018;15(9):1221−9. https://doi.org/10.2174/1567201815666180518124345.

Long Y, Yang Q, Xiang Y, Zhang Y, Wan J, Liu S, Li N, Peng W. Nose to brain drug delivery - a promising strategy for active components from herbal medicine for treating cerebral ischemia reperfusion. Pharmacol Res 2020;159:104795. https://doi.org/10.1016/j.phrs.2020.104795.

Luca SV, Macovei I, Bujor A, Miron A, Skalicka-Woźniak K, Aprotosoaie AC, Trifan A. Bioactivity of dietary polyphenols: the role of metabolites. Crit Rev Food Sci Nutr 2020;60(4):626−59. https://doi.org/10.1080/10408398.2018.1546669.

Luo H, Jiang BH, King SM, Chen YC. Inhibition of cell growth and VEGF expression in ovarian cancer cells by flavonoids. Nutr Cancer 2008;60(6):800−9. https://doi.org/10.1080/01635580802100851.

Matsuo M, Sasaki N, Saga K, Kaneko T. Cytotoxicity of flavonoids toward cultured normal human cells. Biol Pharm Bull 2005;28(2):253—9. https://doi.org/10.1248/bpb.28.253.

Nayki C, Nayki U, Keskin Cimen F, Kulhan M, Yapca OE, Kurt N, Bilgin Ozbek A. The effect of rutin on ovarian ischemia-reperfusion injury in a rat model. Gynecol Endocrinol 2018;34(9):809—14. https://doi.org/10.1080/09513590.2018.1450378.

Negahdari R, Bohlouli S, Sharifi S, Maleki Dizaj S, Rahbar Saadat Y, Khezri K, Jafari S, Ahmadian E, Gorbani Jahandizi N, Raeesi S. Therapeutic benefits of rutin and its nanoformulations. Phytother Res 2021;35(4):1719—38. https://doi.org/10.1002/ptr.6904.

Ninfali P, Antonelli A, Magnani M, Scarpa ES. Antiviral properties of flavonoids and delivery strategies. Nutrients 2020;12(9):2534. https://doi.org/10.3390/nu12092534.

Nouri Z, Fakhri S, Nouri K, Wallace CE, Farzaei MH, Bishayee A. Targeting multiple signaling pathways in cancer: the rutin therapeutic approach. Cancers 2020;12(8):2276. https://doi.org/10.3390/cancers12082276.

Parhi B, Bharatiya D, Swain SK. Application of quercetin flavonoid based hybrid nanocomposites: a review. Saudi Pharmaceut J 2020;28(12):1719—32. https://doi.org/10.1016/j.jsps.2020.10.017.

Rajagopal C, Lankadasari MB, Aranjani JM, Harikumar KB. Targeting oncogenic transcription factors by polyphenols: a novel approach for cancer therapy. Pharmacol Res 2018;130:273—91. https://doi.org/10.1016/j.phrs.2017.12.034.

Sharma S, Ali A, Ali J, Sahni JK, Baboota S. Rutin : therapeutic potential and recent advances in drug delivery. Expet Opin Invest Drugs 2013;22(8):1063—79. https://doi.org/10.1517/13543784.2013.805744.

Simunkova M, Alwasel SH, Alhazza IM, Jomova K, Kollar V, Rusko M, Valko M. Management of oxidative stress and other pathologies in Alzheimer's disease. Arch Toxicol 2019;93(9):2491—513. https://doi.org/10.1007/s00204-019-02538-y.

Sirotkin AV, Radosová M, Tarko A, Fabova Z, Martín-García I, Alonso F. Abatement of the stimulatory effect of copper nanoparticles supported on titania on ovarian cell functions by some plants and phytochemicals. Nanomaterials 2020a;10(9):1859. https://doi.org/10.3390/nano10091859.

Sirotkin AV, Záhoranska Z, Tarko A, Popovska-Percinic F, Alwasel S, Harrath AH. Plant isoflavones can prevent adverse effects of benzene on porcine ovarian activity: an in vitro study. Environ Sci Pollut Res Int 2020b;27(23):29589—98. https://doi.org/10.1007/s11356-020-09260-8.

Sirotkin A, Záhoranska Z, Tarko A, Fabova Z, Alwasel S, Halim Harrath A. Plant polyphenols can directly affect ovarian cell functions and modify toluene effects. J Anim Physiol Anim Nutr 2021;105(1):80—9. https://doi.org/10.1111/jpn.13461.

Slámová K, Kapešová J, Valentová K. "Sweet flavonoids": glycosidase-catalyzed modifications. Int J Mol Sci 2018;19(7):2126. https://doi.org/10.3390/ijms19072126.

Suzuki T, Morishita T, Noda T, Ishiguro K, Otsuka S, Katsu K. Breeding of buckwheat to reduce bitterness and rutin hydrolysis. Plants 2021;10(4):791. https://doi.org/10.3390/plants10040791.

Wang X, Wang GC, Rong J, Wang SW, Ng TB, Zhang YB, Lee KF, Zheng L, Wong HK, Yung KKL, Sze SCW. Identification of steroidogenic components derived from *Gardenia jasminoides* ellis potentially useful for treating postmenopausal syndrome. Front Pharmacol 2018;9:390. https://doi.org/10.3389/fphar.2018.00390.

Yi YS. Regulatory roles of flavonoids on inflammasome activation during inflammatory responses. Mol Nutr Food Res 2018;62(13):e1800147. https://doi.org/10.1002/mnfr.201800147.

Yong DOC, Saker SR, Chellappan DK, Madheswaran T, Panneerselvam J, Choudhury H, Pandey M, Chan YL, Collet T, Gupta G, Oliver BG, Wark P, Hansbro N, Hsu A, Hansbro PM, Dua K, Zeeshan F. Molecular and immunological mechanisms underlying the various pharmacological properties of the potent bioflavonoid, Rutin. Endocr Metab Immune Disord Drug Targets 2020;20(10):1590—6. https://doi.org/10.2174/1871530320666200503053846.

Zhao J, Liu XJ, Ma JW, Zheng RL. DNA damage in healthy term neonate. Early Hum Dev 2004;77(1—2):89—98. https://doi.org/10.1016/j.earlhumdev.2004.02.003.

Conclusion

The present book has summarized the available knowledge concerning provenance, features, effect and practical significance of most common environmental contaminants, medicinal/functional food plants, and their constituents. These substances have different sources and chemical structure and sometimes specific mechanisms of action. They are able either promote or suppress reproductive and non-reproductive processes, as well as to induce or to prevent and to treat reproductive and non-reproductive disorders. Nevertheless, they have some similar targets and mechanisms of action, which define their functional and medicinal interrelationships and applicability.

Usually both contaminants and plant molecules targeting several regulators of reproduction at once—from hypothalamo-pituitary system to ovarian follicular cells, oocytes, embryos, and uterus. The majority of contaminants and plant substances affect oxidative processes and induce or mitigate oxidative stress. Oxidative stress induces inflammatory processes, and via hormones, growth factors and intracellular protein kinases, transcription factors, and other intracellular signaling molecules affect proliferation and apoptosis and angiogenesis, which in turn influence ovarian folikullo-, and oogenesis, embryogenesis, uterus, embryogenesis and their dysfunctions (cancer, polycystic ovarian syndrome, ovarian failure, endometriosis, infertility, etc.). The typical targets and mechanisms of action of these external regulators of female reproduction are present at Fig. 1. It is however to take in mind, that usually one molecule affects several reproductive organs and signaling pathways, therefore the hierarchical interrelationships between they are sometimes difficult to demonstrate.

The ability of plant molecules to affect extra- and intracellular regulators of female reproduction explain their stimulatory or inhibitory action on reproduction and fecundity. Furthermore, both the harmful external factors and medicinal and functional food plant molecules are affecting the same signaling molecules. This fact defines ability of plant molecules to modify the action of environmental contaminants and resulted reproductive illnesses. This ability is a basis of protective and curative effects of phytotherapeutic drugs.

The available data demonstrate the large potential of some plant extracts and molecules for control of female reproductive processes and treatment of their disorders. Some plants or plant molecule could be promising as a functional food or drugs in animal and human nutrition, human and veterinary medicine, assisted reproduction and animal production. Their bioavailability and efficiency could be enhanced by application of their metabolites, chemical analogs and carriers. Nevertheless, the evidence for potential applicability of

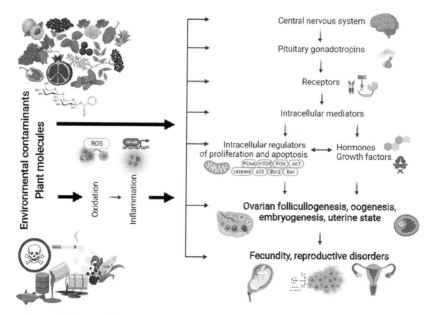

FIGURE 1 The possible targets and mediators of action of environmental factors and plant molecules on female reproductive processes (simplified). Explanations are in the text.

each plant and plant molecule was obtained mainly by studies on cell cultures or laboratory animals. Therefore, applicability of all plants, plant molecules or their analogs listed in this book requires verification by adequate in vivo experiments and clinical trials. The authors hope, that the current state of art described in this book could help to find the adequate approaches to affect animal and human female reproductive processes, to prevent and to treat their disorders.

Index

Note: 'Page numbers followed by "*f*" indicate figures and "*t*" indicate tables.'

A

Acetaminophen (APAP), 111
Acetyl-CoA carboxylase (ACC), 123
Acetylcholine receptors, 235
Activation protein 1 (AP-1), 187, 292
Adenosine monophosphate (AMN), 281
5'adenosine monophosphate-activated protein kinase, 135
Adipocyte protein 2 (aP2), 123
Adipogenesis, 353
Adipogenic transcription factor, 125
Adipokines, 310
Adiponectin receptors, 353
Adrenocorticotropic hormone (ACTH), 13−14, 74
Agavaceae, 317−318
Age-related reproductive insufficiency, 173
Akt
  Akt/GSK3b signaling pathway, 332
  AKT/mTOR/p70S6K, 135
  signaling, 135
Alanine aminotransferase (ALT), 111
Alpha-estrogen receptors, 31
Alternaria, 38−39
Alzheimer's disease, 278
Amaryllidaceae, 317−318
Amelioration, 234
δ-aminolevulinic acid, 30
AMN kinase signaling, 281
AMP-activated protein kinase (AMPK), 125, 290−291, 353
  AMPK/p27 signaling pathway, 342
  AMPKα, 331
Amygdalin, 83, 89, 245−246, 246f
  application in reproductive biology and medicine, 253
  effects on female reproductive processes, 251−252
    effect on ovarian follicular cell functions, 251
    effect on reproductive hormones, 251
    effect on reproductive state, 252

future studies, 253−254
mechanisms of action, 249−250
mechanisms of action on female reproductive processes, 252
physiological and therapeutic actions, 248−249
provenance and properties, 246−248
Amygdalus scoparia kernel oil supplementation, 84−85
Amyloid-β proteins, 354
Androgens, 9−11, 282, 310
  androgen-responsive genes, 112−113
Anelgesic action of *Vitex agnus-cactus*, 233
Angiogenesis, 233, 291−292, 394
Angiotensin II (Ang II), 11−12
Angiotensin-converting enzyme 2 (ACE2), 11−12
Anovulation, 1
Anti-asthmatic extract, 234
Anti-Mullerian hormone (AMH), 177
Anti-obesity effects of capsaicin, 290−291
Antibacterial activity, 234
Antibody, 354
Anticancer effects, 85
Antiglioma, 234
Antihyperprolactinemia, 234
Antiinflammatory
  action of buckwheat, 124
  activity, 233
  properties, 235
Antioxidant, 85, 393
  activities, 226−227
    of extracts, 87
  capacity, 267
  properties, 202, 235
    f buckwheat substances, 123−124
  response element, 354−355
Antioxidative effect, 311
Antioxidative enzymes, 75
Antioxidative properties, 205
Antiproliferative agent, 87
Antiviral drugs, 278

403

404 Index

Antral follicles, 4
Apigenin, 259, 259f
    application in reproductive biology and
        medicine, 270—271
    effect on oocytes and embryos, 265
    effect on ovarian and uterine cell functions,
        264—265
    effect on reproductive hormones, 265—266
    effect on response to adverse external
        factors, 266—267
    effects on female reproductive processes,
        263—264
        effect on ovarian and reproductive
            state, 263—264
    future studies, 271—272
    mechanisms of action, 262—263
    mechanisms of action on female
        reproductive processes, 267—270
    physiological and therapeutical actions,
        260—261
    provenance and properties, 259—260
Apob/apoA-I, 354—355
Apoptosis, 4, 44—48, 61, 73, 87, 112—113,
        115, 124, 134—135, 176, 189,
        226—227, 250, 252, 267—268,
        280—282, 291—292, 295, 310—311,
        331, 352—353, 361
Apoptosis inducing factor (AIF), 331
Apoptotic process, 89
Apricot kernel oil, 83
Apricot seeds, 81—82
    application in reproductive biology and
        medicine, 89—90
    effects on female reproductive processes,
        87—88
        effect on ovarian follicular cell
            functions, 88
        effect on reproductive hormones, 87—88
    future studies, 90
    mechanisms of action, 86—87
    mechanisms of action on female
        reproductive processes, 88—89
    physiological and therapeutic actions,
        83—86
    provenance and properties, 82—83
Apricot tree (*Prunus armeniaca* L.), 81—82
Arginine-vasotocin, 13—14
Armeniacae semen, 84
Aromatic hydrocarbons, 54—55
Arsenic, 21—22
    exposure, 24
Arterial smooth muscle, 202—203

Aspalathus linearis. See Rooibos
    (*Aspalathus linearis*)
Aspartate aminotransferase (AST), 111
Aspergillus, 38—39
    A. terreus, 328—329
ATF6, 115
Atherosclerosis, 248—249
ATP, 236
Atresia, 4
Autophagy, 4, 47—48

B

B-cell lymphoma 1 (Bcl-1), 352—353
B-cell lymphoma 2 (Bcl-2), 13—14, 111, 135,
        250, 331, 341, 344, 361
B-cells inhibitor alpha (IkBα), 341—342
Bacterial diarrhea, 278
Bacteriocide action, 203
Bacteriostatic action, 203
Bad, 361
Bcl-2-associated X protein (Bax),
        13—14, 31, 111—113, 250,
        331, 340—341, 344, 361
Bcl-XL, 341
Beclin 1, 115
Benzene, 57—60, 360
Benzene, toluene, ethyl benzene, xylene
        (BTEX), 54—55, 55f, 60—61
Benzo(a)pyrene (BaP), 72—73
Berberidis radix. See Sankezhen
    (*Berberidis radix*)
Berberine, 276, 277f
    application in reproductive biology and
        medicine, 283
    effects on female reproductive processes,
        280—281
        effect on ovaries, 280—281
        effect of vagina and uterus, 281
    future studies, 283—284
    mechanisms of action, 278—280
    mechanisms of action on female
        reproductive processes, 281—283
        hormones and growth factors, 282—283
        oxidative stress, 282
        proliferation and apoptosis, 281—282
        prostaglandins, 282
    physiological and therapeutic actions,
        277—278
    provenance and properties, 276—277
Berberis, 276
    B. vulgaris, 276

Berries, 109–110
Beta-estrogen receptors, 31
Bid, 361
Bidens
  B. bipinnata L., 331
  B. pilosa L., 332
Bilobalide, 160
Binding, 201–202
1,7-bis (4-hydroxy-3-methoxyphenyl)-1,
    6-eptadiene-3,5-dione, 130–131
Black elder (*Sambucus nigra* L.),
    96, 99–101
  application in reproductive biology and
    medicine, 102
  effects on female reproductive processes,
    100–101
    effect on embryo, 101
    effect on ovarian cell functions,
      100–101
  future studies, 102–103
  mechanisms of action, 99–100
    on female reproductive processes,
      101–102
  physiological and therapeutic actions,
    98–99
  provenance and properties, 96–98
Black elderberry. *See* Black elder
    (*Sambucus nigra* L.)
Black tea, 221
Blood flow, 269
Body weight, 290–291
Bone marrow cells, 330
Bone microstructure of rabbits, 86
Bone morphogenetic protein (BMP), 206
  BMP-15, 134, 207
Bovine endometrial epithelial cells (bEECs),
    343–344
Bovine sperm cells, 330
Brain neurotransmitters, 202, 279–280
Brain structures, 205
Buckthorn, 108–109
  effects on female reproductive processes,
    113–114
    effect on ovarian follicular cell
      functions, 113
    effect on vagina and uterus, 114
  future studies, 116
  mechanisms of action, 111–113
    on female reproductive processes,
      114–115
  physiological and therapeutic actions,
    110–111

  potential for application in reproductive
    biology and medicine, 115–116
  provenance and properties, 109–110
Buckwheat (*Fagopyrum esculentum*),
    63–64, 121
  application in reproductive biology and
    medicine, 127
  effects on female reproductive processes,
    125–126
    effect on ovarian cell functions, 125
    effect on ovarian cell response to
      environmental contaminants, 126
    effect on reproductive hormones,
      125–126
    effect of on ovarian and reproductive
      state, 125
  future studies, 89
  mechanisms of action, 123–125
  physiological and therapeutical actions,
    122–123
  provenance and properties, 121–122

C

c-Fos, 87
C-glucosides, 259–260
c-Jun N-terminal kinase (JNK), 115, 202,
    343–344
  JNK 1, 2, 3, 187
C/EBP homologous protein (CHOP), 115
$C_{15}H_{20}O_6$. *See* Deoxynivalenol (DON)
Ca^{2+} ATPase, 135
Ca^{2+}-sparing effect, 202
Cadmium, 21–22, 24, 27, 72
  accumulation, 29
  ions, 360
Calcitonin gene-related peptide (CGRP),
    290–291
Camelia sinensis L. *See* Tea (*Camelia
    sinensis* L.)
cAMP responsive element binding protein
    1 (CREB-1), 13–14
Cancer, 71–72, 289–290
  cells, 234, 330–331
Capillary permeability, 353
Capsaicin, 287, 288f
  application in reproductive biology and
    medicine, 295–296
  effects on female reproductive processes,
    293–294
    effect on ovarian and reproductive state,
      293–294

406 Index

Capsaicin (*Continued*)
    effect on ovarian cell functions, 294
    effect on reproductive hormones, 294
    future studies, 296
    mechanisms of action, 290–293
    on female reproductive processes, 294–295
    physiological and therapeutic actions, 288–290
    provenance and properties, 287–288
Capsaicinoids, 287, 290–291
Capsicum, 287
    C. *annuum*, 287–288
    C. *frutescens*, 287–288
Cardiovascular diseases, 84–85, 248–249, 278
Cardiovascular systems, 290–291
Casein kinase 2 (CK2α), 268
Caspases, 361
    caspase-3, 30, 112–113, 331, 341–342, 361
        caspase 3-iPLA2-AA-COX-2-PGE2, 282
        downregulation of, 340–341
    caspase-8, 341–342
    caspase-9, 112–113, 331, 341–342
CAT. *See* Catalase (CAT)
Catalase (CAT), 30
β-catenin, 333
    signaling pathway, 344
CCAAT/enhancer binding protein-α (CEBP-α), 123
Cell cycle, 89, 124, 250, 361
    arrest, 111–112
    regulators, 134–135
Cell death, 47–48, 342
Cell division cycle protein two homolog (CDC2 kinase), 13–14
Cell migration, 281
Cell proliferation, 44–45, 47, 164, 281–282, 352–353
Cell survival, 291–292
Cells renewal, 268
Central mechanisms, 164
Central nervous system (CNS), 39–40
Cervical cancer, 174
Checkpoint kinase 1/2 (CHK1/2), 61–62
Chemotherapeutic agent, 360
Chinese traditional medicine, 84
Chromium, 21–22
Chromosomal misalignment, 75
Cis-isomers, 371–372

Cobalt, 28
Colorectal cancer, 330–331
Common buckwheat (*Fagopyrum esculentum*), 121
Congenital hypogonadotropic hypogonadism, 1
CONTAM. *See* Panel on Contaminants in the Food Chain (CONTAM)
Copper
    administration, 28
    nanoparticles, 25
Coptis, 276
Corona radiata, 3–4
Coronary heart disease, 71–72
Corpus albicans, 5–6
Corpus luteum, 4–6, 8–10, 12
Cortisol, 44–45
Cumulus oophorus, 4
Curcuma/turmeric, 130
    effect on female reproductive processes, 132–135
        application in reproductive biology and medicine, 135–136
        effect on ovarian and reproductive state, 132–133
        effect on ovarian cell functions, 133
        effect on reproductive hormones, 133–134
        mechanisms of action on female reproductive processes, 134–135
    mechanisms of action, 132
    physiological and therapeutic actions, 131–132
    provenance and properties, 130–131
    studies, 136–137
Curcumin, 131, 131f
Cyanide, 249
Cyanidin-3-glucoside, 97–98
Cyanidin-3-sambubioside, 97–98
Cyanogenic glycosides, 248
Cyclin-dependent kinase (CDK), 188
Cyclins
    cyclin A, 341
    cyclin B, 236
    cyclin B1, 13–14, 30–31, 341, 352–353, 361
    cyclin D1, 310–311, 341, 344
    cyclin D2, 341
    cyclin E6, 341
Cyclooxygenase-1 (COX-1), 131–132
Cyclooxygenase-2 (COX-2), 187, 202, 279, 282
Cytochrome C, 361

Index **407**

Cytochrome C7, 341
Cytochrome P(450), 354–355
Cytochrome P450–3A (CYP3A), 188
Cytokines, 13, 111–112, 114, 160–161, 201–202, 340
Cytoplasmic maturation, 6
Cytosolic β-glycosidase (CBG), 329–330

D

Daidzein, 303
  effect on female reproductive processes, 307–312
    application in reproductive biology and medicine, 311–312
    effect on ovarian and reproductive state, 307–308
    effect on ovarian cell functions, 308
    effect on reproductive hormones, 308–309
    mechanisms of action on female reproductive processes, 309–311
    studies, 312–313
  mechanisms of action, 306–307
  physiological and therapeutical actions, 305–306
  provenance and properties, 303–305
Decidualization, 174
Delphinidin, 172
Deoxynivalenol (DON), 39–40, 39f, 44
Detoxifying phase II enzymes, 354–355
Diferuloylmethane, 130–131
Dimethylmercury, 23–24
Dioscoreaceae, 317–318
Diosgenin, 317, 317f
  application in reproductive biology and medicine, 323–324
  effect on oocytes and embryos, 321
  effect on ovarian cell functions, 321
  effect on reproductive hormones, 321–322
  effects on female reproductive processes, 320–321
    effect on ovarian and reproductive state, 320–321
  future studies, 324–325
  mechanisms of action, 320
    on female reproductive processes, 322–323
  physiological action, 318–319
  provenance and properties, 317–318
Diploid zygote, 6

DNA, 354
  damage, 47–48, 73
  repair, 164
  repair, 352–353
DNA methyltransferase 3A (DNMT3A), 47–48
Docosahexaenoic acid, 150
Dopamine-2 receptors, 235
Downregulation of promoters and markers of proliferation, 281
Doxorubicin, 360

E

E-cadherin, 342, 361
Elderberries, 97–99
  extract, 99
Elderflowers, 97–99
Ellagitannins, 338
Embryo, 45–46
Endocrine profile, 248–249
Endocrine regulators, 88–89
Endometrial functions, 174
Endometrial stromal cells (ESC), 174
Endometriosis, 7, 14, 174, 359
Endoplasmic reticulum stress, 115, 124, 361
Endothelin (ET-1), 13–14
Environmental contaminants and influence on health and female reproduction
  heavy metals, 21–22
  mycotoxins, 38
  oil-related environmental contaminants, 54–55
  tobacco smoking, 70
Enzymes, 111
Epidermal growth factor (EGF), 13–14
Epigallocatechin gallate, 222
Equol, 306–307
Esterases, 161
17β-estradiol, 44, 332–333
Estradiol, 44–45, 101
Estradiol receptors (ERα), 101
Estrogen receptors (ER), 30, 46, 175, 188, 269, 310, 320, 361
  ERα, 8, 100, 235
Estrogens, 3–4, 9–11, 149, 310
  estrogen-beta receptors, 235
  estrogenic extract, 234
Ethyl benzene, 57–60
European Food Safety Authority (EFSA), 247–248
Extracellular matrix (ECM), 99
Extracellular mediators, 322

408 Index

Extracellular regulators of female
  reproductive processes, 7–14
Extracellular signal-regulated kinase (ERK),
  343–344
  ERK1/2, 61–62, 115, 187

F

Fagopyrum, 121
Fagopyrum dibotrys. *See* Perennial
  buckwheat (*Fagopyrum dibotrys*)
Fagopyrum esculentum. *See* Buckwheat
  (*Fagopyrum esculentum*)
Fagopyrum tataricum. *See* Tartary buckwheat
  (*Fagopyrum tataricum*)
Fallopian tubes, 6–7
Fat metabolism, 361–362
Fat storage, 123
Fatty acid synthase (FAS), 112–113, 123,
  354–355
Fatty acid-binding protein 4 (FABP4), 187
FDA. *See* US Food and Drug Administration
  (FDA)
Female reproductive system and regulation, 1
  extra-and intracellular regulators of female
    reproductive processes, 7–14
  Fallopian tubes, 6–7
  female reproductive disorders, 14–15
  folliculogenesis, 2–4
  luteogenesis and luteolysis, 5–6
  oocytes, 6
  ovary, 1–6
  preovulatory changes and ovulation, 4–5
  uterus, 7
Fennel (*Foeniculum vulgare*), 63–64
Ferroptosis, 31
Flavonoids, 328
Flaxseed (*Linum usitatissimum* L.), 141
  application in reproductive biology and
    medicine, 150–151
  effects on female reproductive processes,
    145–148
    effect on oocytes and embryos, 146–147
    effect on ovarian and reproductive state,
      145–146
    effect on reproductive hormones,
      147–148
  future studies, 151–152
  mechanisms of action, 143–145
    constituents responsible for
      physiological effects, 143–144
    mediators of flaxseed effects, 144–145

mechanisms of action on female
    reproductive processes, 148–150
  constituents responsible for effects on
    female reproductive processes, 148
  mediators of flaxseed effects on female
    reproductive processes, 149–150
  oil, 142–143
  physiological action, 142–143
  provenance and properties, 141–142
Foeniculum vulgare. *See* Fennel
  (*Foeniculum vulgare*)
Follicle growth, 75
Follicle-stimulating hormone (FSH), 8,
  87–88, 280, 282
Follicular waves, 4
Folliculogenesis, 2–4, 11–12, 75
Food/medicinal herbs and influence on health
    and female reproduction
  apricot seeds, 81–82
  black elder, 96
  buckthorn, 108–109
  buckwheat, 121
  curcuma/turmeric, 130
  flaxseed, 141
  ginkgo, 156
  grape, 170
  pomegranate, 183
  puncture vine, 199
  rooibos, 213
  tea, 220
  vitex, 232
Forkhead box protein O1 (FOXO1), 293
Fumonisin B1, 44, 47
Fungi, 41–42
Fusarium, 38–39
  mycotoxins, 44

G

G1/S phase transition of human ovarian
    cancer cells, 344
G2/M phase, 111–112
Gamma-aminobutyric acid (GABA), 59
  receptors, 164
Gastroenteritis, 278
Gastrointestinal system, 289–290
Gastroprotective effect of apricot
    kernel oil, 86
Gene mutations, 73
Genistein, 303–304
Ghrelin, 13–14, 45
Ginkgo (*Ginkgo biloba* L.), 156

application in reproductive biology and medicine, 165

constituents for effects, 159–160

effects on female reproductive processes, 162–163

  effect on oocytes and embryos, 163

  effect on ovarian and reproductive state, 162

  effect on ovarian cell functions, 163

  effect on reproductive hormones, 163

future studies, 165–166

mechanisms of action, 159–162

  on female reproductive processes, 164

mediators of ginkgo and constituents' effects, 160–162

physiological and therapeutic actions, 157–159

provenance and properties, 156–157

Ginkgo biloba L. *See* Ginkgo (*Ginkgo biloba* L.)

Ginkgolide A, 160

Ginkgolides, 160

Glioma-associated oncogene 1 (Gli1), 268

Glucagon-like peptide 1 (GLP-1), 290–291

Glucose, 290–291

  absorption, 353

  utilization, 353

Glucose transporter (GLUT4), 187

Glucose-regulated protein. *See* Protein disulfide-isomerase A3 (PDIA3)

Glucose-regulated protein 78 (GRP78), 115

Glutathione (GSH), 30, 111

Glutathione peroxidase (GSH-Px), 30, 111, 172, 187

Glutathione reductase (GR), 30

Glutathione-S transferase, 30

Glycitein, 303–304

Glycosides, 329–330

Gonadotropin-releasing hormone (GnRH), 8

Gonadotropins, 174

  model, 9–10

  receptors, 310

Graafian follicles, 11

Granulosa cells, 3–4, 9–11, 322

Grape (*Vitis vinifera*), 170

  application in reproductive biology and medicine, 177–178

  effects on female reproductive processes, 172–174

    effect on ovaries, 173–174

    effect on uterus, 174

  future studies, 178

mechanisms of action, 172

  on female reproductive processes, 174–177

physiological and therapeutic actions, 171–172

provenance and properties, 171

seed procyanidin, 171–172

Grape seed flavonoid proanthocyanidin B2 (GSPB2), 171–173, 175

Green rooibos, 213

Green tea, 220–221

Growth arrest, 291–292

Growth differentiation factor (GDF), 206

GDF-9, 134, 207

Growth factors, 160–161, 201–202, 282–283

GSK3β/βcatenin, 202

H

Heat shock protein 70, 135

Heat shock transcription factor, 124

Heat stress, 359

Heavy metals, 21–22

  application in reproductive biology and medicine, 31–32

  effects on female reproductive processes, 27–29

    effect on ovaries, 27–28

    effect on uterus and pregnancy, 29

  future studies, 32

  mechanisms of action, 26–27

    on female reproductive processes, 30–31

  physiological actions, 23–25

  provenance and properties, 22–23

Hematological parameters, 248–249

Heme oxygenase-1 (HO-1), 111, 172, 205, 235, 278–279

Hippophae rhamnoides L. *See* Sea buckthorn (*Hippophae rhamnoides* L.)

Hirsutrin, 328–329

Histone methylation, 361

HMG-CoA reductase, 354–355

Hormonal regulators, 362

Hormones, 3–4, 62, 100, 160–161, 164, 201–202, 234–235, 282–283, 295

HT2 toxin, 39–40, 42f, 45

Huangbo (*Phellodendri chinensis cortex*), 276

Huanglian (*Rhizoma coptidis*), 276

Human chorionic gonadotropin (hCG), 8

410 Index

Human somatic cells, 330
Hydrastis, 276
Hydrogen cyanide (HCN), 247
Hydrogen peroxide (H_2O_2), 176, 331
Hydrolyzable tannins, 183–184
17β-hydroxysteroid dehydrogenase
(17β-OHDH), 175
Hypothalamic monoamines, 295
Hypothalamic neurohormones, 174
Hypothalamus-pituitary-ovary axis, 1, 7
Hypoxia-inducible factor 1α (HIF-1α),
114, 333
Hyptis fasciculata, 331

I

iIkBα, 343–344
IkBα. *See* B-cells inhibitor alpha (IkBα)
Immunomodulatory extract, 234
In vitro studies on DON, 45–46
Inducible nitric oxide synthase (iNOS), 111,
124, 187
Infections, 278
Infertility, 14
Inflammation, 134, 268–269, 353, 393–394
inflammation-related cytokines, 134
Inflammatory bowel disease, 278
Inflammatory cytokines, 13, 87
Inflammatory mediators, 279
Inflammatory processes, 1, 177, 295
Inflammatory response, 30
Inhibition of antioxidant enzymes, 30
Inhibitory effect of PUN, 341–342
Insulin, 12
insulin-sensitizing activities, 353
response to, 282
secretory activities, 353
Insulin-like growth factor binding protein 1
(IGFBP1), 175
Insulin-like growth factor binding protein-3
(IGFBP-3), 201–202
Insulin-like growth factors (IGFs), 12
IGF-I, 11–14, 28, 31, 45, 321–322
IGF-II, 12
insulin/insulin-like growth factor
–dependent signaling pathway, 191
Intercellular adhesion molecule 1 (ICAM-1),
187, 341
Interferon gamma (IFN-γ), 341
Interleukins (IL), 99–100, 262
IL-1β, 87, 187, 250, 262, 340–341,
343–344
IL-2, 262

IL-6, 124, 262, 266, 279, 294, 340–341,
343–344
IL-8, 343–344
IL-17A, 262
Intestinal infection, 278
Intracellular mediators, 164, 322–323
Intracellular reactive oxygen species,
332–333
Intracellular regulators of female
reproductive processes, 7–14
Intracellular signaling pathways, 161
Intracellular steroidogenesis, 44
Intrinsic pathway, 342
Invasion, 341–342
IRE-1, 115
Iron, 24–25, 29
iron-dependent form of cell death, 31
Ischemia/reperfusion, 173
Isoquercetin. *See* Isoquercitrin (IQ)
Isoquercetrin, 328–329
Isoquercitrin (IQ), 328, 329f. *See also*
Quercetin
application in reproductive biology and
medicine, 334
effects on female reproductive processes,
332–333
future studies, 334
mechanisms of action, 331–332
on female reproductive processes, 333
physiological and therapeutic actions,
330–331
provenance and properties, 328–330
"Itai-itai" disease, 24

J

Janus kinase (JAK), 43
pathway, 352–353
JNK. *See* c-Jun N-terminal kinase (JNK)

K

Kaempferol, 160
Kaempherol, 113
Kelch-like ECH-associated protein 1
(Keap1), 30
Kinases, 111–112
Kisspeptin, 8

L

Lactase phlorizin hydrolase (LPH), 329–330
Laetrile, 83, 246–247
LC3II, 115

Lead, 21–22
  exposure, 28
  poisoning, 24
  release of, 29
Learning, 234
Leaves, 110
Leguminosae, 317–318
Leptin, 8, 13–14, 45
Liliaceae, 317–318
Linum usitatissimum L. *See* Flaxseed
    (*Linum usitatissimum* L.)
Lipid
  metabolism, 290–291
  peroxidation, 353
Lipogenesis, 353
Lipopolysaccharide (LPS), 340
  LPS-induced productions of interleukin,
    343–344
Liver, 248–249
  cells, 330–331
Liver kinase B1 (LKB1), 331
Long noncoding RNAs, 135
Low-birth rate in offspring, 71–72
Lung, 248–249
Luteinizing hormone (LH), 8–9, 87–88
Luteogenesis, 5–6
Luteolysis, 5–6
Lysine-specific demethylase 1 (LSD1), 332

M

m-xylene, 59–60
Macrophage, 99–100
Male reproductive functions, 86
Malonaldehyde, 30
Malondialdehyde (MDA), 30, 172, 187, 291
Mammalian target of rapamycin
    (mTOR), 293
  regulators, 13
Manganese, 25
Manganism, 25
Matcha, 221
Matrix metalloproteinase
    (MMP), 269
  MMP-1, 99
  MMP-2, 280, 342
  MMP-9, 135, 280, 342
Medaka (*Oryzias latipes*), 355
Meiosis, 6
Memory, 234
Menopause, 173
Mercury, 21–24

chloride treatment of hamsters, 27
compounds, 29
Metabolic processes, 290–291
Metabolism, 233
Metastasis, 291–292
2-methoxyestradiol, 150
Methylcyclopentadienyl manganese
    tricarbonyl (MMT), 25
4-(methylnitrosamino)-1-(3-pyridyl)-
    1-butanol (NNAL), 70–71
4-(methylnitrosamino)-1-(3-pyridyl)-
    1-butanone (NNK), 70–71
4-(methylnitrosamino)-4-(3-pyridyl)-
    1-butanol (iso-NNAL), 70–71
4-(methylnitrosamino)-4–3-pyridyl butyric
    acid (iso-NNAC), 70–71
MicroRNAs, 135, 161, 354, 361
  let-7a, 176
  miR-101, 262
  miR-138, 262
  miR-423–5p, 262
  miR-520b, 262
Milk production, 234
Mitochondria, 353
  dysfunction, 47–48
  mitochondrial activity, 361
  mitochondrial pathway, 342
Mitogen activating kinases, 310–311
Mitogen-activated protein kinases (MAPK),
    13–14, 43, 111–112, 135,
    187, 235, 331, 343–344. *See also*
    AMP-activated protein kinase (AMPK)
  MAPK p44/42, 361
  MAPK/ERK1/2, 352–353, 361
  pathway, 202
Mitogen-activated protein kinases/activator
    protein 1 (MAPK/AP-1), 99
Mitosis, 46–47
Mixed lineage kinase domain-like pathway
    (MLKL pathway), 281
Molybdenum, 28
Monoamine oxidase, 354
Monocyte chemoattractant protein-1
    (MCP-1), 124, 187, 279, 341
Monounsaturated fatty acids (MUFA),
    82–83, 171
mRNA, 30
Muscarinergic receptors, 235
Mycotoxicosis, 40–41
Mycotoxins, 38
  application in reproductive biology and
    medicine, 48–49

412 Index

Mycotoxins (*Continued*)
  effects on female reproductive processes, 44–46
    effect on ovarian cell functions, 44–45
    effects on oocytes and embryos, 45–46
  mechanism of action, 30
    on female reproductive processes, 46–48
  physiological actions, 43
  provenance and properties, 38–42
Myeloid differentiation 2 (MD-2), 112

N

N-cadherin, 342
N-methyl-D-aspartate (NMDA), 59
Na$^+$-K$^+$-2Cl$^-$cotransporter, 353
Nanoparticles (NPs), 30
Nerve regeneration, 330
Nervous system, 161
Neuronal survival pathway, 202
Neurons, 331
Nickel, 22–23
Nicotiana tabacum L., 70–71
Nicotine, 72
Nitric oxide (NO), 87, 111, 124, 162, 290–291
Nitrosamines, 70–71
Nitrosoanabasine (NAB), 70–71
Nitrosoanatabine (NAT), 70–71
Nitrosonornicotine (NNN), 70–71
NLRP3, 262
Non-reproductive processes, 401
Nonspecific metabolic or toxic effects, 226–227
NOTCH, 352–353
Nuclear factor (erythroid-derived 2)-like 2 (Nrf2), 30, 132, 187, 205, 215, 278–279, 340, 354
  dependent pathways, 111
  Nrf-2/HO-1-SOD-2 signaling pathway, 111
  Nrf2/HO-1, 99, 135, 340
  nuclear erythroid 2-related factor 2/ARE pathway, 172
Nuclear factor kappa B (NF-κB), 99, 111–112, 135, 187, 202, 205, 262, 292, 333, 343–344, 352–354, 361, 374
  signaling pathway, 341–342

O

O-glucosides, 259–260
Obesity, 278, 289–290
Oil-related environmental contaminants
  application in reproductive biology and medicine, 62–64
  effects on female reproductive processes, 57–61
    effect on ovarian cell functions, 58–59
    effects on CNS, 57
    effects on cytogenetics of somatic and generative cells, 60
    effects on oocytes and embryos, 59–60
    effects on ovarian and reproductive state, 57–58
    effects on oviducts, 60
    effects on reproductive hormones, 60–61
  future studies, 64–66
  mechanisms of action, 56
    on female reproductive processes, 61–62
  physiological actions, 55–56
  provenance and properties, 54–55
Oocytes, 1–2, 6
  quality, 45–46
Oogenesis, 75
Oolong, 221
Opioid receptors, 235
Opioidergic extract, 234
Oryzias latipes. See Medaka (*Oryzias latipes*)
Osteogenesis, 234
Ovarian aging, 75, 359
Ovarian angiogenesis, 269
Ovarian cancer, 15, 358–359
  cells, 100–101, 173
Ovarian cell cycle, regulators and markers of, 226–227
Ovarian follicle, 4–6
Ovarian follicullogenesis, 75, 146–147
Ovarian hormones, 134, 207
Ovarian response cells to hormones, 226–227
Ovarian steroidogenesis, 173
Ovary, 1–6
Ovulation, 4–5
Oxidative balance, 248–249
Oxidative stress, 13, 47–48, 62, 73, 75, 175, 235, 282, 291, 295, 331, 353, 360–361
Oxytocin, 13–14, 310

Index **413**

P

p-Akt, 115, 310−311
p-FAK, 310−311
p-GSK3b, 310−311
p-PI3K, 310−311
p-Tau (p-τ), 132
p21, 236, 310−311
p38 mitogen-activated protein kinases
(p38 MAPK), 61−62, 236, 250
p38, 187, 343−344
P450 aromatase, 30
p53 transcription factor, 62, 135, 226−227,
331, 341, 352−353, 361
p65, 343−344
Pancreatic cancer cells, 330−331
Panel on Contaminants in the Food Chain
(CONTAM), 247−248
Para-xylene, 60−61
Penicillium, 38−39
Perennial buckwheat (*Fagopyrum dibotrys*),
121
PERK. *See* RNA-like endoplasmic reticulum
kinase (PERK)
Peroxisome proliferator-activated receptor
α (PPAR α), 290−291
Peroxisome proliferator-activated receptor
γ (PPAR γ), 123, 125, 187
Phayre's leaf monkeys (*Trachypithecus
phayrei crepusculus*), 236
Phellodendri chinensis cortex. *See* Huangbo
(*Phellodendri chinensis cortex*)
Phenolic compounds, 172
Phosphatidylinositide 3-kinases (PI3K),
187−188
Phosphatidylinositol-4,5-bisphosphate
3-kinase/protein kinase B
(PI3K/AKT), 135, 216
intracellular signaling pathway, 282
pathway, 236
PI3K-Akt-mTOR pathway, 111−112,
352−353, 361
PI3K/AKT/PKB, 111−112
signaling pathway, 340
Pituitary gonadotrophs, 8
Pituitary hormones, 134, 207, 234
Plant molecules, 13−14, 401
amygdalin, 245−246
apigenin, 259
berberine, 276
capsaicin, 287
daidzein, 303
diosgenin, 317

isoquercitrin, 328
punicalagin, 338
quercetin, 349
resveratrol, 371
rutin, 387
Platelet aggregation, 353
Poly ADP-ribose polymerase (PARP), 250,
331, 342
PARP-1, 177
Polyamines, 202
Polycystic ovary syndrome (PCOS), 1, 14,
173, 189−190, 359
Polyphenols, 175
Polyunsaturated fatty acids (PUFAs),
82−83, 171
Pomegranate (*Punica granatum* L.),
183−184, 338
application in reproductive biology and
medicine, 191
effects on female reproductive processes,
189−190
flower, 185−186
fruit extract, 185−186
future studies, 192
mechanisms of action, 186−189
on female reproductive processes, 191
physiological and therapeutic actions,
185−186
provenance and properties, 183−185
bioavailability, metabolism and
pharmacokinetics, 184−185
seed oil, leaves, juice and peel, 185−186
Postmenopausal osteoporosis, 333
Prebiotic properties of buckwheat, 124
Premature ovarian failure (POF),
1, 14, 359
Preovulatory changes, 4−5
Primary follicle, 3−4
Primordial follicles, 2
Primordial germ cells, 3
Pro-inflammatory cytokines, 202, 250
Proanthocyanidin, 173
Procyanidin, 171
Progestagen, 310
Progesterone, 5−6, 10, 44−45, 332−333
release, 45
Progesterone receptor (PR),
31, 100, 269
Progestins, 9−10
Prolactin (PRL), 9, 88−89, 175
Proliferating cell nuclear antigen (PCNA),
46−47, 188, 331

414 Index

Proliferation, 61, 73, 100, 111–113, 115, 176, 188, 250, 252, 267, 280–282, 295, 310–311, 331, 341–342
Proportion of live, dead, and apoptotic cells, 332–333
Prostaglandin E2 (PGE2), 87, 202, 279
Prostaglandins (PGs), 12, 174, 282
Protein disulfide-isomerase A3 (PDIA3), 340–341
Protein kinase A (PKA), 13–14
Protein kinase C alpha (PKCα), 374
Protein kinase G, 13–14
Protein kinases, 61–62
Proteins, 40–41
Prunus armeniaca L. *See* Apricot tree (*Prunus armeniaca* L.); Wild apricot (*Prunus armeniaca* L.)
Pu-erh, 221
PUN, 340–341, 343
Puncture vine (*Tribulus terrestris* L.), 199
  application in reproductive biology and medicine, 208
  effect on male reproductive processes, 203–204
  effects on female reproductive processes, 205–207
  future studies, 208–209
  mechanisms of action, 201–203
  mechanisms of effects
    on female reproductive processes, 207–208
    on male reproductive processes, 204–205
  physiological and therapeutical actions, 200–201
  provenance and properties, 199–200
Punica granatum L. *See* Pomegranate (*Punica granatum* L.)
Punicalagin, 338, 339f
  application in reproductive biology and medicine, 344–345
  effects on female reproductive processes, 342–343
    effect on ovarian follicular cell functions, 343
    effect on reproductive health and fertility, 342–343
    effect on uterus and pregnancy, 343
  future studies, 345
  mechanisms of action, 340–342
    on female reproductive processes, 343–344

  physiological and therapeutic actions, 340
  provenance and properties, 338–339

Q

Quercetin, 160, 328, 349, 350f
  application in reproductive biology and medicine, 362–363
  effects on female reproductive processes, 355–360
    effect on oocytes and embryos, 356–357
    effect on ovarian and reproductive state, 355
    effect on ovarian cell functions, 355–356
    effect on ovarian disfunctions, 358–359
    effect on reproductive hormones, 357–358
    effect on response to hazardous factors, 359–360
  future studies, 363–364
  mechanisms of action, 352–355
    on female reproductive processes, 360–362
  physiological actions, 351–352
  provenance and properties, 349–351
Quercetin-3-glucoside (Q3G), 328–330
Quercetin-3-O-β-D-glucoside, 328–329
Quercetin-3-rutinoside, 387

R

R-amygdalin, 246–247
RANKL-mediated osteoclastogenesis, 354
Reactive nitrogen species (RNS), 338–339
Reactive oxygen species (ROS), 13, 31–32, 43, 99, 135, 186–188, 248–249, 338–339
  formation, 75
  generation, 47–48
  production, 333
Reception, 201–202
Receptor-interacting protein kinase 3 (RIPK3), 281
Renin-angiotensin system, 11–12
Reproductive disorders, 1, 7
Reproductive hormones, 204
Reproductive processes, 401
Resveratrol, 171–172, 177, 371
  application in reproductive biology and medicine, 378–380
  in combination with DON, 44–45

effects on female reproductive processes, 374–377
  causes of variability in resveratrol effects, 376–377
  effect on ovarian and reproductive state and on ovarian cell functions, 374–375
  effect on reproductive hormones, 375–376
  mechanisms of action, 374
    on female reproductive processes, 377–378
  physiological and therapeutic actions, 372–373
  provenance and properties, 371–372
  studies, 380–382
Retinal ganglion cells, 330
Rhamnaceae, 317–318
Rheumatoid arthritis, 234
Rhizoma coptidis. See Huanglian (*Rhizoma coptidis*)
Rhizomes of curcumas, 136–137
RIPK3. *See* Receptor-interacting protein kinase 3 (RIPK3)
RIPK3–MLKL pathway, 281
RNA, 40–41
RNA-like endoplasmic reticulum kinase (PERK), 115
Rooibos (*Aspalathus linearis*), 63–64, 213
  application in reproductive biology and medicine, 217
  effects on female reproductive processes, 216
  future studies, 218
  mechanisms of action, 215–216
    on female reproductive processes, 217
  physiological actions, 214–215
  provenance and properties, 213–214
Rutin, 328–329, 387, 388f
  application in reproductive biology and medicine, 395
  effect on reproductive hormones, 393
  effects on female reproductive processes, 391–392
    effect on oocytes and embryos, 392
    effect on ovarian and reproductive state, 391–392
    effect on ovarian cell functions, 392
  future studies, 395–396
  mechanisms of action, 390–391
    on female reproductive processes, 393–394

physiological actions, 389
provenance and properties, 387–389
Rutoside, 387

S

Sambucus nigra L. *See* Black elder (*Sambucus nigra* L.)
Sambunigrin, 97
Sankezhen (*Berberidis radix*), 276
SARS-CoV-2, 278
Saturated fatty acid (SFA), 171
Sciatic nerve, 331
Scrophulariaceae, 317–318
Sea buckthorn (*Hippophae rhamnoides* L.), 108
  H. rhamnoides L. subsp. *mongolica*, 109–110
  H. rhamnoides L. subsp. *sinensis*, 109–110
  H. rhamnoides L. subsp. *turkestanica*, 109–110
  H. rhamnoides L. subsp. *yunnanensis*, 109–110
Secondary follicles, 4
Seed oil, 185–186
Semen armeniacae amarum, 248
Senecio biafrae, 271
Sensory denervation, 295
Sex hormone binding globulin (SHBG), 74, 175
Sexual maturity, 47
Signal transducers and activators of transcription (STAT), 43, 135, 352–353, 361
  STAT3, 292, 294
Silencing information regulator 1 (SIRT1), 177
Sirtuin 1 (SIRT1), 135, 292, 361, 374, 377–378
Sirtuin 3 (SIRT-3), 135
Skin diseases, 84
SMAD, 354
Small RNAs, 13–14
Sodium arsenite, 29
Sodium-dependent glucose transporter 1 (SGLT1), 329–330
Solanaceae, 317–318
Sonic hedgehog pathway (Shh pathway), 30, 352–353
Sophorin, 387
Sperm abnormalities, 113
Spermatogenesis, 205

416 Index

Spermatogonial proliferation, 113
STAY-3, 352–353, 361
Stearoylcoenzyme A desaturase-1
  (SCD-1), 123
Steroid hormones, 9–10, 100–101, 175,
  189–191, 269
  receptors, 134, 269
Steroidogenesis, 75, 188
Steroidogenic acute regulatory protein
  (StAR), 175, 263
Steroids, 174
Superoxide dismutase (SOD), 30, 172, 187
Survivin, 135, 344, 352–353

T

T tyrosine kinase, 13–14
T-2 toxin, 39–40, 41f, 45
Tartary buckwheat (*Fagopyrum tataricum*),
  121
Tea (*Camelia sinensis* L.), 220
  application in reproductive biology and
    medicine, 227–228
  effects on female reproductive processes,
    223–225
    effect on oocytes and embryos, 224–225
    effect on ovarian cell functions, 224
    effect on reproductive hormones, 225
    effect of tea on ovarian and reproductive
      state, 223–224
  future studies, 228
  mechanisms of action, 222–223
    on female reproductive processes,
      226–227
  physiological actions, 221–222
  provenance and properties, 220–221
Testicular cells, 205
Testicular torsion, 289–291
Testosterone, 44–45
Th1/Th2 stability, 354
Theca cells, 9–10
Thyroid hormones, 11
Thyroid-stimulating hormone (TSH), 8
Tissue inhibitor of metalloproteinase
  (TIMP), 344
  TIMP-2 and TIMP-3, 344
Tobacco smoking, 70
  application in reproductive biology and
    medicine, 75–76
  effects on female reproductive processes,
    73–74
    effect on embryos, 74

effect on ovarian and reproductive state,
  73–74
  effect on reproductive hormones, 74
  future studies, 76
  mechanism of action, 73
    on female reproductive processes, 75
  physiological actions, 71–73
  provenance and properties, 70–71
Toll-like receptor 4 (TLR4), 112
Toluene, 57–59
Total superoxide dismutase, 30
Toxicity, 290
Tracheospasmolytic extract, 234
Trachypithecus phayrei crepusculus. See
  Phayre's leaf monkeys
  (*Trachypithecus phayrei crepusculus*)
Trans-isomers, 371–372
Transcription factors, 13–14, 340
Transforming growth factor beta
  (TGF-β), 99, 330–331,
  352–353, 361
  TGF-β2, 332–333
Tribulus, 207
Tribulus terrestris L. *See* Puncture vine
  (*Tribulus terrestris* L.)
TRPV1, 290–291
Tuberculosis, 71–72
Tumor necrosis factor (TNF), 99–100
  TNF-α, 43, 87, 124, 187, 250, 262, 266,
    279, 340–341, 343–344
  TNF-related apoptosis-inducing
    ligand-receptor, 333
Tumor suppression, 164
Tumor/antitumor proteins, 340
Tyrosine kinases, 13–14

U

Uridine diphosphate-
  glucuronosyltransferases (UDP-GT),
  329–330
Urinary frequency, 289–290
US Food and Drug Administration (FDA),
  143
Uterus, 7

V

Vascular endothelial growth factor
  (VEGF), 12, 134, 266,
  282, 333, 352–353
Viability, 267, 332–333
  of human ovarian cells, 190

of human ovarian granulose cells, 100–101
of ovarian cells, 61
Vitamin B17, 83, 246–247
Vitamin P, 387
Vitex (*Vitex agnus-castus* L.), 232
  application in reproductive biology and
    medicine, 238–240
  effects on female reproductive processes,
    236–237
    effect on ovarian and reproductive
      state, 236
    effect on ovarian cell functions,
      236–237
    effect on reproductive hormones, 237
  future studies, 240
  mechanisms of action, 234–236
    on female reproductive processes,
      237–238
  physiological actions, 233–234
  provenance and properties, 232–233
Vitex agnus-castus L. *See* Vitex
    (*Vitex agnus-castus* L.)
Vitis vinifera. See Grape (*Vitis vinifera*)

W

Wild apricot (*Prunus armeniaca* L.), 245–246
Wnt, 354
  Wnt/-catenin, 352–353
  Wnt/β-catenin signaling pathways, 135
World Health Organization (WHO), 7

X

Xenopus laevis oocytes, 265
XIAP, 361
Xylene, 57–58, 60, 360

Y

Yucca schidigera, 64

Z

Zearalenone (ZEA), 39–40, 40f, 44
  and metabolites, 44–45
Zinc, 25
Zinc oxide (ZnO), 30
Zona pellucida, 3–4